P9-EDP-599

FIELD ENGINEER'S MANUAL

BOARD OF CONSULTANTS

Harold J. Cleary, P. E. — Member
Manager, Jump River Electric Co-op
Ladysmith, Wisconsin

Dan S. Kling, P. E. — Member
Consulting Engineer, Cooper Engineering Co.
Rice Lake, Wisconsin

George H. Morgan, P. E. — Member
Consulting Engineer, Morgan & Parmley, Ltd.
Ladysmith, Wisconsin

Robert O. Parmley, P. E. — Chairman
Consulting Engineer, Morgan & Parmley, Ltd.
Ladysmith, Wisconsin

SPECIAL CONSULTANT

Gerald L. Bradshaw, Sr., CET
Certified Engineering Technician
Flambeau Mining Company
Ladysmith, Wisconsin

JOHN BARBER
SILVER PEAK, NEV.

FIELD ENGINEER'S MANUAL

■

ROBERT O. PARMLEY, P. E., Editor
Consulting Engineer, Morgan & Parmley, Ltd.
Ladysmith, Wisconsin

MCGRAW-HILL BOOK COMPANY
New York St. Louis San Francisco Auckland
Bogotá Hamburg Johannesburg London Madrid Mexico
Montreal New Delhi Panama Paris São Paulo
Singapore Sydney Tokyo Toronto

Library of Congress Cataloging in Publication Data

Parmley, Robert O
Field engineer's manual.

Includes index.
1. Civil engineering—Handbooks, manuals, etc.
I. Title.
TA151.P35 624'.0212 80-28187
ISBN 0-07-048513-5

Copyright © 1981 by McGraw-Hill, Inc. All rights reserved. Printed in the United States of America. No part of this publication may be reproduced, stored in a retrieval system, or transmitted, in any form or by any means, electronic, mechanical, photocopying, recording, or otherwise, without the prior written permission of the publisher.

90 KPKP 898

The editors for this book were Harold B. Crawford, Patricia Allen-Browne, and Ruth L. Weine, the designer was Richard A. Roth, and the production supervisor was Thomas G. Kowalczyk. It was set in Times Roman by David E. Seham, Inc.
Printed and bound by The Kingsport Press.

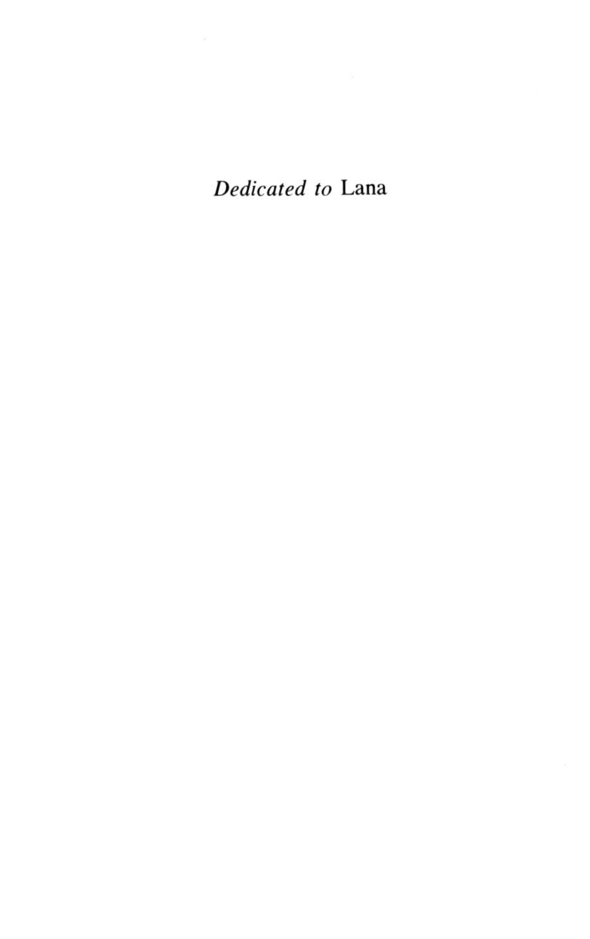

Dedicated to Lana

CONTENTS

Section 4 The Metric System of Units (SI) 103

Section 5 Construction: Materials, Data and Safety 121

Section 11 Welding 333

Section 12 Surveying 355

Section 14 Hydraulics 421

Section 15 Sewage Collection 443

Section 16 Water Supply, Water Storage, and Fire Protection 485

PREFACE

This Manual has been designed to provide you and other field engineers and technical specialists with a compact, accurate, and quick reference to basic engineering data that are needed to assist preliminary surveys, solve field problems, supplement technical material shown on construction plans, and give assurance that proven standards are being used.

Think, for example, of the times when you are caught short for a specific technical answer or formula. In the office, you can consult your technical library or the applicable project file. But when the problem arises in the field you are helpless without basic technical references. Return to the office is out of the question, and perhaps the needed support is not in your briefcase or even in your car. If you could reach into your pocket or into the glove compartment of your car or field vehicle and immediately have a reliable authority, imagine how many problems you can solve on the spot and and how much time loss you can avoid.

Then too, meetings with clients, special public assemblies, and conferences with governmental officials often generate technical questions that can place you in an awkward position. This Manual provides you with a handy compact memory bank of basic accepted data that can help you answer difficult technical questions on the spot. Simply refer to this silent aide at any point without creating a gap in the discussion or your thought patterns.

The general public seems to think that every competent engineer has a perfect memory—with total recall and with the ability to re-peat lengthy data and formulas at a moment's notice. Comparatively few engineers have photographic memories; and, in my opinion, it is fortunate that such is the case, for nature rarely combines excellent memory and recall talent with the practical tact, creative ability, and deep insight so essential to a successful engineer. There have been brilliant exceptions to this rule, but they are very rare indeed. Therefore, when an engineer can reach into a pocket and produce a Manual that will answer some basic technical questions, such an action would enhance a professional presentation and perhaps preserve a client's confidence in the engineer's ability when he or she is trying to break through a technical bottleneck.

All engineers worthy of the title attempt to amass all the technical data related to their respective fields. As an outgrowth of my personal effort to achieve a comprehensive collection, the idea for a general yet all-inclusive manual for field use was germinated. I realize that it is impossible to capture all the technical knowledge of the entire engineering profession and then condense the total data into one small portable manual. However, it is my hope that this volume helps you, and other professionals like you, to develop a more personalized reference tool. For this very reason a blank page has been placed at the end of each section to allow you the opportunity to insert additional technical data you deem important for your specific needs.

The design of the Manual assumes that you have constantly at hand a pocket-sized electronic calculator (slide-rule type). Neither trigonometric and logarithm tables nor functions of numbers have been included here because they are readily recalled from the calculator. Because these and many other memory features are stored within your calculator, much valuable space has been saved in the Manual, thus allowing the inclusion of a substantial amount of additional data.

The material contained in this Manual is so extensive and has been culled from so many different sources that it is impossible to list all of the technical organizations, societies, manufacturing firms, publications, and consultants who contributed to this effort; however, where possible I have noted the source and given credit on the page where their particular material appears.

At the beginning of this effort, a special Board of Consultants was instituted to monitor and periodically review the progress of the manuscript. This Board's wise counsel and analysis of the material has strengthened the main thread of the Manual and ensured the practical format we deemed mandatory in the original concept. Their engineering expertise and advice have been of primary value to me. A list of these board members faces the title page.

Each section of the Manual is devoted to a particular segment of engineering and was organized to include only the most valuable and basic data. If you require a more in-depth study or detailed code, you should select a book or manual from the literature devoted solely to that particular subject. Local, state, and federal codes also should be checked when one is determining a design.

I am sure that some material contained in this Manual will seem very basic and perhaps unimportant to some readers, but bear in mind that just as a dictionary contains all words common to most, this Manual also attempts to be all-inclusive. One never knows who might have need of that minor item common to most, but rare to a few.

I hope you find the FIELD ENGINEER'S MANUAL useful in practice and true to its mission. When it is updated and revised, perhaps some additional information and data can be included—especially those that meet the needs of many peripheral areas. I invite you to suggest any additional entry you feel will enhance the Manual's value and usefulness.

ROBERT O. PARMLEY, P. E.

Section One

■

THE POCKET ELECTRONIC CALCULATOR

Major portions of this section were adapted with permission from "Understanding Calculator Math," which was developed by the staffs of Texas Instruments Learning Center and the University of Denver Mathematics Laboratory. Copyright © 1976, 1978 by Texas Instruments Incorporated, all rights reserved.

1-1 INTRODUCTION

The main purpose of this section is to assemble in one place a wide variety of useful and interesting basic information involving calculators as an aid to the total use of this handbook. It is designed to be a working tool that, when used with your calculator and the following sections, becomes a system for solving field-engineering and design problems as well as a key to discovery.

Today's hand-held or pocket calculators are rugged and inexpensive enough to be a natural and "go-anywhere" tool for the active engineer in a mobile-type practice involving fieldwork. Your calculator can help in exploring and learning about mathematics itself, as well as in solving a variety of number problems from design to construction.

Electronic calculators are a natural part of living today—a technological answer to the real need we all have for quick, accurate calculations. Any calculator, however, is "no more functional than the knowledge of the person who operates it." As with any convenience or tool, a pen, wrench, car, radio or whatever, it's important to get the "feel" of it. Check out all of its features—get familiar with what it will (and will not) do for you.

To help you get better acquainted, this section is a quick tour of the essential features and keys of most calculators. Some information is included on *why* each key is important, as well as *how* each key works. After completing this tour, you'll be better able to use your calculator and the data contained in the following sections.

1-2 THE STORY OF MATHEMATICS

It is said that mathematics began long ago in Egypt. The Nile River would flood on occasion and wash away all landmarks and monuments. People needed a way to know where their land was after these floods, so methods of earth measurement (later to be called *geometry*) were invented. The Greeks, always thinking, picked up those techniques, developed them further, and added new ideas such as *algebra* and *trigonometry.* Math was off and running. It was used in navigation. It was interesting. It was fun. Mathematics was used to help learn about the ways in which the world worked, what it looked like, and how much things cost. Calculus, statistics, and income taxes were invented.

1-3 THE STORY OF CALCULATORS

As mathematics began to grow, people started to notice that there were some parts of it that were not nearly as much fun as others—downright tedious, in fact. Getting answers not only involved looking carefully at nature and people and analyzing them (fun part), but also often involved adding, subtracting, multiplying, and dividing very cumbersome numbers (not-so-much-fun part). People began looking for tools to help them handle the arithmetic part of mathematics more easily.

First, pebbles were used for counting things and keeping track. Then these were placed on a ruled table or strung on a frame to form the *abacus* (a device still widely used in many parts of the world). Calculating tools then evolved—somewhat slowly—and a series of mechanical devices developed, starting in the 1600s with ideas from men such as John Napier. The first real calculating machine was invented by the French mathematician Blaise Pascal—for handling monetary transactions. It was a complex entanglement of gears, wheels, and windows. Next came even more complex whirling and whizzing mechanical units, with buttons, wheels, and hand cranks. Bigger machines using relays and punched cards came into being as electricity was applied to mathematics in helping take the 1890 U.S. Census.

Computers were born and began to grow. Slide rules (easy to use and much more accessible than computers) were invented to help take some of the tedium out of long calculations.

1-4 THE ELECTRONIC HAND-HELD CALCULATOR

Then, about a decade ago, people working in electronics began making some breakthroughs that resulted in the inexpensive, accurate, and reliable hand-held calculator. Calculator math became available to everyone. Now, throughout the world, people are finding these little devices to be powerful allies as they handle numbers and math in their everyday lives.

Math is all around us and is part of many daily activities. Your calculator allows you to handle many of these problems quickly and accurately—without having to wrestle with lengthy, tedious computations. This section consists of an accessible and compact package of the principles you need to easily work your problems with keyboard solutions. The section was designed to work together with your calculator—to open up all its secrets and let you have more complete access to its power. Use them together! Both of them have been designed for you.

The first step is to really get acquainted with your calculator—to put it through its paces and see all aspects of how it operates. We will take a quick tour of the features and keys on most calculators and see why each key or feature is there, as well as how each can be useful. This tour is important—it will familiarize you with the scope and power of your machine. Subsequent sections are packed with data in a variety of fields as well as with the details needed in putting together quick, accurate solutions to your engineering problems.

Remember, too, that although your calculator is packed with the latest in state-of-the-art solid-state technology, it needs love and respect, as well as occasional use as a toy. Don't be afraid to play with it—it's rugged and durable enough to be used anywhere. You may find yourself exploring patterns and relationships which can lead you to a whole new appreciation of the beautiful side of numbers and mathematics.

1-5 THE SLIDE-RULE CALCULATOR

The slide-rule calculator simply has replaced the conventional "slip-stick." The electronic slide rule is faster, more accurate, and certainly more versatile.

These compact tools perform calculations of roots, powers, reciprocals, common and natural logarithms, and trigonometry (in degrees, radians or grads), in addition to basic arithmetic. Versatile memory functions include STORE, RECALL, SUM TO MEMORY, and MEMORY/DISPLAY EXCHANGE. Besides these, answer display to 10 places to the right of the decimal, sign in standard format or in scientific notation, correlation, linear regression, trend-line analysis, and more are available in these units. (See Fig. 1-1.)

1-6 PROGRAMMABLE CALCULATORS

The programmable calculators are truly computers—with capabilities of astronomical proportions. Multi-use memories provide addressable memory locations in which you can store data for recall.

Several hundred program steps or up to 100 memories are available. (See Fig. 1-2).

1-7 MASTER LIBRARY MODULES

The master library module simply plugs into the electronic calculator. It includes many different programs in key areas. An example of its size and ease of installation is shown in Fig. 1-3.

Fig. 1-1 Slide-rule electronic calculator, Model TI-55 *(Texas Instruments Incorporated)*.

Fig. 1-2 Programmable calculator, Model TI-59 *(Texas Instruments Incorporated)*.

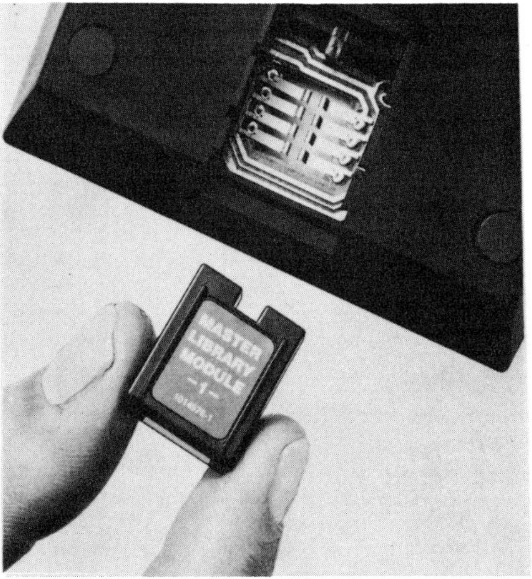

ig. 1-3 Master library module *(Texas Instruments Incorporated).*

1-8 SOLID-STATE SOFTWARE LIBRARIES

Solid-state software is an advance in micromemory technology an provides new programming versatility. It only takes seconds to dro in a module and access a program with a few key strokes. Option solid-state software library modules include surveying, aviatio marine navigation, applied statistics, real estate, and investment.

1-9 PRINTING CALCULATORS

Several calculators are available with printing attachments whi imprint on a tape all actions of the calculator. Figure 1-4 pictures programmable calculator installed on a printer.

Fig. 1-4 Programmable calculator and printer, Model TI-59/P 100A *(Texas Instruments Incorporated).*

1-10 GENERAL OPERATION

The keys and features described in this section and used throughout this book are common to many calculators made by several manufacturers.

For simplicity, the key symbols used here are taken from the TI-30 calculator. (See Fig. 1-5.) They may be slightly different from the keys on your calculator—your owner's manual should clarify which

g. 1-5 Electronic calculator, Model TI-30 *(Texas Instruments In-porated).*

keys perform each function. The keystroke sequences shown assume your calculator has at least one memory, two levels of parentheses, and scientific notation.

Answers may vary slightly in the final decimal place since each calculator has its own technique for rounding the answer. When in doubt about what your calculator does, check your owner's manual.

1-11 BATTERY INSTALLATION AND CONSIDERATIONS

If your calculator is of the TI-30 type, it operates on a standard 9-volt (non-rechargeable) battery. For best results it's recommended that you use an *ALKALINE* battery in your calculator. If you use a non-alkaline battery, it's important to remove it before storing your calculator or at the first signs of discharge. (If you don't – there's danger of damage to the unit from battery leakage.) It's time for a new battery at the *first* signs of erratic display or calculating behavior, and replacing the battery right away is always the best practice.

To install a battery in a TI-30, just insert a small coin into the slot on the back of the calculator case, and pry up the cover gently. Be careful not to tug on the battery wires when handling the snap connectors on the battery. When your new battery's connected, slide the ridged edge of the cover into the case and snap down the edge with the little latch. That's all there is to it!

Check your owner's manual for instructions on replacing batteries and recharging your calculator.

1-12 KEY-TURN ON/CLEARING

The ON key on the calculator keyboard (or ON/OFF switch) turns the calculator on (power "on" is indicated by lighted digits in the display). Just turning most calculators on from the "off" condition clears everything inside to zero, and a **0.** should appear in the display.

Machines vary as to the keys used for clearing them. (Clearing is basically the process of sweeping any previous calculations from your machine in preparation for new ones.) With TI-30 type calculators the [ON/C] key is also used as a clear key. Others may have different arrangements – check your owner's manual.

- *ONE push of the CLEAR key clears the last number you entered into the calculator,* as long as the number wasn't followed by a function or operation key. (So if you hit a 5 instead of a 6 in the middle of a problem, just hit [ON/C] once, and try again.)
- If the CLEAR key is pushed right after an operation or function key (including the [=] key), the display, constant and all operations are cleared.
- *TWO pushes of the CLEAR key clears the entire calculator,* except for what's in memory. (The memory is a storage place for numbers you need to use over and over – we'll say more about it later on.)
- Check your owner's manual for the differences between a *clear* key and a *CLEAR ENTRY* or *CLEAR DISPLAY* key.

1-13 THE DISPLAY

To check out your calculator's display, turn it on and then push the [8] key until the display is filled with 8's. (In the TI-30 there should be 8* of them, and check to be sure that all of the parts of all your 8's are lit up.) You can enter up to eight* digits into the calculator at any one time; digit keys pressed after the 8th* key are ignored. (Internally, however, a calculator may work with more digits than shown in the display—see "Data Entry" section.)

Now press the CLEAR key twice and press the decimal point key [•] and the change sign key [+/-]. The change sign key changes the sign of any displayed number, and allows you to enter *negative numbers*—those numbers *less than zero* that we all have to deal with from time to time (such as money owed, etc.).

Your calculator's display really has a powerful bundle of technology behind it. If your display is made up of small, bright red digits, each segment is a little diode made of Gallium arsenide (GaAs), a substance which emits light when electric current is passed through it under just the right conditions. If your display is bright green, your digits are of the *vacuum fluorescent* (VF) variety, where a gas discharge provides the light for the display segments. If your display is made up of darker digits on a silver or silver/green background it's of the *liquid crystal* variety. Liquid Crystal Displays (LCD's) use an electro-chemical action to change the physical appearance of the display digits, and use very little electric power.

*these vary from calculator to calculator, so check your owner's manual for specifics.

Each number you see in your display isn't lit up or activated all at once either. All the display segments are turned on and off very rapidly ("strobed") by the electronic "brain" of your calculator, so fast that your eye puts the segments together into numbers. (Shake your calculator slightly while watching the display, and you'll begin to catch the display in the act.) Once the number is lit up, specially designed lens systems and colored windows are often used to get the light to you as a clean, easy to read display.

1-14 KEY AND POWER SAVER

The OFF key or ON/OFF switch just turns off the calculator, plain and simple. Turning the calculator off and on will clear most calculators completely – including the memory.

TI-30 type calculators have an added bonus feature – automatic shutoff power saving circuitry. If you leave the calculator on without activating any keys, the calculator takes these steps to save power:

1) After typically 25 to 30 seconds the display (which consumes the most power) shuts down to a single traveling decimal point which moves across the display from left to right. As soon as any key is depressed the display comes "back to life". (A good way to reactivate the display is to press the EXC key twice. This makes sure that any calculations in progress or data in the machine are all unaffected.)

2) If the calculator continues in the traveling decimal mode uninterrupted for 7 to 14 minutes, it quietly shuts itself off!

Working together these two features can save up to 50% on battery life. The common problem of inadvertently leaving your calculator on when not in use is solved! To see if your calculator has a power saving feature, refer to your owner's manual.

1-15 DATA ENTRY KEYS

Most of today's calculators provide a full floating decimal point, and numbers are entered into the machine with the data entry keys: [0] - [9] [•] [+/-] . As you press number keys, the decimal point remains to the right of your entry until the decimal point key [•] is pressed. The fractional part of the number is then keyed in and the decimal point floats to the left with it. To change the sign of a number in the display just push the change sign key [+/-] once. (Pressing [+/-] again changes the sign back again).

ACCURACY:

With most calculators you can directly enter numbers up to 8 digits in length. Your calculator may hold and work internally with more digits* — for extra accuracy. In these cases, you can enter numbers with more digits than appear in the display as the sum of two numbers, as shown below. (As an example, results of calculations are computed to eleven digits in the TI-30, and then rounded off to 8 digits in the display).

* These numbers vary from calculator to calculator, so check your owner's manual for specifics.

Example: To enter the maximum number of digits, for instance, an eleven-digit number – 418413.23106.

Press	**Display/Comments**
418413 [+]	**418413.** The whole part of the number.
.23106 [=]	**418413.23**

1-16 THE BASIC KEYS

If you're looking at buying a new car, there are certain basic things you must have before you start checking out the extra options. Calculators have "basics", too – the four basic operations.

When you press the [=] key, all pending operations (things waiting to happen inside the calculator) are completed, you get your result, and the calculator is cleared – ready to start on the next problem. Here are a couple of quick examples:

[+] [-]

You start with $150 in your checking account, and write checks for $10, $45.25, $15, and then make a $50 deposit; what's your balance?

Press	**Display/Comments**
150 [-] 10 [-] 45.25 [-] 15 [+] 50 [=]	**129.75** Your balance.

[×] [÷] : If pencils cost $6.48 per gross, how much will 47 of them cost? (A "gross" of anything is 144 of them.)

Press	**Display/Comments**
6.48 [÷] 144 [×]	$6.48 [÷] 144 is how much 1 pencil costs
47 [=]	**2.115** Cost of 47 pencils (You'd probably pay $2.12.)

1-17 ALGEBRAIC OPERATING SYSTEM™

Mathematics is a science which is persnickety about some things. One of them is that it never permits two different answers to the same series of computations. Because of this requirement – one solution for any computation – mathematicians have established a set of accepted (and universal) rules when mixed operations are in one calculation. For example, the problem:

$$3 + 10 - 2 \times 14 \div 7 = ?$$

has only one right answer! (Know what it is? It's 9.)

You can key the above problem directly, left to right, into your TI-30 type calculator with AOS entry method and you'll get the correct answer. *Not all calculators will do this.* A calculator with AOS (Algebraic Operating System™) entry method, receives all the numbers and operations in the problem, automatically sorts them out for you, and applies them all according to the correct rules of mathematics – letting you see intermediate steps along the way. A calculator with AOS entry method automatically performs operations in the following order:

1) Single Variable Functions (the keys [sin] [cos] [tan] [log] [lnx] [x²] [√x] [1/x] [%] and [INV] working with them) act on the displayed number immediately. (All these keys are discussed for you later in this chapter). Then:

2) Exponentiation [y^x], and roots, $\sqrt[x]{y}$ ([INV] [y^x]), are performed as soon as single variable functions are completed. (More on these functions later.)
3) Multiplications and divisions are completed next, then
4) Additions and subtractions are completed.

Finally, the equals key completes all operations.

In elementary school you may have heard the memory aid "My Dear Aunt Sally" (MDAS) applied to help you remember the last part of this hierarchy (multiplications and divisions first – in order left to right – then additions and subtractions in the same way). In a calculator equipped with AOS entry method – all of this is remembered for you. If your calculator does not have AOS entry method, you must apply the MDAS rule yourself.

There are cases in problem solving when *you* want to specify the order in which an expression is evaluated, or the way in which numbers and operations are grouped. In these cases you'd use the parentheses keys: [( ] [ )] , which are discussed in the next section. Parentheses always demand a special first level of attention in mathematics, and they are treated that way by your calculator. (Parentheses say "Do Me First".)

1-18 PARENTHESES KEYS

In a variety of problems, you may need to specify the exact order in which expressions are evaluated, or the way in which numbers are grouped, as a problem is solved. The parentheses keys allow you to do this. Parentheses give you a way to "cluster" numbers and operations. By putting a series of numbers and operations in parentheses you tell the calculator: "Evaluate this little problem first – down to a single number result, then use this result for the rest of the calculation." You should make use of parentheses whenever you need the calculator to make an "intermediate" calculation, or if you have any doubts in your mind about how the calculator is going to reduce an expression.

Note that different calculators have different limits as to the number of parentheses that can be opened at any one time, and how many "pending" operations that can be handled. In a calculator of the TI-30 type, you can open up to 15 sets of parentheses at any one time, with up to 4 operations pending. Exceeding these limits results in an *Error* indication. (In this book, we will assume your calculation has two levels of parentheses.)

Notice the following important point when using parentheses. Often you'll see an equation or expression with parentheses indicating an *implied multiplication:* $(2 + 1)(3 + 2) = 15$. *Your calculator will not perform implied multiplications.* You have to key in the operation between parentheses:

[(] 2 [+] 1 [)] [×] [(] 3 [+] 2 [)] [=] **15.**

Here's an example on the use of parentheses:

Evaluate: $\dfrac{(8 \times 4) + (9 \times -19)}{(3 + 10 \div 7) \times 2} =$

Solution: Here it's important that the calculator evaluates the entire numerator, and then divides by the entire denominator. In problems of this type, you can be sure of this by placing an extra set of parentheses around the numerator and denominator as you key in the problem.

Press	**Display/Comments**
[( ] [( ] 8 [×] 4 [ )] [+]	**32.** (8 × 4) displayed
[( ] 9 [×] 19 [+/−] [ )] [ )] [÷]	**−139.** This is the value of the entire numerator
[( ] [( ] 10 [÷] 7 [+] 3 [ )]	**4.4285714** This is (3 + 10 ÷ 7)
[×] 2 [ )]	**8.8571429** This is the denominator.
[=]	**−15.693548** The final result. The [=] key completes the division and the entire problem.

Note: The order of the denominator was changed to follow the MDAS rule of algebraic hierarchy.

1-19 SCIENTIFIC NOTATION

Very often, particularly in problems that relate to science or engineering, you find yourself needing to handle an astronomically huge or really small number. Such numbers are easily handled (by you and your calculator) using *scientific notation.* A number in scientific notation is expressed as a base number (or "mantissa") times ten raised to some power (or "exponent").

$$\text{Mantissa} \times 10^{\text{power}}$$

(Raising a number to a power or exponent just means to multiply it by itself the number of times the power indicates: $10^3 = 10 \times 10 \times 10$.) To enter a number in scientific notation:

- Enter the mantissa (then press [+/−] if it's negative)
- Press [EE↓] (EE stands for "Enter Exponent")
- Enter the power of ten (then press [+/−] if it's negative.)

A number such as -3.6089×10^{32} will look like this in the calculator display:

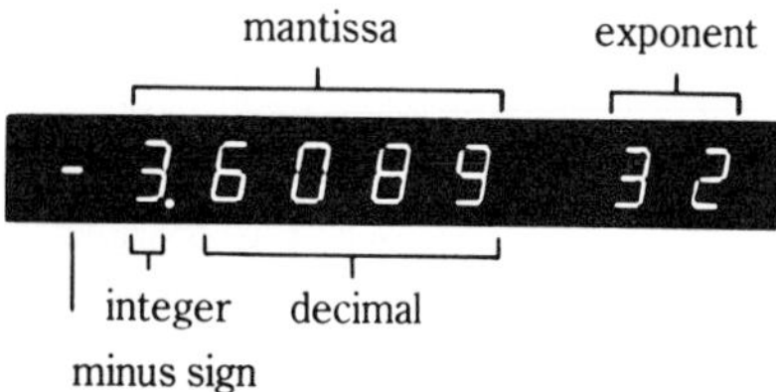

In scientific notation the power of *ten* tells you where the decimal point would have to be if you were going to write the number out longhand.

A positive exponent tells you how many places the deci mal point should be shifted to the right.

Example: 2.9979×10^{8} equals
299,790,000.
Move decimal 8 places right, add zeros as needed.

A negative exponent tells you how many places the decimal point should be shifted left.

Example: 1.6021×10^{-19} equals
.000 000 000 000 000 000 16021
Move decimal 19 places left, add zeros as needed.

It's easy to see why most folks prefer to handle very large and small numbers in Scientific Notation!

A few points on your calculator and scientific notation:

- No matter how you enter a mantissa the calculator will usually convert it to standard scientific form, with one digit to the left of the decimal point, when any function or operation key is pressed.
- Numbers in scientific notation can be mixed with numbers in standard form in any calculation.

Example: How many 0.5 meter footsteps are there from here to the moon? (Moon-Earth distance is approximately 3.8×10^8 meters)

Press	**Display/Comments**
3.8 [EE↓] 8 [÷]	
.5 [=]	**7.6 08** 760 million footsteps!

1-20 EXPONENTIAL SHIFT

Some calculators have an [EE↓] key that allows you to shift the position of the decimal point when in scientific notation. Once you've entered a number in scientific notation, and pressed [=] or some other operation key, each push of the [EE↓] key *decreases* the exponent by 1, and moves the decimal point one place to the right. (This doesn't change the value of the number —just the way it looks.) In this way, you can explore how the position of the decimal point is related to the value of the exponent.

If you press [INV] [EE↓], just the reverse happens. The exponent goes up by one, and the decimal moves one place to the left. (The inverse key [INV], when present, works with several keys on your calculator, and will be discussed later on in this chapter.)

As an example, you might try exploring with the following numbers:

The speed of light is 2.9979250×10^8 m/sec. How many million meters per second is that?

Press	**Display/Comments**
2.9979250 [EE↓] 8 [=]	**2.9979 08**
[EE↓]	**29.979 07**
[EE↓]	**299.79 06**
	Since one million is 1×10^6, the speed of light is over 299 million meters per second. Try moving things the other way with the [INV] [EE↓] sequence:
[INV] [EE↓]	**29.979 07**
[INV] [EE↓]	**2.9979 08**
[INV] [EE↓]	**.29979 09**
[INV] [EE↓]	**.02997 10**

With the [EE↓] and [INV] keys, you can put the decimal just about anywhere you'd want to in the mantissa. The exponent automatically changes to keep the value of the displayed number the same.

1-21 INVERSE FUNCTION OR RECIPROCAL

The [1/x] key just takes the number in the display and divides it *into* 1. (By the way—the letter "x", used in calculator keys just means "any number that may be in the display.") The [1/x] key can be used at any time: it acts immediately on whatever number is in the display, and doesn't affect other calculations in progress.

Now – why have a whole key just for [1/x] ? Well, this operation is useful and important in a variety of problem solving situations:

Example:
You're trying to fill a swimming pool and want to speed up the process. You turn on a main fill faucet that would fill the pool by itself in 10 hours, set up a garden hose that would do it in 28 hours, and a fire hose that would take 6 hours on its own. How long does it take with all three working?

Solution:

$$\frac{1}{\text{Time Total}} = \frac{1}{T_1} + \frac{1}{T_2} + \frac{1}{T_3}$$

where T_1, T_2, and T_3 are the times for the faucet, garden hose and fire hose, respectively.

Press	**Display/Comments**
10 [1/x] [+] 28 [1/x] [+]	
6 [1/x] [=] [1/x]	**3.3070866** hours, or about 3 hours, 18 minutes.

Note that the [1/x] key "inverts" or flips over fractions; and this process can be useful in evaluating expressions you'll find in many situations – especially in basic science.

1-22 MEMORY KEYS

The memory in your calculator is a special place in the machine to store numbers you may need to use in calculations later on. It's sort of a "calculator within a calculator" since you can store numbers, or add to what's in memory, without affecting any other calculations you may have in progress. The CLEAR key will not clear out what's in memory.

[STO] – The "Store" key just "stores" the displayed number in the memory, without removing it from the display. (Any number previously stored in memory is *cleared out first.*)

By pressing
[RCL] – (Recall key) any time after a number is stored in memory, the number reappears on the display and can be used in operations and calculations. The number remains in the memory after you press the [RCL] key, and can be recalled as many times as you need it any time after that. The number will remain in memory until you alter it with one of the memory keys, or until you turn the calculator off. (Turning the calculator [OFF] and [ON/C] completely clears everything!)

Here's a quick example on the use of memory:

$$a = \frac{3}{8}(44 - 16)$$

$b = 144 - 9a$, and you need to find b:

Press	**Display/Comments**
3 [÷] 8 [×] [(] 44 [−] 16 [)] [=] [STO]	**10.5** The value for a, store it.
144 [−] [(] 9 [×] [RCL] [)] [=]	**49.5** The value of b.

Another way of looking at memory is to consider that [RCL] is a key that has a number value you can decide. If you need some weird number a lot of times in different calculations, just key it into your calculator and press [STO] . Every time you need it later on press [RCL] , and there it is!

[SUM] or [M+] – Key allows you to algebraically add whatever's in the display directly to what's in the memory. (On most calculators this doesn't affect any calculation in progress – check your owner's manual.) This key comes in handy when you want to keep a running total on something (say your grocery bill), while keeping the rest of the calculator clear for other things (such as calculating best unit prices or discounts).

[EXC] – "Exchange" Key. This key "swaps" what's stored in memory with what's in the display. (The display value gets stored, while the stored number is recalled.) On most calculators this key doesn't affect any calculation in progress, and can come in handy in a variety of problem situations. Not all calculators have an Exchange key.

(If your calculator has multiple memories, your owner's manual will describe how to use them.)

1-23 SQUARE AND SQUARE ROOT KEYS

The square key, [x^2] , just takes the number in the display and multiplies it by itself.

This process, *squaring*, is a very handy one in a whole variety of situations in mathematics and in problems from everyday life – so your calculator has a whole key just for it.

One place squaring is handy is in calculating *areas.* For example, if you have a square field that is 5 kilometers on a side, its area is 5 squared (5^2) or 25 square kilometers.

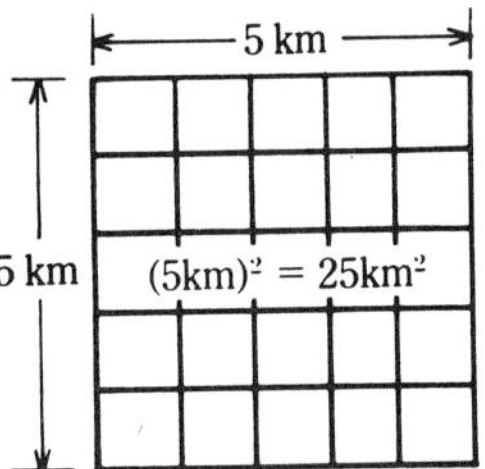

This illustration also suggests the origin of the term "square root" ($\sqrt{x}$). Let's say you have a *square* field that covers 25 square kilometers (25 km^2) area and need to know how long it is on a side. Looking at the figure, you can see that the answer is 5 km. Well, the *square root* of 25 equals 5. You might say that the 5 km sides of the field give rise to its 25 km^2 area, in much the same way a "root" gives rise to a plant. Notice that the "$\sqrt{\ }$" symbol means square root, so you write the phrase "the square root of 25 is 5" simply: $\sqrt{25} = 5$.

The square root key [√x] takes the square root of the number in the display. The square root of any number (say x) is another number ($\sqrt{x}$) that when multiplied by itself gives you back the original number. (Got that?)

Example: Evaluate:

$$\frac{(3 \times 8)}{\sqrt{2}} + 6 \times 3.1^2$$

Press	**Display/Comments**
3.1 [x^2] [×] 6 [=] [+]	
[(] [(] 3 [×] 8 [)] [÷] 2 [$\sqrt{x}$] [)]	
[=]	**74.630563**

In expressions like those above, again notice that the [x^2] and [$\sqrt{x}$] functions do not act to complete a calculation that is not complete (such as 6 × 3.1) but *only* act on the number in the display. An "Error" indication will result if the [$\sqrt{x}$] key is pressed with a negative number in the display. Square roots of negative numbers are called "imaginary" numbers, and your calculator is not equipped to handle these (see "Error" Indications section).

1-24 POWERS AND ROOTS

A power (or "exponent") is a number that's written above and to the right of another number (called the base).

base y^x power or exponent

y^x just means:
Take the number y, and multiply it by itself x times – and that's just what the [y^x] key does for you! This process is quite often required in problem solving, and can be a tedious process prone to all sorts of errors. With the [y^x] key helping you, however, much of the hassle is eliminated.

To use the [y^x] key just:

- Enter the base (y)
- Press [y^x]
- Enter the power (x)
- Press [=]

Example: You have a cubic bin that's exactly 3.21413 meters on a side. What's its volume in m^3?

Solution:

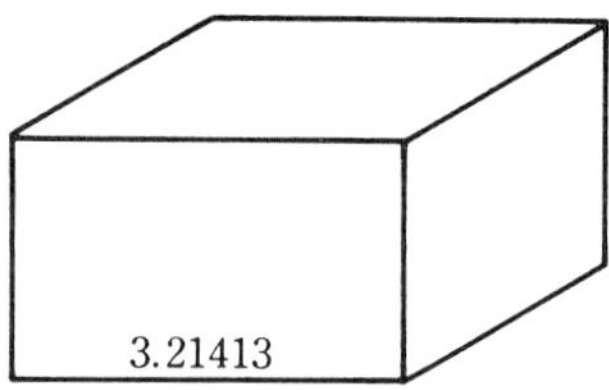

The volume of the bin is 3.21413^3.

Press	**Display/Comments**
3.21413 [y^x] 3 [=]	**33.203993** cubic meters

Important Note: On some calculators the right-most digit flashes during the short time the calculator is "grinding out" the result; on other calculators, the display goes blank. Be sure the calculator *has completely finished an operation* before pressing the next key!

Using the key sequence [INV] [y^x] gives you the "xth root of y", which is often written as $\sqrt[x]{y}$. The xth root of a number y, is another number ($\sqrt[x]{y}$), that when multiplied by itself x times, gives you back y. (Got that one?) Mathematically you would write:

$$(\sqrt[x]{y})^x = y.$$

This process of taking roots also crops up quite often in various applications of mathematics, and it's a downright "bear" of a task in many situations! With a calculator it's easy and accurate. To compute the xth root of y:

- Enter the base (y)
- Press [INV] [y^x]
- Enter the root (x)
- Press [=]

Example: Compute $\sqrt[3.12]{1460}$

(This would be tough, without your calculator)

Press	**Display/Comments**
1460 [INV] [y^x] 3.12 [=]	**10.332744**

If your calculator does not have an [INV] key, here is the procedure for taking the $\sqrt[x]{y}$:

Press	**Display/Comments**
1460 [y^x] 3.12 [1/x] [=]	**10.332744**

Additional important notes on [y^x] and [INV] [y^x] :
These two functions are the only special functions that do not act on the displayed value immediately. The second number (x value) must be entered in each case, before the function can be completed. The [=] key completes the calculation. (Closing a parenthesis that contains either of these functions will complete the operation also.)

On most calculators there is a restriction on these functions – the variable y should *not* be negative (this has to do with the way the calculator goes about computing the functions). If you try either of these with a negative number – you'll get an "Error" indication. (Trying to calculate a "zeroth" root also gives you an error!)

1-25 CALCULATIONS WITH A CONSTANT

Here's a real labor-saving feature – one that can increase accuracy and reduce tedium when you've got to do a whole load of repetitive calculations. If your calculator has a [K] constant key you can store *a number and an operation sequence,* and then these can be used by the calculator to operate on any displayed number. This type of feature is really handy if you have to "mark down" all the items in a store, or multiply all the ingredients in a recipe by 3, or in any repetitive situation!

To use the [K] feature:

- Enter the repetitive number, m.
- Enter the desired operation.
- Press [K] .

From then on in, all you do is enter the string of numbers you want to operate on, and press [=] after each entry to complete each calculation. The table below summarizes how the [K] feature will work in each case:

Key sequence	Result
m [+] [K]	adds m to each subsequent entry, when [=] is pressed.
m [−] [K]	subtracts m from each subsequent entry.
m [×] [K]	multiplies each subsequent entry by m.
m [÷] [K]	divides each subsequent entry by m.
m [y^x] [K]	raises each subsequent entry to the m power.
m [INV] [y^x] [K]	takes the mth root of each subsequent entry.

Example: Multiply the numbers 81, 67, 21, 32 by .69174385.

Press	Display/Comments
.69174385 [×] [K]	Enter the repetitive number m and the operation ([×]), then press [K]
81 [=]	**56.031252**
67 [=]	**46.346838**
21 [=]	**14.526621**
32 [=]	**22.135803**

Note: Clearing the calculator or entering any of the arithmetic functions clears the [K] constant feature.

Some calculators have an automatic constant function that is activated by the [=] key. Check your owner's manual for details.

1-26 PI (π) KEY

The [π] key, when pressed, displays the value of "π". (That's the Greek letter Pi, pronounced "pie".) The number you'll see in the display is 3.14159265... The quantity may actually be entered into your calculator correct to more digits, than shown in your display.

Pi is a very special number that represents a relationship that is found in *all circles.* The Greeks were the first to discover this relationship.

The Greeks probably studied thousands of circles before they determined the following fact: In any circle if you take the *distance around its edge* (called the circumference), and divide that by the *distance across its middle* (the diameter), the result is always the *same number.* That number is π, or about 3.1415927.

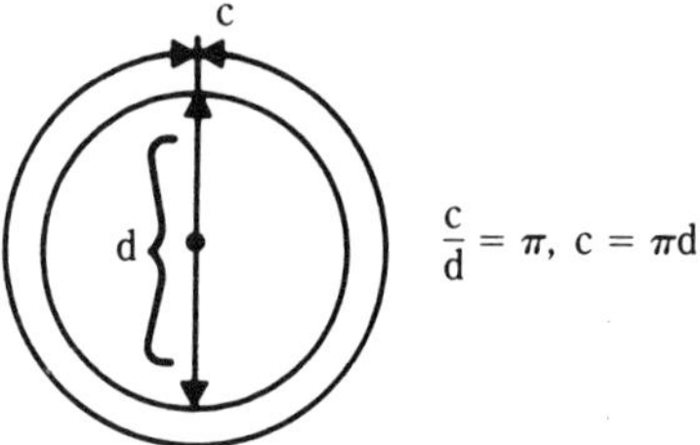

π is found in almost any problem or calculation involving circles, and since circles are pretty common things a whole key on your calculator is devoted to remembering π for you. The [π] key displays π immediately, does not affect calculations in progress, and can be used at any time in a calculation.

1-27 PERCENT (%) KEY

Taxes, discounts, inflation, etc. all involve percentages – the number of "cents per dollar" or parts per hundred. The [%] key on your calculator is a genuine labor-saving device that handles more than it appears to at first glance.

When the [%] key is pressed, the number in the display is immediately converted to a decimal percent (multiplied by 0.01). If you enter 43.9 and press [%] , **0.439** appears in the display.

The real power of an add-on [%] key is turned on when it's used in combination with an operation key, which allows a wide variety of percentage problems to be tackled. The following key sequences, each operating on the displayed number, will handle the most common problems involving percentage.

[+] n [%] [=] adds n% to the number displayed.

Example: What will the cost of a new $75 jacket be with a 6.5% sales tax?

Press	**Display/Comments**
75 [+] 6.5 [%]	**4.875** *Note:* At this point 6.5 per cent of 75 is computed and displayed. Pressing [=] adds this amount to 75 and completes the calculation.
[=]	**79.875** (You'd pay $79.88).

Note: To work this problem without an *add-on* percent key follow this procedure.

75 [+] [(] 75 [X] .065 [)] [=] **79.875**

[−] n [%] [=] subtracts n% from the number displayed.

Example: You want to buy a stereo headset for $35; and a sale sign states 38% off. What's the actual price?

Press	**Display/Comments**
35 [−] 38 [%]	**13.3** At this point, 38% of 35 is computed and displayed. Pressing [=] subtracts it from 35 and completes the calculation.

[=]	**21.7** (You'd get the phones for $21.70)

[X] n [%] [=] multiplies the number in the display by n%.

This sequence is for straight percentage calculations: What's 31.258% of $270.00?

Press	**Display/Comments**
270 [X] 31.258 [%] [=]	**84.3966**

[÷] n [%] [=] divides the displayed number by n%.

This key sequence helps to solve those "inverted" percentage problems.

Example: 25 is 15% of what number?

Press	**Display/Comments**
25 [÷] 15 [%] [=]	**166.66667**

To figure percent without a [%] key, simply multiply the number by 0.01.

Press	**Display/Comments**
43.9 [X] .01 [=]	**0.439**

1-28 ANGULAR MEASURE KEY

Angles are measurements that describe how 2 lines or surfaces meet each other. As discussed later on in the trigonometry section, angles are an important part of your life – and if you look around you you'll see angles everywhere.

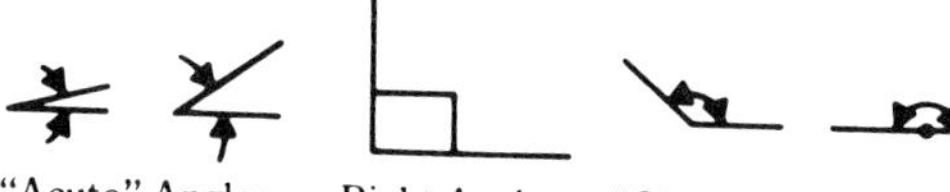

One common angle is the *right angle* (the angle at which walls meet floors), shown above. Angles smaller than right angles are called "acute", larger angles are called "obtuse."

There are 3 common sets of mathematical units you can use to specify angular measure, and they're all related to how they divide up a circle as follows:

The *degree:* $1^\circ = 1/360$ of a circle

The *radian:* $1 \text{ rad} = \dfrac{1}{2\pi}$ of a circle

The *grad:* $1 \text{ grad} = \dfrac{1}{400}$ of a circle

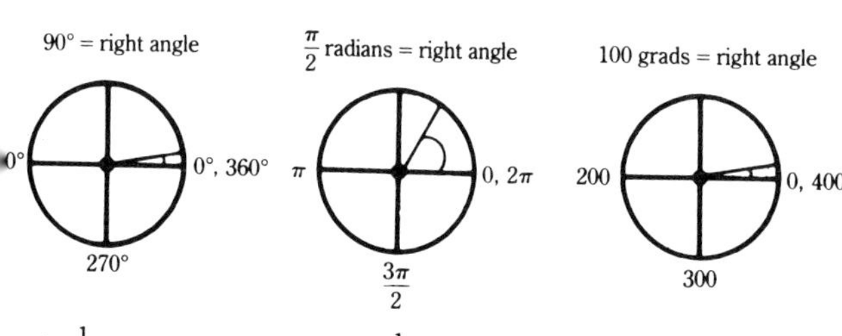

Many calculators can handle all three of these units and a DRG or similar key specifies which units the calculator assumes will be entered into its display. The DRG key works with the Trig functions (sin cos tan) and can be used with them to convert between systems of angular units.

The DRG key on a TI-30 type calculator is actually a 3-position switch that works as follows:

When your calculator is turned on, it is automatically in the degree mode.

Pressing DRG once puts the calculator in radian mode: an apostrophe (') at the left of the calculator display indicates this.

Pressing DRG once again puts the calculator in grad mode: a set of quotation marks (") at the left of the display indicates this mode.

(If you don't see an apostrophe or quotation marks at the left of the display – you're in degree mode.)

Be sure DRG is set correctly when solving Trig problems!

1-29 LOGARITHM KEYS

Logarithms, or "Logs" as they are commonly called, were originally worked up by mathematicians to make computations easier (much as calculators have been devised today). Logarithmic functions have gone on to be involved in man's mathematical descriptions of many natural phenomena. They are found to be helpful describing many natural effects – so you'll be seeing them in a variety of situations.

Logarithms are related to exponential functions, and work like this: If you pick a number called a base (b) then any other number, (say x) can be expressed as b raised to some power (y).

$$x = b^y$$

The logarithm is the *inverse* of this exponential function, and may be written:

$$y = \log_b x.$$

which is stated: "y equals log to the base b of x."

Now why all of this? Logs are very useful (your calculator even uses them internally) in handling complex problems, because using logs allows multiplication, division, and exponentiation (raising to a power) to be replaced by the simpler operations of addition, subtraction and multiplication. The rules for math with logs are as follows;

$\log_b(xy) = \log_b(x) + \log_b(y)$
$\log_b (x/y) = \log_b x - \log_b (y)$
$\log_b (x^n) = n \log_b x$

To do math with logs (before your calculator came along) you would first look up logarithms of the numbers x and y (in bulky tables), perform the operations you need, longhand, according to the rules above, then take the *antilogarithm* (using tables again) of your results to get your final answers.

There are two common bases used for logarithms, and hence your calculator's two log keys. Logarithms to the base 10 are convenient for use in calculating, and are called the *common* logarithms. The [log] key on your calculator immediately displays the common logarithm (base 10) of the number in the display.

The other common base for logarithms is a special number called "e", whose value is 2.7182818 . . . . Logs with this base are called the *natural* logarithms, and occur in many formulas in higher mathematics. (Natural logs are often abbreviated "ln x".) The [lnx] key on your calculator immediately displays the natural logarithm (base e) of the number in the display.

Note: When calculating logarithms with either the [log] or [lnx] keys, the number in the display must be *positive*, or an "Error" indication will result.

The [INV] key works with the [log] and [lnx] keys to calculate *antilogarithms:*

The common antilogarithm (10 to the x power) sequence [INV] [log] calculates the common antilogarithm of the displayed value. This sequence raises 10 to the displayed power.

The natural antilogarithm (e to the x power) sequence [INV] [lnx] calculates the natural antilogarithm of the number in the display. This sequence raises the number e to the displayed power.

Examples:

Calculate: log 15.32
ln 203.451
$e^{-.693}$
10^{π}

Press	Display/Comments
15.32 [log]	1.1852588
203.451 [lnx]	5.3154252
.693 [+/-] [INV] [lnx]	0.5000736
[π] [INV] [log]	1385.4557

1-30 INVERSE KEY SUMMARY

The inverse key has been mentioned throughout this chapter in each situation where it can work in a sequence with other keys. Here is a summary of where it works and what it does.

[INV] [EE↓] adds one to the exponent and moves the decimal one place to the left.

[INV] [y^x] takes the xth root of the displayed value y. Order of entry is y [INV] [y^x] x. y cannot be negative, but both x and y can be fractional.

[INV] [sin] Arcsine ($\sin^{-1}$) Sequence – Calculates the smallest angle whose sine is in the display (first or fourth quadrant).

[INV] [cos] Arccosine ($\cos^{-1}$) Sequence – Calculates the smallest angle whose cosine is in the display (first or second quadrant).

[INV] [tan] Arctangent ($\tan^{-1}$) Sequence – Calculates the smallest angle whose tangent is in the display (first or fourth quadrant).

[INV] [log] Common Antilogarithm (10 to the x power) Sequence – Calculates the common antilogarithm of the displayed value (raises 10 to the displayed power).

[INV] [lnx] Natural Antilogarithm (e to the x power) Sequence – Calculates the natural antilogarithm of the number in the display (raises e to the displayed power).

1-31 ERROR INDICATIONS

If your calculator displays an "Error" indication, you probably asked it to try to do something it couldn't do. (It tries to do everything you ask it to; when it can't, it signals for help with the "Error" Signal.)

When this occurs, no entry from the keyboard is accepted until your calculator is cleared or turned off and on again. This clears the error condition and all pending operations. Then, you have to begin the problem from scratch—starting right at the top. Here are typical reasons why you'll get an "Error" message:

1. Number entry or calculation result (including summation into memory) outside the range of the calculator.
2. Dividing a number by zero.
3. The mantissa is zero and [log], [lnx] or [1/x] is pressed.
4. The mantissa is negative and [log], [√x], [yˣ], [lnx] or [INV] [yˣ] is pressed.
5. Inverse of sine or cosine (arcsine, arccosine) when the mantissa is greater than 1.
6. Tangent of 90°, 270°, $\pi/2$, $3\pi/2$, 100 grads, 300 grads or their rotation multiples like 450°, etc.

Check your owner's manual for "Error" messages unique to your calculator.

1-32 TRIGONOMETRIC FUNCTION KEYS

Triangles! There are whole courses of study devoted to them. Why? Well, triangles are very common shapes—found in a variety of natural situations, they're ideal for certain construction and architectural applications, part of inclined planes and roadbeds, etc. In addition, the relationships between the sides and angles of a right triangle (that's a triangle with one right angle), crop up again and again in nature. The three "trigonometric functions", called the *sine, cosine,* and *tangent,* are an important part of the way scientists describe electrical and wave phenomena, as well as many kinds of *periodic* motion of mechanical systems.

The relationships are related to the right triangle as shown below:

Hypotenuse
Opposite side
Θ
Adjacent side

The angle we'll focus on is Θ, and notice the side of the triangle opposite to Θ is labelled "*Opposite*", while the side next to Θ is labelled "*Adjacent*". The side opposite to the right angle is called the *Hypotenuse.* The three trigonometric functions are defined as follows:

$$\text{SIN } \Theta = \frac{\text{Length of Opposite Side}}{\text{Length of Hypotenuse}} = \frac{O}{H}$$

$$\text{COS } \Theta = \frac{\text{Length of Adjacent Side}}{\text{Length of Hypotenuse}} = \frac{A}{H}$$

$$\text{TAN } \Theta = \frac{\text{Length of Opposite Side}}{\text{Length of Adjacent Side}} = \frac{O}{A}$$

The [sin] [cos] and [tan] keys each assume that there is an angle in the display, *in units specified by the setting of the* [DRG] *key.* When any of these 3 keys are pressed, the appropriate function (sine, cosine, or tangent) of the displayed angle is computed and displayed immediately. These **3 keys** do not affect calculations in progress, and can be used at any time.

Example: Compute the sine, cosine, and tangent of 90 degrees and 90 grads.

Press	Display/Comments
[OFF] [ON/C]	This makes certain the calculator is in degree mode.
90 [sin]	**1.**
90 [cos]	**0.**
90 [tan]	**Error.** (The tangent of 90° is undefined.)
[ON/C]	Clears "Error" condition.
[DRG] [DRG]	" **0.** Converts to grads mode.
90 [sin]	**".98768834**
90 [cos]	**".15643447**
90 [tan]	**"6.3137515**

The "ARC" trigonometric functions – ARCSINE, ARCCOSINE and ARCTANGENT,

are the trigonometric functions in "reverse." The term *Arcsine* (often written $\sin^{-1}$) means "The Angle whose SINE is." You calculate arcsines, arccosines, and arctangents using the [INV] key, with the [sin] [cos] and [tan] keys on your calculator. The result of an "arc" calculation is an angle, that will be in the units specified by the setting of the [DRG] key.

[INV] [sin] – Calculates the smallest angle whose sine is in the display (first or fourth quadrant).

[INV] [cos] – Calculates the smallest angle whose cosine is in the display (first or second quadrant).

[INV] [tan] – Calculates the smallest angle whose tangent is in the display (first or fourth quadrant).

Examples: Calculate the arcsin of 0.514
arccos of 1.4
and arctan of 15, all in degrees.

Press	**Display/Comments**
[OFF] [ON/C]	This makes sure you're in degree mode.
.514 [INV] [sin]	**30.930637** degrees
1.4 [INV] [cos]	**Error.** *Note:* The argument of the sine or cosine is always between plus and minus 1.
15 [INV] [tan]	**86.185925** degrees.

NOTES

Section Two

▪

MATHEMATICAL DATA AND FORMULAS

TABLE 2-1 Area Formulas

FORM		METHOD OF FINDING AREAS
TRIANGLE		Base × ½ perpendicular height. $\sqrt{s(s-a)(s-b)(s-c)}$, s = ½ sum of the three sides a, b, c.
TRAPEZIUM		Sum of area of the two triangles
TRAPEZOID		½ sum of parallel sides × perpendicular height.
PARALLELOGRAM		Base × perpendicular height.
REG. POLYGON		½ sum of sides × inside radius.
CIRCLE		$\pi r^2 = 0.78540 \times \text{diam}^2 = 0.07958 \times \text{circumference}^2$.
SECTOR OF A CIRCLE		$\frac{\pi r^2 A^\circ}{360} = 0.0087266\, r^2 A^\circ$, = arc × ½ radius
SEGMENT OF A CIRCLE		$\frac{r^2}{2}\left(\frac{\pi A^\circ}{180} - \sin A^\circ\right)$
CIRCLE of same area as a square		Diameter = side × 1.12838
SQUARE of same area as a circle		Side = diameter × 0.88623
ELLIPSE		Long diameter × short diameter × 0.78540
PARABOLA		Base × ⅔ perpendicular height.

SOURCE: N. Foster, *Practical Tables for Building Construction*. Copyright © 1963 by McGraw-Hill, Inc. New York. (Used with permission of McGraw-Hill Book Co.)

TABLE 2-2 Finding Surfaces and Volumes of Solids
(S = lateral or convex surface; V = volume)

SHAPE	FORMULAS
Parallelepiped 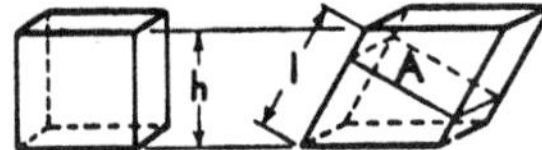	S = perimeter, P, perp. to sides × lat. length l : Pl. V = area of base, B, × perpendicular height, h : Bh. V = area of section, A, perp. to sides, × lat. length l : Al.
Prism right, or oblique, regular or irregular 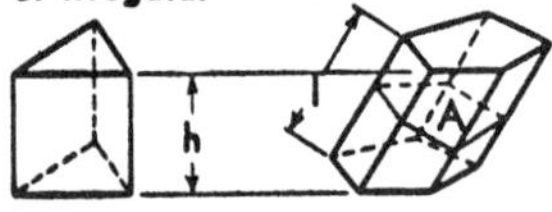	S = perimeter, P, perp. to sides × lat. length l : Pl. V = area of base, B, × perpendicular height, h : Bh. V = area of section, A, perp. to sides, × lat. length l : Al.
Cylinder, right or oblique, circular or elliptic etc. 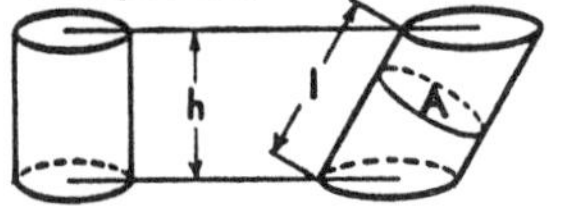	S = perimeter of base, P, × perp. height, h:Ph. S_1 = perimeter, P_1, perp., × lat. length, l : P_1 l. V = area of base, B, × perp. height, h:Bh. V = area of section, A, perp. to sides × lat. length l:Al.
Frustum of any prism or cylinder 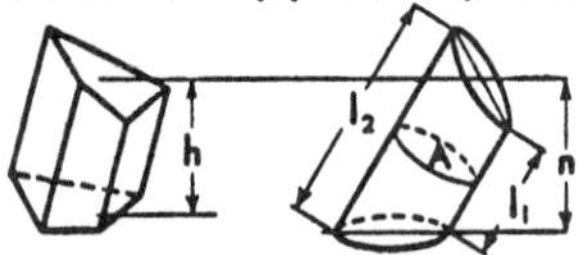	V = area of base, B, × perpendicular distance h, from base to centre of gravity of opposite face: Bh. for cylinder, $\frac{1}{2}$ A $(l_1 + l_2)$
Pyramid or Cone, right and regular 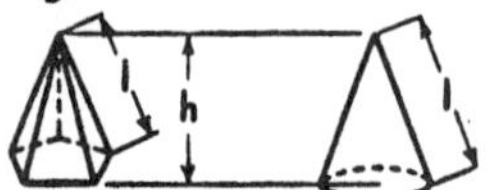	S = perimeter of base, P, × $\frac{1}{2}$ slant height l : $\frac{1}{2}$ Pl. V = area of base, B, × $\frac{1}{3}$ perpendicular ht., h : $\frac{1}{3}$ Bh.

TABLE 2-2 Finding Surfaces and Volumes of Solids *(cont.)*
(S = lateral or convex surface; V = volume)

SHAPE	FORMULAS
Pyramid or Cone, right or oblique, regular or irregular 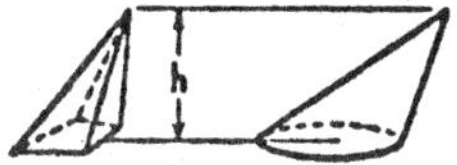	V = area of base, B, × $\frac{1}{3}$ perp. height, h : $\frac{1}{3}Bh$. V = $\frac{1}{3}$ vol. of prism or cylinder of same base & perp. height. V = $\frac{1}{2}$ vol. of hemisphere of same base and perp. height.
Frustum of pyramid or cone, right and regular, parallel ends 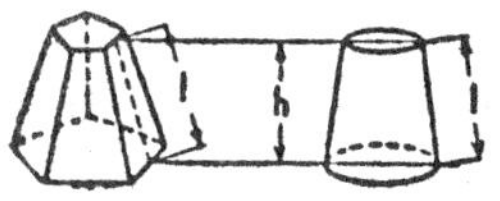	S = (sum of perimeter of base, P, and top, p) × $\frac{1}{2}$ slant height l : $\frac{1}{2}l(P+p)$. V = (sum of areas of base, B, and top, b, + sq. root of their products) × $\frac{1}{3}$ perp. height, h : $\frac{1}{3}h(B+b+\sqrt{Bb})$.
Frustum of any pyramid or cone, parallel ends 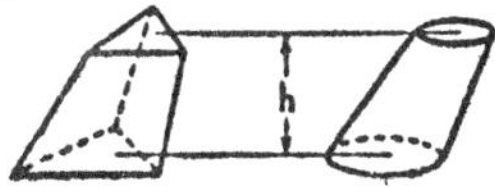	V = (sum of areas of base, B, and top, b, + sq. root of their products) × $\frac{1}{3}$ perpendicular height, h : $\frac{1}{3}h(B+b+\sqrt{Bb})$.
Wedge, parallelogram face 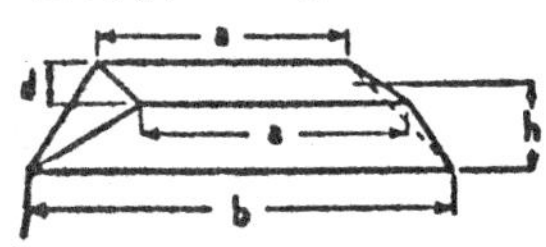	V = $\frac{1}{6}$ (sum of three edges, a b a, × perpendicular height, h, × perpendicular width, d) : $\frac{1}{6}dh(2a+b)$

SOURCE: N. Foster, *Practical Tables for Building Construction*. Copyright © 1963 by McGraw-Hill, Inc. New York. (Used with permission of McGraw-Hill Book Co.)

TABLE 2-2 Finding Surfaces and Volumes of Solids *(cont.)*
(S = lateral or convex surface; V = volume)

SHAPE	FORMULAS

Ungula of right, regular cylinder

1. Base = segment, bab. 2. Base = half circle

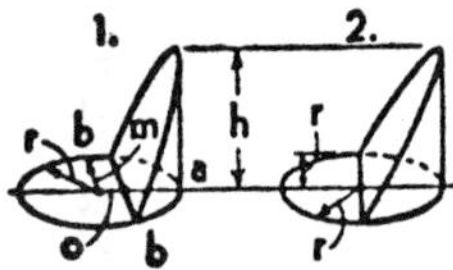

1. $S = (2rm - o \times \text{arc, } bab)\,\frac{h}{r-o}$.
$V = (\frac{2}{3}m^3 - o \times \text{area, } bab)\,\frac{h}{r-o}$.

2. $S = 2rh$.
$V = \frac{2}{3}r^2h$.

1. Base = segment, cac. 2. Base = circle

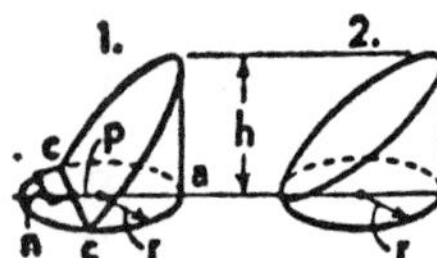

1. $S = (2rn + p \times \text{arc, } cac)\,\frac{h}{r+p}$.
$V = (\frac{2}{3}n^3 + p \times \text{area, } cac)\,\frac{h}{r+p}$.

2. $S = \pi rh$.
$V = \frac{1}{2}r^2\pi h$.

Ellipsoid

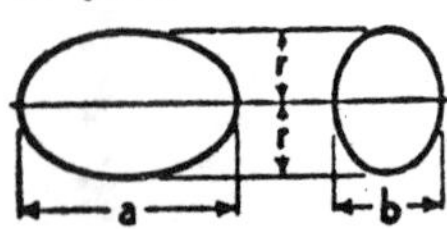

$V = \frac{1}{3}\pi rab$.

Paraboloid

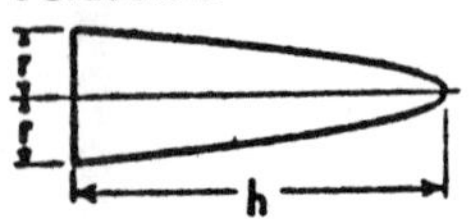

$V = \frac{1}{2}\pi r^2h$.

Ratio of corresponding volume of a Cone, Paraboloid, Sphere & Cylinder of equal height: $\frac{1}{3}$, $\frac{1}{2}$, $\frac{2}{3}$, 1.

TABLE 2-2 Finding Surfaces and Volumes of Solids *(cont.)*
(S = lateral or convex surface; V = volume)

SHAPE	FORMULAS
Sphere 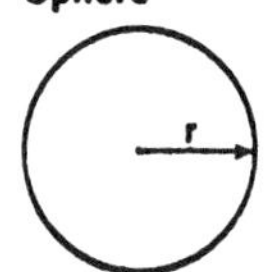	$S = 4\pi r^2 = \pi d^2 = 3.14159265\, d^2.$ $V = \frac{4}{3}\pi r^3 = \frac{1}{6}\pi d^3 = 0.52359878\, d^3.$
Spherical Sector 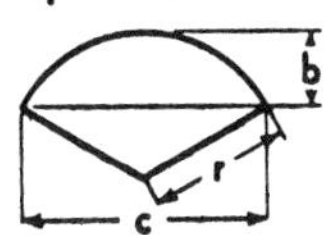	$S = \frac{1}{2}\pi r(4b + c).$ $V = \frac{2}{3}\pi r^2 b.$
Spherical Segment 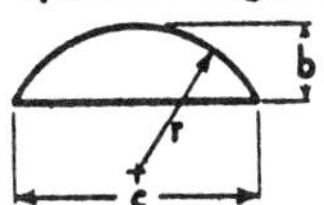	$S = 2\pi rb = \frac{1}{4}\pi(4b^2 + c^2).$ $V = \frac{1}{3}\pi b^2(3r - b) = \frac{1}{24}\pi b(3c^2 + 4b^2).$
Spherical Zone 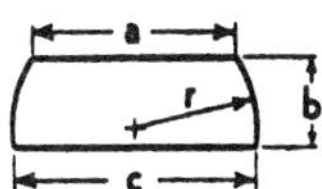	$S = 2\pi rb.$ $V = \frac{1}{24}\pi b(3a^2 + 3c^2 + 4b^2).$
Circular Ring 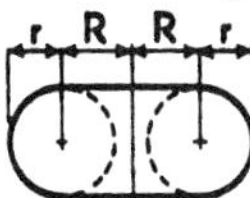	$S = 4\pi^2 Rr.$ $V = 2\pi^2 Rr^2.$

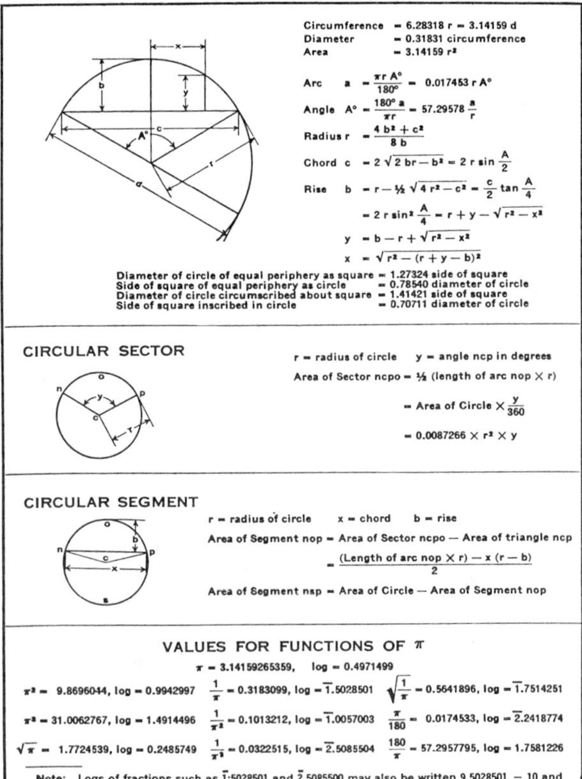

Circumference $= 6.28318\ r = 3.14159\ d$

Diameter $= 0.31831$ circumference

Area $= 3.14159\ r^2$

Arc $a = \dfrac{\pi r A^\circ}{180^\circ} = 0.017453\ r A^\circ$

Angle $A^\circ = \dfrac{180^\circ a}{\pi r} = 57.29578 \dfrac{a}{r}$

Radius $r = \dfrac{4b^2 + c^2}{8b}$

Chord $c = 2\sqrt{2br - b^2} = 2r \sin \dfrac{A}{2}$

Rise $b = r - \frac{1}{2}\sqrt{4r^2 - c^2} = \dfrac{c}{2} \tan \dfrac{A}{4}$

$= 2r \sin^2 \dfrac{A}{4} = r + y - \sqrt{r^2 - x^2}$

$y = b - r + \sqrt{r^2 - x^2}$

$x = \sqrt{r^2 - (r + y - b)^2}$

Diameter of circle of equal periphery as square $= 1.27324$ side of square

Side of square of equal periphery as circle $= 0.78540$ diameter of circle

Diameter of circle circumscribed about square $= 1.41421$ side of square

Side of square inscribed in circle $= 0.70711$ diameter of circle

CIRCULAR SECTOR

r = radius of circle y = angle ncp in degrees

Area of Sector ncpo $= \frac{1}{2}$ (length of arc nop $\times$ r)

$=$ Area of Circle $\times \dfrac{y}{360}$

$= 0.0087266 \times r^2 \times y$

CIRCULAR SEGMENT

r = radius of circle x = chord b = rise

Area of Segment nop = Area of Sector ncpo — Area of triangle ncp

$= \dfrac{(\text{Length of arc nop} \times r) - x(r - b)}{2}$

Area of Segment nsp = Area of Circle — Area of Segment nop

VALUES FOR FUNCTIONS OF π

$\pi = 3.14159265359$, log $= 0.4971499$

$\pi^2 = 9.8696044$, log $= 0.9942997$	$\dfrac{1}{\pi} = 0.3183099$, log $= \bar{1}.5028501$	$\sqrt{\dfrac{1}{\pi}} = 0.5641896$, log $= \bar{1}.7514251$
$\pi^3 = 31.0062767$, log $= 1.4914496$	$\dfrac{1}{\pi^2} = 0.1013212$, log $= \bar{1}.0057003$	$\dfrac{\pi}{180} = 0.0174533$, log $= \bar{2}.2418774$
$\sqrt{\pi} = 1.7724539$, log $= 0.2485749$	$\dfrac{1}{\pi^3} = 0.0322515$, log $= \bar{2}.5085504$	$\dfrac{180}{\pi} = 57.2957795$, log $= 1.7581226$

Note: Logs of fractions such as $\bar{1}$:5028501 and $\bar{2}$.5085500 may also be written 9.5028501 — 10 and 8.5085500 — 10 respectively.

Fig. 2-1 Properties of the circle *(American Institute of Steel Construction, Chicago, Ill.)*

TABLE 2-3 Length of Circular Arcs for Unit Radius

By the use of this table, the length of any arc may be found if the length of the radius and the angle of the segment are known.

Example: Required the length of arc of segment 32° 15′ 27″ with radius of 24 feet 3 inches.

From table: Length of arc (Radius 1) for 32° = .5585054
15′ = .0043633
27″ = .0001309
.5629996

.5629996 × 24.25 (length of radius) = 13.65 feet

DEGREES						MINUTES		SECONDS	
1	.017 4533	61	1.064 6508	121	2.111 8484	1	.000 2909	1	.000 0048
2	.034 9066	62	1.082 1041	122	2.129 3017	2	.000 5818	2	.000 0097
3	.052 3599	63	1.099 5574	123	2.146 7550	3	.000 8727	3	.000 0145
4	.069 8132	64	1.117 0107	124	2.164 2083	4	.001 1636	4	.000 0194
5	.087 2665	65	1.134 4640	125	2.181 6616	5	.001 4544	5	.000 0242
6	.104 7198	66	1.151 9173	126	2.199 1149	6	.001 7453	6	.000 0291
7	.122 1730	67	1.169 3706	127	2.216 5682	7	.002 0362	7	.000 0339
8	.139 6263	68	1.186 8239	128	2.234 0214	8	.002 3271	8	.000 0388
9	.157 0796	69	1.204 2772	129	2.251 4747	9	.002 6180	9	.000 0436
10	.174 5329	70	1.221 7305	130	2.268 9280	10	.002 9089	10	.000 0485
11	.191 9862	71	1.239 1838	131	2.286 3813	11	.003 1998	11	.000 0533
12	.209 4395	72	1.256 6371	132	2.303 8346	12	.003 4907	12	.000 0582
13	.226 8928	73	1.274 0904	133	2.321 2879	13	.003 7815	13	.000 0630
14	.244 3461	74	1.291 5436	134	2.338 7412	14	.004 0724	14	.000 0679
15	.261 7994	75	1.308 9969	135	2.356 1945	15	.004 3633	15	.000 0727
16	.279 2527	76	1.326 4502	136	2.373 6478	16	.004 6542	16	.000 0776
17	.296 7060	77	1.343 9035	137	2.391 1011	17	.004 9451	17	.000 0824
18	.314 1593	78	1.361 3568	138	2.408 5544	18	.005 2360	18	.000 0873
19	.331 6126	79	1.378 8101	139	2.426 0077	19	.005 5269	19	.000 0921
20	.349 0659	80	1.396 2634	140	2.443 4610	20	.005 8178	20	.000 0970
21	.366 5191	81	1.413 7167	141	2.460 9142	21	.006 1087	21	.000 1018
22	.383 9724	82	1.431 1700	142	2.478 3675	22	.006 3995	22	.000 1067
23	.401 4257	83	1.448 6233	143	2.495 8208	23	.006 6904	23	.000 1115
24	.418 8790	84	1.466 0766	144	2.513 2741	24	.006 9813	24	.000 1164
25	.436 3323	85	1.483 5299	145	2.530 7274	25	.007 2722	25	.000 1212
26	.453 7856	86	1.500 9832	146	2.548 1807	26	.007 5631	26	.000 1261
27	.471 2389	87	1.518 4364	147	2.565 6340	27	.007 8540	27	.000 1309
28	.488 6922	88	1.535 8897	148	2.583 0873	28	.008 1449	28	.000 1357
29	.506 1455	89	1.553 3430	149	2.600 5406	29	.008 4358	29	.000 1406
30	.523 5988	90	1.570 7963	150	2.617 9939	30	.008 7266	30	.000 1454
31	.541 0521	91	1.588 2496	151	2.635 4472	31	.009 0175	31	.000 1503
32	.558 5054	92	1.605 7029	152	2.652 9005	32	.009 3084	32	.000 1551
33	.575 9587	93	1.623 1562	153	2.670 3538	33	.009 5993	33	.000 1600
34	.593 4119	94	1.640 6095	154	2.687 8070	34	.009 8902	34	.000 1648
35	.610 8652	95	1.658 0628	155	2.705 2603	35	.010 1811	35	.000 1697
36	.628 3185	96	1.675 5161	156	2.722 7136	36	.010 4720	36	.000 1745
37	.645 7718	97	1.692 9694	157	2.740 1669	37	.010 7629	37	.000 1794
38	.663 2251	98	1.710 4227	158	2.757 6202	38	.011 0538	38	.000 1842
39	.680 6784	99	1.727 8760	159	2.775 0735	39	.011 3446	39	.000 1891
40	.698 1317	100	1.745 3293	160	2.792 5268	40	.011 6355	40	.000 1939
41	.715 5850	101	1.762 7825	161	2.809 9801	41	.011 9264	41	.000 1988
42	.733 0383	102	1.780 2358	162	2.827 4334	42	.012 2173	42	.000 2036
43	.750 4916	103	1.797 6891	163	2.844 8867	43	.012 5082	43	.000 2085
44	.767 9449	104	1.815 1424	164	2.862 3400	44	.012 7991	44	.000 2133
45	.785 3982	105	1.832 5957	165	2.879 7933	45	.013 0900	45	.000 2182
46	.802 8515	106	1.850 0490	166	2.897 2466	46	.013 3809	46	.000 2230
47	.820 3047	107	1.867 5023	167	2.914 6999	47	.013 6717	47	.000 2279
48	.837 7580	108	1.884 9556	168	2.932 1531	48	.013 9626	48	.000 2327
49	.855 2113	109	1.902 4089	169	2.949 6064	49	.014 2535	49	.000 2376
50	.872 6646	110	1.919 8622	170	2.967 0597	50	.014 5444	50	.000 2424
51	.890 1179	111	1.937 3155	171	2.984 5130	51	.014 8353	51	.000 2473
52	.907 5712	112	1.954 7688	172	3.001 9663	52	.015 1262	52	.000 2521
53	.925 0245	113	1.972 2221	173	3.019 4196	53	.015 4171	53	.000 2570
54	.942 4778	114	1.989 6753	174	3.036 8729	54	.015 7080	54	.000 2618
55	.959 9311	115	2.007 1286	175	3.054 3262	55	.015 9989	55	.000 2666
56	.977 3844	116	2.024 5819	176	3.071 7795	56	.016 2897	56	.000 2715
57	.994 8377	117	2.042 0352	177	3.089 2328	57	.016 5806	57	.000 2763
58	1.012 2910	118	2.059 4885	178	3.106 6861	58	.016 8715	58	.000 2812
59	1.029 7443	119	2.076 9418	179	3.124 1394	59	.017 1624	59	.000 2860
60	1.047 1976	120	2.094 3951	180	3.141 5927	60	.017 4533	60	.000 2909

SOURCE: *Manual of Steel Construction,* 7th ed., American Institute of Steel Construction, Chicago, Ill., 1970 and 1973.

TABLE 2-4 Trigonometric Formulas

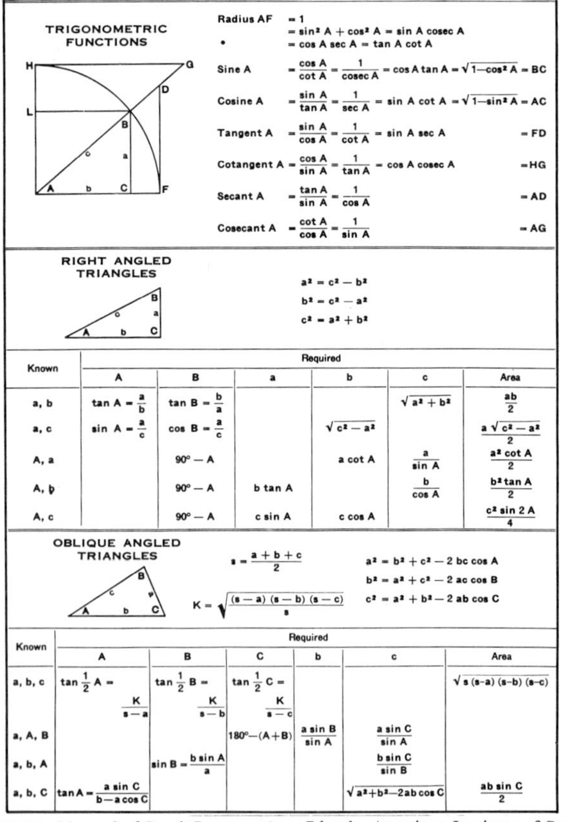

TRIGONOMETRIC FUNCTIONS

Radius AF $= 1$
$= \sin^2 A + \cos^2 A = \sin A \operatorname{cosec} A$
$= \cos A \sec A = \tan A \cot A$

Sine A $= \dfrac{\cos A}{\cot A} = \dfrac{1}{\operatorname{cosec} A} = \cos A \tan A = \sqrt{1-\cos^2 A} = BC$

Cosine A $= \dfrac{\sin A}{\tan A} = \dfrac{1}{\sec A} = \sin A \cot A = \sqrt{1-\sin^2 A} = AC$

Tangent A $= \dfrac{\sin A}{\cos A} = \dfrac{1}{\cot A} = \sin A \sec A = FD$

Cotangent A $= \dfrac{\cos A}{\sin A} = \dfrac{1}{\tan A} = \cos A \operatorname{cosec} A = HG$

Secant A $= \dfrac{\tan A}{\sin A} = \dfrac{1}{\cos A} = AD$

Cosecant A $= \dfrac{\cot A}{\cos A} = \dfrac{1}{\sin A} = AG$

RIGHT ANGLED TRIANGLES

$a^2 = c^2 - b^2$

$b^2 = c^2 - a^2$

$c^2 = a^2 + b^2$

Known	Required: A	B	a	b	c	Area
a, b	$\tan A = \frac{a}{b}$	$\tan B = \frac{b}{a}$			$\sqrt{a^2+b^2}$	$\frac{ab}{2}$
a, c	$\sin A = \frac{a}{c}$	$\cos B = \frac{a}{c}$		$\sqrt{c^2-a^2}$		$\frac{a\sqrt{c^2-a^2}}{2}$
A, a		$90° - A$		$a \cot A$	$\frac{a}{\sin A}$	$\frac{a^2 \cot A}{2}$
A, b		$90° - A$	$b \tan A$		$\frac{b}{\cos A}$	$\frac{b^2 \tan A}{2}$
A, c		$90° - A$	$c \sin A$	$c \cos A$		$\frac{c^2 \sin 2A}{4}$

OBLIQUE ANGLED TRIANGLES

$s = \dfrac{a+b+c}{2}$

$K = \sqrt{\dfrac{(s-a)(s-b)(s-c)}{s}}$

$a^2 = b^2 + c^2 - 2bc \cos A$

$b^2 = a^2 + c^2 - 2ac \cos B$

$c^2 = a^2 + b^2 - 2ab \cos C$

Known	Required: A	B	C	b	c	Area
a, b, c	$\tan \frac{1}{2} A = \frac{K}{s-a}$	$\tan \frac{1}{2} B = \frac{K}{s-b}$	$\tan \frac{1}{2} C = \frac{K}{s-c}$			$\sqrt{s(s-a)(s-b)(s-c)}$
a, A, B			$180° - (A+B)$	$\frac{a \sin B}{\sin A}$	$\frac{a \sin C}{\sin A}$	
a, b, A		$\sin B = \frac{b \sin A}{a}$			$\frac{b \sin C}{\sin B}$	
a, b, C	$\tan A = \frac{a \sin C}{b - a \cos C}$				$\sqrt{a^2+b^2-2ab \cos C}$	$\frac{ab \sin C}{2}$

SOURCE: *Manual of Steel Construction,* 7th ed., American Institute of Steel Construction, Chicago, Ill., 1970 and 1973.

TABLE 2-5 Properties of Geometric Sections

SQUARE Axis of moments through center d, c, d	$A = d^2$ $c = \frac{d}{2}$ $I = \frac{d^4}{12}$ $S = \frac{d^3}{6}$ $r = \frac{d}{\sqrt{12}} = .288675\ d$ $Z = \frac{d^3}{4}$
SQUARE Axis of moments on base d, c, d	$A = d^2$ $c = d$ $I = \frac{d^4}{3}$ $S = \frac{d^3}{3}$ $r = \frac{d}{\sqrt{3}} = .577350\ d$
SQUARE Axis of moments on diagonal c, d, d	$A = d^2$ $c = \frac{d}{\sqrt{2}} = .707107\ d$ $I = \frac{d^4}{12}$ $S = \frac{d^3}{6\sqrt{2}} = .117851\ d^3$ $r = \frac{d}{\sqrt{12}} = .288675\ d$ $Z = \frac{2c^3}{3} = \frac{d^3}{3\sqrt{2}} = .235702d^3$
RECTANGLE Axis of moments through center d, c, b	$A = bd$ $c = \frac{d}{2}$ $I = \frac{bd^3}{12}$ $S = \frac{bd^2}{6}$ $r = \frac{d}{\sqrt{12}} = .288675\ d$ $Z = \frac{bd^2}{4}$

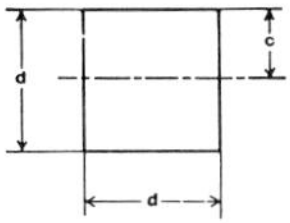

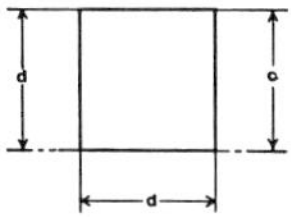

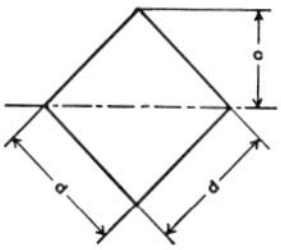

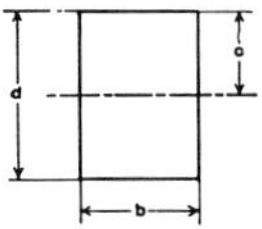

SOURCE: *Manual of Steel Construction,* 7th ed., American Institute of Steel Construction, Chicago, Ill., 1970 and 1973.

TABLE 2-5 Properties of Geometric Sections *(cont.)*

RECTANGLE

Axis of moments on base

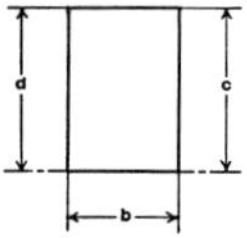

$$A = bd$$
$$c = d$$
$$I = \frac{bd^3}{3}$$
$$S = \frac{bd^2}{3}$$
$$r = \frac{d}{\sqrt{3}} = .577350\ d$$

RECTANGLE

Axis of moments on diagonal

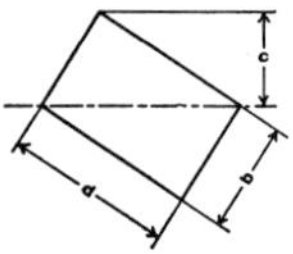

$$A = bd$$
$$c = \frac{bd}{\sqrt{b^2 + d^2}}$$
$$I = \frac{b^3d^3}{6\,(b^2 + d^2)}$$
$$S = \frac{b^2d^2}{6\sqrt{b^2 + d^2}}$$
$$= \frac{bd}{\sqrt{6\,(b^2 + d^2)}}$$

RECTANGLE

Axis of moments any line through center of gravity

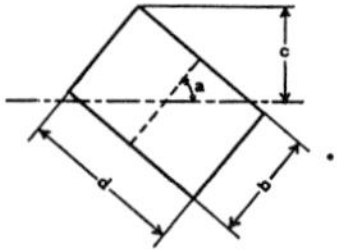

$$A = bd$$
$$c = \frac{b \sin a + d \cos a}{2}$$
$$I = \frac{bd\,(b^2 \sin^2 a + d^2 \cos^2 a)}{12}$$
$$S = \frac{bd\,(b^2 \sin^2 a + d^2 \cos^2 a)}{6\,(b \sin a + d \cos a)}$$
$$r = \sqrt{\frac{b^2 \sin^2 a + d^2 \cos^2 a}{12}}$$

HOLLOW RECTANGLE

Axis of moments through center

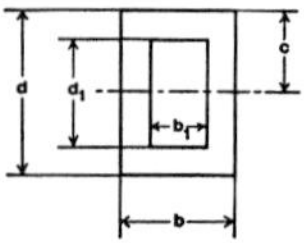

$$A = bd - b_1d_1$$
$$c = \frac{d}{2}$$
$$I = \frac{bd^3 - b_1d_1^3}{12}$$
$$S = \frac{bd^3 - b_1d_1^3}{6d}$$
$$r = \sqrt{\frac{bd^3 - b_1d_1^3}{12\,A}}$$
$$Z = \frac{bd^2}{4} - \frac{b_1d_1^2}{4}$$

TABLE 2-5 Properties of Geometric Sections *(cont.)*

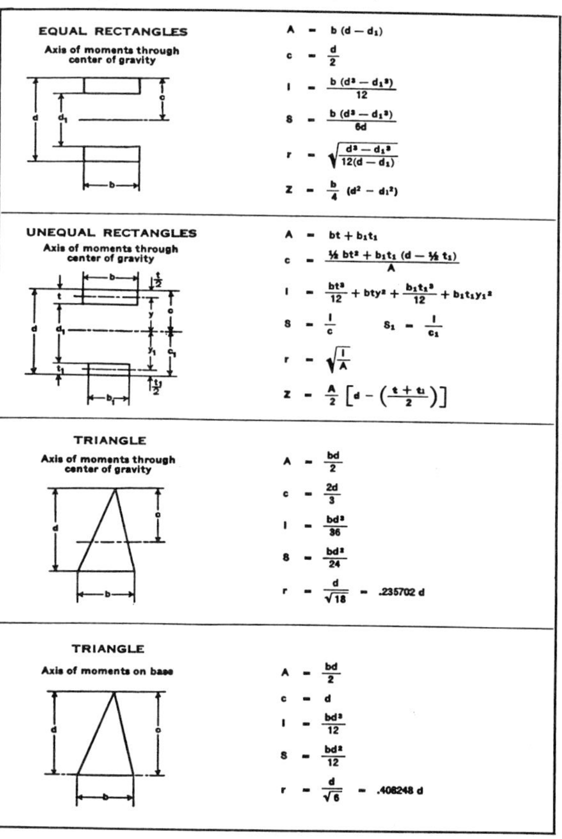

EQUAL RECTANGLES

Axis of moments through center of gravity

$$A = b(d - d_1)$$

$$c = \frac{d}{2}$$

$$I = \frac{b(d^3 - d_1^3)}{12}$$

$$S = \frac{b(d^3 - d_1^3)}{6d}$$

$$r = \sqrt{\frac{d^3 - d_1^3}{12(d - d_1)}}$$

$$Z = \frac{b}{4}(d^2 - d_1^2)$$

UNEQUAL RECTANGLES

Axis of moments through center of gravity

$$A = bt + b_1t_1$$

$$c = \frac{\frac{1}{2}bt^2 + b_1t_1(d - \frac{1}{2}t_1)}{A}$$

$$I = \frac{bt^3}{12} + bty^2 + \frac{b_1t_1^3}{12} + b_1t_1y_1^2$$

$$S = \frac{I}{c} \qquad S_1 = \frac{I}{c_1}$$

$$r = \sqrt{\frac{I}{A}}$$

$$Z = \frac{A}{2}\left[d - \left(\frac{t + t_1}{2}\right)\right]$$

TRIANGLE

Axis of moments through center of gravity

$$A = \frac{bd}{2}$$

$$c = \frac{2d}{3}$$

$$I = \frac{bd^3}{36}$$

$$S = \frac{bd^2}{24}$$

$$r = \frac{d}{\sqrt{18}} = .235702\,d$$

TRIANGLE

Axis of moments on base

$$A = \frac{bd}{2}$$

$$c = d$$

$$I = \frac{bd^3}{12}$$

$$S = \frac{bd^2}{12}$$

$$r = \frac{d}{\sqrt{6}} = .408248\,d$$

TABLE 2-5 Properties of Geometric Sections *(cont.)*

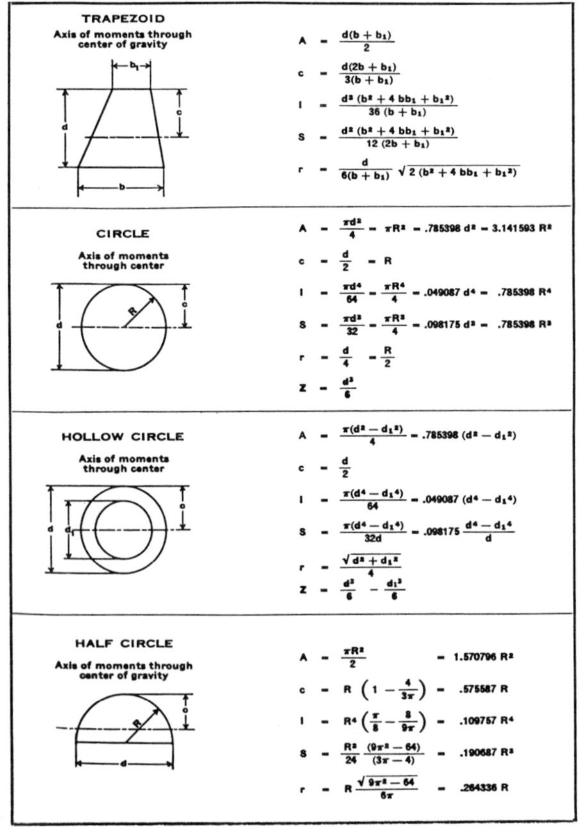

TRAPEZOID Axis of moments through center of gravity	A	$= \frac{d(b + b_1)}{2}$
	c	$= \frac{d(2b + b_1)}{3(b + b_1)}$
	I	$= \frac{d^3 (b^2 + 4\,bb_1 + b_1^2)}{36\,(b + b_1)}$
	S	$= \frac{d^2 (b^2 + 4\,bb_1 + b_1^2)}{12\,(2b + b_1)}$
	r	$= \frac{d}{6(b + b_1)} \sqrt{2\,(b^2 + 4\,bb_1 + b_1^2)}$
CIRCLE Axis of moments through center	A	$= \frac{\pi d^2}{4} = \pi R^2 = .785398\,d^2 = 3.141593\,R^2$
	c	$= \frac{d}{2} = R$
	I	$= \frac{\pi d^4}{64} = \frac{\pi R^4}{4} = .049087\,d^4 = .785398\,R^4$
	S	$= \frac{\pi d^3}{32} = \frac{\pi R^3}{4} = .098175\,d^3 = .785398\,R^3$
	r	$= \frac{d}{4} = \frac{R}{2}$
	Z	$= \frac{d^3}{6}$
HOLLOW CIRCLE Axis of moments through center	A	$= \frac{\pi(d^2 - d_1^2)}{4} = .785398\,(d^2 - d_1^2)$
	c	$= \frac{d}{2}$
	I	$= \frac{\pi(d^4 - d_1^4)}{64} = .049087\,(d^4 - d_1^4)$
	S	$= \frac{\pi(d^4 - d_1^4)}{32d} = .098175\,\frac{d^4 - d_1^4}{d}$
	r	$= \frac{\sqrt{d^2 + d_1^2}}{4}$
	Z	$= \frac{d^3}{6} - \frac{d_1^3}{6}$
HALF CIRCLE Axis of moments through center of gravity	A	$= \frac{\pi R^2}{2} = 1.570796\,R^2$
	c	$= R\left(1 - \frac{4}{3\pi}\right) = .575587\,R$
	I	$= R^4\left(\frac{\pi}{8} - \frac{8}{9\pi}\right) = .109757\,R^4$
	S	$= \frac{R^3}{24}\,\frac{(9\pi^2 - 64)}{(3\pi - 4)} = .190687\,R^3$
	r	$= R\,\frac{\sqrt{9\pi^2 - 64}}{6\pi} = .264336\,R$

TABLE 2-5 Properties of Geometric Sections *(cont.)*

*** HALF ELLIPSE**

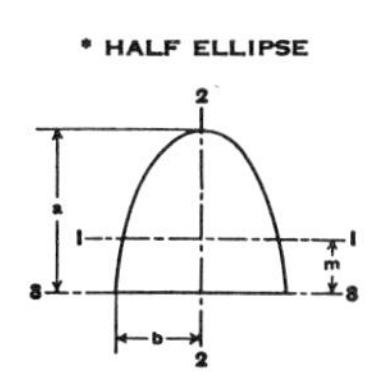

$$A = \frac{1}{2}\pi ab$$

$$m = \frac{4a}{3\pi}$$

$$I_1 = a^3b\left(\frac{\pi}{8} - \frac{8}{9\pi}\right)$$

$$I_2 = \frac{1}{8}\pi ab^3$$

$$I_3 = \frac{1}{8}\pi a^3b$$

*** QUARTER ELLIPSE**

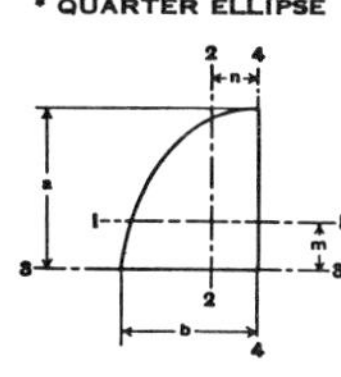

$$A = \frac{1}{4}\pi ab$$

$$m = \frac{4a}{3\pi}$$

$$n = \frac{4b}{3\pi}$$

$$I_1 = a^3b\left(\frac{\pi}{16} - \frac{4}{9\pi}\right)$$

$$I_2 = ab^3\left(\frac{\pi}{16} - \frac{4}{9\pi}\right)$$

$$I_3 = \frac{1}{16}\pi a^3b$$

$$I_4 = \frac{1}{16}\pi ab^3$$

*** ELLIPTIC COMPLEMENT**

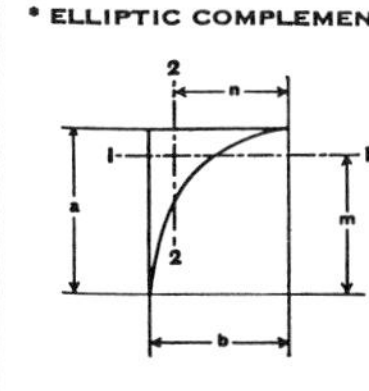

$$A = ab\left(1 - \frac{\pi}{4}\right)$$

$$m = \frac{a}{6\left(1 - \frac{\pi}{4}\right)}$$

$$n = \frac{b}{6\left(1 - \frac{\pi}{4}\right)}$$

$$I_1 = a^3b\left(\frac{1}{3} - \frac{\pi}{16} - \frac{1}{36\left(1 - \frac{\pi}{4}\right)}\right)$$

$$I_2 = ab^3\left(\frac{1}{3} - \frac{\pi}{16} - \frac{1}{36\left(1 - \frac{\pi}{4}\right)}\right)$$

* To obtain properties of half circle, quarter circle and circular complement substitute $a = b = R$.

TABLE 2-5 Properties of Geometric Sections *(cont.)*

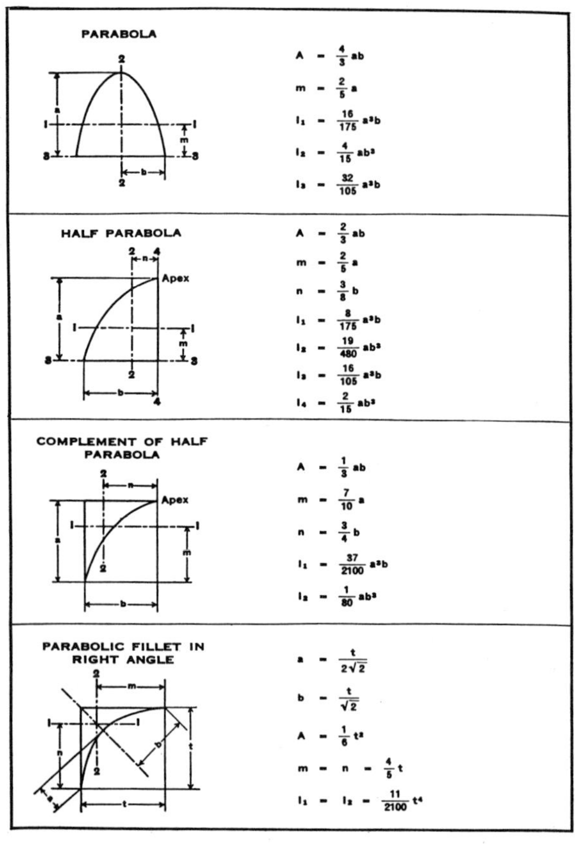

PARABOLA

$A = \frac{4}{3}ab$

$m = \frac{2}{5}a$

$I_1 = \frac{16}{175}a^3b$

$I_2 = \frac{4}{15}ab^3$

$I_3 = \frac{32}{105}a^3b$

HALF PARABOLA

$A = \frac{2}{3}ab$

$m = \frac{2}{5}a$

$n = \frac{3}{8}b$

$I_1 = \frac{8}{175}a^3b$

$I_2 = \frac{19}{480}ab^3$

$I_3 = \frac{16}{105}a^3b$

$I_4 = \frac{2}{15}ab^3$

COMPLEMENT OF HALF PARABOLA

$A = \frac{1}{3}ab$

$m = \frac{7}{10}a$

$n = \frac{3}{4}b$

$I_1 = \frac{37}{2100}a^3b$

$I_2 = \frac{1}{80}ab^3$

PARABOLIC FILLET IN RIGHT ANGLE

$a = \frac{t}{2\sqrt{2}}$

$b = \frac{t}{\sqrt{2}}$

$A = \frac{1}{6}t^2$

$m = n = \frac{4}{5}t$

$I_1 = I_2 = \frac{11}{2100}t^4$

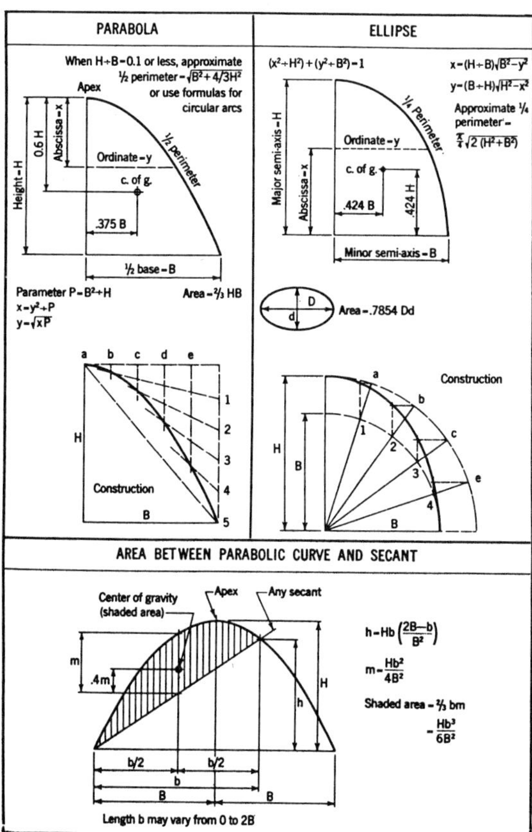

Fig. 2-2 Properties of parabola and ellipse *(American Institute of Steel Construction, Chicago, Ill.)*

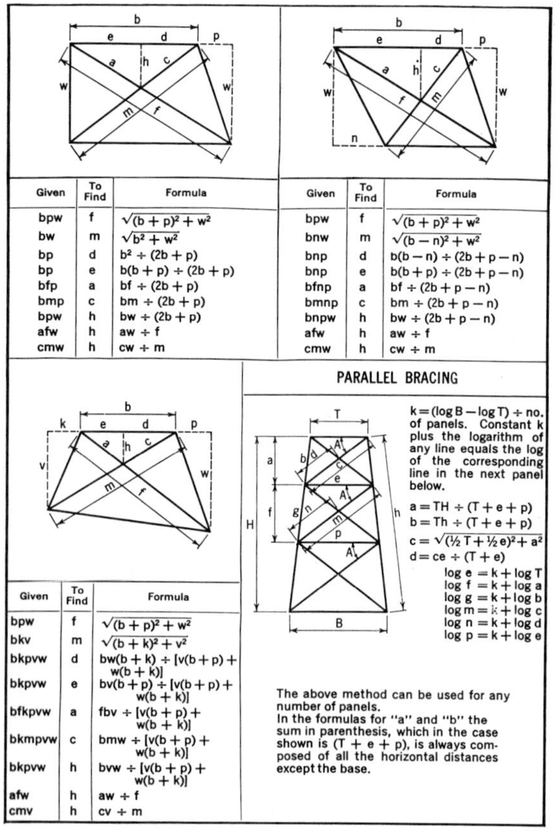

Given	To Find	Formula
bpw	f	$\sqrt{(b+p)^2+w^2}$
bw	m	$\sqrt{b^2+w^2}$
bp	d	$b^2 \div (2b+p)$
bp	e	$b(b+p) \div (2b+p)$
bfp	a	$bf \div (2b+p)$
bmp	c	$bm \div (2b+p)$
bpw	h	$bw \div (2b+p)$
afw	h	$aw \div f$
cmw	h	$cw \div m$

Given	To Find	Formula
bpw	f	$\sqrt{(b+p)^2+w^2}$
bnw	m	$\sqrt{(b-n)^2+w^2}$
bnp	d	$b(b-n) \div (2b+p-n)$
bnp	e	$b(b+p) \div (2b+p-n)$
bfnp	a	$bf \div (2b+p-n)$
bmnp	c	$bm \div (2b+p-n)$
bnpw	h	$bw \div (2b+p-n)$
afw	h	$aw \div f$
cmw	h	$cw \div m$

Given	To Find	Formula
bpw	f	$\sqrt{(b+p)^2+w^2}$
bkv	m	$\sqrt{(b+k)^2+v^2}$
bkpvw	d	$bw(b+k) \div [v(b+p)+w(b+k)]$
bkpvw	e	$bv(b+p) \div [v(b+p)+w(b+k)]$
bfkpvw	a	$fbv \div [v(b+p)+w(b+k)]$
bkmpvw	c	$bmw \div [v(b+p)+w(b+k)]$
bkpvw	h	$bvw \div [v(b+p)+w(b+k)]$
afw	h	$aw \div f$
cmv	h	$cv \div m$

PARALLEL BRACING

$k = (\log B - \log T) \div$ no. of panels. Constant k plus the logarithm of any line equals the log of the corresponding line in the next panel below.

$a = TH \div (T + e + p)$
$b = Th \div (T + e + p)$
$c = \sqrt{(\frac{1}{2}T + \frac{1}{2}e)^2 + a^2}$
$d = ce \div (T + e)$
$\log e = k + \log T$
$\log f = k + \log a$
$\log g = k + \log b$
$\log m = k + \log c$
$\log n = k + \log d$
$\log p = k + \log e$

The above method can be used for any number of panels.
In the formulas for "a" and "b" the sum in parenthesis, which in the case shown is $(T + e + p)$, is always composed of all the horizontal distances except the base.

Fig. 2-3 Bracing formulas *(American Institute of Steel Construction, Chicago, Ill.)*

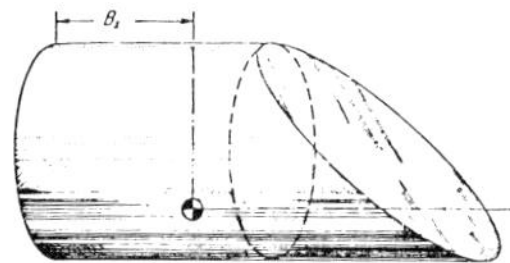

VOLUME equations are included for all cases. Where the equation for the CG (center of gravity) is not given, you can easily obtain it by looking up the volume and CG equations for portions of the shape and then combining values. For example, for the shape above, use the equations for a cylinder, Fig 1, and a truncated cylinder, Fig 10 (subscripts C and T, respectively, in the equations below). Hence taking moments

$$B_x = \frac{V_C B_C + V_T(B_T + L_C)}{V_C + V_T}$$

or

$$B_x = \frac{\left(\frac{\pi}{4} D^2 L_C\right)\left(\frac{L_C}{2}\right) + \frac{\pi}{8} D^2 L_T\left(\frac{5}{16} L_T + L_C\right)}{\frac{\pi}{4} D^2 L_C + \frac{\pi}{8} D^2 L_T}$$

$$B_x = \frac{L_C^2 + L_T\left(\frac{5}{16} L_T + L_C\right)}{2L_C + L_T}$$

In the equations to follow, angle θ can be either in degrees or in radians. Thus θ (rad) $= \pi\theta/180$ (deg) $= 0.01745\ \theta$ (deg). For example, if $\theta = 30$ deg in Case 3, then $\sin\theta = 0.5$ and

$$B = \frac{2R\,(0.5)}{3\,(30)\,(0.01745)} = 0.637R$$

Symbols used are: B = distance from CG to reference plane, V = volume, D and d = diameter, R and r = radius, H = height, L = length.—*Nicholas P. Chironis*

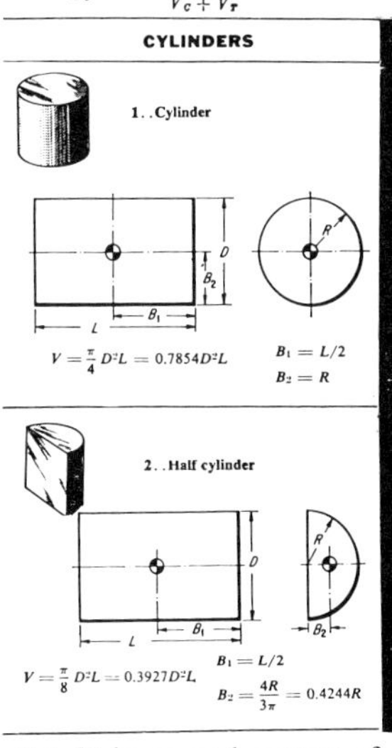

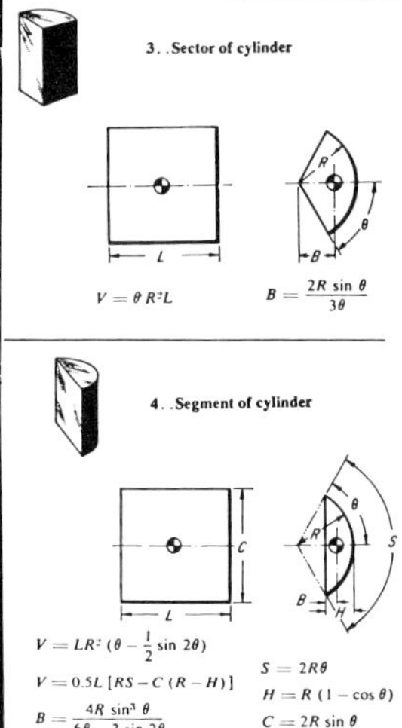

3. . Sector of cylinder
L
R
θ
B
$V = \theta R^2L$
$B = \frac{2R \sin\theta}{3\theta}$
4. . Segment of cylinder
C
L
θ
R
S
B
H
$V = LR^2 (\theta - \frac{1}{2} \sin 2\theta)$
$V = 0.5L\,[RS - C\,(R - H)]$
$B = \frac{4R \sin^3\theta}{6\theta - 3 \sin 2\theta}$
$S = 2R\theta$
$H = R\,(1 - \cos\theta)$
$C = 2R \sin\theta$

Fig. 2-4 Volume and center of gravity (CG). Equations for 55 shapes, many of which are the result of drilled holes, bosses, and fillets in machined and cast parts *(From an original article by E. W. Jenkins, reprinted from "Product Engineering" by permission of Morgan-Grampian Publishing Co., New York.)*

5. . Quadrant of cylinder

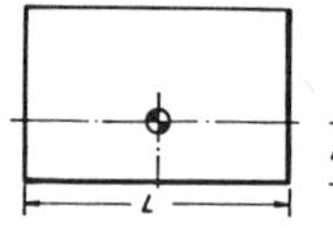

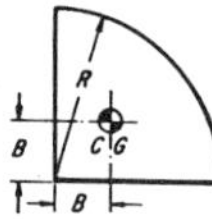

$V = \frac{\pi}{4} R^2 L = 0.7854 R^2 L \qquad B = \frac{4R}{3\pi} = 0.4244R$

6. . Fillet or spandrel

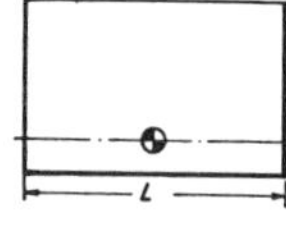

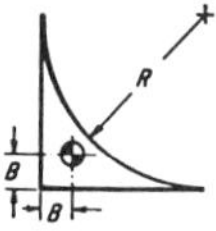

$V = \left(1 - \frac{\pi}{4}\right) R^2 L = 0.2146 R^2 L$

$B = \frac{10 - 3\pi}{12 - 3\pi} R = 0.2234R$

7. . Hollow cylinder

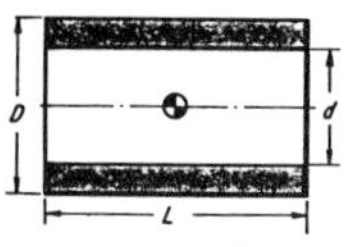

$V = \frac{\pi L}{4} (D^2 - d^2)$

CG at center of part

8. . Half hollow cylinder

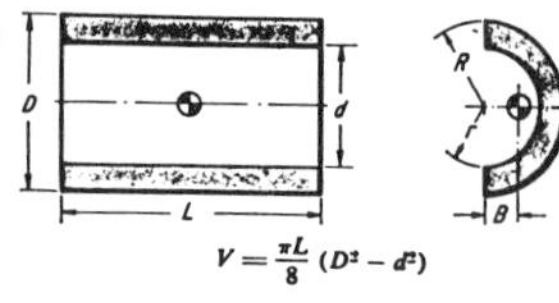

$V = \frac{\pi L}{8} (D^2 - d^2)$

$B = \frac{4}{3\pi} \left[\frac{R^3 - r^3}{R^2 - r^2} \right]$

Fig. 2-4 (*cont.*)

9. . Sector of hollow cylinder

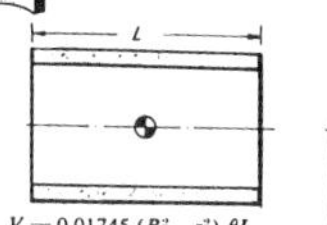

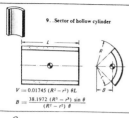

$$V = 0.01745\,(R^2 - r^2)\,\theta L$$

$$B = \frac{38.1972\,(R^3 - r^3)\sin\theta}{(R^2 - r^2)\,\theta}$$

10. . Truncated cylinder (with full circle base)

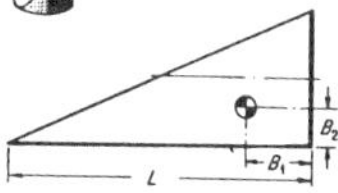

$$V = \frac{\pi}{8}D^2L = 0.3927D^2L \qquad B_1 = 0.3125L \qquad B_2 = 0.375D$$

11. . Truncated cylinder
(with partial circle base)

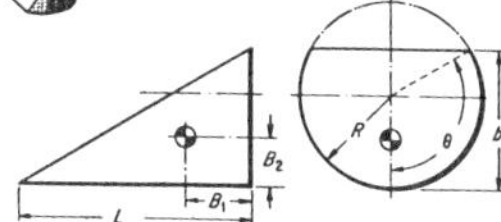

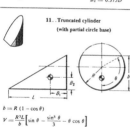

$$b = R\,(1 - \cos\theta)$$

$$V = \frac{R^3L}{b}\left[\sin\theta - \frac{\sin^3\theta}{3} - \theta\cos\theta\right]$$

$$B_1 = \frac{L\left[\frac{\theta\cos^2\theta}{2} - \frac{5\sin\theta\cos\theta}{8} + \frac{\sin^3\theta\cos\theta}{12} + \frac{\theta}{8}\right]}{\left[1 - \cos\theta\right]\left[\sin\theta - \frac{\sin^3\theta}{3} - \theta\cos\theta\right]}$$

$$B_2 = \frac{2R\left[-\frac{\theta\cos\theta}{2} + \frac{\sin\theta}{2} - \frac{\theta}{8} + \frac{\sin\theta\cos\theta}{8} - N\right]}{\left[\sin\theta - \frac{\sin^3\theta}{3} - \theta\cos\theta\right]}$$

where $N = \frac{5\sin^3\theta}{6} - \frac{\sin^3\theta\cos\theta}{12}$

12. . Oblique cylinder
(or circular hole at oblique angle)

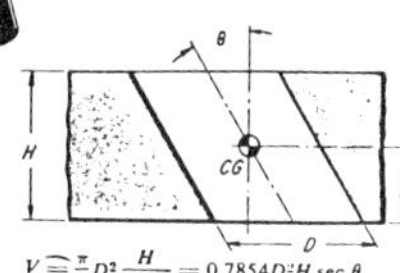

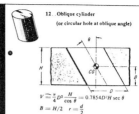

$$V = \frac{\pi}{4}D^2\frac{H}{\cos\theta} = 0.7854D^2H\sec\theta$$

$$B = H/2 \qquad r = \frac{d}{2}$$

13. . Bend in cylinder

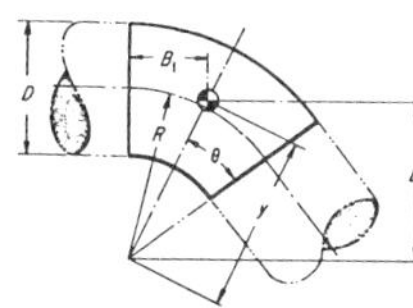

$$V = \frac{\pi^2}{360}D^2R\theta = 0.0274D^2R\theta$$

$$y = R\left[1 + \frac{r^2}{4R^2}\right] \qquad B_1 = y\tan\theta \qquad B_2 = y\cot\theta$$

14. . Curved groove in cylinder

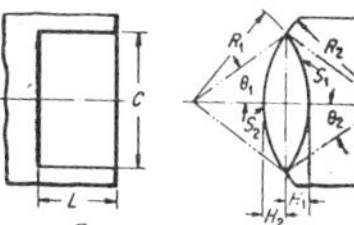

$$\sin\theta_1 = \frac{C}{2R_1} \qquad \sin\theta_2 = \frac{C}{2R_2} \qquad S = 2R\theta$$

$$H_1 = R_1\,(1 - \cos\theta_1) \qquad H_2 = R_2\,(1 - \cos\theta_2)$$

$$V = L\left[R_1^2\left(\theta_1 - \frac{1}{2}\theta_1\sin 2\theta_1\right) + R_2^2\left(\theta_2 - \frac{1}{2}\theta_2\sin 2\theta_2\right)\right]$$

Compute CG of each part separately

ig. 2-4 (*cont.*)

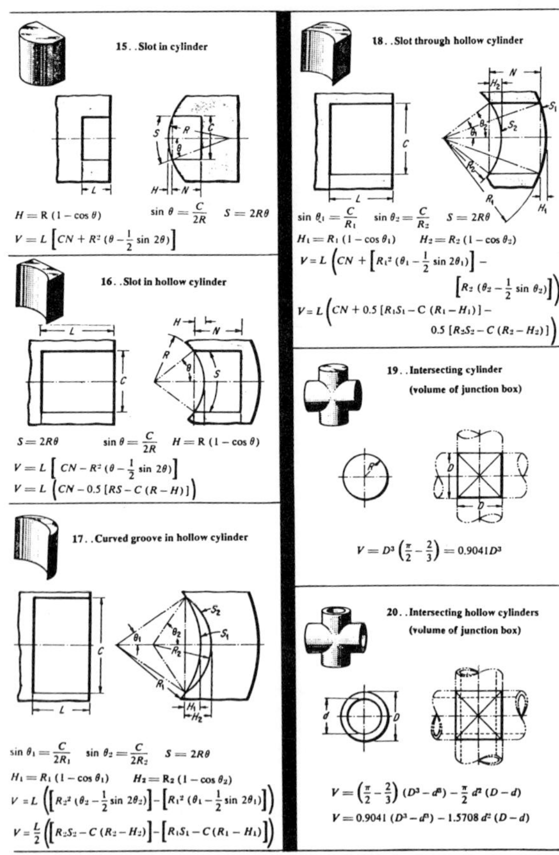
15. . Slot in cylinder
S
R
C
θ
L
H
N
$H = R(1 - \cos\theta)$
$\sin\theta = \frac{C}{2R}$
$S = 2R\theta$
$V = L\left[CN + R^2\left(\theta - \frac{1}{2}\sin 2\theta\right)\right]$
16. . Slot in hollow cylinder
L
C
H
N
R
θ
S
$S = 2R\theta$
$\sin\theta = \frac{C}{2R}$
$H = R(1 - \cos\theta)$
$V = L\left[CN - R^2\left(\theta - \frac{1}{2}\sin 2\theta\right)\right]$
$V = L\left(CN - 0.5\left[RS - C(R - H)\right]\right)$
17. . Curved groove in hollow cylinder
C
L
S_2
S_1
θ_1
θ_2
R_2
R_1
H_1
H_2
$\sin\theta_1 = \frac{C}{2R_1}$
$\sin\theta_2 = \frac{C}{2R_2}$
$S = 2R\theta$
$H_1 = R_1(1 - \cos\theta_1)$
$H_2 = R_2(1 - \cos\theta_2)$
$V = L\left(\left[R_2^2\left(\theta_2 - \frac{1}{2}\sin 2\theta_2\right)\right] - \left[R_1^2\left(\theta_1 - \frac{1}{2}\sin 2\theta_1\right)\right]\right)$
$V = \frac{L}{2}\left(\left[R_2S_2 - C(R_2 - H_2)\right] - \left[R_1S_1 - C(R_1 - H_1)\right]\right)$
18. . Slot through hollow cylinder
N
H_2
S_1
S_2
θ_2
θ_1
C
R_2
R_1
L
H_1
$\sin\theta_1 = \frac{C}{R_1}$
$\sin\theta_2 = \frac{C}{R_2}$
$S = 2R\theta$
$H_1 = R_1(1 - \cos\theta_1)$
$H_2 = R_2(1 - \cos\theta_2)$
$V = L\left(CN + \left[R_1^2\left(\theta_1 - \frac{1}{2}\sin 2\theta_1\right)\right] - \left[R_2\left(\theta_2 - \frac{1}{2}\sin\theta_2\right)\right]\right)$
$V = L\left(CN + 0.5\left[R_1S_1 - C(R_1 - H_1)\right] - 0.5\left[R_2S_2 - C(R_2 - H_2)\right]\right)$
19. . Intersecting cylinder
(volume of junction box)
R
D
D
$V = D^3\left(\frac{\pi}{2} - \frac{2}{3}\right) = 0.9041D^3$
20. . Intersecting hollow cylinders
(volume of junction box)
d
D
$V = \left(\frac{\pi}{2} - \frac{2}{3}\right)(D^3 - d^3) - \frac{\pi}{2}d^2(D - d)$
$V = 0.9041(D^3 - d^3) - 1.5708d^2(D - d)$

Fig. 2-4 (*cont.*)

21. . Intersecting parallel cylinders

$(M<R_1)$

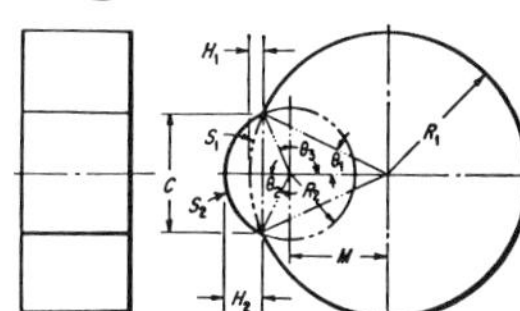

$\theta_2 = 180° - \theta_3 \quad \cos\theta_3 = \dfrac{R_2^2 + M^2 - R_1^2}{2MR_2}$

$\cos\theta_1 = \dfrac{R_1^2 + M^2 - R_2^2}{2MR_1} \qquad H_1 = R_1(1-\cos\theta_1)$

$S_1 = 2R_1\theta_1$

$V = L\left(\pi R_1^2 + \left[R_2^2\left(\theta_2 - \frac{1}{2}\sin 2\theta_2\right)\right] - \left[R_1^2\left(\theta_1 - \frac{1}{2}\sin 2\theta_1\right)\right]\right)$

22. . Intersecting parallel cylinders

$(M>R_1)$

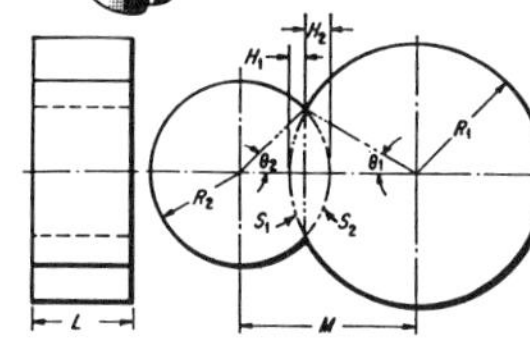

$H_1 = R_1(1-\cos\theta_1) \qquad \cos\theta_1 = \dfrac{R_1^2 + M^2 - R_2^2}{2MR_1}$

$S_1 = 2R_1\theta_1$

$V = L\left(\left[\pi(R_1^2 + R_2^2)\right] - \left[R_1^2\left(\theta_1 - \frac{1}{2}\sin 2\theta_1\right)\right] - \left[R_2^2\left(\theta_2 - \frac{1}{2}\sin 2\theta_2\right)\right]\right)$

SPHERES

23. . Sphere

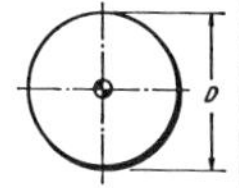

$V = \dfrac{\pi D^3}{6} = 0.5236D^3$

24. . Hemisphere

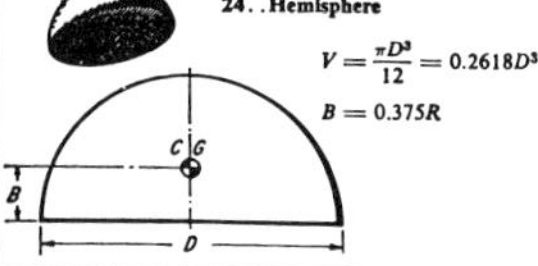

$V = \dfrac{\pi D^3}{12} = 0.2618D^3$

$B = 0.375R$

25. . Spherical segment

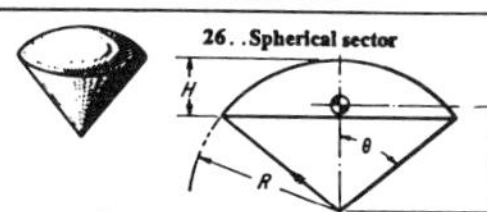

$V = \pi H^2\left(R - \dfrac{H}{3}\right)$

$B_1 = \dfrac{H(4R-H)}{4(3R-H)}$

$B_2 = \dfrac{3(2R-H)^2}{4(3R-H)}$

26. . Spherical sector

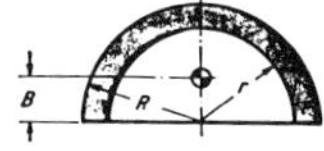

$V = \dfrac{2\pi}{3}R^2H = 2.0944R^2H$

$B = 0.375(1+\cos\theta)R = 0.375(2R-H)$

27. . Shell of hollow hemisphere

$V = \dfrac{2\pi}{3}(R^3 - r^3) \qquad B = 0.375\left(\dfrac{R^4 - r^4}{R^3 - r^3}\right)$

28. . Hollow sphere

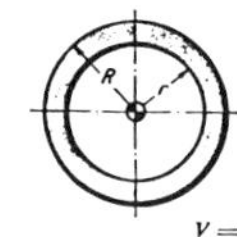

$V = \dfrac{4\pi}{3}(R^3 - r^3)$

ig. 2-4 (*cont.*)

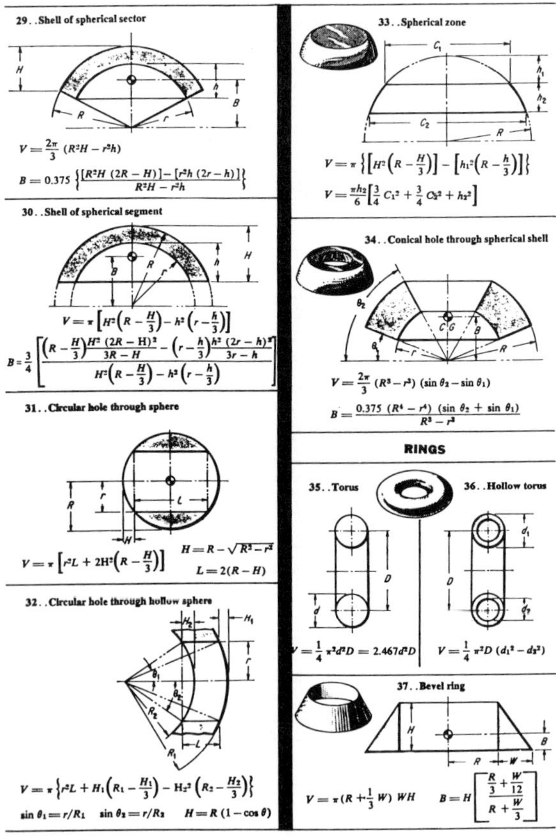
29. .Shell of spherical sector
$V = \frac{2\pi}{3}(R^2H - r^2h)$
$B = 0.375\left\{\frac{[R^2H(2R-H)]-[r^2h(2r-h)]}{R^2H - r^2h}\right\}$
30. .Shell of spherical segment
$V = \pi\left[H^2\left(R - \frac{H}{3}\right) - h^2\left(r - \frac{h}{3}\right)\right]$
$B = \frac{3}{4}\left[\frac{\left(R-\frac{H}{3}\right)\frac{H^2(2R-H)^2}{3R-H} - \left(r-\frac{h}{3}\right)\frac{h^2(2r-h)^2}{3r-h}}{H^2\left(R-\frac{H}{3}\right) - h^2\left(r-\frac{h}{3}\right)}\right]$
31. .Circular hole through sphere
$V = \pi\left[r^2L + 2H^2\left(R - \frac{H}{3}\right)\right]$
$H = R - \sqrt{R^2 - r^2}$
$L = 2(R - H)$
32. .Circular hole through hollow sphere
$V = \pi\left\{r^2L + H_1\left(R_1 - \frac{H_1}{3}\right) - H_2^2\left(R_2 - \frac{H_2}{3}\right)\right\}$
$\sin\theta_1 = r/R_1$ $\sin\theta_2 = r/R_2$ $H = R(1 - \cos\theta)$
33. .Spherical zone
$V = \pi\left\{\left[H^2\left(R - \frac{H}{3}\right)\right] - \left[h_1^2\left(R - \frac{h}{3}\right)\right]\right\}$
$V = \frac{\pi h_2}{6}\left[\frac{3}{4}C_1^2 + \frac{3}{4}C_2^2 + h_2^2\right]$
34. .Conical hole through spherical shell
$V = \frac{2\pi}{3}(R^3 - r^3)(\sin\theta_2 - \sin\theta_1)$
$B = \frac{0.375(R^4 - r^4)(\sin\theta_2 + \sin\theta_1)}{R^3 - r^3}$
RINGS
35. .Torus
$V = \frac{1}{4}\pi^2d^2D = 2.467d^2D$
36. .Hollow torus
$V = \frac{1}{4}\pi^2D(d_1^2 - d_2^2)$
37. .Bevel ring
$V = \pi\left(R + \frac{1}{3}W\right)WH$
$B = H\left[\frac{\frac{R}{3} + \frac{W}{12}}{R + \frac{W}{3}}\right]$

Fig. 2-4 (*cont.*)

38. Bevel ring

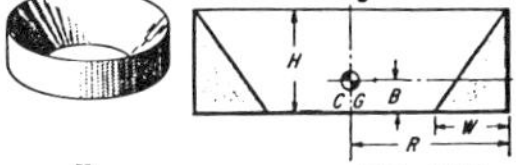

$$B > \frac{H}{3}$$

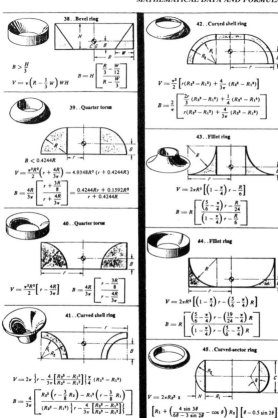

$$V = \pi\left(R - \frac{1}{3}W\right)WH \qquad B = H\left[\frac{\frac{R}{3} - \frac{W}{12}}{R - \frac{W}{3}}\right]$$

39. Quarter torus

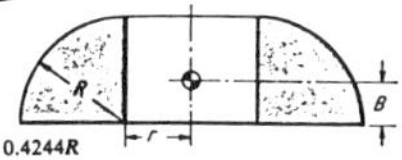

$$B < 0.4244R$$

$$V = \frac{\pi^2 R^2}{2}\left(r + \frac{4R}{3\pi}\right) = 4.9348R^2\,(r + 0.4244R)$$

$$B = \frac{4R}{3\pi}\left[\frac{r + \frac{3R}{8}}{r + \frac{4R}{3\pi}}\right] = \frac{0.4244Rr + 0.1592R^2}{r + 0.4244R}$$

40. Quarter torus

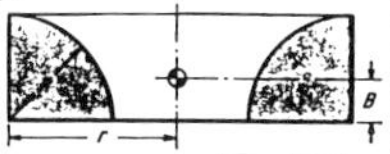

$$V = \frac{\pi^2 R^2}{2}\left[r - \frac{4R}{3\pi}\right] \qquad B = \frac{4R}{3\pi}\left[\frac{r - \frac{3R}{8}}{r - \frac{4R}{3\pi}}\right]$$

41. Curved shell ring

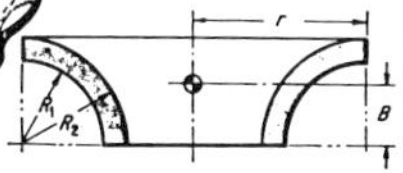

$$V = 2\pi\left\{r - \frac{4}{3\pi}\left[\frac{R_2^3 - R_1^3}{R_2^2 - R_1^2}\right]\right\}\frac{\pi}{4}\,(R_2^2 - R_1^2)$$

$$B = \frac{4}{3\pi}\left[\frac{R_2^3\left(r - \frac{3}{8}R_2\right) - R_1^3\left(r - \frac{3}{8}R_1\right)}{(R_2^2 - R_1^2)\left\{r - \frac{4}{3\pi}\left[\frac{R_2^3 - R_1^3}{R_2^2 - R_1^2}\right]\right\}}\right]$$

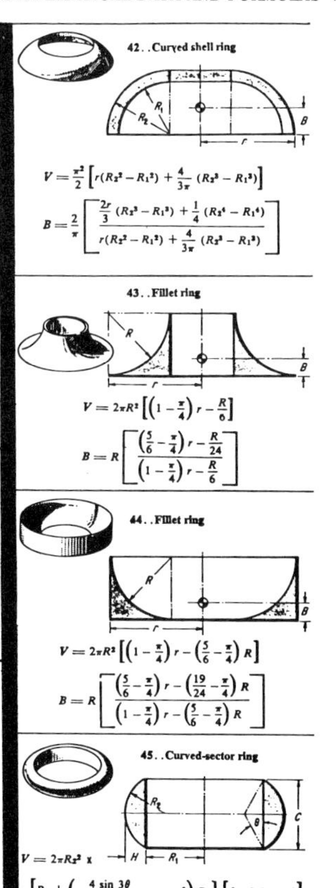

42. Curved shell ring

$$V = \frac{\pi^2}{2}\left[r(R_2^2 - R_1^2) + \frac{4}{3\pi}(R_2^3 - R_1^3)\right]$$

$$B = \frac{2}{\pi}\left[\frac{\frac{2r}{3}(R_2^3 - R_1^3) + \frac{1}{4}(R_2^4 - R_1^4)}{r(R_2^2 - R_1^2) + \frac{4}{3\pi}(R_2^3 - R_1^3)}\right]$$

43. Fillet ring

$$V = 2\pi R^2\left[\left(1 - \frac{\pi}{4}\right)r - \frac{R}{6}\right]$$

$$B = R\left[\frac{\left(\frac{5}{6} - \frac{\pi}{4}\right)r - \frac{R}{24}}{\left(1 - \frac{\pi}{4}\right)r - \frac{R}{6}}\right]$$

44. Fillet ring

$$V = 2\pi R^2\left[\left(1 - \frac{\pi}{4}\right)r - \left(\frac{5}{6} - \frac{\pi}{4}\right)R\right]$$

$$B = R\left[\frac{\left(\frac{5}{6} - \frac{\pi}{4}\right)r - \left(\frac{19}{24} - \frac{\pi}{4}\right)R}{\left(1 - \frac{\pi}{4}\right)r - \left(\frac{5}{6} - \frac{\pi}{4}\right)R}\right]$$

45. Curved-sector ring

$$V = 2\pi R_2^2 \times \left[R_1 + \left(\frac{4\sin^3\theta}{6\theta - 3\sin 2\theta} - \cos\theta\right)R_2\right]\left[\theta - 0.5\sin 2\theta\right]$$

ig. 2-4 (*cont.*)

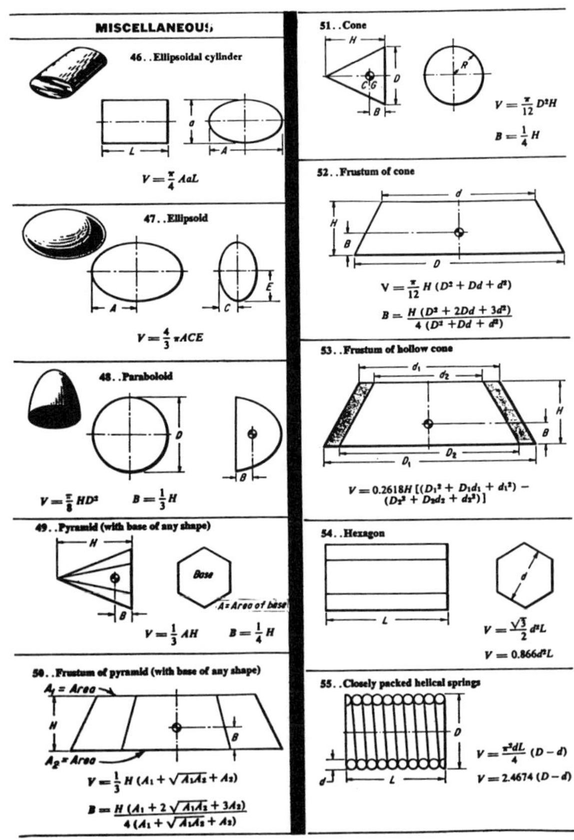
MISCELLANEOUS
46. . Ellipsoidal cylinder
$V = \frac{\pi}{4} AaL$
47. . Ellipsoid
$V = \frac{4}{3} \pi ACE$
48. . Paraboloid
$V = \frac{\pi}{8} HD^2$
$B = \frac{1}{3} H$
49. . Pyramid (with base of any shape)
Base
A = Area of base
$V = \frac{1}{3} AH$
$B = \frac{1}{4} H$
50. . Frustum of pyramid (with base of any shape)
A_1 = Area
A_2 = Area
$V = \frac{1}{3} H (A_1 + \sqrt{A_1 A_2} + A_2)$
$B = \frac{H (A_1 + 2\sqrt{A_1 A_2} + 3A_2)}{4 (A_1 + \sqrt{A_1 A_2} + A_2)}$
51. . Cone
C.G
$V = \frac{\pi}{12} D^2 H$
$B = \frac{1}{4} H$
52. . Frustum of cone
$V = \frac{\pi}{12} H (D^2 + Dd + d^2)$
$B = \frac{H (D^2 + 2Dd + 3d^2)}{4 (D^2 + Dd + d^2)}$
53. . Frustum of hollow cone
$V = 0.2618H [(D_1^2 + D_1 d_1 + d_1^2) - (D_2^2 + D_2 d_2 + d_2^2)]$
54. . Hexagon
$V = \frac{\sqrt{3}}{2} d^2 L$
$V = 0.866 d^2 L$
55. . Closely packed helical springs
$V = \frac{\pi^2 dL}{4} (D - d)$
$V = 2.4674 (D - d)$

Fig. 2-4 (*cont.*)

TABLE 2-6 Important Constants

N	Log N	N	Log N
π = 3.14159265	0.4971499	π^2 = 9.86960440	0.9942997
2π = 6.28318531	0.7981799	$\frac{1}{\pi^2}$ = 0.10132118	9.0057003–10
4π = 12.56637061	1.0992099	$\sqrt{\pi}$ = 1.77245385	0.2485749
$\frac{\pi}{2}$ = 1.57079633	0.1961199	$\frac{1}{\sqrt{\pi}}$ = 0.56418958	9.7514251–10
$\frac{\pi}{3}$ = 1.04719755	0.0200286	$\sqrt{\frac{3}{\pi}}$ = 0.97720502	9.9899857–10
$\frac{4\pi}{3}$ = 4.18879020	0.6220886	$\sqrt{\frac{4}{\pi}}$ = 1.12837917	0.0524551
$\frac{\pi}{4}$ = 0.78539816	9.8950899–10	$\sqrt[3]{\pi}$ = 1.46459189	0.1657166
$\frac{\pi}{6}$ = 0.52359878	9.7189986–10	$\frac{1}{\sqrt[3]{\pi}}$ = 0.68278406	9.8342834–10
$\frac{1}{\pi}$ = 0.31830989	9.5028501–10	$\sqrt[3]{\pi^2}$ = 2.14502940	0.3314332
$\frac{1}{2\pi}$ = 0.15915494	9.2018201–10		
$\frac{3}{\pi}$ = 0.95492966	9.9799714–10	$\sqrt[3]{\frac{3}{4\pi}}$ = 0.62035049	9.7926371–10
$\frac{4}{\pi}$ = 1.27323954	0.1049101	$\sqrt[3]{\frac{\pi}{6}}$ = 0.80599598	9.9063329–10

e = Napierian base	= 2.71828183	0.43429448
$M = \log_{10} e$	= 0.43429448	9.63778431—10
$1/M = \log_e 10$	= 2.30258509	0.36221569
$180/\pi$ = degrees in 1 radian	= 57.2957795	1.75812263
$\pi/180$ = radians in 1°	= 0.01745329	8.24187737—10
$\pi/10800$ = radians in 1′	= 0.0002908882	6.46372612—10
$\pi/648000$ = radians in 1″	= 0.000004848136811095	4.68557487—10
sin 1″	= 0.000004848136811076	4.68557487—10
tan 1″	= 0.000004848136811133	4.68557487—10
centimeters in 1 (U.S.) ft	= 30.4800	1.4840150
feet in 1 cm	= 0.0328084	8.5159850—10
inches in 1 m	= 39.3700	1.5951663
pounds in 1 kg	= 2.20462	0.3433343
kilograms in 1 lb	= 0.453592	9.6566657—10
cubic inches in 1 (U.S.) gal	= 231	2.3636120
g (average value)	= 32.16 ft/sec/sec	1.5073
g (legal)	= 980.665 cm/sec²	2.9915207
weight of 1 cu ft of water	= 62.425 lb (max. density)	1.7953586
weight of 1 cu ft of air	= 0.0807 lb (at 32°F)	8.907—10
ft-lb per sec in 1 hp	= 550	2.7403627
kg-m per sec in 1 hp	= 76.0404	1.8810445
watts in 1 hp (legal)	= 745.70	2.8725649

π = 3.14159 26535 89793 23846
e = 2.71828 18284 59045 23536
M = 0.43429 44819 03251 82765
$1/M$ = 2.30258 50929 94045 68402
$\log_{10} \pi$ = 0.49714 98726 94133 85435
$\log_{10} M$ = 9.63778 43113 00536 78912

SOURCE: R. S. Burington, *Handbook of Mathematical Tables and Formulas,* 5th ed. Copyright © 1972 by McGraw-Hill, Inc., New York. (Used with permission of McGraw-Hill Book Co.)

TABLE 2-7 Complete Elliptic Integrals, K and E for Different Values of the Modulus k

$\sin^{-1} k$	K	E	$\sin^{-1} k$	K	E	$\sin^{-1} k$	K	E
0°	1.5708	1.5708	50°	1.9356	1.3055	81°.0	3.2553	1.0338
1	1.5709	1.5707	51	1.9539	1.2963	81.2	3.2771	1.0326
2	1.5713	1.5703	52	1.9729	1.2870	81.4	3.2995	1.0314
3	1.5719	1.5697	53	1.9927	1.2776	81.6	3.3223	1.0302
4	1.5727	1.5689	54	2.0133	1.2681	81.8	3.3458	1.0290
5	1.5738	1.5678	55	2.0347	1.2587	82.0	3.3699	1.0278
6	1.5751	1.5665	56	2.0571	1.2492	82.2	3.3946	1.0267
7	1.5767	1.5649	57	2.0804	1.2397	82.4	3.4199	1.0256
8	1.5785	1.5632	58	2.1047	1.2301	82.6	3.4460	1.0245
9	1.5805	1.5611	59	2.1300	1.2206	82.8	3.4728	1.0234
10	1.5828	1.5589	60	2.1565	1.2111	83.0	3.5004	1.0223
11	1.5854	1.5564	61	2.1842	1.2015	83.2	3.5288	1.0213
12	1.5882	1.5537	62	2.2132	1.1920	83.4	3.5581	1.0202
13	1.5913	1.5507	63	2.2435	1.1826	83.6	3.5884	1.0192
14	1.5946	1.5476	64	2.2754	1.1732	83.8	3.6196	1.0182
15	1.5981	1.5442	65	2.3088	1.1638	84.0	3.6519	1.0172
16	1.6020	1.5405	65.5	2.3261	1.1592	84.2	3.6852	1.0163
17	1.6061	1.5367	66.0	2.3439	1.1545	84.4	3.7198	1.0153
18	1.6105	1.5326	66.5	2.3622	1.1499	84.6	3.7557	1.0144
19	1.6151	1.5283	67.0	2.3809	1.1453	84.8	3.7930	1.0135
20	1.6200	1.5238	67.5	2.4001	1.1408	85.0	3.8317	1.0127
21	1.6252	1.5191	68.0	2.4198	1.1362	85.2	3.8721	1.0118
22	1.6307	1.5141	68.5	2.4401	1.1317	85.4	3.9142	1.0110
23	1.6365	1.5090	69.0	2.4610	1.1272	85.6	3.9583	1.0102
24	1.6426	1.5037	69.5	2.4825	1.1228	85.8	4.0044	1.0094
25	1.6490	1.4981	70.0	2.5046	1.1184	86.0	4.0528	1.0086
26	1.6557	1.4924	70.5	2.5273	1.1140	86.2	4.1037	1.0079
27	1.6627	1.4864	71.0	2.5507	1.1096	86.4	4.1574	1.0072
28	1.6701	1.4803	71.5	2.5749	1.1053	86.6	4.2142	1.0065
29	1.6777	1.4740	72.0	2.5998	1.1011	86.8	4.2744	1.0059
30	1.6858	1.4675	72.5	2.6256	1.0968	87.0	4.3387	1.0053
31	1.6941	1.4608	73.0	2.6521	1.0927	87.2	4.4073	1.0047
32	1.7028	1.4539	73.5	2.6796	1.0885	87.4	4.4811	1.0041
33	1.7119	1.4469	74.0	2.7081	1.0844	87.6	4.5609	1.0036
34	1.7214	1.4397	74.5	2.7375	1.0804	87.8	4.6477	1.0031
35	1.7312	1.4323	75.0	2.7681	1.0764	88.0	4.7427	1.0026
36	1.7415	1.4248	75.5	2.7998	1.0725	88.2	4.8478	1.0021
37	1.7522	1.4171	76.0	2.8327	1.0686	88.4	4.9654	1.0017
38	1.7633	1.4092	76.5	2.8669	1.0648	88.6	5.0988	1.0014
39	1.7748	1.4013	77.0	2.9026	1.0611	88.8	5.2527	1.0010
40	1.7868	1.3931	77.5	2.9397	1.0574	89.0	5.4349	1.0008
41	1.7992	1.3849	78.0	2.9786	1.0538	89.1	5.5402	1.0006
42	1.8122	1.3765	78.5	3.0192	1.0502	89.2	5.6579	1.0005
43	1.8256	1.3680	79.0	3.0617	1.0468	89.3	5.7914	1.0004
44	1.8396	1.3594	79.5	3.1064	1.0434	89.4	5.9455	1.0003
45	1.8541	1.3506	80.0	3.1534	1.0401	89.5	6.1278	1.0002
46	1.8691	1.3418	80.2	3.1729	1.0388	89.6	6.3509	1.0001
47	1.8848	1.3329	80.4	3.1928	1.0375	89.7	6.6385	1.0001
48	1.9011	1.3238	80.6	3.2132	1.0363	89.8	7.0440	1.0000
49	1.9180	1.3147	80.8	3.2340	1.0350	89.9	7.7371	1.0000

$$K = \int_0^{\pi/2} \frac{dx}{\sqrt{1 - k^2 \sin^2 x}} \qquad E = \int_0^{\pi/2} \sqrt{1 - k^2 \sin^2 x}\, dx$$

SOURCE: R. S. Burington, *Handbook of Mathematical Tables and Formulas* 5th ed. Copyright © 1972 by McGraw-Hill, Inc., New York. (Used with permission of McGraw-Hill Book Co.)

TABLE 2-8 Glossary of Mathematical Symbols

Symbol	*Name or meaning*
Algebra	
$\|a\|$	absolute value of a
$n!$	n factorial
$0!$	zero factorial, $0! = 1$
$\binom{n}{r} = \frac{n!}{(n-r)!r!}$	binomial coefficient; $\binom{n}{r}$ is coefficient of $a^{n-r}b^r$ in expansion of $(a+b)^n$
$\sqrt[n]{a} = a^{1/n}$	the positive nth root of a, $a > 0$
$\sqrt[q]{a^p} = a^{p/q}$	the positive qth root of a^p, $a > 0$
$\sum_{i=1}^{n} a_i$	$\sum_{i=1}^{n} a_i = a_1 + a_2 + \cdots + a_n$
Sets	
$\in$	belongs to; is an element of; is a member of
$\notin$	does not belong to; is not an element of; is not a member of
$\supset$	contains
$\subseteq$	is included in; is a subset of
$\subset$	is a proper subset of; is properly included in; *inclusion*
$\not\subset$	is not a proper subset of; is not included in
$\cup$	union; cup; join; logical sum
$\cap$	intersection; cap; logical product
A' or $C_U A$	complement of A relative to universe U
$\emptyset$	the null set; the empty set
$=$	is equal to; equals; is identical with; is equivalent to
$\neq$	is not equal to; not equal
$\approx$ or $\leftrightarrow$	is equivalent to
$\not\approx$	is not equivalent to
$\{\ \}$	used to represent a set
$\{x\}$	the set of objects x
$\|$ or $:$	such that; for which
$\{x \mid x \text{ has property } p\}$	the set of objects x having the property p
$\{x \mid \quad \}$	a set builder, that is, a set consisting of all x such that
$[a,b]$	the set $a \leqq x \leqq b$
$[a,b)$	the set $a \leqq x < b$

TABLE 2-8 Glossary of Mathematical Symbols ***(cont.)***

Symbol	Meaning
$(a,b]$	the set $a < x \leqq b$
(a,b)	the set $a < x < b$
$\langle x, y \rangle$	ordered pair
$A \times B$	cartesian product of A and B, A cross B
$\cup A_r$ or $\bigcup_{r=1}^{n} A_r = A_1 \cup A_2 \cup \cdots \cup A_n$	
$\cap A_r$ or $\bigcap_{r=1}^{n} A_r = A_1 \cap A_2 \cap \cdots \cap A_n$	
$\aleph_0$	*cardinal number* of set of all natural numbers 1, 2, 3, . . . ; aleph null
$\aleph$	*cardinal number* of set of all real numbers; aleph
Logic	
$\wedge$, &	and; *conjunction*
$\vee$	inclusive or; *inclusive disjunction*
$\underline{\vee}$	exclusive or; *exclusive disjunction*
$\rightarrow$ or $\Rightarrow$	implies; *implication*
$\leftrightarrow$ or $\Leftrightarrow$	is equivalent to; *equivalence*
$\sim$ or $'$	not; *negation*
$\forall$	for all, for every; *universal quantifier*
$\forall_x$	for all x
$\exists$	there exists at least one; for some; *existential quantifier*
$\exists_x$	there exists an x such that; for some x
$\exists \vert$	there exists uniquely
E!v	there exists exactly one v such that; *uniqueness quantifier*
$\vdash .$	it is asserted that
Relations	
R	has the *relation* R to
$\not R$ or R'	does not have the relation R to
$\breve{R}$	the *converse* of relation R
$\approx$ or $=$	is identical (or equivalent) to; *identity* or *equivalence* relation
$\not\approx$ or $\neq$	is not identical (or not equivalent) to; *diversity relation*
$\subset$	is included in
R^{-1}	the *inverse* of relation R
$x\,R\,y$	x stands in the (binary) relation R to y

TABLE 2-8 Glossary of Mathematical Symbols *(cont.)*

$x \not R y$	x does not stand in the (binary) relation R to y
$\leftrightarrow$	one-to-one reciprocal correspondence
$\rightarrow$ or $\hookrightarrow$	corresponds to; gives; is transformed into; has for an image
$\xrightarrow{T}$ or $\overset{T}{\hookrightarrow}$	passage by means of T; is transformed by T into
$F: S \rightarrow T$	F maps S into T
Functions	
$f(x)$	*value* of function f at x
f^{-1}	*inverse* of function f
$f \circ g$	*composite* of function f and function g
$f^{-1} \circ f = f \circ f^{-1} = E$	
$f: (x, y)$	the function f whose ordered pairs are (x, y)
Probabilities	
$P(E)$	probability of E
$P(E \cap F)$ or $P(E \wedge F)$	probability of both E and F
$P(E \mid F)$	probability of E, given F
$P(E \cup F)$ or $P(E \vee F)$	probability of E or F or both
$P(E \veebar F)$	probability of E or F, but not both E and F
Algebraic Structures	
$\circ, \odot, \oplus, \otimes, \star, \cup, \cap, +, -$	symbols for *operations*
z or 0	zero
u or 1	unity
a^{-1}	*inverse* of a
$\otimes$ (circled <)	symbol used in ordering
$\otimes$ (circled >)	symbol used in ordering
Vectors	
$V_1 \cdot V_2$	scalar product
$V_1 \times V_2$	vector product
∇S, grad S	gradient of S
div V, $\nabla \cdot V$	divergence of V

TABLE 2-8 Glossary of Mathematical Symbols ***(cont.)***

curl $\boldsymbol{V}$, rot $\boldsymbol{V}$, $\nabla \times \boldsymbol{V}$	curl of $\boldsymbol{V}$
$\nabla^2 S$, div grad S	divergence of grad S
Matrices	
$\begin{pmatrix} a_{11} \cdots a_{1n} \\ \cdots\cdots \\ a_{n1} \cdots a_{nn} \end{pmatrix}$ or $[a_{rs}]$ or (a_{ij}) or A	*matrix*
$(x_1, x_2, \ldots, x_n)$ or $\overrightarrow{x_p}$ or $\lVert x_1, x_2, \ldots, x_n \rVert$	row matrix, row vector
$\begin{Bmatrix} y_1 \\ \cdots \\ y_n \end{Bmatrix}$ or $y_p\uparrow$ or $\begin{Vmatrix} y_1 \\ \cdots \\ y_n \end{Vmatrix}$	column matrix, column vector
$\det A = \lvert A \rvert = \begin{vmatrix} a_{11} \cdots a_{1n} \\ \cdots\cdots \\ a_{n1} \cdots a_{nn} \end{vmatrix}$	*determinate* of matrix A

SOURCE: R. S. Burington, *Handbook of Mathematical Tables and Formulas*, 5th ed. Copyright © 1972 by McGraw-Hill, Inc., New York. (Used with permission of McGraw-Hill Book Co.)

NOTES

NOTES

Section Three

■

GENERAL CONVERSION TABLES

TABLE 3-1 Length

Ordinary Units

1 foot = 12 inches
1 yard = 3 feet
1 mile = 5280 feet
1 nautical mi = 1.1516 statute mi
1° of latitude at the equator = 69.16 statute mi
= 60 nautical mi
1 acre = 208.71 ft on one side of square

Metric Units

10 millimeters (mm) = 1 centimeter (cm)
100 cm = 1 meter (m)
1000 m = 1 kilometer (km) (about ⅝ mi)

Equivalents

1 inch	= 2.5400 centimeters
1 foot	= 0.3048 meter
1 statute mi	= 1.60935 kilometers
1 nautical mi	= 1.853 kilometers
1 centimeter	= 0.39370 inch
1 meter	= 3.28 feet
1 kilometer	= 3280.83 feet = 0.62137 mile

URCE: *Handbook of Steel Drainage and Highway Construction Products*, 2d ed., copyright © 1971 by American Iron and Steel Institute, Washington, D.C.

TABLE 3-2 Area

Ordinary Units

1 square foot	=	144 square inches
1 square yard	=	9 sq ft
	≐	1296 sq in.
1 acre	=	43,560 sq ft
	=	4840 sq yds
1 sq mile	=	640 acres
	=	1 section of land (U.S.)

Equivalents

1 square centimeter	=	0.155 square inch
1 square meter	=	10.76 square feet
	=	1.196 square yards
1 square kilometer	=	0.386 square mile
1 square inch	=	6.45 square centimeters
1 square foot	=	0.0929 square meter
1 square yard	=	0.836 square meter
1 square mile	=	2.59 square kilometers

SOURCE: *Handbook of Steel Drainage and Highway Construction Products* 2d ed., copyright © 1971 by American Iron and Steel Institute Washington, D.C.

TABLE 3-3 Volume and Capacity

Ordinary Units

1 cu ft of water at 39.1° F = 62.425 lbs
1 United States gallon = 231 cu in.
1 imperial gallon = 277.274 cu in.
1 cubic foot of water = 1728 cu in.
= 7.480519 U. S. gallons
= 6.232103 imperial gallons
1 cubic yard = 27 cu ft = 46,656 cu in.
1 quart = 2 pints
1 gallon = 4 quarts
1 U. S. gallon = 231 cu in.
= 0.133681 cu ft
= 0.83311, imperial gallon
= 8.345 lbs
1 barrel = 31.5 gallons = 4.21 cu ft
1 U. S. bushel = 1.2445 cu ft
1 fluid ounce = 1.8047 cu in.
1 acre foot = 43,560 cu ft
= 1,613.3 cu yds
1 acre inch = 3,630 cu ft
1 million U. S. gallons = 133,681 cu ft
= 3.0689 acre-ft
1 ft depth on 1 sq mi = 27,878, 400 cu ft
= 640 acre-ft

Equivalents

1 cu in. = 16.387 cu cm
1 cu ft = 0.0283 cu m
1 cu yd = 0.765 cu m
1 cu cm = 0.0610 cu in.
1 cu m = 35.3 cu ft
= 1.308 cu yds
1 liter = 61.023378 cu in. (about 1 quart)
= 0.264170 U. S. liquid gallon
= 0.2201 imperial gallon
1 U. S. liquid quart = 0.946 liter
1 U. S. liquid gallon = 3.785 liters

SOURCE: *Handbook of Steel Drainage and Highway Construction Products*, 2d ed., copyright © 1971 by American Iron and Steel Institute, Washington, D.C.

TABLE 3-4 Weight (Mass)

Ordinary Units

1 pound = 16 ounces (avoirdupois)
1 ton = 2000 lbs
1 long ton = 2240 lbs
1 lb of water (39.1° F) = 27.681217 cu in.
= 0.016019 cu ft
= 0.119832 U. S. gallon
= 0.453617 liter

Equivalents

1 kilogram	**=**	**2.205 avoirdupois pounds**
1 metric ton	**=**	**0.984 gross or long ton**
	=	**1.102 net or short tons**
1 avoirdupois ounce	**=**	**28.35 grams**
1 avoirdupois pound	**=**	**0.4536 kilogram**

SOURCE: *Handbook of Steel Drainage and Highway Construction Products*, 2d ed., copyright © 1971 by American Iron and Steel Institute, Washington, D.C.

TABLE 3-5 Pressure

Comparison of Heads of Water in Feet with Pressures in Various Units

One foot of water at 39.1° F = 62.425 pounds per square foot (psf)
= 0.4335 pound per square inch
= 0.0295 atmosphere
= 0.8826 inch of mercury at 30° F
= 773.3 feet of air at 32° F and atmospheric pressure
One foot of water at 62° F = 62.355 pounds per square foot
= 0.43302 pound per square inch
One pound of water on the square inch at 62° F = 2.3094 feet of water
One ounce of water on the square inch at 62° F = 1.732 inches of water
1 atmosphere at sea level (32° F) = 14.697 lbs per sq in.
= 29.921 in. of mercury
1 inch of mercury (32° F) = 0.49119 lb per sq in.

TABLE 3-6 Flowing Water

cfs = cubic feet per second, or second feet
gpm = gallons per minute
1 cfs = 60 cu ft per min
= 86,400 cu ft per 24 hrs
= 448.83 U. S. gals per min
= 646,317 U. S. gals per 24 hrs
= 1.9835 acre-foot per 24 hrs (usually taken as 2)
= 1 acre-inch per hour (approximate)
= .028317 cu meter per second
1 U. S. gpm = 1440 U. S. gals per 24 hrs
= 0.00442 acre-foot per 24 hrs
= 0.0891 Miners inches, Ariz., Calif.
1 million U. S. gal per day = 1.5472 cfs
= 3.07 acre-feet
= 2.629 cu meters per min

SOURCE: *Handbook of Steel Drainage and Highway Construction Products*, 2d ed., copyright © 1971 by American Iron and Steel Institute, Washington, D.C.

TABLE 3-7 Miscellaneous Measurements
(Temperature, Circular Measure, Time Measure, Ice and Snow)

Temperature

Freezing point of water = 32° Fahrenheit
= 0° Celsius

Boiling point of water (at normal air pressure) = 212° Fahrenheit
= 100° Celsius

1 degree Fahrenheit = 0.5556 degree (Celsius)
1 degree Celsius = 1.8 degrees Fahrenheit

Circular Measure

1 minute (′) = 60 seconds (″)
1 degree (°) = 60 minutes
1 right angle = 90 degrees
1 circumference = 360 degrees

Time Measure

1 minute = 60 seconds
1 hour = 60 minutes = 3600 seconds
1 day = 24 hours = 1440 minutes
1 week = 7 days
1 year = 365 days, 5 hr, 48 min, 48 sec

Ice and Snow

1 cubic foot of ice at 32° F weighs 57.50 pounds; 1 pound of ice at 32° F has a volume of 0.0174 cubic foot = 30.067 cubic inches (Clark).

1 cubic foot of fresh snow, according to humidity of atmosphere, weighs 5 pounds to 12 pounds. 1 cubic foot of snow moistened and compacted by rain weighs 15 pounds to 50 pounds (Trautwine).

SOURCE: *Handbook of Steel Drainage and Highway Construction Products*, 2d ed., copyright © 1971 by American Iron and Steel Institute, Washington, D.C.

TABLE 3-8 Areas of Plane Figures

Triangle:	Base × ½ perpendicular height $\sqrt{s(s-a)\ (s-b)\ (s-c)}$ s = ½ sum of the three sides a, b and c
Trapezium:	Sum of areas of the two triangles
Trapezoid:	½ sum of parallel sides × perpendicular height
Parallelogram:	Base × perpendicular height
Regular Polygon:	½ sum of sides × inside radius
Circle:	$\pi r^2 = 0.78540 \times \text{dia}^2 = 0.07958 \times \text{circumference}^2$
Sector of Circle:	$\dfrac{\pi r^2 A^\circ}{360} = 0.0087266\ r^2 A^\circ = \text{arc} \times \tfrac{1}{2}\ \text{radius}$
Segment of Circle:	$\dfrac{r^2}{2}\left(\dfrac{\pi A^\circ}{180} - \sin A^\circ\right)$

Circle of same area as square: diameter = side × 1.12838

Square of same area as circle: side = diameter × 0.88623

Ellipse:	Long diameter × short diameter × 0.78540
Parabola:	Base × ⅔ perpendicular height

SOURCE: *Handbook of Steel Drainage and Highway Construction Products*, 2d ed., copyright © 1971 by American Iron and Steel Institute, Washington, D.C.

TABLE 3-9 Trigonometric Formulas

$$\text{Radius, } 1 = \sin^2 A + \cos^2 A = \sin A \operatorname{cosec} A = \cos A \sec A = \tan A \cot A$$

Sine $$A = \frac{\cos A}{\cot A} = \frac{1}{\operatorname{cosec} A} = \cos A \tan A = \sqrt{1 - \cos^2 A}$$

Cosine $$A = \frac{\sin A}{\tan A} = \frac{1}{\sec A} = \sin A \cot A = \sqrt{1 - \sin^2 A}$$

Tangent $$A = \frac{\sin A}{\cos A} = \frac{1}{\cot A} = \sin A \sec A$$

Cotangent $$A = \frac{\cos A}{\sin A} = \frac{1}{\tan A} = \cos A \operatorname{cosec} A$$

Secant $$A = \frac{\tan A}{\sin A} = \frac{1}{\cos A}$$

Cosecant $$A = \frac{\cot A}{\cos A} = \frac{1}{\sin A}$$

SOURCE: *Handbook of Steel Drainage and Highway Construction Products*, 2d ed., copyright © 1971 by American Iron and Steel Institute, Washington, D.C.

TABLE 3-10 Areas of Circles

Diameter in Inches	Area		Diameter in Inches	Area		Diameter in Inches	Area	
	Square Inches	Square Feet		Square Inches	Square Feet		Square Inches	Square Feet
1.0	0.7854	.005454	4.0	12.5664	.087266	7.0	38.4845	.267254
.1	0.9503	.006599	.1	13.2025	.091684	.1	39.5919	.274944
.2	1.1310	.007854	.2	13.8544	.096211	.2	40.7150	.282743
.25	1.2272	.008522	.25	14.1863	.098516	.25	41.2825	.286684
.3	1.3273	.009218	.3	14.5220	.100847	.3	41.8539	.290652
.4	1.5394	.010690	.4	15.2053	.105592	.4	43.0084	.298669
.5	1.7671	.012272	.5	15.9043	.110446	.5	44.1786	.306796
.6	2.0106	.013963	.6	16.6190	.115410	.6	45.3646	.315032
.7	2.2698	.015763	.7	17.3494	.120482	.7	46.5663	.323377
.75	2.4053	.016703	.75	17.7205	.123059	.75	47.1730	.327590
.8	2.5447	.017671	.8	18.0956	.125664	.8	47.7336	.331831
.9	2.8353	.019689	.9	18.8574	.130954	.9	49.0167	.340394
2.0	3.1416	.021816	5.0	19.6350	.136354	8.0	50.2655	.349066
.1	3.4636	.024053	.1	20.4282	.141863	.1	51.5300	.357847
.2	3.8013	.026398	.2	21.2372	.147480	.2	52.8102	.366737
.25	3.9761	.027612	.25	21.6475	.150330	.25	53.4562	.371223
.3	4.1548	.028852	.3	22.0618	.153207	.3	54.1061	.375736
.4	4.5239	.031416	.4	22.9022	.159043	.4	55.4177	.384845
.5	4.9087	.034088	.5	23.7583	.164988	.5	56.7450	.394063
.6	5.3093	.036870	.6	24.6301	.171042	.6	58.0880	.403389
.7	5.7256	.039760	.7	25.5176	.177205	.7	59.4468	.412825
.75	5.9396	.041247	.75	25.9672	.180328	.75	60.1320	.417584
.8	6.1575	.042760	.8	26.4208	.183477	.8	60.8212	.422370
.9	6.6052	.045869	.9	27.3397	.189859	.9	62.2114	.432024
3.0	7.0686	.049087	6.0	28.2743	.196350	9.0	63.6173	.441786
.1	7.5477	.052414	.1	29.2247	.202949	.1	65.0388	.451658
.2	8.0425	.055851	.2	30.1907	.209658	.2	66.4761	.461640
.25	8.2958	.057609	.25	30.6796	.213053	.25	67.2006	.466671
.3	8.5530	.059396	.3	31.1725	.216475	.3	67.9291	.471730
.4	9.0792	.063050	.4	32.1699	.223402	.4	69.3978	.481929
.5	9.6211	.066813	.5	33.1831	.230438	.5	70.8822	.492237
.6	10.1788	.070686	.6	34.2119	.237583	.6	72.3823	.502654
.7	10.7521	.074667	.7	35.2565	.244837	.7	73.8981	.513181
.75	11.0447	.076699	.75	35.7847	.248505	.75	74.6619	.518486
.8	11.3411	.078758	.8	36.3168	.252200	.8	75.4296	.523817
.9	11.9459	.082958	.9	37.3928	.259672	.9	76.9769	.534561

The above table may be used for finding the areas of circles whose diameters are not within the limits of the table. Since the areas vary as the squares of their diameters, the given diameter may be divided (or multiplied) by 10, and the area found from the table under the resulting diameter corrected by moving the decimal point two places to the right (or left). Thus to find the area of a 22-inch circle:

From table, area of 2.2-inch circle = 3.8013 sq in. = .026398 sq ft

Therefore area of 22-inch circle = 380.13 sq in. = 2.64 sq ft

Again, to find the area of a 0.75-inch circle:

From table, area of 7.5-inch circle = 44.1786 sq in. = 0.306796 sq ft

Therefore area of 0.75-inch circle = 0.4418 sq in. = 0.00307 sq ft

It will also be apparent that the *first two* columns in the table may be used for any unit of measure.

SOURCE: *Handbook of Steel Drainage and Highway Construction Products*, 2d ed., copyright © 1971 by American Iron and Steel Institute, Washington, D.C.

TABLE 3-11 Slope in Inches Reduced to Feet

In. per 100 Ft	Ft per 100 Ft	Ft per Mile	In. per 100 Ft	Ft per 100 Ft	Ft per Mile
¼	.0208	1.098	¼	.5208	27.498
½	.0417	2.202	½	.5417	28.602
¾	.0625	3.300	¾	.5625	29.700
1	.0833	4.398	7	.5833	30.798
¼	.1042	5.502	¼	.6042	31.902
½	.1250	6.600	½	.6250	33.000
¾	.1458	7.698	¾	.6458	34.098
2	.1667	8.802	8	.6667	35.202
¼	.1875	9.900	¼	.6875	36.300
½	.2083	10.998	½	.7083	37.398
¾	.2292	12.102	¾	.7292	38.502
3	.2500	13.200	9	.7500	39.600
¼	.2708	14.298	¼	.7708	40.698
½	.2917	15.402	½	.7917	41.802
¾	.3125	16.500	¾	.8125	42.900
4	.3333	17.598	10	.8333	43.998
¼	.3542	18.702	¼	.8542	45.102
½	.3750	19.800	½	.8750	46.200
¾	.3958	20.898	¾	.8958	47.298
5	.4167	22.002	11	.9167	48.402
¼	.4375	23.100	¼	.9375	49.500
½	.4583	24.198	½	.9583	50.598
¾	.4792	25.302	¾	.9792	51.702
6	.5000	26.400	12	1.0000	52.800

SOURCE: *Handbook of Steel Drainage and Highway Construction Products*, 2d ed., copyright © 1971 by American Iron and Steel Institute, Washington, D.C.

TABLE 3-12 Frequently Used Conversion Factors

Multiply	By	To Get
Acre-feet	43,560	Cubic feet
Atmosphere	29.92	Inches of mercury
"	33.90	Feet of water
"	14.70	Lb per sq in.
Barrels, oil	42	Gallons, oil
Cubic feet	1728	Cubic inches
" "	7.48	Gallons
" "	28.32	Liters
Cubic feet per sec	448.83	Gallons per min
Feet	30.48	Centimeters
"	0.3048	Meters
Gallons	231	Cubic inches
"	3.785	Liters
" (Imperial)	1.20095	U.S. gallons
" (U.S.)	0.83267	Imp. gallons
" water	8.3453	Pounds of water
" per min	2.228×10^{-3}	Cubic ft/sec
Grains per U.S. gal	17.118	Parts per million
Grams	10^{3}	Milligrams
" per liter	10^{3}	Parts per million
Horsepower	0.7457	Kilowatts
"	745.7	Watts
" hours	0.7457	Kilowatt-hours
Kilograms	10^{3}	Grams
Kilowatts	1.341	Horsepower
" hours	1.341	Horsepower-hours
Liters	0.2642	Gallons
Meters	39.37	Inches
"	3.281	Feet
Microns	10^{-6}	Meters
Milligrams	10^{-3}	Grams
Milliliters	10^{-3}	Liters
Milligrams per liter	1	Parts per million
Million gallons per day	1.54723	Cubic feet per sec
Pounds	453.59	Grams
Square miles	640	Acres
Temp. (°C) + 17.78	1.8	Temp. (°F)
Temp. (°F) − 32	5/9	Temp. (°C)
Watts	1.341×10^{-3}	Horsepower
"	10^{-3}	Kilowatts
Watt hours	1.341×10^{-3}	Horsepower hours

The figures 10^{-1}, 10^{-2}, 10^{-3}, etc., denote 0.10, 0.01, 0.001, etc.
The figures 10^{1}, 10^{2}, 10^{3}, etc., denote 10, 100, 1000, etc.

SOURCE: R. O. Parmley, *Standard Handbook of Fastening and Joining*. Copyright © McGraw-Hill, Inc. New York, 1977. (Used with permission of McGraw-Hill Book Co.)

Table 3-13 General Conversion Factors

Multiply	*by*	*to obtain*
acres	43,560	square feet
acres	4047	square meters
acres	1.562×10^{-3}	square miles
acres	5645.38	square varas
acres	4840	square yards
amperes	1/10	abamperes
amperes	3×10^{9}	statamperes
atmospheres	76.0	cms. of mercury
atmospheres	29.92	inches of mercury
atmospheres	33.90	feet of water
atmospheres	10,333	kgs. per sq. meter
atmospheres	14.70	pounds per sq. inch
atmospheres	1.058	tons per sq. foot
British thermal units	0.2520	kilogram-calories
British thermal units	777.5	foot-pounds
British thermal units	3.927×10^{-4}	horse-power-hours
British thermal units	1054	joules
British thermal units	107.5	kilogram-meters
British thermal units	2.928×10^{-4}	kilowatt-hours
B.t.u. per min.	12.96	foot-pounds per sec.
B.t.u. per min.	0.02356	horse-power
B.t.u. per min.	0.01757	kilowatts
B.t.u. per min.	17.57	watts
B.t.u. per sq. ft. per min.	0.1220	watts per sq. inch
bushels	1.244	cubic feet
bushels	2150	cubic inches
bushels	0.03524	cubic meters
bushels	4	pecks
bushels	64	pints (dry)
bushels	32	quarts (dry)
centimeters	0.3397	inches
centimeters	0.01	meters
centimeters	393.7	mils
centimeters	10	millimeters
centimeter-grams	980.7	centimeter-dynes
centimeter-grams	10^{-5}	meter-kilograms
centimeter-grams	7.233×10^{-5}	pound-feet

SOURCE: Dietzgen Corporation, Des Plaines, Ill.

Table 3-13 General Conversion Factors (*cont.*)

Multiply	*by*	*to obtain*
centimeters of mercury	0.01316	atmospheres
centimeters of mercury	0.4461	feet of water
centimeters of mercury	136.0	kgs. per sq. meter
centimeters of mercury	27.85	pounds per sq. foot
centimeters of mercury	0.1934	pounds per sq. inch
centimeters per second	1.969	feet per minute
centimeters per second	0.03281	feet per second
centimeters per second	0.036	kilometers per hour
centimeters per second	0.6	meters per minute
centimeters per second	0.02237	miles per hour
centimeters per second	3.728×10^{-4}	miles per minute
cubic centimeters	3.531×10^{-5}	cubic feet
cubic centimeters	6.102×10^{-2}	cubic inches
cubic centimeters	10^{-6}	cubic meters
cubic centimeters	1.308×10^{-6}	cubic yards
cubic centimeters	2.642×10^{-4}	gallons
cubic centimeters	10^{-3}	liters
cubic centimeters	2.113×10^{-3}	pints (liq.)
cubic centimeters	1.057×10^{-3}	quarts (liq.)
cubic feet	62.43	pounds of water
cubic feet	2.832×10^{4}	cubic cms.
cubic feet	1728	cubic inches
cubic feet	0.02832	cubic meters
cubic feet	0.03704	cubic yards
cubic feet	7.481	gallons
cubic feet	28.32	liters
cubic feet	59.84	pints (liq.)
cubic feet	29.92	quarts (liq.)
cubic feet per minute	472.0	cubic cms. per sec.
cubic feet per minute	0.1247	gallons per sec.
cubic feet per minute	0.4720	liters per second
cubic feet per minute	62.4	lbs. of water per min.
cubic inches	16.39	cubic centimeters
cubic inches	5.787×10^{-4}	cubic feet
cubic inches	1.639×10^{-5}	cubic meters
cubic inches	2.143×10^{-5}	cubic yards
cubic inches	4.329×10^{-3}	gallons
cubic inches	1.639×10^{-2}	liters

Table 3-13 General Conversion Factors (*cont.*)

Multiply	*by*	*to obtain*
cubic inches	0.03463	pints (liq.)
cubic inches	0.01732	quarts (liq.)
cubic yards	7.646×10^{5}	cubic centimeters
cubic yards	27	cubic feet
cubic yards	46,656	cubic inches
cubic yards	0.7646	cubic meters
cubic yards	202.0	gallons
cubic yards	764.6	liters
cubic yards	1616	pints (liq.)
cubic yards	807.9	quarts (liq.)
cubic yards per minute	0.45	cubic feet per sec.
cubic yards per minute	3.367	gallons per second
cubic yards per minute	12.74	liters per second
degrees (angle)	60	minutes
degrees (angle)	0.01745	radians
degrees (angle)	3600	seconds
dynes	1.020×10^{-3}	grams
dynes	7.233×10^{-5}	poundals
dynes	2.248×10^{-6}	pounds
ergs	9.486×10^{-11}	British thermal units
ergs	1	dyne-centimeters
ergs	7.376×10^{-8}	foot-pounds
ergs	1.020×10^{-3}	gram-centimeters
ergs	10^{-7}	joules
ergs	2.390×10^{-11}	kilogram-calories
ergs	1.020×10^{-8}	kilogram-meters
feet	30.48	centimeters
feet	12	inches
feet	0.3048	meters
feet	.36	varas
feet	1/3	yards
feet of water	0.02950	atmospheres
feet of water	0.8826	inches of mercury
feet of water	304.8	kgs. per sq. meter
feet of water	62.43	pounds per sq. ft.
feet of water	0.4335	pounds per sq. inch
foot-pounds	1.286×10^{-3}	British thermal units

Table 3-13 General Conversion Factors (*cont.*)

Multiply	*by*	*to obtain*
foot-pounds	1.356×10^{7}	ergs
foot-pounds	5.050×10^{-7}	horse-power-hours
foot-pounds	1.356	joules
foot-pounds	3.241×10^{-4}	kilogram-calories
foot-pounds	0.1383	kilogram-meters
foot-pounds	3.766×10^{-7}	kilowatt-hours
foot-pounds per min.	1.286×10^{-3}	B.t. units per minute
foot-pounds per min.	0.01667	foot-pounds per sec.
foot-pounds per min.	3.030×10^{-5}	horse-power
foot-pounds per min.	3.241×10^{-4}	kg.-calories per min.
foot-pounds per min.	2.260×10^{-5}	kilowatts
foot-pounds per sec.	7.717×10^{-2}	B.t. units per minute
foot-pounds per sec.	1.818×10^{-3}	horse-power
foot-pounds per sec.	1.945×10^{-2}	kg-calories per min.
foot-pounds per sec.	1.356×10^{-3}	kilowatts
gallons	8.345	pounds of water
gallons	3785	cubic centimeters
gallons	0.1337	cubic feet
gallons	231	cubic inches
gallons	3.785×10^{-3}	cubic meters
gallons	4.951×10^{-3}	cubic yards
gallons	3.785	liters
gallons	8	pints (liq.)
gallons	4	quarts (liq.)
gallons per minute	2.228×10^{-3}	cubic ft. per second
gallons per minute	0.06308	liters per second
grains (troy)	1	grains (av.)
grains (troy)	0.06480	grams
grains (troy)	0.04167	pennyweights (troy)
grams	980.7	dynes
grams	15.43	grains (troy)
grams	10^{-3}	kilograms
grams	10^{3}	milligrams
grams	0.03527	ounces
grams	0.03215	ounces (troy)
grams	0.07093	poundals
grams	2.205×10^{-3}	pounds
horse-power	42.44	B.t. units per min.

Table 3-13 General Conversion Factors (*cont.*)

Multiply	*by*	*to obtain*
horse-power	33,000	foot-pounds per min.
horse-power	550	foot-pounds per sec.
horse-power	1.014	horse-power (metric)
horse-power	10.70	kg.-calories per min.
horse-power	0.7457	kilowatts
horse-power	745.7	watts
horse-power (boiler)	33,520	B.t.u. per hour
horse-power (boiler)	9.804	kilowatts
horse-power-hours	2547	British thermal units
horse-power-hours	1.98×10^6	foot-pounds
horse-power-hours	2.684×10^6	joules
horse-power-hours	641.7	kilogram-calories
horse-power-hours	2.737×10^5	kilogram-meters
horse-power-hours	0.7457	kilowatt-hours
inches	2.540	centimeters
inches	10^3	mils
inches	.03	varas
inches of mercury	0.03342	atmospheres
inches of mercury	1.133	feet of water
inches of mercury	345.3	kgs. per sq. meter
inches of mercury	70.73	pounds per sq. ft.
inches of mercury	0.4912	pounds per sq. in.
inches of water	0.002458	atmospheres
inches of water	0.07355	inches of mercury
inches of water	25.40	kgs. per sq. meter
inches of water	0.5781	ounces per sq. in.
inches of water	5.204	pounds per sq. ft.
inches of water	0.03613	pounds per sq. in.
kilograms	980,665	dynes
kilograms	10^3	grams
kilograms	70.93	poundals
kilograms	2.2046	pounds
kilograms	1.102×10^{-3}	tons (short)
kilogram-calories	3.968	British thermal units
kilogram-calories	3086	foot-pounds
kilogram-calories	1.558×10^{-3}	horse-power-hours
kilogram-calories	4183	joules

Table 3-13 General Conversion Factors (*cont.*)

Multiply	*by*	*to obtain*
kilogram-calories	426.6	kilogram-meters
kilogram-calories	1.162×10^{-3}	kilowatt-hours
kg.-calories per min.	51.43	foot-pounds per sec.
kg.-calories per min.	0.09351	horse-power
kg.-calories per min.	0.06972	kilowatts
kilometers	10^5	centimeters
kilometers	3281	feet
kilometers	10^3	meters
kilometers	0.6214	miles
kilometers	1093.6	yards
kilowatts	56.92	B.t. units per min.
kilowatts	4.425×10^4	foot-pounds per min.
kilowatts	737.6	foot-pounds per sec.
kilowatts	1.341	horse-power
kilowatts	14.34	kg.-calories per min.
kilowatts	10^3	watts
kilowatt-hours	3415	British thermal units
kilowatt-hours	2.655×10^6	foot-pounds
kilowatt-hours	1.341	horse-power-hours
kilowatt-hours	3.6×10^6	joules
kilowatt-hours	860.5	kilogram-calories
kilowatt-hours	3.671×10^5	kilogram-meters
$\log^{10} N$	2.303	$\log_\epsilon N$ or $ln\ N$
$\log^\epsilon N$ or $ln\ N$	0.4343	$\log_{10} N$
meters	100	centimeters
meters	3.2808	feet
meters	39.37	inches
meters	10^{-3}	kilometers
meters	10^3	millimeters
meters	1.0936	yards
miles	1.609×10^5	centimeters
miles	5280	feet
miles	1.6093	kilometers
miles	1760	yards
miles	1900.8	varas
miles per hour	44.70	centimeters per sec.
miles per hour	88	feet per minute
miles per hour	1.467	feet per second

Table 3-13 General Conversion Factors (*cont.*)

Multiply	*by*	*to obtain*
miles per hour	1.6093	kilometers per hour
miles per hour	0.8684	knots per hour
miles per hour	26.82	meters per minute
miles per hour per sec.	44.70	cms. per sec. per sec.
miles per hour per sec.	1.467	ft. per sec. per sec.
miles per hour per sec.	1.6093	kms. per hr. per sec.
miles per hour per sec.	0.4470	M. per sec. per sec.
months	30.42	days
months	730	hours
months	43,800	minutes
months	2.628×10^{6}	seconds
ounces	8	drams
ounces	437.5	grains
ounces	28.35	grams
ounces	.0625	pounds
ounces per square inch	0.0625	pounds per sq. inch
pints (dry)	33.60	cubic inches
pints (liq.)	28.87	cubic inches
pounds	444,823	dynes
pounds	7000	grains
pounds	453.6	grams
pounds	16	ounces
pounds	32.17	poundals
pounds of water	0.01602	cubic feet
pounds of water	27.68	cubic inches
pounds of water	0.1198	gallons
pounds of water per min.	2.669×10^{-4}	cubic feet per sec.
pounds per cubic foot	0.01602	grams per cubic cm.
pounds per cubic foot	16.02	kgs. per cubic meter
pounds per cubic foot	5.787×10^{-4}	pounds per cubic in.
pounds per cubic foot	5.456×10^{-9}	pounds per mil foot
pounds per square foot	0.01602	feet of water
pounds per square foot	4.882	kgs. per sq. meter
pounds per square foot	6.944×10^{-3}	pounds per sq. inch
pounds per square inch	0.06804	atmospheres
pounds per square inch	2.307	feet of water
pounds per square inch	2.036	inches of mercury

Table 3-13 General Conversion Factors (*cont.*)

Multiply	*by*	*to obtain*
pounds per square inch	703.1	kgs. per sq. meter
pounds per square inch	144	pounds per sq. foot
quarts	32	fluid ounces
quarts (dry)	67.20	cubic inches
quarts (liquid)	57.75	cubic inches
rods	16.5	feet
square centimeters	1.973×10^{5}	circular mils
square centimeters	1.076×10^{-3}	square feet
square centimeters	0.1550	square inches
square centimeters	10^{-4}	square meters
square centimeters	100	square millimeters
square feet	2.296×10^{-5}	acres
square feet	929.0	square centimeters
square feet	144	square inches
square feet	0.09290	square meters
square feet	3.587×10^{-8}	square miles
square feet	.1296	square varas
square feet	1/9	square yards
square inches	1.273×10^{6}	circular mils
square inches	6.452	square centimeters
square inches	6.944×10^{-3}	square feet
square inches	10^{6}	square mils
square inches	645.2	square millimeters
square miles	640	acres
square miles	27.88×10^{6}	square feet
square miles	2.590	square kilometers
square miles	3,613,040.45	square varas
square miles	3.098×10^{6}	square yards
square yards	2.066×10^{-4}	acres
square yards	9	square feet
square yards	0.8361	square meters
square yards	3.228×10^{-7}	square miles
square yards	1.1664	square varas
temp. (degs. C.) +17.8	1.8	temp. (degs. Fahr.)
temp. (degs. F.) −32	5/9	temp. (degs. Cent.)
tons (long)	2240	pounds
tons (short)	2000	pounds
yards	.9144	meters

NOTES

Section Four

THE METRIC SYSTEM OF UNITS (SI)

TABLE 4-1 Basic SI Units

Quantity	*Unit*
length	meter (m)
mass	kilogram (kg)
time	second (s)
electric current	ampere (A)
temperature (thermodynamic)	kelvin (K)
amount of substance	mole (mol)
luminous intensity	candela (cd)

TABLE 4-2 Prefixes for SI Units

Multiple and submultiple	*Prefix*	*Symbol*
$1{,}000{,}000{,}000{,}000 = 10^{12}$	tera	T
$1{,}000{,}000{,}000 = 10^{9}$	giga	G
$1{,}000{,}000 = 10^{6}$	mega	M
$1{,}000 = 10^{3}$	kilo	k
$100 = 10^{2}$	hecto	h
$10 = 10$	deka	da
$0.1 = 10^{-1}$	deci	d
$0.01 = 10^{-2}$	centi	c
$0.001 = 10^{-3}$	milli	m
$0.000\ 001 = 10^{-6}$	micro	μ
$0.000\ 000\ 001 = 10^{-9}$	nano	n
$0.000\ 000\ 000\ 001 = 10^{-12}$	pico	p
$0.000\ 000\ 000\ 000\ 001 = 10^{-15}$	femto	f
$.000\ 000\ 000\ 000\ 000\ 001 = 10^{-18}$	atto	a

TABLE 4-3 Derived Units of the International System

Quantity	*Name of unit*	*Unit symbol or abbreviation, where differing from basic form*	*Unit expressed in terms of basic or supplementary units**
area	square meter		m^2
volume	cubic meter		m^3
frequency	hertz, cycle per second†	Hz	s^{-1}
density	kilogram per cubic meter		kg/m^3
velocity	meter per second		m/s
angular velocity	radian per second		rad/s
acceleration	meter per second squared		m/s^2
angular acceleration	radian per second squared		rad/s^2
volumetric flow rate	cubic meter per second		m^3/s
force	newton	N	$kg \cdot m/s^2$
surface tension	newton per meter, joule per square meter	N/m, J/m^2	kg/s^2
pressure	newton per square meter, pascal†	N/m^2, Pa†	$kg/m \cdot s^2$
viscosity, dynamic	newton-second per square meter, poiseuille†	N s/m^2, Pl†	$kg/m \cdot s$
viscosity, kinematic	meter squared per second		m^2/s
work, torque, energy, [illegible]	joule, newton meter, watt-second	J, N·m, W·s	$kg \cdot m^2/s^2$

	watt, joule per second	W, J/s	kg·m²/s³
heat flux density	watt per square meter	W/m²	kg/s³
volumetric heat release rate	watt per cubic meter	W/m³	kg/m·s³
heat transfer coefficient	watt per square meter degree	W/m²·deg	kg/s³·deg
heat capacity (specific)	joule per kilogram degree	J/kg·deg	m²/s²·deg
capacity rate	watt per degree	W/deg	kg·m²/s³·deg
thermal conductivity	watt per meter degree	W/m·deg, $\frac{Jm}{s \cdot m^2 \cdot deg}$	kg·m/s³·deg
quantity of electricity	coulomb	C	A·s
electromotive force	volt	V, W/A	kg·m²/A·s³
electric field strength	volt per meter		V/m
electric resistance	ohm	Ω, V/A	kg·m²/A²·s³
electric conductivity	ampere per volt meter	A/V·m	A²s³/kg·m³
electric capacitance	farad	F, A·s/V	A³s¹/kg·m²
magnetic flux	weber	Wb, V·s	kg·m²/A·s²
inductance	henry	H, V·s/A	kg·m²/A²s²
magnetic permeability	henry per meter	H/m	kg·m/A²s²
magnetic flux density	tesla, weber per square meter	T, Wb/m²	kg/A·s²
magnetic field strength	ampere per meter		A/m
magnetomotive force	ampere		A
luminous flux	lumen	lm	cd sr
luminance	candela per square meter		cd/m²
illumination	lux, lumen per square meter	lx, lm/m²	cd·sr/m²

*Supplementary units are plane angle, radian (rad); solid angle, steradian (sr).
†Not used in all countries.

SOURCE: "McGraw-Hill Metrication Manual." Copyright © 1971 by McGraw-Hill, Inc., New York. (Used with permission of McGraw-Hill Book Co.)

TABLE 4-4 SI Derived Units Used in Civil and Mechanical Engineering

Quantity	*Usual units*	*Symbol*
water usage	liter per person day	l/person·d
runoff	cubic meter per square kilometer	m^3/km^2
precipitation	millimeter per hour	mm/h
river flow	cubic meter per second	m^3/s
discharge	cubic meter per day	m^3/d
hydraulic load per unit area, e.g., filtration rates	cubic meter per square meter day	$m^3/m^2 \cdot d$
concentration	gram per cubic meter	g/m^3
	milligram per liter	mg/l
BOD produced	kilogram per day	kg/d
BOD loading	kilogram per cubic meter day	$kg/m^3 \cdot d$
hydraulic load per unit volume, e.g., biological filters	cubic meter per cubic meter day	$m^3/m^3 \cdot d$
sludge treatment	square meter per person	m^2/person
flow in pipes, channels, etc., peak demands; surface water runoff	cubic meters per second	m^3/s
flow in small pipes; demands of sanitary fittings; pumping rates	liters per second	l/s
water demand; water supply	cubic meters per day	m^3/d
force	kilonewtons	kN
moment of force	newton meter	Nm
	kilonewton meter	kNm

SOURCE: "McGraw-Hill Metrication Manual." Copyright © 1971 by Mc-Graw-Hill, Inc., New York. (Used with permission of McGraw-Hill Book Co.)

The following tables express the definitions of various units of measure as exact numerical multiples of coherent SI units, and provide multiplying factors for converting numbers and miscellaneous units to corresponding new numbers and SI units.

The first two digits of each numerical entry represent a power of 10. An asterisk follows each number which expresses an exact definition. For example, the entry "–02 2.54*" expresses the fact that 1 inch = 2.54×10^{-2} meter, exactly, by definition. Most of the definitions are extracted from National Bureau of Standards documents. Numbers not followed by an asterisk are only approximate representations of definitions or are the results of physical measurements. The conversion factors are listed alphabetically and by physical quantity.

The Listing by Physical Quantity includes only relationships which are frequently encountered and deliberately omits the great multiplicity of combinations of units which are used for more specialized purposes. Conversion factors for combinations of units are easily generated from numbers given in the Alphabetical Listing by the technique of direct substitution or by other well-known rules for manipulating units. These rules are adequately discussed in many science and engineering textbooks and are not repeated here.

TABLE 4-5 Conversion Factors as Exact Numerical Multiples of SI Units

ALPHABETICAL LISTING

To convert from	*to*	*multiply by*
abampere	ampere	+01 1.00*
abcoulomb	coulomb	+01 1.00*
abfarad	farad	+09 1.00*
abhenry	henry	–09 1.00*
abmho	mho	+09 1.00*
abohm	ohm	–09 1.00*
abvolt	volt	–08 1.00*
acre	meter^2	+03 4.046 856 422 4*

TABLE 4-5 Conversion Factors as Exact Numerical Multiples of SI Units *(cont.)*

To convert from	*to*	*multiply by*
ampere (international of 1948)	ampere	−01 9.998 35
angstrom	meter	−10 1.00*
are	meter²	+02 1.00*
astronomical unit	meter	+11 1.495 978 9
atmosphere	newton/meter²	+05 1.013 25*
bar	newton/meter²	+05 1.00*
barn	meter²	−28 1.00*
barrel (petroleum, 42 gallons)	meter³	−01 1.589 873
barye	newton/meter²	−01 1.00*
British thermal unit (ISO/TC 12)	joule	+03 1.055 06
British thermal unit (International Steam Table)	joule	+03 1.055 04
British thermal unit (mean)	joule	+03 1.055 87
British thermal unit (thermochemical)	joule	+03 1.054 350 264 488
British thermal unit (39° F)	joule	+03 1.059 67
British thermal unit (60° F)	joule	+03 1.054 68
bushel (U.S.)	meter³	−02 3.523 907 016 688*
cable	meter	+02 2.194 56*
caliber	meter	−04 2.54*
calorie (International Steam Table)	joule	+00 4.1868
calorie (mean)	joule	+00 4.190 02
calorie (thermochemical)	joule	+00 4.184*
calorie (15° C)	joule	+00 4.185 80
calorie (20° C)	joule	+00 4.181 90
calorie (kilogram, International Steam Table)	joule	+03 4.1868
calorie (kilogram, mean)	joule	+03 4.190 02
calorie (kilogram, thermochemical)	joule	+03 4.184*
carat (metric)	kilogram	−04 2.00*
Celsius (temperature)	kelvin	$t_K = t_C + 273.15$
centimeter of mercury (0° C)	newton/meter²	+03 1.333 22
centimeter of water (4° C)	newton/meter²	+01 9.806 38
chain (engineer or ramden)	meter	+01 3.048*
chain (surveyor or gunter)	meter	+01 2.011 68*
circular mil	meter²	−10 5.067 074 8
cord	meter³	+00 3.624 556 3
coulomb (international of 1948)	coulomb	−01 9.998 35
cubit	meter	−01 4.572*
cup	meter³	−04 2.365 882 365*
curie	disintegration/second	+10 3.70*
day (mean solar)	second (mean solar)	+04 8.64*
day (sidereal)	second (mean solar)	+04 8.616 409 0
degree (angle)	radian	−02 1.745 329 251 994 3
denier (international)	kilogram/meter	−07 1.00*
dram (avoirdupois)	kilogram	−03 1.771 845 195 312 5*
dram (troy or apothecary)	kilogram	−03 3.887 934 6*
dram (U.S. fluid)	meter³	−06 3.696 691 195 312 5*
dyne	newton	−05 1.00*
electron volt	joule	−19 1.602 10
erg	joule	−07 1.00*
Fahrenheit (temperature)	kelvin	$t_K = (5/9)\ (t_F + 459.67)$
Fahrenheit (temperature)	Celsius	$t_C = (5/9)\ (t_F - 32)$
farad (international of 1948)	farad	−01 9.995 05
faraday (based on carbon 12)	coulomb	+04 9.648 70

TABLE 4-5 Conversion Factors as Exact Numerical Multiples of SI Units *(cont.)*

To convert from	*to*	*multiply by*
faraday (chemical)	coulomb	+04 9.649 57
faraday (physical)	coulomb	+04 9.652 19
fathom	meter	+00 1.828 8*
fermi (femtometer)	meter	−15 1.00*
fluid ounce (U.S.)	meter³	−05 2.957 352 956 25*
foot	meter	−01 3.048*
foot (U.S. survey)	meter	+00 1200/3937*
foot (U.S. survey)	meter	−01 3.048 006 096
foot of water (39.2° F)	newton/meter²	+03 2.988 98
foot-candle	lumen/meter²	+01 1.076 391 0
foot-lambert	candela/meter²	+00 3.426 259
furlong	meter	+02 2.011 68*
gal (galileo)	meter/second²	−02 1.00*
gallon (U.K. liquid)	meter³	−03 4.546 087
gallon (U.S. dry)	meter³	−03 4.404 883 770 86*
gallon (U.S. liquid)	meter³	−03 3.785 411 784*
gamma	tesla	−09 1.00*
gauss	tesla	−04 1.00*
gilbert	ampere turn	−01 7.957 747 2
gill (U.K.)	meter³	−04 1.420 652
gill (U.S.)	meter³	−04 1.182 941 2
grad	degree (angular)	−01 9.00*
grad	radian	−02 1.570 796 3
grain	kilogram	−05 6.479 891*
gram	kilogram	−03 1.00*
hand	meter	−01 1.016*
hectare	meter²	+04 1.00*
henry (international of 1948)	henry	+00 1.000 495
hogshead (U.S.)	meter³	−01 2.384 809 423 92*
horsepower (550 foot lbf/second)	watt	+02 7.456 998 7
horsepower (boiler)	watt	+03 9.809 50
horsepower (electric)	watt	+02 7.46*
horsepower (metric)	watt	+02 7.354 99
horsepower (U.K.)	watt	+02 7.457
horsepower (water)	watt	+02 7.460 43
hour (mean solar)	second (mean solar)	+03 3.60*
hour (sidereal)	second (mean solar)	+03 3.590 170 4
hundredweight (long)	kilogram	+01 5.080 234 544*
hundredweight (short)	kilogram	+01 4.535 923 7*
inch	meter	−02 2.54*
inch of mercury (32° F)	newton/meter²	+03 3.386 389
inch of mercury (60° F)	newton/meter²	+03 3.376 85
inch of water (39.2° F)	newton/meter²	+02 2.490 82
inch of water (60° F)	newton/meter²	+02 2.4884
joule (international of 1948)	joule	+00 1.000 165
kayser	1/meter	+02 1.00*
kilocalorie (International Steam Table)	joule	+03 4.186 74
kilocalorie (mean)	joule	+03 4.190 02
kilocalorie (thermochemical)	joule	+03 4.184*
kilogram mass	kilogram	+00 1.00*
kilogram force (kgf)	newton	+00 9.806 65*
kilopond force	newton	+00 9.806 65*
kip	newton	+03 4.448 221 615 260 5*
knot (international)	meter/second	−01 5.144 444 444

TABLE 4-5 Conversion Factors as Exact Numerical Multiples of SI Units *(cont.)*

To convert from	*to*	*multiply by*
lambert	candela/meter²	+04 1/π*
lambert	candela/meter²	+03 3.183 098 8
langley	joule/meter²	+04 4.184*
lbf (pound force, avoirdupois)	newton	+00 4.448 221 615 260 5*
lbm (pound mass, avoirdupois)	kilogram	−01 4.535 923 7*
league (British nautical)	meter	+03 5.559 552*
league (international nautical)	meter	+03 5.556*
league (statute)	meter	+03 4.828 032*
light year	meter	+15 9.460 55
link (engineer or ramden)	meter	−01 3.048*
link (surveyor or gunter)	meter	−01 2.011 68*
liter	meter³	−03 1.00*
lux	lumen/meter²	+00 1.00*
maxwell	weber	−08 1.00*
meter	wavelengths Kr 86	+06 1.650 763 73*
micron	meter	−06 1.00*
mil	meter	−05 2.54*
mile (U.S. statute)	meter	+03 1.609 344*
mile (U.K. nautical)	meter	+03 1.853 184*
mile (international nautical)	meter	+03 1.852*
mile (U.S. nautical)	meter	+03 1.852*
millibar	newton/meter²	+02 1.00*
millimeter of mercury (0° C)	newton/meter²	+02 1.333 224
minute (angle)	radian	−04 2.908 882 086 66
minute (mean solar)	second (mean solar)	+01 6.00*
minute (sidereal)	second (mean solar)	+01 5.983 617 4
month (mean calendar)	second (mean solar)	+06 2.628*
nautical mile (international)	meter	+03 1.852*
nautical mile (U.S.)	meter	+03 1.852*
nautical mile (U.K.)	meter	+03 1.853 184*
oersted	ampere/meter	+01 7.957 747 2
ohm (international of 1948)	ohm	+00 1.000 495
ounce force (avoirdupois)	newton	−01 2.780 138 5
ounce mass (avoirdupois)	kilogram	−02 2.834 952 312 5*
ounce mass (troy or apothecary)	kilogram	−02 3.110 347 68*
ounce (U.S. fluid)	meter³	−05 2.957 352 956 25*
pace	meter	−01 7.62*
parsec	meter	+16 3.083 74
pascal	newton/meter²	+00 1.00*
peck (U.S.)	meter³	−03 8.809 767 541 72*
pennyweight	kilogram	−03 1.555 173 84*
perch	meter	+00 5.0292*
phot	lumen/meter²	+04 1.00
pica (printers)	meter	−03 4.217 517 6*
pint (U.S. dry)	meter³	−04 5.506 104 713 575*
pint (U.S. liquid)	meter³	−04 4.731 764 73*
point (printers)	meter	−04 3.514 598*
poise	newton second/meter²	−01 1.00*
pole	meter	+00 5.0292*
pound force (lbf avoirdupois)	newton	+00 4.448 221 615 260 5*
pound mass (lbm avoirdupois)	kilogram	−01 4.535 923 7*
pound mass (troy or apothecary)	kilogram	−01 3.732 417 216*
poundal	newton	−01 1.382 549 543 76*
quart (U.S. dry)	meter³	−03 1.101 220 942 715*
quart (U.S. liquid)	meter³	−04 9.463 529 5

TABLE 4-5 Conversion Factors as Exact Numerical Multiples of SI Units *(cont.)*

To convert from	*to*	*multiply by*
rad (radiation dose absorbed)	joule/kilogram	−02 1.00*
Rankine (temperature)	kelvin	$t_K = (5/9)t_R$
rayleigh (rate of photon emission)	1/second meter²	+10 1.00*
rhe	meter²/newton second	+01 1.00*
rod	meter	+00 5.0292*
roentgen	coulomb/kilogram	−04 2.579 76*
rutherford	disintegration/second	+06 1.00*
second (angle)	radian	−06 4.848 136 811
second (ephemeris)	second	+00 1.000 000 000
second (mean solar)	second (ephemeris)	Consult American Ephemeris and Nautical Almanac
second (sidereal)	second (mean solar)	−01 9.972 695 7
section	meter²	+06 2.589 988 110 336*
scruple (apothecary)	kilogram	−03 1.295 978 2*
shake	second	−08 1.00
skein	meter	+02 1.097 28*
slug	kilogram	+01 1.459 390 29
span	meter	−01 2.286*
statampere	ampere	−10 3.335 640
statcoulomb	coulomb	−10 3.335 640
statfarad	farad	−12 1.112 650
stathenry	henry	+11 8.987 554
statmho	mho	−12 1.112 650
statohm	ohm	+11 8.987 554
statute mile (U.S.)	meter	+03 1.609 344*
statvolt	volt	+02 2.997 925
stere	meter³	+00 1.00*
stilb	candela/meter²	+04 1.00
stoke	meter²/second	−04 1.00*
tablespoon	meter³	−05 1.478 676 478 125*
teaspoon	meter³	−06 4.928 921 593 75*
ton (assay)	kilogram	−02 2.916 666 6
ton (long)	kilogram	+03 1.016 046 908 8*
ton (metric)	kilogram	+03 1.00*
ton (nuclear equivalent of TNT)	joule	+09 4.20
ton (register)	meter³	+00 2.831 684 659 2*
ton (short, 2000 pound)	kilogram	+02 9.071 847 4*
tonne	kilogram	+03 1.00*
torr (0° C)	newton/meter²	+02 1.333 22
township	meter²	+07 9.323 957 2
unit pole	weber	−07 1.256 637
volt (international of 1948)	volt	+00 1.000 330
watt (international of 1948)	watt	+00 1.000 165
yard	meter	−01 9.144*
year (calendar)	second (mean solar)	+07 3.1536*
year (sidereal)	second (mean solar)	+07 3.155 815 0
year (tropical)	second (mean solar)	+07 3.155 692 6
year 1900, tropical, Jan., day 0, hour 12	second (ephemeris)	+07 3.155 692 597 47*
year 1900, tropical, Jan., day 0, hour 12	second	+07 3.155 692 597 47

SOURCE: "McGraw-Hill Metrication Manual." Copyright © 1971 by McGraw-Hill, Inc., New York. (Used with permission of McGraw-Hill Book Co.)

TABLE 4-6 Listing of Conversion Factors by Physical Quantity

ACCELERATION

To convert from	to	Multiply by
foot/second^2	meter/second^2	−01 3.048*
free fall, standard	meter/second^2	+00 9.806 65*
gal (galileo)	meter/second^2	−02 1.00*
inch/second^2	meter/second^2	−02 2.54*

AREA

To convert from	to	Multiply by
acre	meter^2	+03 4.046 856 422 4*
are	meter^2	+02 1.00*
barn	meter^2	−28 1.00*
circular mil	meter^2	−10 5.067 074 8
foot^2	meter^2	−02 9.290 304*
hectare	meter^2	+04 1.00*
inch^2	meter^2	−04 6.4516*
mile^2 (U.S. statute)	meter^2	+06 2.589 988 110 336*
section	meter^2	+06 2.589 988 110 336*
township	meter^2	+07 9.323 957 2
yard^2	meter^2	−01 8.361 273 6*

DENSITY

To convert from	to	Multiply by
gram/centimeter^3	kilogram/meter^3	+03 1.00*
lbm/inch^3	kilogram/meter^3	+04 2.767 990 5
lbm/foot^3	kilogram/meter^3	+01 1.601 846 3
slug/foot^3	kilogram/meter^3	+02 5.153 79

ENERGY

To convert from	to	Multiply by
British thermal unit (ISO/TC 12)	joule	+03 1.055 06
British thermal unit (International Steam Table)	joule	+03 1.055 04
British thermal unit (mean)	joule	+03 1.055 87
British thermal unit (thermochemical)	joule	+03 1.054 350 264 488
British thermal unit (39° F)	joule	+03 1.059 67
British thermal unit (60° F)	joule	+03 1.054 68
calorie (International Steam Table)	joule	+00 4.1868
calorie (mean)	joule	+00 4.190 02
calorie (thermochemical)	joule	+00 4.184*
calorie (15° C)	joule	+00 4.185 80
calorie (20° C)	joule	+00 4.181 90
calorie (kilogram, International Steam Table)	joule	+03 4.1868
calorie (kilogram, mean)	joule	+03 4.190 02
calorie (kilogram, thermochemical)	joule	+03 4.184*
electron volt	joule	−19 1.602 10
erg	joule	−07 1.00*
foot lbf	joule	+00 1.355 817 9
foot poundal	joule	−02 4.214 011 0
joule (international of 1948)	joule	+00 1.000 165
kilocalorie (International Steam Table)	joule	+03 4.1868
kilocalorie (mean)	joule	+03 4.190 02
kilocalorie (thermochemical)	joule	+03 4.184*

TABLE 4-6 Listing of Conversion Factors by Physical Quantity *(cont.)*

To convert from	*to*	*multiply by*
kilowatt hour	joule	+06 3.60*
kilowatt hour (international of 1948)	joule	+06 3.600 59
ton (nuclear equivalent of TNT)	joule	+09 4.20
watt hour	joule	+03 3.60*
	ENERGY/AREA TIME	
Btu (thermochemical)/foot² second	watt/meter²	+04 1.134 893 1
Btu (thermochemical)/foot² minute	watt/meter²	+02 1.891 488 5
Btu (thermochemical)/foot² hour	watt/meter²	+00 3.152 480 8
Btu (thermochemical)/inch² second	watt/meter²	+06 1.634 246 2
calorie (thermochemical)/cm² minute	watt/meter²	+02 6.973 333 3
erg/centimeter² second	watt/meter²	−03 1.00*
watt/centimeter²	watt/meter²	+04 1.00*
	FORCE	
dyne	newton	−05 1.00*
kilogram force (kgf)	newton	+00 9.806 65*
kilopond force	newton	+00 9.806 65*
kip	newton	+03 4.448 221 615 260 5*
lbf (pound force, avoirdupois)	newton	+00 4.448 221 615 260 5*
ounce force (avoirdupois)	newton	−01 2.780 138 5
pound force, lbf (avoirdupois)	newton	+00 4.448 221 615 260 5*
poundal	newton	−01 1.382 549 543 76*
	LENGTH	
angstrom	meter	−10 1.00*
astronomical unit	meter	+11 1.495 978 9
cable	meter	+02 2.194 56*
caliber	meter	−04 2.54*
chain (surveyor or gunter)	meter	+01 2.011 68*
chain (engineer or ramden)	meter	+01 3.048*
cubit	meter	−01 4.572*
fathom	meter	+00 1.8288*
fermi (femtometer)	meter	−15 1.00*
foot	meter	−01 3.048*
foot (U.S. survey)	meter	+00 1200/3937*
foot (U.S. survey)	meter	−01 3.048 006 096
furlong	meter	+02 2.011 68*
hand	meter	−01 1.016*
inch	meter	−02 2.54*
league (U.K. nautical)	meter	+03 5.559 552*
league (international nautical)	meter	+03 5.556*
league (statute)	meter	+03 4.828 032*
light year	meter	+15 9.460 55
link (engineer or ramden)	meter	−01 3.048*
link (surveyor or gunter)	meter	−01 2.011 68*
meter	wavelengths Kr 86	+06 1.650 763 73*
micron	meter	−06 1.00*
mil	meter	−05 2.54*
mile (U.S. statute)	meter	+03 1.609 344*
mile (U.K. nautical)	meter	+03 1.853 184*
mile (international nautical)	meter	+03 1.852*
mile (U.S. nautical)	meter	+03 1.852*
nautical mile (U.K.)	meter	+03 1.853 184*
nautical mile (international)	meter	+03 1.852*
nautical mile (U.S.)	meter	+03 1.852*

TABLE 4-6 Listing of Conversion Factors by Physical Quantity *(cont.)*

To convert from	*to*	*multiply by*
pace	meter	−01 7.62*
parsec	meter	+16 3.083 74
perch	meter	+00 5.0292*
pica (printers)	meter	−03 4.217 517 6*
point (printers)	meter	−04 3.514 598*
pole	meter	+00 5.0292*
rod	meter	+00 5.0292*
skein	meter	+02 1.097 28*
span	meter	−01 2.286*
statute mile (U.S.)	meter	+03 1.609 344*
yard	meter	−01 9.144*
	MASS	
carat (metric)	kilogram	−04 2.00*
dram (avoirdupois)	kilogram	−03 1.771 845 195 312 5*
dram (troy or apothecary)	kilogram	−03 3.887 934 6*
grain	kilogram	−05 6.479 891*
gram	kilogram	−03 1.00*
hundredweight (long)	kilogram	+01 5.080 234 544*
hundredweight (short)	kilogram	+01 4.535 923 7*
kgf second² meter (mass)	kilogram	+00 9.806 65*
kilogram mass	kilogram	+00 1.00*
lbm (pound mass, avoirdupois)	kilogram	−01 4.535 923 7*
ounce mass (avoirdupois)	kilogram	−02 2.834 952 312 5*
ounce mass (troy or apothecary)	kilogram	−02 3.110 347 68*
pennyweight	kilogram	−03 1.555 173 84*
pound mass, lbm (avoirdupois)	kilogram	−01 4.535 923 7*
pound mass (troy or apothecary)	kilogram	−01 3.732 417 216*
scruple (apothecary)	kilogram	−03 1.295 978 2*
slug	kilogram	+01 1.459 390 29
ton (assay)	kilogram	−02 2.916 666 6
ton (long)	kilogram	+03 1.016 046 908 8*
ton (metric)	kilogram	+03 1.00*
ton (short, 2000 pound)	kilogram	+02 9.071 847 4*
tonne	kilogram	+03 1.00*
	POWER	
Btu (thermochemical)/second	watt	+03 1.054 350 264 488
Btu (thermochemical)/minute	watt	+01 1.757 250 4
calorie (thermochemical)/second	watt	+00 4.184*
calorie (thermochemical)/minute	watt	−02 6.973 333 3
foot lbf/hour	watt	−04 3.766 161 0
foot lbf/minute	watt	−02 2.259 696 6
foot lbf/second	watt	+00 1.355 817 9
horsepower (550 foot lbf/second)	watt	+02 7.456 998 7
horsepower (boiler)	watt	+03 9.809 50
horsepower (electric)	watt	+02 7.46*
horsepower (metric)	watt	+02 7.354 99
horsepower (U.K.)	watt	+02 7.457
horsepower (water)	watt	+02 7.460 43
kilocalorie (thermochemical)/minute	watt	+01 6.973 333 3
kilocalorie (thermochemical)/second	watt	+03 4.184*
watt (international of 1948)	watt	+00 1.000 165
	PRESSURE	
atmosphere	newton/meter²	+05 1.013 25*
bar	newton/meter²	+05 1.00*

TABLE 4-6 Listing of Conversion Factors by Physical Quantity *(cont.)*

To convert from	*to*	*multiply by*
barye	newton/meter²	−01 1.00*
centimeter of mercury (0° C)	newton/meter²	+03 1.333 22
centimeter of water (4° C)	newton/meter²	+01 9.806 38
dyne/centimeter²	newton/meter²	−01 1.00*
foot of water (39.2° F)	newton/meter²	+03 2.988 98
inch of mercury (32° F)	newton/meter²	+03 3.386 389
inch of mercury (60° F)	newton/meter²	+03 3.376 85
inch of water (39.2° F)	newton/meter²	+02 2.490 82
inch of water (60° F)	newton/meter²	+02 2.4884
kgf/centimeter²	newton/meter²	+04 9.806 65*
kgf/meter²	newton/meter²	+00 9.806 65*
lbf/foot²	newton/meter²	+01 4.788 025 8
lbf/inch² (psi)	newton/meter²	+03 6.894 757 2
millibar	newton/meter²	+02 1.00*
millimeter of mercury (0° C)	newton/meter²	+02 1.333 224
pascal	newton/meter²	+00 1.00*
psi (lbf/inch²)	newton/meter²	+03 6.894 757 2
torr (0° C)	newton/meter²	+02 1.333 22

SPEED

foot/hour	meter/second	−05 8.466 666 6
foot/minute	meter/second	−03 5.08*
foot/second	meter/second	−01 3.048*
inch/second	meter/second	−02 2.54*
kilometer/hour	meter/second	−01 2.777 777 8
knot (international)	meter/second	−01 5.144 444 444
mile/hour (U.S. statute)	meter/second	−01 4.4704*
mile/minute (U.S. statute)	meter/second	+01 2.682 24*
mile/second (U.S. statute)	meter/second	+03 1.609 344*

TEMPERATURE

Celsius	kelvin	$t_K = t_C + 273.15$
Fahrenheit	kelvin	$t_K = (5/9)(t_F + 459.67)$
Fahrenheit	Celsius	$t_C = (5/9)(t_F - 32)$
Rankine	kelvin	$t_K = (5/9)t_R$

TIME

day (mean solar)	second (mean solar)	+04 8.64*
day (sidereal)	second (mean solar)	+04 8.616 409 0
hour (mean solar)	second (mean solar)	+03 3.60*
hour (sidereal)	second (mean solar)	+03 3.590 170 4
minute (mean solar)	second (mean solar)	+01 6.00*
minute (sidereal)	second (mean solar)	+01 5.983 617 4
month (mean calendar)	second (mean solar)	+06 2.628*
second (ephemeris)	second	+00 1.000 000 000
second (mean solar)	second (ephemeris)	Consult American Ephemeris and Nautical Almanac
second (sidereal)	second (mean solar)	−01 9.972 695 7
year (calendar)	second (mean solar)	+07 3.1536*
year (sidereal)	second (mean solar)	+07 3.155 815 0
year (tropical)	second (mean solar)	+07 3.155 692 6
year 1900, tropical, Jan., day 0, hour 12	second (ephemeris)	+07 3.155 692 597 47*
year 1900, tropical, Jan., day 0, hour 12	second	+07 3.155 692 597 47

TABLE 4-6 Listing of Conversion Factors by Physical Quantity *(cont.)*

To convert from	*to*	*multiply by*
	VISCOSITY	
centistoke	meter^2/second	−06 1.00*
stoke	meter^2/second	−04 1.00*
foot^2/second	meter^2/second	−02 9.290 304*
centipoise	newton second/meter^2	−03 1.00*
lbm/foot second	newton second/meter^2	+00 1.488 163 9
lbf second/foot^2	newton second/meter^2	+01 4.788 025 8
poise	newton second/meter^2	−01 1.00*
poundal second/foot^2	newton second/meter^2	+00 1.488 163 9
slug/foot second	newton second/meter^2	+01 4.788 025 8
rhe	meter^2/newton second	+01 1.00*
	VOLUME	
acre foot	meter^3	+03 1.233 481 9
barrel (petroleum, 42 gallons)	meter^3	−01 1.589 873
board foot	meter^3	−03 2.359 737 216*
bushel (U.S.)	meter^3	−02 3.523 907 016 688*
cord	meter^3	+00 3.624 556 3
cup	meter^3	−04 2.365 882 365*
dram (U.S. fluid)	meter^3	−06 3.696 691 195 312 5*
fluid ounce (U.S.)	meter^3	−05 2.957 352 956 25*
foot^3	meter^3	−02 2.831 684 659 2*
gallon (U.K. liquid)	meter^3	−03 4.546 087
gallon (U.S. dry)	meter^3	−03 4.404 883 770 86*
gallon (U.S. liquid)	meter^3	−03 3.785 411 784*
gill (U.K.)	meter^3	−04 1.420 652
gill (U.S.)	meter^3	−04 1.182 941 2
hogshead (U.S.)	meter^3	−01 2.384 809 423 92*
inch^3	meter^3	−05 1.638 706 4*
liter	meter^3	−03 1.00*
ounce (U.S. fluid)	meter^3	−05 2.957 352 956 25*
peck (U.S.)	meter^3	−03 8.809 767 541 72*
pint (U.S. dry)	meter^3	−04 5.506 104 713 575*
pint (U.S. liquid)	meter^3	−04 4.731 764 73*
quart (U.S. dry)	meter^3	−03 1.101 220 942 715*
quart (U.S. liquid)	meter^3	−04 9.463 529 5
stere	meter^3	+00 1.00*
tablespoon	meter^3	−05 1.478 676 478 125*
teaspoon	meter^3	−06 4.928 921 593 75*
ton (register)	meter^3	+00 2.831 684 659 2*
yard^3	meter^3	−01 7.645 548 579 84*

SOURCE: "McGraw-Hill Metrication Manual." Copyright © 1971 by McGraw-Hill, Inc., New York. (Used with permission of McGraw-Hill Book Co.)

NOTES

NOTES

Section Five

■

CONSTRUCTION: MATERIALS, DATA, AND SAFETY

TABLE 5-1 Weights and Specific Gravities

Substance	Weight Lb. per Cu. Ft.	Specific Gravity
ASHLAR MASONRY		
Granite, syenite, gneiss	165	2.3-3.0
Limestone, marble	160	2.3-2.8
Sandstone, bluestone	140	2.1-2.4
MORTAR RUBBLE MASONRY		
Granite, syenite, gneiss	155	2.2-2.8
Limestone, marble	150	2.2-2.6
Sandstone, bluestone	130	2.0-2.2
DRY RUBBLE MASONRY		
Granite, syenite, gneiss	130	1.9-2.3
Limestone, marble	125	1.9-2.1
Sandstone, bluestone	110	1.8-1.9
BRICK MASONRY		
Pressed brick	140	2.2-2.3
Common brick	120	1.8-2.0
Soft brick	100	1.5-1.7
CONCRETE MASONRY		
Cement, stone, sand	144	2.2-2.4
Cement, slag, etc.	130	1.9-2.3
Cement, cinder, etc.	100	1.5-1.7
VARIOUS BUILDING MATERIALS		
Ashes, cinders	40-45	....
Cement, portland, loose	90	....
Cement, portland, set	183	2.7-3.2
Lime, gypsum, loose	53-64	....
Mortar, set	103	1.4-1.9
Slags, bank slag	67-72	....
Slags, bank screenings	98-117	....
Slags, machine slag	96	....
Slags, slag sand	49-55	....
EARTH, ETC., EXCAVATED		
Clay, dry	63	....
Clay, damp, plastic	110	....
Clay and gravel, dry	100	....
Earth, dry, loose	76	....
Earth, dry, packed	95	....
Earth, moist, loose	78	....
Earth, moist, packed	96	....
Earth, mud, flowing	108	....
Earth, mud, packed	115	....
Riprap, limestone	80-85	....
Riprap, sandstone	90	....
Riprap, shale	105	....
Sand, gravel, dry, loose	90-105	....
Sand, gravel, dry, packed	100-120	....
Sand, gravel, wet	118-120	....
EXCAVATIONS IN WATER		
Sand or gravel	60	....
Sand or gravel and clay	65	....
Clay	80	....
River mud	90	....
Soil	70	....
Stone riprap	65	....

Substance	Weight Lb. per Cu. Ft.	Specific Gravity
MINERALS		
Asbestos	153	2.1-2.8
Barytes	281	4.50
Basalt	184	2.7-3.2
Bauxite	159	2.55
Borax	109	1.7-1.8
Chalk	137	1.8-2.6
Clay, marl	137	1.8-2.6
Dolomite	181	2.9
Feldspar, orthoclase	159	2.5-2.6
Gneiss, serpentine	159	2.4-2.7
Granite, syenite	175	2.5-3.1
Greenstone, trap	187	2.8-3.2
Gypsum, alabaster	159	2.3-2.8
Hornblende	187	3.0
Limestone, marble	165	2.5-2.8
Magnesite	187	3.0
Phosphate rock, apatite	200	3.2
Porphyry	172	2.6-2.9
Pumice, natural	40	0.37-0.90
Quartz, flint	165	2.5-2.8
Sandstone, bluestone	147	2.2-2.5
Shale, slate	175	2.7-2.9
Soapstone, talc	169	2.6-2.8
STONE, QUARRIED, PILED		
Basalt, granite, gneiss	96	....
Limestone, marble, quartz	95	....
Sandstone	82	....
Shale	92	....
Greenstone, hornblende	107	....
BITUMINOUS SUBSTANCES		
Asphaltum	81	1.1-1.5
Coal, anthracite	97	1.4-1.7
Coal, bituminous	84	1.2-1.5
Coal, lignite	78	1.1-1.4
Coal, peat, turf, dry	47	0.65-0.85
Coal, charcoal, pine	23	0.28-0.44
Coal, charcoal, oak	33	0.47-0.57
Coal, coke	75	1.0-1.4
Graphite	131	1.9-2.3
Paraffine	56	0.87-0.91
Petroleum	54	0.87
Petroleum, refined	50	0.79-0.82
Petroleum, benzine	46	0.73-0.75
Petroleum, gasoline	42	0.66-0.69
Pitch	69	1.07-1.15
Tar, bituminous	75	1.20
COAL AND COKE, PILED		
Coal, anthracite	47-58	....
Coal, bituminous, lignite	40-54	....
Coal, peat, turf	20-26	....
Coal, charcoal	10-14	....
Coal, coke	23-32	....

The specific gravities of solids and liquids refer to water at 4°C., those of gases to air at 0°C. and 760 mm. pressure. The weights per cubic foot are derived from average specific gravities, except where stated that weights are for bulk, heaped or loose material, etc.

SOURCE: *Manual of Steel Construction,* 7th ed., American Institute of Steel Construction, Chicago, Ill., 1973.

TABLE 5-1 Weights and Specific Gravities *(cont.)*

Substance	Weight Lb. per Cu. Ft.	Specific Gravity
METALS, ALLOYS, ORES		
Aluminum, cast, hammered	165	2.55-2.75
Brass, cast, rolled	534	8.4-8.7
Bronze, 7.9 to 14% Sn	509	7.4-8.9
Bronze, aluminum	481	7.7
Copper, cast, rolled	556	8.8-9.0
Copper ore, pyrites	262	4.1-4.3
Gold, cast, hammered	1205	19.25-19.3
Iron, cast, pig	450	7.2
Iron, wrought	485	7.6-7.9
Iron, spiegel-eisen	468	7.5
Iron, ferro-silicon	437	6.7-7.3
Iron ore, hematite	325	5.2
Iron ore, hematite in bank	160-180	....
Iron ore, hematite loose	130-160	....
Iron ore, limonite	237	3.6-4.0
Iron ore, magnetite	315	4.9-5.2
Iron slag	172	2.5-3.0
Lead	710	11.37
Lead ore, galena	465	7.3-7.6
Magnesium, alloys	112	1.74-1.83
Manganese	475	7.2-8.0
Manganese ore, pyrolusite	259	3.7-4.6
Mercury	849	13.6
Monel Metal	556	8.8-9.0
Nickel	565	8.9-9.2
Platinum, cast, hammered	1330	21.1-21.5
Silver, cast, hammered	656	10.4-10.6
Steel, rolled	490	7.85
Tin, cast, hammered	459	7.2-7.5
Tin ore, cassiterite	418	6.4-7.0
Zinc, cast, rolled	440	6.9-7.2
Zinc ore, blende	253	3.9-4.2
VARIOUS SOLIDS		
Cereals, oats .... bulk	32	....
Cereals, barley .... bulk	39	....
Cereals, corn, rye .... bulk	48	....
Cereals, wheat .... bulk	48	....
Hay and Straw .... bales	20	....
Cotton, Flax, Hemp	93	1.47-1.50
Fats	58	0.90-0.97
Flour, loose	28	0.40-0.50
Flour, pressed	47	0.70-0.80
Glass, common	156	2.40-2.60
Glass, plate or crown	161	2.45-2.72
Glass, crystal	184	2.90-3.00
Leather	59	0.86-1.02
Paper	58	0.70-1.15
Potatoes, piled	42	....
Rubber, caoutchouc	59	0.92-0.96
Rubber goods	94	1.0-2.0
Salt, granulated, piled	48	....
Saltpeter	67	....
Starch	96	1.53
Sulphur	125	1.93-2.07
Wool	82	1.32

Substance	Weight Lb. per Cu. Ft.	Specific Gravity
TIMBER, U. S. SEASONED		
Moisture Content by Weight:		
Seasoned timber 15 to 20%		
Green timber up to 50%		
Ash, white, red	40	0.62-0.65
Cedar, white, red	22	0.32-0.38
Chestnut	41	0.66
Cypress	30	0.48
Fir, Douglas spruce	32	0.51
Fir, eastern	25	0.40
Elm, white	45	0.72
Hemlock	29	0.42-0.52
Hickory	49	0.74-0.84
Locust	46	0.73
Maple, hard	43	0.68
Maple, white	33	0.53
Oak, chestnut	54	0.86
Oak, live	59	0.95
Oak, red, black	41	0.65
Oak, white	46	0.74
Pine, Oregon	32	0.51
Pine, red	30	0.48
Pine, white	26	0.41
Pine, yellow, long-leaf	44	0.70
Pine, yellow, short-leaf	38	0.61
Poplar	30	0.48
Redwood, California	26	0.42
Spruce, white, black	27	0.40-0.46
Walnut, black	38	0.61
Walnut, white	26	0.41
VARIOUS LIQUIDS		
Alcohol, 100%	49	0.79
Acids, muriatic 40%	75	1.20
Acids, nitric 91%	94	1.50
Acids, sulphuric 87%	112	1.80
Lye, soda 66%	106	1.70
Oils, vegetable	58	0.91-0.94
Oils, mineral, lubricants	57	0.90-0.93
Water, 4°C. max. density	62.428	1.0
Water, 100°C.	59.830	0.9584
Water, ice	56	0.88-0.92
Water, snow, fresh fallen	8	.125
Water, sea water	64	1.02-1.03
GASES		
Air, 0°C. 760 mm.	.08071	1.0
Ammonia	.0478	0.5920
Carbon dioxide	.1234	1.5291
Carbon monoxide	.0781	0.9673
Gas, illuminating	.028-.036	0.35-0.45
Gas, natural	.038-.039	0.47-0.48
Hydrogen	.00559	0.0693
Nitrogen	.0784	0.9714
Oxygen	.0892	1.1056

The specific gravities of solids and liquids refer to water at 4°C., those of gases to air at 0°C. and 760 mm. pressure. The weights per cubic foot are derived from average specific gravities, except where stated that weights are for bulk, heaped or loose material, etc.

TABLE 5-2 Weights of Building Materials

Materials	Weight Lb. per Sq. Ft.
CEILINGS	
Channel suspended system	1
Lathing and plastering	See Partitions
Acoustical fiber tile	1
FLOORS	
Steel Deck	See Manufacturer
Concrete-Reinforced 1 in.	
Stone	12½
Slag	11½
Lightweight	6 to 10
Concrete-Plain 1 in.	
Stone	12
Slag	11
Lightweight	3 to 9
Fills 1 inch	
Gypsum	6
Sand	8
Cinders	4
Finishes	
Terrazzo 1 in.	13
Ceramic or Quarry Tile ¾ in.	10
Linoleum ¼ in.	1
Mastic ¾ in.	9
Hardwood ⅞ in.	4
Softwood ¾ in.	2½
ROOFS	
Copper or tin	1
Corrugated steel	See page 6 - 5
3-ply ready roofing	1
3-ply felt and gravel	5½
5-ply felt and gravel	6
Shingles	
Wood	2
Asphalt	3
Clay tile	9 to 14
Slate ¼	10
Sheathing	
Wood ¾ in.	3
Gypsum 1 in.	4
Insulation 1 in.	
Loose	½
Poured in place	2
Rigid	1½

Materials	Weight Lb. per Sq. Ft.
PARTITIONS	
Clay Tile	
3 in.	17
4 in.	18
6 in.	28
8 in.	34
10 in.	40
Gypsum Block	
2 in.	9½
3 in.	10½
4 in.	12½
5 in.	14
6 in.	18½
Wood Studs 2 × 4 12–16 in. o.c.	2
Steel partitions	4
Plaster 1 inch	
Cement	10
Gypsum	5
Lathing	
Metal	½
Gypsum Board ½ in.	2
WALLS	
Brick	
4 in.	40
8 in.	80
12 in.	120
Hollow Concrete Block (Heavy Aggregate)	
4 in.	30
6 in.	43
8 in.	55
12½ in.	80
Hollow Concrete Block (Light Aggregate)	
4 in.	21
6 in.	30
8 in.	38
12 in.	55
Clay tile (Load Bearing)	
4 in.	25
6 in.	30
8 in.	33
12 in.	45
Stone 4 in.	55
Glass Block 4 in.	18
Windows, Glass, Frame & Sash	8
Curtain Walls	See Manufacturer
Structural Glass 1 in.	15
Corrugated Cement Asbestos ¼ in.	3

SOURCE: *Manual of Steel Construction*, 7th ed., American Institute of Steel Construction, Chicago, Ill., 1973.

TABLE 5-3 Wire and Sheet Metal Gages

Name of Gage	*United States Standard Gage		The United States Steel Wire Gage	American or Brown & Sharpe Wire Gage	New Birmingham Standard Sheet & Hoop Gage	British Imperial or English Legal Standard Wire Gage	Birmingham or Stubs Iron Wire Gage	Name of Gage
Principal Use	Uncoated Steel Sheets and Light Plates		Steel Wire except Music Wire	Non-Ferrous Sheets and Wire	Iron and Steel Sheets and Hoops	Wire	Strips, Bands, Hoops and Wire	**Principal Use**
Gage No.	Weight Oz. per Sq. Ft.	Approx. Thickness Inches	Thickness, Inches					**Gage No.**
7/0's			.4900		.6666	.500		7/0's
6/0's			.4615	.5800	.625	.464		6/0's
5/0's			.4305	.5165	.5883	.432	.500	5/0's
4/0's			.3938	.4600	.5416	.400	.454	4/0's
3/0's			.3625	.4096	.500	.372	.425	3/0's
2/0's			.3310	.3648	.4452	.348	.380	2/0's
1/0			.3065	.3249	.3964	.324	.340	1/0
1			.2830	.2893	.3532	.300	.300	1
2			.2625	.2576	.3147	.276	.284	2
3	160	.2391	.2437	.2294	.2804	.252	.259	3
4	150	.2242	.2253	.2043	.250	.232	.238	4
5	140	.2092	.2070	.1819	.2225	.212	.220	5
6	130	.1943	.1920	.1620	.1981	.192	.203	6
7	120	.1793	.1770	.1443	.1764	.176	.180	7
8	110	.1644	.1620	.1285	.1570	.160	.165	8
9	100	.1495	.1483	.1144	.1398	.144	.148	9
10	90	.1345	.1350	.1019	.1250	.128	.134	10
11	80	.1196	.1205	.0907	.1113	.116	.120	11
12	70	.1046	.1055	.0808	.0991	.104	.109	12
13	60	.0897	.0915	.0720	.0882	.092	.095	13
14	50	.0747	.0800	.0641	.0785	.080	.083	14
15	45	.0673	.0720	.0571	.0699	.072	.072	15
16	40	.0598	.0625	.0508	.0625	.064	.065	16
17	36	.0538	.0540	.0453	.0556	.056	.058	17
18	32	.0478	.0475	.0403	.0495	.048	.049	18
19	28	.0418	.0410	.0359	.0440	.040	.042	19
20	24	.0359	.0348	.0320	.0392	.036	.035	20
21	22	.0329	.0317	.0285	.0349	.032	.032	21
22	20	.0299	.0286	.0253	.0313	.028	.028	22
23	18	.0269	.0258	.0226	.0278	.024	.025	23
24	16	.0239	.0230	.0201	.0248	.022	.022	24
25	14	.0209	.0204	.0179	.0220	.020	.020	25
26	12	.0179	.0181	.0159	.0196	.018	.018	26
27	11	.0164	.0173	.0142	.0175	.0164	.016	27
28	10	.0149	.0162	.0126	.0156	.0148	.014	28
29	9	.0135	.0150	.0113	.0139	.0136	.013	29
30	8	.0120	.0140	.0100	.0123	.0124	.012	30
31	7	.0105	.0132	.0089	.0110	.0116	.010	31
32	6.5	.0097	.0128	.0080	.0098	.0108	.009	32
33	6	.0090	.0118	.0071	.0087	.0100	.008	33
34	5.5	.0082	.0104	.0063	.0077	.0092	.007	34
35	5	.0075	.0095	.0056	.0069	.0084	.005	35
36	4.5	.0067	.0090	.0050	.0061	.0076	.004	36
37	4.25	.0064	.0085	.0045	.0054	.0068		37
38	4	.0060	.0080	.0040	.0048	.0060		38
39			.0075	.0035	.0043	.0052		39
40			.0070	.0031	.0039	.0048		40

* U. S. Standard Gage is officially a weight gage, in oz. per sq. ft. as tabulated. The Approx. Thickness shown is the "Manufacturers' Standard" of the American Iron and Steel Institute, based on steel as weighing 501.81 lb. per cu. ft. (489.6 true weight plus 2.5 per cent for average over-run in area and thickness). The AISI standard nomenclature for flat rolled carbon steel is as follows:

Thickness (Inches)	Width (Inches)					
	To 3½ incl.	Over 3½ To 6	Over 6 To 8	Over 8 To 12	Over 12 To 48	Over 48
0.2300 & thicker	Bar	Bar	Bar	Plate	Plate	Plate
0.2299 to 0.2031	Bar	Bar	Strip	Strip	Sheet	Plate
0.2030 to 0.1800	Strip	Strip	Strip	Strip	Sheet	Plate
0.1799 to 0.0449	Strip	Strip	Strip	Strip	Sheet	Sheet
0.0448 to 0.0344	Strip	Strip	Hot rolled sheet and strip not generally produced in these widths and thicknesses			
0.0343 to 0.0255	Strip					
0.0254 & thinner						

SOURCE: *Manual of Steel Construction,* 7th ed., American Institute of Steel Construction, Chicago, Ill., 1973.

TABLE 5-4 Square and Round Bars—Weight and Area

Size Inches	Weight Lb. per Foot		Area Square Inches		Size Inches	Weight Lb. per Foot		Area Square Inches	
	■ (square)	● (round)	▨ (square)	◍ (round)		■ (square)	● (round)	▨ (square)	◍ (round)
0					3	30.60	24.03	9.000	7.069
1/16	.013	.010	.0039	.0031	1/16	31.89	25.05	9.379	7.366
1/8	.053	.042	.0156	.0123	1/8	33.20	26.08	9.766	7.670
3/16	.120	.094	.0352	.0276	3/16	34.54	27.13	10.160	7.980
1/4	.213	.167	.0625	.0491	1/4	35.91	28.21	10.563	8.296
5/16	.332	.261	.0977	.0767	5/16	37.31	29.30	10.973	8.618
3/8	.478	.376	.1406	.1105	3/8	38.73	30.42	11.391	8.946
7/16	.651	.511	.1914	.1503	7/16	40.18	31.55	11.816	9.281
1/2	.850	.668	.2500	.1963	1/2	41.65	32.71	12.250	9.621
9/16	1.076	.845	.3164	.2485	9/16	43.15	33.89	12.691	9.968
5/8	1.328	1.043	.3906	.3068	5/8	44.68	35.09	13.141	10.321
11/16	1.607	1.262	.4727	.3712	11/16	46.23	36.31	13.598	10.680
3/4	1.913	1.502	.5625	.4418	3/4	47.81	37.55	14.063	11.045
13/16	2.245	1.763	.6602	.5185	13/16	49.42	38.81	14.535	11.416
7/8	2.603	2.044	.7656	.6013	7/8	51.05	40.10	15.016	11.793
15/16	2.988	2.347	.8789	.6903	15/16	52.71	41.40	15.504	12.177
1	3.400	2.670	1.0000	.7854	4	54.40	42.73	16.000	12.566
1/16	3.838	3.015	1.1289	.8866	1/16	56.11	44.07	16.504	12.962
1/8	4.303	3.380	1.2656	.9940	1/8	57.85	45.44	17.016	13.364
3/16	4.795	3.766	1.4102	1.1075	3/16	59.62	46.83	17.535	13.772
1/4	5.313	4.172	1.5625	1.2272	1/4	61.41	48.23	18.063	14.186
5/16	5.857	4.600	1.7227	1.3530	5/16	63.23	49.66	18.598	14.607
3/8	6.428	5.049	1.8906	1.4849	3/8	65.08	51.11	19.141	15.033
7/16	7.026	5.518	2.0664	1.6230	7/16	66.95	52.58	19.691	15.466
1/2	7.650	6.008	2.2500	1.7671	1/2	68.85	54.07	20.250	15.904
9/16	8.301	6.519	2.4414	1.9175	9/16	70.78	55.59	20.816	16.349
5/8	8.978	7.051	2.6406	2.0739	5/8	72.73	57.12	21.391	16.800
11/16	9.682	7.604	2.8477	2.2365	11/16	74.71	58.67	21.973	17.257
3/4	10.413	8.178	3.0625	2.4053	3/4	76.71	60.25	22.563	17.721
13/16	11.170	8.773	3.2852	2.5802	13/16	78.74	61.85	23.160	18.190
7/8	11.953	9.388	3.5156	2.7612	7/8	80.80	63.46	23.766	18.665
15/16	12.763	10.024	3.7539	2.9483	15/16	82.89	65.10	24.379	19.147
2	13.600	10.681	4.0000	3.1416	5	85.00	66.76	25.000	19.635
1/16	14.463	11.359	4.2539	3.3410	1/16	87.14	68.44	25.629	20.129
1/8	15.353	12.058	4.5156	3.5466	1/8	89.30	70.14	26.266	20.629
3/16	16.270	12.778	4.7852	3.7583	3/16	91.49	71.86	26.910	21.135
1/4	17.213	13.519	5.0625	3.9761	1/4	93.71	73.60	27.563	21.648
5/16	18.182	14.280	5.3477	4.2000	5/16	95.96	75.36	28.223	22.166
3/8	19.178	15.062	5.6406	4.4301	3/8	98.23	77.15	28.891	22.691
7/16	20.201	15.866	5.9414	4.6664	7/16	100.53	78.95	29.566	23.221
1/2	21.250	16.690	6.2500	4.9087	1/2	102.85	80.78	30.250	23.758
9/16	22.326	17.534	6.5664	5.1572	9/16	105.20	82.62	30.941	24.301
5/8	23.428	18.400	6.8906	5.4119	5/8	107.58	84.49	31.641	24.850
11/16	24.557	19.287	7.2227	5.6727	11/16	109.98	86.38	32.348	25.406
3/4	25.713	20.195	7.5625	5.9396	3/4	112.41	88.29	33.063	25.967
13/16	26.895	21.123	7.9102	6.2126	13/16	114.87	90.22	33.785	26.535
7/8	28.103	22.072	8.2656	6.4918	7/8	117.35	92.17	34.516	27.109
15/16	29.338	23.042	8.6289	6.7771	15/16	119.86	94.14	35.254	27.688
3	30.600	24.033	9.0000	7.0686	6	122.40	96.13	36.000	28.274

TABLE 5-4 Square and Round Bars—Weight and Area *(cont.)*

Size Inches	Weight Lb. per Foot		Area Square Inches		Size Inches	Weight Lb. per Foot		Area Square Inches	
	■	●	▨	◍		■	●	▨	◍
6	122.40	96.13	36.000	28.274	9	275.40	216.30	81.000	63.617
1/16	124.96	98.15	36.754	28.866	1/16	279.24	219.31	82.129	64.504
1/8	127.55	100.18	37.516	29.465	1/8	283.10	222.35	83.266	65.397
3/16	130.17	102.23	38.285	30.069	3/16	286.99	225.41	84.410	66.296
1/4	132.81	104.31	39.063	30.680	1/4	290.91	228.48	85.563	67.201
5/16	135.48	106.41	39.848	31.296	5/16	294.86	231.58	86.723	68.112
3/8	138.18	108.53	40.641	31.919	3/8	298.83	234.70	87.891	69.029
7/16	140.90	110.66	41.441	32.548	7/16	302.83	237.84	89.066	69.953
1/2	143.65	112.82	42.250	33.183	1/2	306.85	241.00	90.250	70.882
9/16	146.43	115.00	43.066	33.824	9/16	310.90	244.18	91.441	71.818
5/8	149.23	117.20	43.891	34.472	5/8	314.98	247.38	92.641	72.760
11/16	152.06	119.43	44.723	35.125	11/16	319.08	250.61	93.848	73.708
3/4	154.91	121.67	45.563	35.785	3/4	323.21	253.85	95.063	74.662
13/16	157.79	123.93	46.410	36.450	13/16	327.37	257.12	96.285	75.622
7/8	160.70	126.22	47.266	37.122	7/8	331.55	260.40	97.516	76.589
15/16	163.64	128.52	48.129	37.800	15/16	335.76	263.71	98.754	77.561
7	166.60	130.85	49.000	38.485	10	340.00	267.04	100.000	78.540
1/16	169.59	133.19	49.879	39.175	1/16	344.26	270.38	101.254	79.525
1/8	172.60	135.56	50.766	39.871	1/8	348.55	273.75	102.516	80.516
3/16	175.64	137.95	51.660	40.574	3/16	352.87	277.14	103.785	81.513
1/4	178.71	140.36	52.563	41.282	1/4	357.21	280.55	105.063	82.516
5/16	181.81	142.79	53.473	41.997	5/16	361.58	283.99	106.348	83.525
3/8	184.93	145.24	54.391	42.718	3/8	365.98	287.44	107.641	84.541
7/16	188.07	147.71	55.316	43.445	7/16	370.40	290.91	108.941	85.563
1/2	191.25	150.21	56.250	44.179	1/2	374.85	294.41	110.250	86.590
9/16	194.45	152.72	57.191	44.918	9/16	379.33	297.92	111.566	87.624
5/8	197.68	155.26	58.141	45.664	5/8	383.83	301.46	112.891	88.664
11/16	200.93	157.81	59.098	46.415	11/16	388.36	305.02	114.223	89.710
3/4	204.21	160.39	60.063	47.173	3/4	392.91	308.59	115.563	90.763
13/16	207.52	162.99	61.035	47.937	13/16	397.49	312.19	116.910	91.821
7/8	210.85	165.60	62.016	48.707	7/8	402.10	315.81	118.266	92.886
15/16	214.21	168.24	63.004	49.483	15/16	406.74	319.45	119.629	93.957
8	217.60	170.90	64.000	50.265	11	411.40	323.11	121.000	95.033
1/16	221.01	173.58	65.004	51.054	1/16	416.09	326.80	122.379	96.116
1/8	224.45	176.29	66.016	51.849	1/8	420.80	330.50	123.766	97.205
3/16	227.92	179.01	67.035	52.649	3/16	425.54	334.22	125.160	98.301
1/4	231.41	181.75	68.063	53.456	1/4	430.31	337.97	126.563	99.402
5/16	234.93	184.52	69.098	54.269	5/16	435.11	341.73	127.973	100.510
3/8	238.48	187.30	70.141	55.088	3/8	439.93	345.52	129.391	101.623
7/16	242.05	190.11	71.191	55.914	7/16	444.78	349.33	130.816	102.743
1/2	245.65	192.93	72.250	56.745	1/2	449.65	353.16	132.250	103.869
9/16	249.28	195.78	73.316	57.583	9/16	454.55	357.00	133.691	105.001
5/8	252.93	198.65	74.391	58.426	5/8	459.48	360.87	135.141	106.139
11/16	256.61	201.54	75.473	59.276	11/16	464.43	364.76	136.598	107.284
3/4	260.31	204.45	76.563	60.132	3/4	469.41	368.68	138.063	108.434
13/16	264.04	207.38	77.660	60.994	13/16	474.42	372.61	139.535	109.591
7/8	267.80	210.33	78.766	61.863	7/8	479.45	376.56	141.016	110.754
15/16	271.59	213.31	79.879	62.737	15/16	484.51	380.54	142.504	111.923
9	275.40	216.30	81.000	63.617	12	489.60	384.53	144.000	113.098

SOURCE: *Manual of Steel Construction,* 7th ed., American Institute of Steel Construction, Chicago, Ill., 1973.

TABLE 5-5 Weights and Areas of Steel Bars

Size, Inches	Weight, Lbs. per Foot		Area, Square Ins.	
	Square	Round	Square	Round
1/16	.013	.010	.0039	.0031
1/8	.053	.042	.0156	.0123
3/16	.120	.094	.0352	.0276
1/4	.213	.167	.0625	.0491
5/16	.332	.261	.0977	.0767
3/8	.478	.376	.1406	.1105
7/16	.651	.511	.1914	.1503
1/2	.850	.668	.2500	.1963
9/16	1.076	.845	.3164	.2485
5/8	1.328	1.043	.3906	.3068
11/16	1.607	1.262	.4727	.3712
3/4	1.913	1.502	.5625	.4418
13/16	2.245	1.763	.6602	.5185
7/8	2.603	2.044	.7656	.6013
15/16	2.988	2.347	.8789	.6903
1	3.400	2.670	1.0000	.7854
1-1/16	3.838	3.015	1.1289	.8866
1-1/8	4.303	3.380	1.2656	.9940
1-3/16	4.795	3.766	1.4102	1.1075
1-1/4	5.313	4.172	1.5625	1.2272
1-5/16	5.857	4.600	1.7227	1.3530
1-3/8	6.428	5.049	1.8906	1.4849
1-7/16	7.026	5.518	2.0664	1.6230
1-1/2	7.650	6.008	2.2500	1.7671
1-9/16	8.301	6.519	2.4414	1.9175
1-5/8	8.978	7.051	2.6406	2.0739
1-11/16	9.682	7.604	2.8477	2.2365
1-3/4	10.413	8.178	3.0625	2.4053
1-13/16	11.170	8.773	3.2852	2.5802
1-7/8	11.953	9.388	3.5156	2.7612
1-15/16	12.763	10.024	3.7539	2.9483
2	13.600	10.681	4.0000	3.1416
2-1/4	17.213	13.519	5.0625	3.9761
2-1/2	21.250	16.690	6.2500	4.9087
2-3/4	25.713	20.195	7.5625	5.9396
3	30.600	24.033	9.0000	7.0686
3-1/2	41.65	32.71	12.250	9.621
4	54.40	42.73	16.000	12.566
5	85.00	66.76	25.000	19.635
6	122.40	96.13	36.000	28.274

SOURCE: Pittsburgh-Des Moines Steel Company, Pittsburgh, Pa.

TABLE 5-6 Weight of Steel Bars or Plates in Pounds per Linear Foot

Width	THICKNESS INCHES															
Ins.	1/16	1/8	3/16	1/4	5/16	3/8	7/16	1/2	9/16	5/8	11/16	3/4	13/16	7/8	15/16	1
1/4	.053	.106	.159	.213	.266	.319	.372	.425	.478	.531	.584	.638	.691	.744	.797	.850
1/2	.106	.213	.319	.425	.531	.638	.744	.850	.956	1.063	1.169	1.275	1.381	1.488	1.594	1.700
3/4	.159	.319	.478	.638	.797	.956	1.116	1.275	1.434	1.594	1.753	1.913	2.072	2.231	2.391	2.550
1	.213	.425	.638	.850	1.063	1.275	1.488	1.700	1.913	2.125	2.338	2.550	2.763	2.975	3.188	3.400
2	.425	.850	1.275	1.700	2.125	2.550	2.975	3.400	3.825	4.250	4.675	5.100	5.525	5.950	6.375	6.800
3	.638	1.275	1.913	2.550	3.188	3.825	4.463	5.100	5.738	6.375	7.013	7.650	8.288	8.925	9.563	10.20
4	.850	1.700	2.550	3.400	4.250	5.100	5.950	6.800	7.650	8.500	9.350	10.20	11.05	11.90	12.75	13.60
5	1.063	2.125	3.188	4.250	5.313	6.375	7.438	8.500	9.563	10.63	11.69	12.75	13.81	14.88	15.94	17.00
6	1.275	2.550	3.825	5.100	6.375	7.650	8.925	10.20	11.48	12.75	14.03	15.30	16.58	17.85	19.13	20.40
7	1.488	2.975	4.463	5.950	7.438	8.925	10.41	11.90	13.39	14.88	16.36	17.85	19.34	20.83	22.31	23.80
8	1.700	3.400	5.100	6.800	8.500	10.20	11.90	13.60	15.30	17.00	18.70	20.40	22.10	23.80	25.50	27.20
9	1.913	3.825	5.738	7.650	9.563	11.48	13.39	15.30	17.21	19.13	21.04	22.95	24.86	26.78	28.69	30.60
10	2.125	4.250	6.375	8.500	10.63	12.75	14.88	17.00	19.13	21.25	23.38	25.50	27.63	29.75	31.88	34.00
11	2.338	4.675	7.013	9.350	11.69	14.03	16.36	18.70	21.04	23.38	25.71	28.05	30.39	32.73	35.06	37.40
12	2.550	5.100	7.650	10.20	12.75	15.30	17.85	20.40	22.95	25.50	28.05	30.60	33.15	35.70	38.25	40.80

SOURCE: Pittsburgh-Des Moines Steel Company, Pittsburgh, Pa.

TABLE 5-7 Wide Flange Beams

W is weight in pounds per linear foot of beam.
S is greatest section modulus of beam of given weight

Size	W	S
8x5¼	17	14.1
	20	17.0
8x6½	24	20.8
	28	24.3
8x8	31	27.4
	35	31.3
	40	35.5
	48	43.2
	58	52.0
	67	60.4
10x5¾	21	21.5
	25	26.4
	29	30.8
10x8	33	35.0
	39	42.2
	45	49.1
10x10	49	54.6
	54	60.4
	60	67.1
	66	73.7
	72	80.1
	77	86.1
	89	99.7
	100	112.4
	112	126.3
12x6½	27	34.1
	31	39.4
	36	45.9
12x8	40	51.9
	45	58.2
	50	64.7
12x10	53	70.7
	58	78.1
12x12	65	88.0
	72	97.5
	79	107.1
	85	115.7
	92	125.0
	99	134.7
	106	144.5
	120	163.4
	133	182.5
12x12 Cont'd	161	222.2
	190	263.2
14x6¾	30	41.8
	34	48.5
	38	54.6
14x8	43	62.7
	48	70.2
	53	77.8
14x10	61	92.2
	68	103.0
	74	112.3
14x12	78	121.1
	84	130.9
14x14½	87	138.1
	95	150.6
	103	163.6
	111	176.3
	119	189.4
	127	202.0
	136	216.0
14x16	320	492.8
	142	226.7
	150	240.2
	158	253.4
	167	267.3
	176	281.9
	184	295.8
	193	310.0
	202	324.9
	211	339.2
	219	352.6
	228	367.8
	237	382.2
	246	397.4
	264	427.4
	287	465.5
	314	511.9
	342	559.4
	370	608.1
	398	656.9
	426	707.4

TABLE 5-7 Wide Flange Beams *(cont.)*

W is weight in pounds per linear foot of beam.
S is greatest section modulus of beam of given weight

Size	W	S
16x7	36	56.3
	40	64.4
	45	72.4
	50	80.7
16x8½	58	94.1
	64	104.2
	71	115.9
	78	127.8
16x11½	88	151.3
	96	166.1
18x7½	50	89.0
	55	98.2
	60	107.8
18x8¾	64	117.0
	70	128.2
	77	141.7
	85	156.1
18x11¾	96	184.4
	105	202.2
	114	220.1
21x8¼	62	126.4
	68	139.9
	73	150.7
21x9	82	168.0
	96	197.6
21x13	112	249.6
	127	284.1
	142	317.2
24x9	76	175.4
	84	196.3
	94	220.9
24x12	100	248.9
	110	274.4
	120	299.1
24x14	130	330.7
	145	372.5
	160	413.5
27x10	94	242.8
	102	266.3
	114	299.2
27x14	145	402.9
	160	444.5
	177	492.8
30x10½	108	299.2
	116	327.9
	124	354.6
	132	379.7
30x15	172	528.2
	190	586.1
	210	649.9
33x11½	130	404.8
	141	446.8
	152	486.4
33x15¾	200	669.6
	220	740.6
	240	811.1
36x12	150	502.9
	160	541.0
	170	579.1
	182	621.2
	194	663.6
36x16½	230	835.5
	245	892.5
	260	951.1
	280	1031.2
	300	1105.1

SOURCE: Pittsburgh-Des Moines Steel Company, Pittsburgh, Pa.

TABLE 5-8 Safe Loads for Timber Columns

Size of post, in.	Length of post							
	8 ft	10 ft	12 ft	14 ft	16 ft	18 ft	20 ft	22 ft
	Safe load, lb							
4 × 4	12,109	7,794	5,412	3,976	3,044			
5 × 5	23,980	18,920	13,214	9,708	7,433	5,783	4,757	
6 × 6	37,145	33,660	27,245	20,131	15,412	12,178	9,864	
8 × 8	68,992	67,021	63,430	57,517	48,435	38,488	31,175	25,764
10 × 10	110,000	107,800	105,600	101,750	95,920	87,450	75,680	62,902
12 × 12		158,400	155,232	152,698	148,579	142,718	134,640	123,552
14 × 12			215,600	211,288	208,485	204,173	198,352	189,728
16 × 16				281,600	275,968	278,870	268,083	261,888
18 × 18					356,400	349,272	346,064	340,718
20 × 20						440,000	431,200	427,240

The figures are based on 1,100 psi compressive strength lumber. Columns are standing plumb and supported at ends only.

For 1,000 psi use 91 per cent of loads shown.
For 1,200 psi use 109 per cent of loads shown.
For 1,400 psi use 127 per cent of loads shown.
For 1,600 psi use 145 per cent of loads shown.

Safe Capacity of Spruce Timbers Used as Gin Poles

Section, in.	Length, ft	Safe capacity, tons
6 × 6	20	7
6 × 6	25	5
6 × 6	30	3
8 × 8	25	16
8 × 8	30	12
8 × 8	35	9
8 × 8	40	5
8 × 8	45	2
10 × 10	25	30
10 × 10	30	25
10 × 10	35	20
10 × 10	40	16
10 × 10	45	12
10 × 10	50	8
12 × 12	30	40
12 × 12	40	30
12 × 12	50	20
12 × 12	60	10

SOURCE: N. Foster, *Practical Tables for Building Construction*. Copyright © 1963 by McGraw-Hill, Inc., New York. (Used with permission of McGraw-Hill Book Co.)

TABLE 5-9 Properties of Section of Solid-Sawn Wood

Nominal size, in.	Standard dressed size, in. (S4S)	Area of section sq in.	Moment of inertia in.^{4}	Section modulus, in.^{3}	Weight, lb per lin ft, of piece when weight of wood, lb per cu ft, equals					
					25	30	35	40	45	50
1 × 3	¾ × 2½	1.875	0.977	0.781	0.326	0.391	0.456	0.521	0.586	0.651
1 × 4	¾ × 3½	2.625	2.680	1.531	0.456	0.547	0.638	0.729	0.820	0.911
1 × 6	¾ × 5½	4.125	10.398	3.781	0.716	0.859	1.003	1.146	1.289	1.432
1 × 8	¾ × 7¼	5.438	23.817	6.570	0.944	1.133	1.322	1.510	1.699	1.888
1 × 10	¾ × 9¼	6.938	49.466	10.695	1.204	1.445	1.686	1.927	2.168	2.409
1 × 12	¾ × 11¼	8.438	88.989	15.820	1.465	1.758	2.051	2.344	2.637	2.930
2 × 3*	1½ × 2½	3.750	1.953	1.563	0.651	0.781	0.911	1.042	1.172	1.302
2 × 4	1½ × 3½	5.250	5.359	3.063	0.911	1.094	1.276	1.458	1.641	1.823
2 × 6	1½ × 5½	8.250	20.797	7.563	1.432	1.719	2.005	2.292	2.578	2.865
2 × 8	1½ × 7¼	10.875	47.635	13.141	1.888	2.266	2.643	3.021	3.398	3.776
2 × 10	1½ × 9¼	13.875	98.932	21.391	2.409	2.891	3.372	3.854	4.336	4.818
2 × 12	1½ × 11¼	16.875	177.979	31.641	2.930	3.516	4.102	4.688	5.273	5.859
2 × 14	1½ × 13¼	19.875	290.775	43.891	3.451	4.141	4.831	5.521	6.211	6.901
3 × 1	2½ × ¾	1.875	0.088	0.234	0.326	0.391	0.456	0.521	0.586	0.651
3 × 2	2½ × 1½	3.750	0.703	0.938	0.651	0.781	0.911	1.042	1.172	1.302
3 × 4	2½ × 3½	8.750	8.932	5.104	1.519	1.823	2.127	2.431	2.734	3.038
3 × 6	2½ × 5½	13.750	34.661	12.604	2.387	2.865	3.342	3.819	4.297	4.774
3 × 8	2½ × 7¼	18.125	79.391	21.901	3.147	3.776	4.405	5.035	5.664	6.293
3 × 10	2½ × 9¼	23.125	164.886	35.651	4.015	4.818	5.621	6.424	7.227	8.030
3 × 12	2½ × 11¼	28.125	296.631	52.734	4.883	5.859	6.836	7.813	8.789	9.766
3 × 14	2½ × 13¼	33.125	484.625	73.151	5.751	6.901	8.051	9.201	10.352	11.502
3 × 16	2½ × 15¼	38.125	738.870	96.901	6.619	7.943	9.266	10.590	11.914	13.238
4 × 1	3½ × ¾	2.625	0.123	0.328	0.456	0.547	0.638	0.729	0.820	0.911
4 × 2	3½ × 1½	5.250	0.984	1.313	0.911	1.094	1.276	1.458	1.641	1.823
4 × 3	3½ × 2½	8.750	4.557	3.646	1.519	1.823	2.127	2.431	2.734	3.038
4 × 4	3½ × 3½	12.250	12.505	7.146	2.127	2.552	2.977	3.403	3.828	4.253
4 × 6	3½ × 5½	19.250	48.526	17.646	3.342	4.010	4.679	5.347	6.016	6.684
4 × 8	3½ × 7¼	25.375	111.148	30.661	4.405	5.286	6.168	7.049	7.930	8.811
4 × 10	3½ × 9¼	32.375	230.840	49.911	5.621	6.745	7.869	8.933	10.117	11.241
4 × 12	3½ × 11¼	39.375	415.283	73.828	6.836	8.203	9.570	10.938	12.305	13.672
4 × 14	3½ × 13½	47.250	717.609	106.313	8.203	9.844	11.484	13.125	14.766	16.406
4 × 16	3½ × 15½	54.250	1,086.130	140.146	9.418	11.302	13.186	15.069	16.953	18.837

6 × 1	5½ × ¾	4.125	0.193	0.516	0.716	0.859	1.003	1.146	1.289	1.432
6 × 2	5½ × 1½	8.250	1.547	2.063	1.432	1.719	2.005	2.292	2.578	2.865
6 × 3	5½ × 2½	13.750	7.161	5.729	2.387	2.865	3.342	3.819	4.297	4.774
6 × 4	5½ × 3½	19.250	19.651	11.229	3.342	4.010	4.679	5.347	6.016	6.684
6 × 6	5½ × 5½	30.250	76.255	27.729	5.252	6.302	7.352	8.403	9.453	10.503
6 × 8	5½ × 7½	41.250	193.359	51.563	7.161	8.594	10.026	11.458	12.891	14.323
6 × 10	5½ × 9½	52.250	392.963	82.729	9.071	10.885	12.700	14.514	16.328	18.142
6 × 12	5½ × 11½	63.250	697.068	121.229	10.981	13.177	15.373	17.569	19.766	21.962
6 × 14	5½ × 13½	74.250	1,127.672	167.063	12.891	15.469	18.047	20.625	23.203	25.781
6 × 16	5½ × 15½	85.250	1,706.776	220.229	14.800	17.760	20.720	23.681	26.641	29.601
6 × 18	5½ × 17½	96.250	2,456.380	280.729	16.710	20.052	23.394	26.736	30.078	33.420
6 × 20	5½ × 19½	107.250	3,398.484	348.563	18.620	22.344	26.068	29.792	33.516	37.240
6 × 22	5½ × 21½	118.250	4,555.086	423.729	20.530	24.635	28.741	32.847	36.953	41.059
6 × 24	5½ × 23½	129.250	5,948.191	506.229	22.439	26.927	31.415	35.903	40.391	44.878
8 × 1	7½ × ¾	5.438	0.255	0.680	0.944	1.133	1.322	1.510	1.699	1.888
8 × 2	7½ × 1½	10.875	2.039	2.719	1.888	2.266	2.643	3.021	3.398	3.776
8 × 3	7½ × 2½	18.125	9.440	7.552	3.147	3.776	4.405	5.035	5.664	6.293
8 × 4	7½ × 3½	25.375	25.904	14.802	4.405	5.286	6.168	7.049	7.930	8.811
8 × 6	7½ × 5½	41.250	103.984	37.813	7.161	8.594	10.026	11.458	12.891	14.323
8 × 8	7½ × 7½	56.250	263.672	70.313	9.766	11.719	13.672	15.625	17.578	19.531
8 × 10	7½ × 9½	71.250	535.859	112.813	12.370	14.844	17.318	19.792	22.266	24.740
8 × 12	7½ × 11½	86.250	950.547	165.313	14.974	17.969	20.964	23.958	26.953	29.948
8 × 14	7½ × 13½	101.250	1,537.734	227.813	17.578	21.094	24.609	28.125	31.641	35.156
8 × 16	7½ × 15½	116.250	2,327.422	300.313	20.182	24.219	28.255	32.292	36.328	40.365
8 × 18	7½ × 17½	131.250	3,349.609	382.813	22.786	27.344	31.901	36.458	41.016	45.573
8 × 20	7½ × 19½	146.250	4,634.297	475.313	25.391	30.469	35.547	40.625	45.703	50.781
8 × 22	7½ × 21½	161.250	6,211.484	577.813	27.995	33.594	39.193	44.792	50.391	55.990
8 × 24	7½ × 23½	176.250	8,111.172	690.313	30.599	36.719	42.839	48.958	55.078	61.198
10 × 1	9½ × ¾	6.938	0.325	0.867	1.204	1.445	1.686	1.927	2.168	2.409
10 × 2	9½ × 1½	13.875	2.602	3.469	2.409	2.891	3.372	3.854	4.336	4.818
10 × 3	9½ × 2½	23.125	12.044	9.635	4.015	4.818	5.621	6.424	7.227	8.030
10 × 4	9½ × 3½	32.375	33.049	18.885	5.621	6.745	7.869	8.993	10.117	11.241
10 × 6	9½ × 5½	52.250	131.714	47.896	9.071	10.885	12.700	14.514	16.328	18.142
10 × 8	9½ × 7½	71.250	333.984	89.063	12.370	14.844	17.318	19.792	22.266	24.740
10 × 10	9½ × 9½	90.250	678.755	142.896	15.668	18.802	21.936	25.069	28.203	31.337
10 × 12	9½ × 11½	109.250	1,204.026	209.396	18.967	22.760	26.554	30.347	34.141	37.934
10 × 14	9½ × 13½	128.250	1,947.797	288.563	22.266	26.719	31.172	35.625	40.078	44.531
10 × 16	9½ × 15½	147.250	2,948.068	380.396	25.564	30.677	35.790	40.903	46.016	51.128
10 × 18	9½ × 17½	166.250	4,242.836	484.896	28.863	34.635	40.408	46.181	51.953	57.726
10 × 20	9½ × 19½	185.250	5,870.109	602.063	32.161	38.594	45.026	51.458	57.891	64.323
10 × 22	9½ × 21½	204.250	7,867.879	731.896	35.460	42.552	49.644	56.736	63.828	70.920
10 × 24	9½ × 23½	223.250	10,274.148	874.396	38.759	46.510	54.262	62.014	69.766	77.517
12 × 1	11½ × ¾	8.438	0.396	1.055	1.465	1.758	2.051	2.344	2.637	2.930
12 × 2	11½ × 1½	16.875	3.164	4.219	2.930	3.516	4.102	4.688	5.273	5.859
12 × 3	11½ × 2½	28.125	14.648	11.719	4.883	5.859	6.836	7.813	8.789	9.766
12 × 4	11½ × 3½	39.375	40.195	22.969	6.836	8.203	9.570	10.938	12.305	13.672
12 × 6	11½ × 5½	63.250	159.443	57.979	10.981	13.177	15.373	17.569	19.766	21.962
12 × 8	11½ × 7½	86.250	404.297	107.813	14.974	17.969	20.964	23.958	26.953	29.948
12 × 10	11½ × 9½	109.250	821.651	172.979	18.967	22.760	26.554	30.347	34.141	37.934
12 × 12	11½ × 11½	132.250	1,457.505	253.479	22.960	27.552	32.144	36.736	41.328	45.920
12 × 14	11½ × 13½	155.250	2,357.859	349.313	26.953	32.344	37.734	43.125	48.516	53.906

TABLE 5-9 Properties of Section of Solid-Sawn Wood *(cont.)*

Nominal size, in.	Standard dressed size, in. (S4S)	Area of section, sq in.	Moment of inertia in.^{4}	Section modulus, in.^{3}	Weight, lb per lin ft, of piece when weight of wood, lb per cu ft, equals					
					25	30	35	40	45	50
12 × 16	11½ × 15½	178.250	3,568.713	460.479	30.946	37.135	43.325	49.514	55.703	61.892
12 × 18	11½ × 17½	201.250	5,136.066	586.979	34.939	41.927	48.915	55.903	62.891	69.878
12 × 20	11½ × 19½	224.250	7,105.922	728.813	38.932	46.719	54.505	62.292	70.078	77.865
12 × 22	11½ × 21½	247.250	9,524.273	885.979	42.925	51.510	60.095	68.681	77.266	85.851
12 × 24	11½ × 23½	270.250	12,437.129	1,058.479	46.918	56.302	65.686	75.069	84.453	93.837
14 × 2	13½ × 1½	19.875	3.727	4.969	3.451	4.141	4.831	5.521	6.211	6.901
14 × 3	13½ × 2½	33.125	17.253	13.802	5.751	6.901	8.051	9.201	10.352	11.502
14 × 4	13½ × 3½	46.375	47.34	27.052	8.047	9.657	11.266	12.877	14.485	16.094
14 × 6	13½ × 5½	74.250	187.172	68.063	12.891	15.469	18.047	20.625	23.203	25.781
14 × 8	13½ × 7½	101.250	474.609	126.563	17.578	21.094	24.609	28.125	31.641	35.156
14 × 10	13½ × 9½	128.250	964.547	203.063	22.266	26.719	31.172	35.625	40.078	44.531
14 × 12	13½ × 11½	155.250	1,710.984	297.563	26.953	32.344	37.734	43.125	48.516	53.906
14 × 16	13½ × 15½	209.250	4,189.359	540.563	36.328	43.594	50.859	58.125	65.391	72.656
14 × 18	13½ × 17½	236.250	6,029.297	689.063	41.016	49.219	57.422	65.625	73.828	82.031
14 × 20	13½ × 19½	263.250	8,341.734	855.563	45.703	54.844	63.984	73.125	82.266	91.406
14 × 22	13½ × 21½	290.250	11,180.672	1,040.063	50.391	60.469	70.547	80.625	90.703	100.781
14 × 24	13½ × 23½	317.250	14,600.109	1,242.563	55.078	66.094	77.109	88.125	99.141	110.156
16 × 3	15½ × 2½	38.125	19.857	15.885	6.619	7.944	9.267	10.592	11.915	13.240
16 × 4	15½ × 3½	53.375	54.487	31.135	9.267	11.121	12.975	14.828	16.682	18.536
16 × 6	15½ × 5½	85.250	214.901	78.146	14.800	17.760	20.720	23.681	26.641	29.601
16 × 8	15½ × 7½	116.250	544.922	145.313	20.182	24.219	28.255	32.292	36.328	40.365
16 × 10	15½ × 9½	147.250	1,107.443	233.146	25.564	30.677	35.790	40.903	46.016	51.128
16 × 12	15½ × 11½	178.250	1,964.463	341.646	30.946	37.135	43.325	49.514	55.703	61.892
16 × 14	15½ × 13½	209.250	3,177.984	470.813	36.328	43.594	50.859	58.125	65.391	72.656
16 × 16	15½ × 15½	240.250	4,810.004	620.646	41.710	50.052	58.394	66.736	75.078	83.420
16 × 18	15½ × 17½	271.250	6,922.523	791.146	47.092	56.510	65.929	75.347	84.766	94.184
16 × 20	15½ × 19½	302.250	9,577.547	982.313	52.474	62.969	73.464	83.958	94.453	104.948
16 × 22	15½ × 21½	333.250	12,837.066	1,194.146	57.856	69.427	80.998	92.569	104.141	115.712
16 × 24	15½ × 23½	364.250	16,763.086	1,426.646	63.238	75.885	88.533	101.181	113.828	126.476
18 × 6	17½ × 5½	96.250	242.630	88.229	16.710	20.052	23.394	26.736	30.078	33.420
18 × 8	17½ × 7½	131.250	615.234	164.063	22.786	27.344	31.901	36.458	41.016	45.573
18 × 10	17½ × 9½	166.250	1,250.338	263.229	28.863	34.635	40.408	46.181	51.953	57.726
18 × 12	17½ × 11½	201.250	2,217.943	385.729	34.939	41.927	48.915	55.903	62.891	69.878
18 × 14	17½ × 13½	236.250	3,588.047	531.563	41.016	49.219	57.422	65.625	73.828	82.031
18 × 16	17½ × 15½	271.250	5,430.648	700.729	47.092	56.510	65.929	75.347	84.766	94.184
18 × 18	17½ × 17½	306.250	7,815.754	893.229	53.168	63.802	74.436	85.069	95.703	106.337
18 × 20	17½ × 19½	341.250	10,813.359	1,109.063	59.245	71.094	82.943	94.792	106.641	118.490
18 × 22	17½ × 21½	376.250	14,493.461	1,348.229	65.321	78.385	91.450	104.514	117.578	130.642
18 × 24	17½ × 23½	411.250	18,926.066	1,610.729	71.398	85.677	99.957	114.236	128.516	142.795

20 × 6	19½ × 5½	107.250	270.359	98.313	18.620	22.344	26.068	29.792	33.516	37.240
20 × 8	19½ × 7½	146.250	685.547	182.813	25.391	30.469	35.547	40.625	45.703	50.781
20 × 10	19½ × 9½	185.250	1,393.234	293.313	32.161	38.594	45.026	51.458	57.891	64.323
20 × 12	19½ × 11½	224.250	2,471.422	429.813	38.932	46.719	54.505	62.292	70.078	77.865
20 × 14	19½ × 13½	263.250	3,998.109	592.313	45.703	54.844	63.984	73.125	82.266	91.406
20 × 16	19½ × 15½	302.250	6,051.297	780.813	52.474	62.969	73.464	83.958	94.453	104.948
20 × 18	19½ × 17½	341.250	8,708.984	995.313	59.245	71.094	82.943	94.792	106.641	118.490
20 × 20	19½ × 19½	380.250	12,049.172	1,235.813	66.016	79.219	92.422	105.625	118.828	132.031
20 × 22	19½ × 21½	419.250	16,149.859	1,502.313	72.786	87.344	101.901	116.458	131.016	145.573
20 × 24	19½ × 23½	458.250	21,089.047	1,794.813	79.557	95.469	111.380	127.292	243.203	159.115
22 × 6	21½ × 5½	118.250	298.088	108.396	20.530	24.635	28.741	32.847	36.953	41.059
22 × 8	21½ × 7½	161.250	755.859	201.563	27.995	33.594	39.193	44.792	50.391	55.990
22 × 10	21½ × 9½	204.250	1,536.130	323.396	35.460	42.552	49.644	56.736	63.828	70.920
22 × 12	21½ × 11½	247.250	2,724.901	473.896	42.925	51.510	60.095	68.681	77.266	85.851
22 × 14	21½ × 13½	290.250	4,408.172	653.063	50.391	60.469	70.547	80.625	90.703	100.781
22 × 16	21½ × 15½	333.250	6,671.941	860.896	57.856	69.427	80.998	92.569	104.141	115.712
22 × 18	21½ × 17½	376.250	9,602.211	1,097.396	65.321	78.385	91.450	104.514	117.578	130.642
22 × 20	21½ × 19½	419.250	13,284.984	1,362.563	72.786	87.344	101.901	116.458	131.016	145.573
22 × 22	21½ × 21½	462.250	17,806.254	1,656.396	80.252	96.302	112.352	128.403	144.453	160.503
22 × 24	21½ × 23½	505.250	23,252.023	1,978.896	87.717	105.260	122.804	140.347	157.891	175.434
24 × 6	23½ × 5½	129.250	325.818	118 479	22.439	26.927	31.415	35.903	40.391	44.878
24 × 8	23½ × 7½	176.250	826.172	220.313	30.599	36.719	42.839	48.958	55.078	61.198
24 × 10	23½ × 9½	223.250	1,679.026	353.479	38.759	46.510	54.262	62.014	69.766	77.517
24 × 12	23½ × 11½	270.250	2,978.380	517.979	46.918	56.302	65.686	75.069	84.453	93.837
24 × 14	23½ × 13½	317.250	4,818.234	713.813	55.078	66.094	77.109	88.125	99.141	110.156
24 × 16	23½ × 15½	364.250	7,292.586	940.979	63.238	75.885	88.533	101.181	113.828	126.476
24 × 18	23½ × 17½	411.250	10,495.441	1,199.479	71.398	85.677	99.957	114.236	128.516	142.795
24 × 20	23½ × 19½	458.250	14,520.797	1,489.313	79.557	95.469	111.380	127.292	143.203	159.115
24 × 22	23½ × 21½	505.250	19,462.648	1,810.479	87.717	105.260	122.804	140.347	157.891	175.434
24 × 24	23½ × 23½	552.250	25,415.004	2,162.979	95.877	115.052	134.227	153.403	172.578	191.753

* For lumber surfaced 1⅝ in. thick, instead of 1½ in., the area, moment of inertia, and section modulus may be increased 8.33%.

SOURCE: F. S. Merritt, *Standard Handbook for Civil Engineers*, 2d ed. Copyright © 1975 by McGraw-Hill, Inc., New York. (Used with permission of McGraw-Hill Book Co.)

TABLE 5-10 Lumber Converted to Board Feet

One board foot equals one lineal foot of 1 in. × 12 in. board or its equivalent.

Lumber sizes, in.	Board feet per lin ft	Board feet for standard lengths						
		8 ft	10 ft	12 ft	14 ft	16 ft	18 ft	20 ft
1 × 2	0.166	1.33	1.67	2.0	2.33	2.67	3.0	3.33
1 × 3	0.25	2.00	2.50	3.0	3.5	4.0	4.5	5.0
1 × 4; 2 × 2	0.333	2.67	3.33	4.00	4.67	5.33	6.0	6.67
1 × 6; 2 × 3	0.5	4.0	5.0	6.0	7.0	8.0	9.0	10.0
1 × 8; 2 × 4	0.667	5.34	6.67	8.0	9.34	10.67	12.0	13.34
1 × 12; 2 × 6; 3 × 4	1.0	8.0	10.0	12.0	14.0	16.0	18.0	20.0
2 × 8; 4 × 4	1.333	10.66	13.33	16.0	18.66	21.33	24.0	26.66
2 × 10	1.667	13.34	16.67	20.0	23.34	26.67	30.0	33.34
2 × 12; 3 × 8; 4 × 6	2.0	16.0	20.0	24.0	28.0	32.0	36.0	40.0
2 × 14	2.333	18.66	23.33	28.0	32.66	37.33	42.0	46.66
2½ × 12; 3 × 10	2.5	20.0	25.0	30.0	35.0	40.0	45.0	50.0
2½ × 14	2.917	23.34	29.17	35.0	40.84	46.67	52.50	58.34
2½ × 16; 4 × 10	3.333	26.66	33.33	40.0	46.62	53.33	60.0	66.66
3 × 6	1.5	12.0	15.0	18.0	21.0	24.0	27.0	30.0
3 × 12; 6 × 6	3.0	24.0	30.0	36.0	42.0	48.0	54.0	60.0
3 × 14	3.5	28.0	35.0	42.0	49.0	56.0	63.0	70.0
3 × 16; 4 × 12; 6 × 8	4.0	32.0	40.0	48.0	56.0	64.0	72.0	80.0
4 × 8	2.667	21.34	26.67	32.0	37.34	42.67	48.00	53.34
4 × 14	4.667	37.34	46.67	56.0	65.34	74.67	84.00	93.34
4 × 16; 8 × 8	5.333	42.66	53.33	64.0	74.66	85.33	96.0	106.66
6 × 10	5.0	40.0	50.0	60.0	70.0	80.0	90.0	100.0
6 × 12	6.0	48.0	60.0	72.0	84.0	96.0	108.0	120.0
6 × 14	7.0	56.0	70.0	84.0	98.0	112.0	126.0	140.0
6 × 16; 8 × 12	8.0	64.0	80.0	96.0	112.0	128.0	144.0	160.0
8 × 10	6.667	53.34	66.67	80.0	93.34	106.67	120.00	133.34
8 × 14	9.333	74.66	93.33	112.0	130.67	149.33	168.0	186.66
8 × 16	10.667	85.34	106.67	128.00	149.34	170.67	192.00	213.34

SOURCE: N. Foster, *Practical Tables for Building Construction*. Copyright © 1963 by McGraw-Hill, Inc., New York. (Used with permission of McGraw-Hill Book Co.)

TABLE 5-11 Safe Loads for Structural Lumber

Span, ft	Width × depth, in.									
	1 × 4	1 × 6	1 × 7	1 × 8	1 × 9	1 × 10	1 × 12	1 × 14	1 × 15	1 × 16
	Safe load uniformly distributed, lb*									
6	533	1,200	1,633	2,133	2,700	3,333	4,800	6,533	7,500	8,533
8	400	900	1,225	1,600	2,025	2,500	3,600	4,900	5,633	6,400
10	320	720	980	1,280	1,620	2,000	2,880	3,920	4,500	5,120
12	267	600	816	1,066	1,350	1,666	2,400	3,266	3,750	4,266
14	228	514	700	914	1,157	1,428	2,056	2,800	3,214	3,656
15	213	480	653	853	1,080	1,333	1,920	2,613	3,000	3,412
16	200	450	612	800	1,012	1,250	1,800	2,450	2,816	3,200
17	188	423	576	753	953	1,176	1,694	2,306	2,653	3,012
18	178	400	544	711	900	1,111	1,600	2,177	2,500	2,844
20	160	360	490	640	810	1,000	1,440	1,960	2,250	2,560
22	145	327	445	582	736	909	1,309	1,782	2,045	2,327
23	139	313	426	556	704	869	1,252	1,704	1,956	2,226
24	133	300	408	533	675	833	1,200	1,633	1,875	2,133
25	128	288	392	512	648	800	1,152	1,568	1,800	2,048
26	123	277	377	492	623	769	1,107	1,507	1,730	1,969
27	119	267	363	474	600	740	1,066	1,451	1,666	1,896
28	114	257	350	457	578	714	1,028	1,400	1,607	1,828
30	107	240	326	426	540	667	960	1,306	1,500	1,706
32	100	225	306	400	506	625	900	1,225	1,406	1,600
34				376	476	588	847	1,153	1,323	1,506
36				355	450	555	800	1,088	1,250	1,422
38				337	426	526	757	1,031	1,184	1,347
40				320	405	500	720	980	1,125	1,280

The loads shown are based on 1,800 psi structural lumber per inch of width.

For 1,600 psi use 89 per cent of loads shown.
For 1,400 psi use 78 per cent of loads shown.
For 1,200 psi use 66 per cent of loads shown.
For 1,000 psi use 55 per cent of loads shown.

*These figures are for lumber on edge—simple end supported. For other thicknesses, multiply above load by the actual thickness.

SOURCE: N. Foster, *Practical Tables for Building Construction*. Copyright © 1963 by McGraw-Hill, Inc., New York. (Used with permission of McGraw-Hill Book Co.)

TABLE 5-12 Nails (Lathing; Wire Spikes; Flooring Brads; Shingle; Double-Headed; and Plasterboard)

Size	Length, in.	Type		
		Common	Casing	Finishing
		Approx. number per lb		
2d	1	830		1,351
3d	1¼	528		807
4d	1½	316	473	584
5d	1¾	271		500
6d	2	168	236	309
7d	2¼	150	210	
8d	2½	106	145	189
9d	2¾	96		
10d	3	69	94	121
12d	3¼	63		
16d	3½	49	71	90
20d	4	31	52	62
30d	4½	24		
40d	5	18		
50d	5½	14		
60d	6	11		

Lathing Nails

Size	Gauge	Approx. number per lb
2d	16½	1,265
3d	15	752
3d	16	976

Wire Spikes

Size, in.	Gauge	Approx. number per lb
6	1	9
7	5/16	6
8	3/8	4
9	3/8	3½
10	3/8	3
12	3/8	2½

SOURCE: N. Foster, *Practical Tables for Building Construction*. Copyright © 1963 by McGraw-Hill, Inc., New York. (Used with permission of McGraw-Hill Book Co.)

TABLE 5-12 Nails *(cont.)*

Flooring Brads

Size	Gauge	Approx. number per lb
7d	11	139
8d	10	99
10d	9	69
8d	11	123

Shingle Nails

Size	Gauge	Approx. number per lb
3d	13	429
3½d	12½	345
4d	12	274
5d	12	235

Double-headed Nails

Size, in.	Over-all length, in.	Approx. number per lb
1¾	3	156
2¼	2½	90
2¾	3 1/16	59
3	3⅜	45
3½	3⅞	28

Plasterboard Nails

Size, in.	Approx. number per lb
1	469
1⅛	448
1¼	387
1½	339
1⅝ (⅜ in. head)	419

TABLE 5-13 Concrete Materials per Cubic Yard

Proportions (cem.:sand: stone)	¾" max. coarse aggregate			1" max. coarse aggregate			2½" coarse aggregate			Mixing water per cu yd, gal
	Cement, bags	Sand, cu yd	Stone, cu yd	Cement, bags	Sand, cu yd	Stone, cu yd	Cement, bags	Sand, cu yd	Stone, cu yd	
1:1 :2	9.2	0.36	0.74	10.3	0.39	0.78	10.5	0.40	0.80	38
1:1½:3	6.9	0.40	0.80	7.4	0.42	0.84	7.6	0.44	0.88	36
1:2 :3	6.2	0.48	0.74	6.8	0.53	0.80	6.9	0.54	0.80	34
1:2 :4	5.4	0.42	0.82	5.9	0.45	0.89	6.0	0.46	0.90	33
1:2½:4	5.0	0.47	0.75	5.4	0.53	0.83	5.5	0.54	0.85	34
1:2½:5	4.5	0.42	0.84	4.8	0.47	0.92	4.9	0.47	0.93	32½
1:3 :4	4.6	0.53	0.73	5.1	0.59	0.78	5.2	0.58	0.79	32
1:3 :5	4.2	0.47	0.78	4.5	0.52	0.86	4.6	0.53	0.87	32½
1:3 :6	4.0	0.43	0.85	4.1	0.47	0.92	4.1	0.48	0.94	33

The quantities in the table assume damp sand with a moisture content of not over 5 per cent by volume. The quantity of sand in any mix must be increased by the amount its moisture content exceeds 5 per cent, and the mixing water must be proportionately decreased. The gradations of the coarse aggregate will affect the amount of sand in a mix; for example, if there are no "fines-to-dust" in the coarse aggregate, a slight increase in the volume of both the sand and coarse aggregate is necessary.

SOURCE: N. Foster, *Practical Tables for Building Construction*. Copyright © 1963 by McGraw-Hill, Inc., New York. (Used with permission of McGraw-Hill Book Co.)

TABLE 5-14 Bearing Capacity of Soils

This table is approximate only. The actual bearing capacity of soils depends on the composition, the moisture content, and the extent of the strata. Local building codes usually give the allowable bearing capacity of soils.

Material	Bearing value	
	Tons per sq ft	Lb per sq ft
Clay, soft	1	2,000
Clay, hard	6	12,000
Clay, medium	4	8,000
Sand and clay mixed	2	4,000
Sand, fine, loose	1	2,000
Sand, coarse, loose	3	6,000
Sand (dry), fine, compact	3	6,000
Sand (dry), coarse, compact	4	8,000
Sand and gravel mixed (loose)	4	8,000
Sand and gravel mixed (compact)	5	10,000
Gravel, compact	6	12,000
Rock, soft	8	16,000
Rock, medium	15	30,000
Rock, hard	35	70,000
Hardpan	10	20,000
Shale, in sound condition	10	20,000

SOURCE: N. Foster, *Practical Tables for Building Construction*. Copyright © 1963 by McGraw-Hill, Inc., New York. (Used with permission of McGraw-Hill Book Co.)

TABLE 5-15 Wire Rope (6-strand, 19 wires)

Dia., in.	Approx. wt. per 100 ft, lb	Mild plow steel	Plow steel	Improved plow steel
		Breaking strength, lb		
1/4	10	4,140	4,780	5,480
3/8	23	10,000	11,000	12,600
1/2	40	17,000	18,800	21,600
5/8	63	26,200	28,800	33,200
3/4	90	37,400	41,200	47,400
7/8	123	50,800	56,000	64,400
1	160	66,000	73,000	84,000
1 1/8	203	83,000	92,000	106,000
1 1/4	250	102,000	113,000	130,000
1 1/2	360	145,000	161,000	185,000

Use of Wire Rope	Factor of Safety
Derricks	3.5
Guys	6.0
Hauling	6.0
Hoisting	5.0
Slings	8.0
Track cables	3.2

SOURCE: N. Foster, *Practical Tables for Building Construction*. Copyright © 1963 by McGraw-Hill, Inc., New York. (Used with permission of McGraw-Hill Book Co.)

TABLE 5-16 Safe Capacity of Chains

Dia. of iron, in.	Safe capacity, lb			
	Used straight	Used at 60-deg angle	Used at 45-deg angle	Used at 30-deg angle
1/4	1,330	1,000	850	600
3/8	2,660	2,050	1,700	1,200
1/2	5,330	4,100	3,400	2,400
5/8	8,330	6,800	5,600	4,000
3/4	12,000	9,400	7,800	5,500
7/8	16,330	12,800	10,400	7,400
1	20,830	16,000	13,000	9,400

SOURCE: N. Foster, *Practical Tables for Building Construction*. Copyright © 1963 by McGraw-Hill, Inc., New York. (Used with permission of McGraw-Hill Book Co.)

TABLE 5-17 Fiber Rope

These values are for new rope. For old rope reduce values by half. For safe loads, use a safety factor of 4.

Threads	Dia., in.	Circumference, in.	Approx. length of full coils, ft	Approx. weight per coil, lb	Feet per lb	Breaking strength, lb	
						Manila	Sisal
6(fine)	3/16	5/8	3,000	45	66.6	450	360
6	1/4	3/4	2,750	55	50.0	600	480
9	5/16	1	2,250	65	34.5	1,100	800
12	3/8	1 1/8	1,620	66	24.4	1,350	1,000
15	7/16	1 1/4	1,200	63	19.0	1,750	1,400
18	15/32	1 3/8	1,200	75	16.0	2,250	1,800
21	1/2	1 1/2	1,200	90	13.3	2,650	2,120
	9/16	1 3/4	1,200	125	9.61	3,450	2,760
	5/8	2	1,200	160	7.25	4,400	3,520
	3/4	2 1/4	1,200	200	6.00	5,400	4,320
	13/16	2 1/2	1,200	234	5.13	6,500	5,200
	7/8	2 3/4	1,200	270	4.45	7,700	6,160
	1	3	1,200	324	3.71	9,000	7,200
	1 1/16	3 1/4	1,200	375	3.20	10,500	8,400
	1 1/8	3 1/2	1,200	432	2.78	12,000	9,600
	1 1/4	3 3/4	1,200	502	2.40	13,500	10,800
	1 5/16	4	1,200	576	2.09	15,000	12,000
	1 1/2	4 1/2	1,200	720	1.67	18,500	14,800
	1 5/8	5	1,200	893	1.34	22,500	18,000
	1 3/4	5 1/2	1,200	1,074	1.12	26,500	21,200
	2	6	1,200	1,290	0.930	31,000	24,800
	2 1/8	6 1/2	1,200	1,503	0.800	36,000	28,800
	2 1/4	7	1,200	1,752	0.685	41,000	32,800
	2 1/2	7 1/2	1,200	2,004	0.600	46,500	37,200
	2 5/8	8	1,200	2,290	0.524	52,000	41,600
	2 7/8	8 1/2	1,200	2,580	0.465	58,000	46,400
	3	9	1,200	2,900	0.414	64,000	51,200
	3 1/8	9 1/2	1,200	3,225	0.372	71,000	56,800
	3 1/4	10	1,200	3,590	0.335	77,000	61,600
	3 5/8	11	1,200	4,400	0.273	91,000	72,800
	4	12	1,200	5,225	0.230	105,000	84,000

SOURCE: N. Foster, *Practical Tables for Building Construction*. Copyright © 1963 by McGraw-Hill, Inc., New York. (Used with permission of McGraw-Hill Book Co.)

Heavy Impact Goggle

Protection from impact, dust, smoke, cinders, and welding glare—May be worn with or without spectacles. Not suitable for continuous close-up exposure to welding glare.

Burner Goggle

Protection from light and sparks from burning torch, impact, dust, cinders, and weld glare. May be worn with or without spectacles.

Welder Helper Spectacle

Protection from welding glare. Particularly suitable for continuous close-up exposure to welding operations.

Fig. 5-1 Eye-protection equipment *(Pittsburgh-Des Moines Steel Company, Pittsburgh, Pa.)*

General Purpose Spectacle-Clear

Protection from impact, dust, cinders, and welding glare. Not adequate for continuous close-up exposure to welding operations.

Note: *Also available in shaded lens and frames.*

General Purpose Spectacle-Shaded

Similar to above except shaded lens and frames. Protection from sun glare. Suitable for rivet heaters and affords some protection from welding glare.

Cover-All Plastic Monogoggle-Clear

Protection from impact, dust, and cinders. Soft Vinyl frame readily adjusts to face so this is particularly suitable for dust protection. May be worn with or without spectacles. Does not protect from weld glare.

Note: *Also available in shaded lens and frames.*

Fig. 5-1 Eye-protection equipment *(cont.) (Pittsburgh-Des Moines Steel Company, Pittsburgh, Pa.)*

Fig. 5-2 Crane signals *(Pittsburgh-Des Moines Steel Company, Pittsburgh, Pa.)*

TABLE 5-18 Slings, Wire Rope

Capacity Tons	Rope Dia. Inches	Length Feet	Type
½	⅜	4- 6	Eye and Eye
2	⅝	8	Eye and Eye
5	¾	6-10-14	Eye and Eye
10	1	14-20	Eye and Eye
15	1¼	20-30-36	Eye and Eye
22	1½	20-32	Eye and Eye
2	⅝	16	• Pointed hook, sorting
7	⅞	18	Hook
12	1	4	Tail Hook

Capacities are for straight pull on eye or hook.

*For pointed hook sling with point in open hole, the capacity is ONE TON.

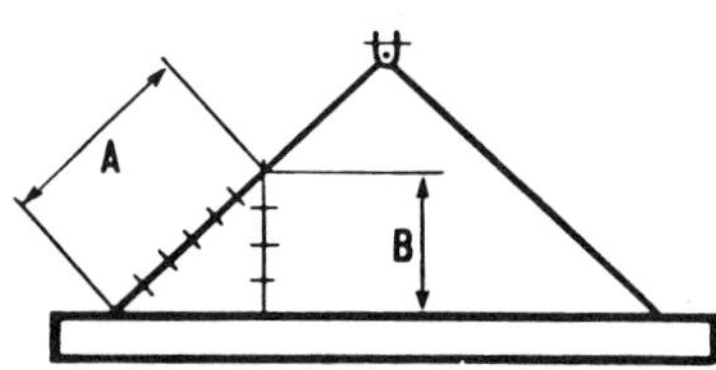

If slings are used as Twin Slings, the carrying capacity decreases as the spread between the slings increases.

To determine the capacity of each sling in a bridle hitch:

1. Measure distance along one sling for the same number of units as the capacity of the sling in tons. (Distance A)

2. Measure vertical distance from the point located in (1) to the load. (Distance B)

The number of units in this vertical distance (B) is the working load for the sling.

NOTE: Use the same unit for measuring both distances. The unit may be feet, half-feet, or any other convenient unit.

SOURCE: *PDM Safety Manual–Field Erection and Construction,* 2d ed., Pittsburgh-Des Moines Steel Company, Pittsburgh, Pa., 1978.

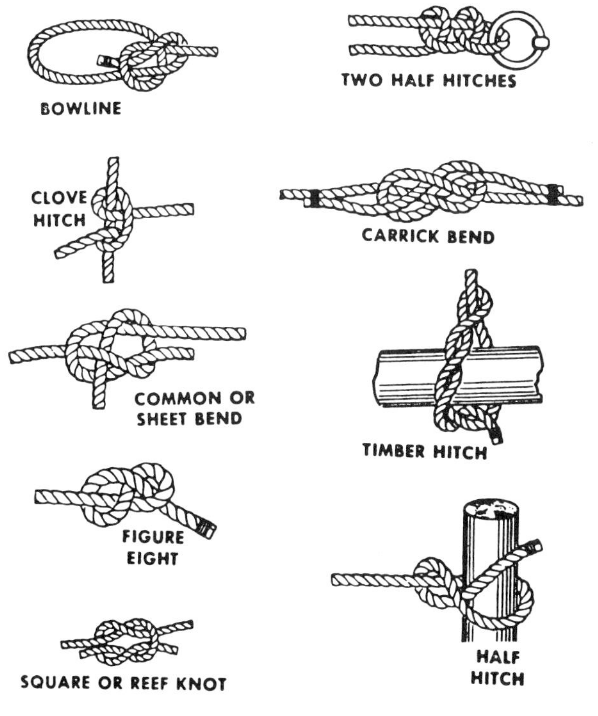

Fig. 5-3 Manila rope knots *(Pittsburgh-Des Moines Steel Company, Pittsburgh, Pa.)*.

TABLE 5-19 Wire Rope Clips

Dia. of Wire Rope	For Splicing to Develop Working Strength of Wire Rope	Distribute Clips Over	For Clamping an Eye in Wire Rope (*)	Distribute Clips Over
½″	Use 5 Clips	3′-0″	Use 3 Clips	2′-0″
⅝″	Use 6 Clips	4′-0″	Use 4 Clips	3′-0″
¾″	Use 6 Clips	4′-0″	Use 4 Clips	3′-0″
⅞″	Use 8 Clips	6′-0″	Use 5 Clips	4′-0″
1″	Use 8 Clips	6′-0″	Use 5 Clips	4′-0″
1¼″	Use 10 Clips	10′-0″	Use 6 Clips	6′-0″

When clamping an eye in wire rope, Crosby (U-Bolt) Clips should be turned so saddle bears on working part of rope. When splicing two wire ropes where both lines will be working, each end clip should be placed with its saddle on the working part of the rope. The interior clips should alternate.

After making one or two lifts, the clips should be gone over and tightened again to take any looseness that may have developed while clips were getting set to the wire rope.

(*) These are enough clips to make an eye either over a pin or over a movable sheave.

SOURCE: *PDM Safety Manual–Field Erection and Construction,* 2d ed., Pittsburgh-Des Moines Steel Company, Pittsburgh, Pa., 1978.

TABLE 5-20 Wire Rope Capacities

Diameter in Inches	Weight Per Foot Pounds	Working Capacity Pounds
3/8″	.23	4070
1/2″	.40	7130
5/8″	.63	11130
3/4″	.90	15870
7/8″	1.23	21470
1″	1.60	27870
1 1/8″	2.03	35070

Working capacities are for new rope. For used rope in good condition they should be reduced 25%.

SOURCE: *PDM Safety Manual–Field Erection and Construction,* 2d ed., Pittsburgh-Des Moines Steel Company, Pittsburgh, Pa., 1978.

TABLE 5-21 Turnbuckles

Type and Name	Capacity Tons	Screw Dia.	Length C Min.	Length C Max.	Wt. Lbs.	Remarks
Guy	12*	1½	60	96	66	For 1″ Guys
	15*	1¾	60	96	96	For 1⅛″ Guys
	19*	2	60	96	118	For 1¼″ Guys
	27*	2¼	66	102	185	For 1½″ Guys
Kicker	9*	1¼	39	63	26	For G4 Dk
	12*	1½	41	65	40	For G3 Dk
	27*	2¼	50	74	133	For G2 Dk
Plumbing	5*	1	36	60	25	For ⅝″ Wire Rope
	9*	1¼	51	87	47	For ¾″ Wire Rope
Standard	10	1½	76	112	275	With 10¼″ Sheave
	25	2¼	80	116	420	With 10¼″ Sheave
	60	3½	92	128	1,045	With 16″ Sheave

***Capacities shown are for holding loads only.**
Do not adjust these turnbuckles under capacity loads.

SOURCE: *PDM Safety Manual–Field Erection and Construction,* 2d ed., Pittsburgh-Des Moines Steel Company, Pittsburgh, Pa., 1978.

TABLE 5-22 Strength of Wire Rope Connections

Properly applied spelter socket	100% of cable
Eye splice with thimble or spool	90% of cable
Eye splice in sling	85% of cable
Cable clamps properly installed	80% of cable

)URCE: *PDM Safety Manual–Field Erection and Construction*, 2d ed., Pittsburgh-Des Moines Steel Company, Pittsburgh, Pa., 1978.

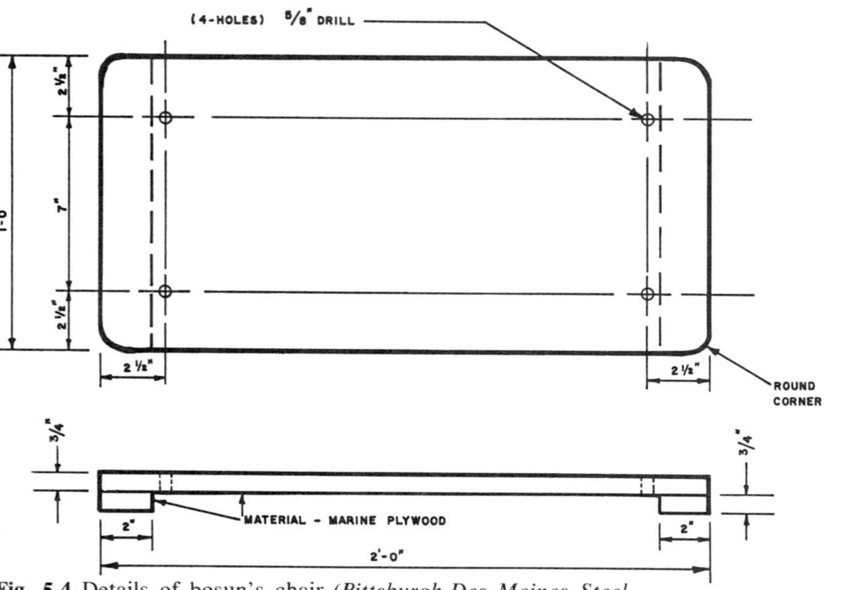

Fig. 5-4 Details of bosun's chair (*Pittsburgh-Des Moines Steel*

GENERAL SAFETY RULES*

1 Keep working areas clean and in an orderly condition. Equipment and material should be neatly arranged.

2 Do not allow drift pins, bolts, nuts, key plates, or other small erection items or tools to lie loose on the structure, planks, or scaffolds. These articles should be kept in erection baskets or in pails. Watch that hand tools or other objects are not placed where they may fall overboard.

3 Be constantly on the alert for unsafe conditions and report such conditions to your foreman.

4 All shaftways and other openings should be barricaded or roped off.

5 Place air lines, welding leads, and burning hose so as to avoid tripping hazards.

6 Always keep clear of suspended or swinging loads.

7 Construction safety signs are available from the insurance company. Additional posters and educational matter are also available from the local insurance company office.

8 Watch where you are walking.
Be sure scaffolds have proper bearing and look out for overhanging ends that may tip.

9 Take the safest way of travel from one place to another. Never jump between scaffolds. Use ladders or walkways.

10 Remove foreign material such as grease, oil, wet paint, frost, or ice from your work area. Also remove the mud from your shoes.

11 If winds are strong or gusty, do not put yourself in a position where you may be blown off the structure. If it is necessary to be on a structure under these conditions, a safety belt and lanyard should be used. The tail line should be as short as practicable and

*From *PDM Safety Manual–Field Erection and Construction,* 2d ed., 1978. Used by permission of Pittsburgh-Des Moines Steel Company, Pittsburgh, Pa.

it should be made fast to a substantial anchorage above the worker's waist.

12 Never slide down rope lines, wire rope, guys, or falls.

13 Riding up on loads being handled by crane or derrick is forbidden.

14 When carrying or handling pieces with other workers, make sure you understand each other, so as to avoid injury that might be caused by unexpected shifting of the load or by dropping it without warning.

15 Know the tools you are using—use the right tool for the right purpose.

16 Report to your foreman all tools which are burred, mushroomed, or defective in any way.

17 When assigned to grinding, inspect the stones for flaws. Be sure to use the correct stone for the speed of the grinder. Never throw stones into a tool box.

18 Where workers are working near power lines, the foreman must check conditions and issue specific instructions. Generally accepted safe working distances from high-voltage wires under normal conditions are as follows:

300 – 8,700 volts— 6 feet
8,700 – 15,000 volts— 8 feet
15,000 – 35,000 volts—10 feet
35,000 – 50,000 volts—12 feet
50,000 – 100,000 volts—15 feet
100,000 – 132,000 volts—17 feet

At all times a minimum distance of 17 feet should be maintained between moving equipment or loads and any power line where the current cannot be shut off.

19 Only qualified welders shall weld scaffold bracket clips, ear plates, erection nuts, and lifting lugs, and each weld shall be initialed by the welder.

20 For personnel assigned to x-ray work. Specific safety regulations are listed in the PDM Corporate "Radiographer's Manual." This covers safety procedures.

21 When conducting hydrostatic tests, install pressure gages so that they can be read at a safe distance. Each project will have a specific set of instructions.

22 When conducting pressure tests on vessels, only qualified personnel will be used. Each project will have a specific set of instructions for conducting a safe efficient method of leak testing.

23 Do not stand in the bight of a line or in the angle formed by a snatch block.

RULES FOR EYE PROTECTION*

1 Eye protection will be worn for protection against impact, glare, and harmful radiation. See chart, Fig. 5-1.

2 For burning, brazing, cutting, or gas welding use regular cup-style goggles with No. 4 shade lenses.

3 For chipping and grinding use cup-style goggles with clear lenses and fitted with clear plastic back-up disks.

4 For fitters and personnel working near welding operations, use goggles with green medium Calobar lenses to protect your eyes from glare and invisible radiation.

5 Employees requiring prescription safety goggles should be encouraged to make arrangements to obtain and wear them.

6 Cover goggles are to be worn over personal spectacles by wearers of same who do not have prescription safety glasses.

7 Eye protection will be furnished by Erection Tool Houses and should not be purchased locally on the job.

*From *PDM Safety Manual–Field Erection and Construction,* 2d ed., 1978. Used by permission of Pittsburgh-Des Moines Steel Company, Pittsburgh, Pa.

RULES FOR UNLOADING STEEL*

1 When moving rail cars, make sure everyone is in the clear. Check the brakes.
2 Chock car wheels on both sides—do not depend on brakes.
3 Crane operators and hoist operators will take directions from only one signalman for each lift and at each landing. Use signals shown in Fig. 5-2. Signalers should wear luminous gloves or vests in dark areas for long lifts.
4 See that steel is taken out of car in proper sequence so that remaining pieces will not fall.
5 Be alert to block up car when necessary to prevent tipping. Do not unbalance car by unloading progressively from one side.
6 Make sure you are in the clear when a load is lifted. Get to the end of the car or off the car. Watch out for bad end-gates.
7 Keep your eye on the load as it is being raised from the car.
8 Use thought in piling steel so that it will be in proper sequence when it is taken to the site.
9 When unloading trucks make sure the truck is as level as possible. Get off the truck before the load is lifted.
10 Keep hands off loads that may nest and pinch your fingers. Use a rope tagline to help control suspended loads.
11 Remember that the capacity of spreaders or twin slings decreases as the spread between the slings increases. (See Table 5-18.) (Check cable, chockers, slings for kinks before using.)
12 Use softeners where necessary to protect the slings from sharp edges or to protect the steel from damage from the slings. Tie softeners to prevent them from falling.

*From *PDM Safety Manual–Field Erection and Construction,* 2d ed., 1978. Used by permission of Pittsburgh-Des Moines Steel Company, Pittsburgh, Pa.

SAFETY RULES FOR ROPE, SLINGS, SHACKLES, AND SHEAVES*

1 Ropes and cables should be properly cared for and handled.
2 Avoid kinks in all hoisting lines. Remember that a loop in a wire rope will lead to a kink and kinks cause permanent damage.
3 Ropes, slings, and shackles must be used within their rated capacity. Consult Tables 5-18, 5-20, and 5-22.
4 Remember that a sudden jerk on a line will result in strain far beyond the weight of the load.
5 Use the sheaves and blocks of the correct sizes for the diameter of the rope. Check with Construction Tool House.
6 Never use wire rope in a block designed for manila line.
7 Never use manila rope on power equipment for hoisting or pulling without approval from the construction manager's office.
8 Manila rope is to be used only for hand falls, scaffold lines, and taglines.
9 To uncoil a new cable, the reel should be mounted on a horizontal shaft and [the cable should be] taken off the bottom.
10 Wire rope should be carefully inspected for wear of the crown wires, corrosion, broken wires, kinking, and high strands.
11 Place cable clamps with the U bolt on the dead end of the wire rope. Follow Tables 5-19 to 5-22.
12 Mouse hooks wherever there is a possibility of a sling eye or shackle jumping out.

SAFETY RULES FOR SCAFFOLDS*

1 Scaffold lumber will be inspected before use in the field to be sure it is free of splits.
2 The ends of each scaffold plank will be banded or bolted or S wedges used to avoid splitting.

*From *PDM Safety Manual–Field Erection and Construction*, 2d ed., 1978. Used by permission of Pittsburgh-Des Moines Steel Company, Pittsburgh, Pa.

3. Each scaffold shall be constructed of not less than three 2″×10″ or two 2″×12″ planks. On special-shaped vessels where space is restricted, the size and shape of scaffolds will be governed by local conditions.
4. Brackets will have a maximum spacing of 8′-0″ and have a minimum overhang of 12 in on each bracket.
5. Scaffolds shall be kept free of ice, snow, mud, oil, and other materials which create a hazard.
6. Excessive storage or accumulation of materials on scaffolds will not be permitted. One-half barrel of fitting equipment weighs approximately 350 lb. This is too much weight to place on a scaffold.
7. Scaffold planks should overlap in *one* direction. This overlap should be a minimum of 2′-0″. For strength of plank see chart (Table 5-10).
8. Keep scaffold brackets plumb. If a bracket is installed only 5 in. out of plumb, it will support only half of its load. Check all brackets before using them.
9. Regardless of the type of scaffolding, some form of guardrail must be used at heights of 36 to 42 in above planks.
10. Never jack from a scaffold.
11. Safety lines are to be secured at each post and must be a minimum of ½-in rope or equivalent. Larger-diameter rope is required in some states, and cable is required in some industrial plants.
12. Use 1-in manila line to hang rivet floats. In fastening the scaffold lines around steel, take one round turn and tie with two half-hitches.
13. When girder hoks are used, make sure the hooks are securely placed. Hooks made of reinforcing rod are not permitted.
14. When carrying floats, be careful of the wind.
15. Check the float prior to each time you get on it, paying attention to the rope and knots.

16 When using tubular steel scaffolding, review general rules and requirements for scaffolds.

17 If casters or wheels are used on tubular steel scaffolding, they shall be locked in position. Check for stable footing. Tubular steel scaffolding should never be placed directly on the ground, but on steel sheets, blocks, masonry, or wood plank.

18 The height of the scaffold shall not exceed four times the smallest base dimension unless it is tied to the structure.

19 All locking pins must be provided and used.

20 Guardrails are required regardless of height.

21 No one is permitted to remain on a tubular steel scaffold while it is being moved.

22 Free-standing tubular steel towers must be guyed every 30 ft of height.

23 For rolling tank scaffolds, review general requirements and rules for scaffolds.

24 When inspecting the scaffold, check for broken wheel flanges. Make sure the boards in the platform are secured.

25 Caution must be used to prevent the scaffold from rolling off the end of the sheet.

26 Safe access must be provided to and from the scaffold.

27 When moving or changing location, make certain the wheels are properly seated on the sheets before cutting loose the rig.

28 When using a bosun's chair, check to see that ropes and chair are in satisfactory condition.

29 Bosun's chair should be constructed as shown. See detail (Fig. 5-4).

NOTES

Section Six

■

ATHLETIC FACILITIES

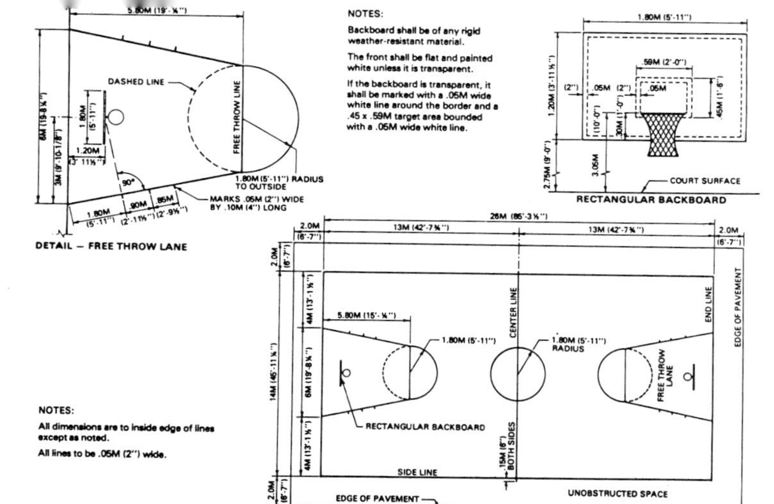

Fig. 6-1 AAU basketball court with free throw lane and rectangular backboard [*U.S. Department of the Army, Office of the Chief of Engineers (TM 5-803-10/NAVFAC P-457/AFR 88-33)*].

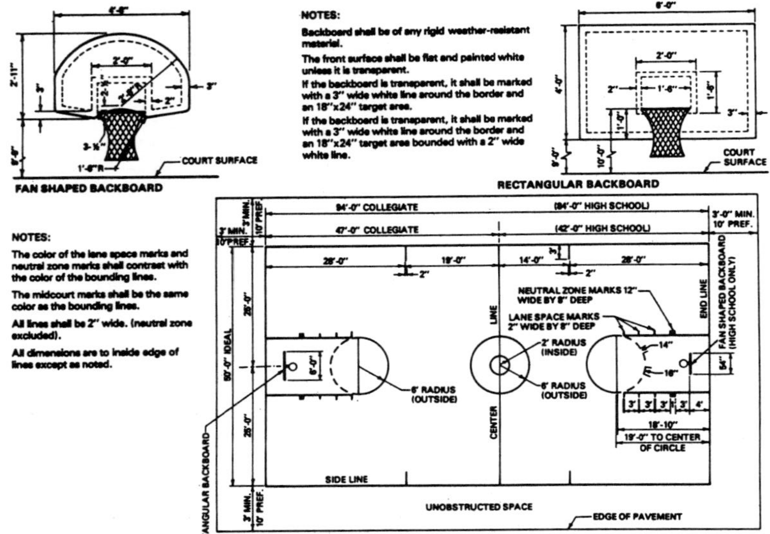
FAN SHAPED BACKBOARD
COURT SURFACE
NOTES:
Backboard shall be of any rigid weather-resistant material.
The front surface shall be flat and painted white unless it is transparent.
If the backboard is transparent, it shall be marked with a 3" wide white line around the border and an 18"x24" target area.
If the backboard is transparent, it shall be marked with a 3" wide white line around the border and an 18"x24" target area bounded with a 2" wide white line.
RECTANGULAR BACKBOARD
COURT SURFACE
NOTES:
The color of the lane space marks and neutral zone marks shall contrast with the color of the bounding lines.
The midcourt marks shall be the same color as the bounding lines.
All lines shall be 2" wide. (neutral zone excluded).
All dimensions are to inside edge of lines except as noted.
94'-0" COLLEGIATE
(84'-0" HIGH SCHOOL)
47'-0" COLLEGIATE
(42'-0" HIGH SCHOOL)
3'-0" MIN.
10' PREF.
50'-0" IDEAL
NEUTRAL ZONE MARKS 12" WIDE BY 8" DEEP
LANE SPACE MARKS 2" WIDE BY 8" DEEP
2' RADIUS (INSIDE)
6' RADIUS (OUTSIDE)
6' RADIUS (OUTSIDE)
CENTER LINE
END LINE
FAN SHAPED BACKBOARD (HIGH SCHOOL ONLY)
18'-10"
19'-0" TO CENTER OF CIRCLE
SIDE LINE
UNOBSTRUCTED SPACE
EDGE OF PAVEMENT
RECTANGULAR BACKBOARD
COURT LAYOUT

Fig. 6-2 NCAA basketball court with backboard details (*U.S. De-*

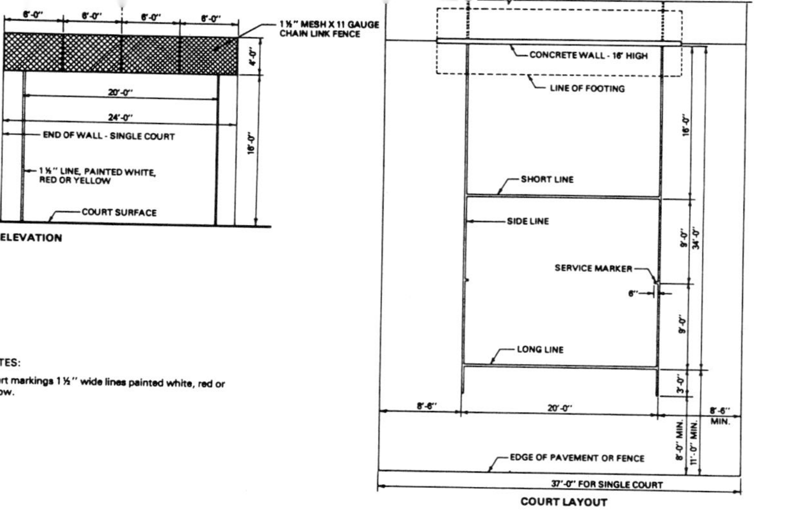

Fig. 6-3 One-wall handball court layout with elevation view (*U.S. Department of the Army, Office of the Chief of Engineers*).

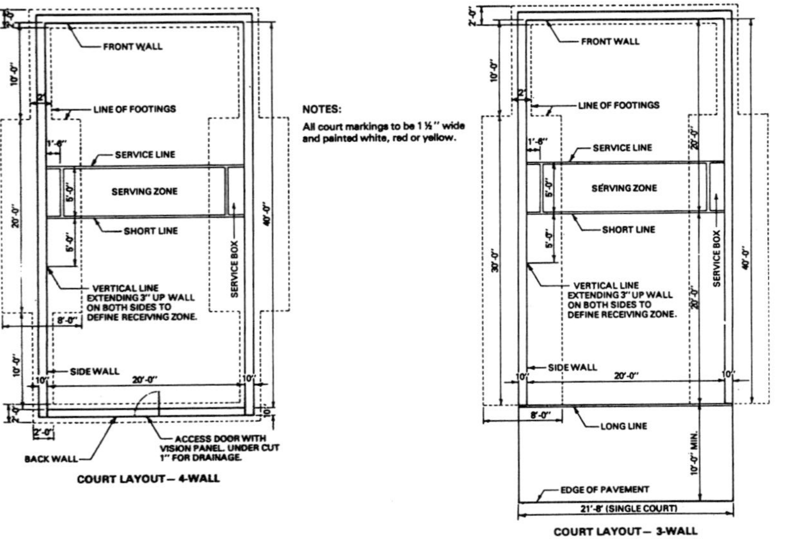

Fig. 6-4 Three- and four-wall handball court layouts (*U.S. Depart-*

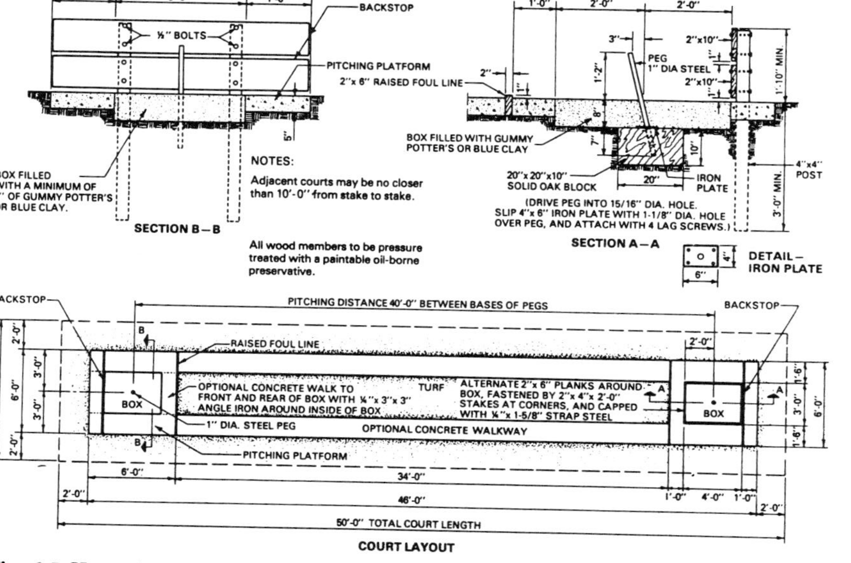

Fig. 6-5 Horseshoe court layout with details (*U.S. Department of the Army, Office of the Chief of Engineers*).

NOTES: The rink must be at least 185'x85', ideal 200'x85'.

The rink shall be surrounded by a wooden wall or fence known as the "BOARDS" which shall extend not less than 40" nor more than 48" above the level of the ice surface. Ideal 42".

The surface of the boards facing the ice shall be smooth and free from obstructions. All access doors to the playing surface must swing away from the ice surface.

A protective screening of heavy gauge wire or safety glass is recommended above the boards, except for the bench areas, for the protection of spectators around the rink.

The center line and the two blue lines shall extend across the rink and vertically to the top of the boards.

Surface to be flooded may be sand-clay or bituminous material

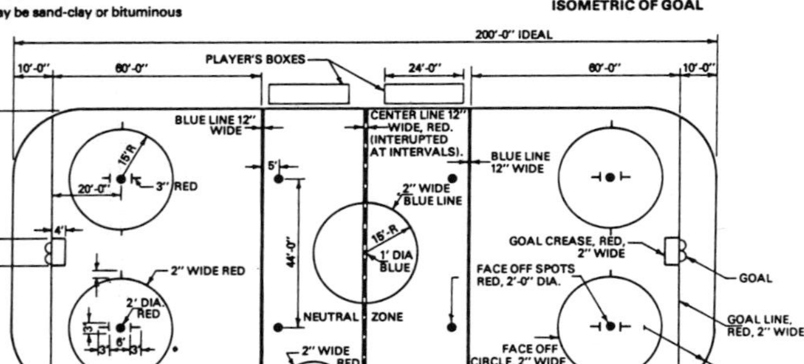

Fig. 6-6 Ice hockey rink layout with isometric of goal (*U.S. Department of the Army, Office of the Chief of Engineers*).

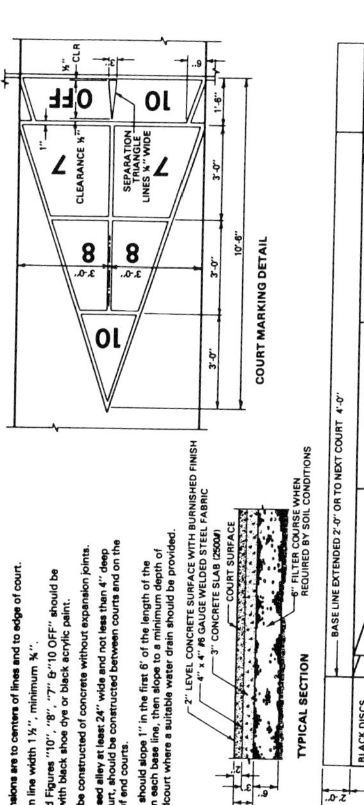

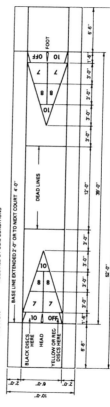

Fig. 6-7 Shuffleboard court layout with marking details (*U.S. Department of the Army, Office of the Chief of Engineers*).

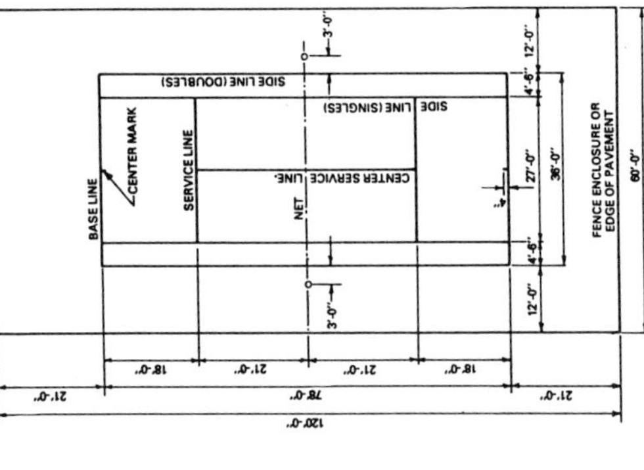

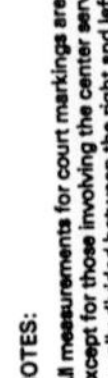

NOTES:
All measurements for court markings are to the outside of lines except for those involving the center service line which is equally divided between the right and left service courts.
All court markings to be 2" wide.
Fence enclosure, if provided, should be 10' high, 11 gauge, 1¾" mesh chain link. For fence details see figure 63.
Minimum distance between sides of parallel courts to be 12'-0".

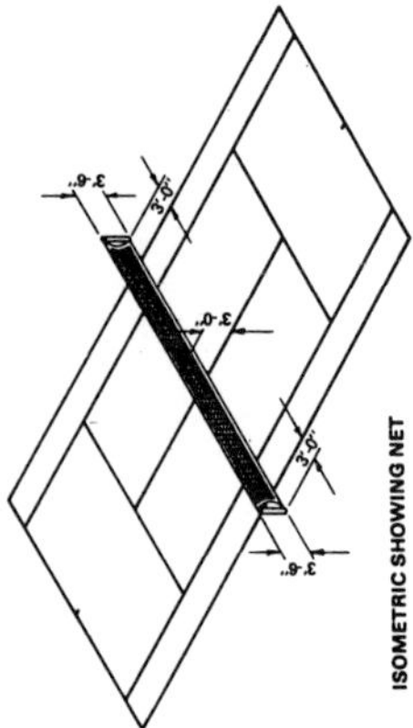

Fig. 6-8 Tennis court layout with net detail (*U.S. Department of the*

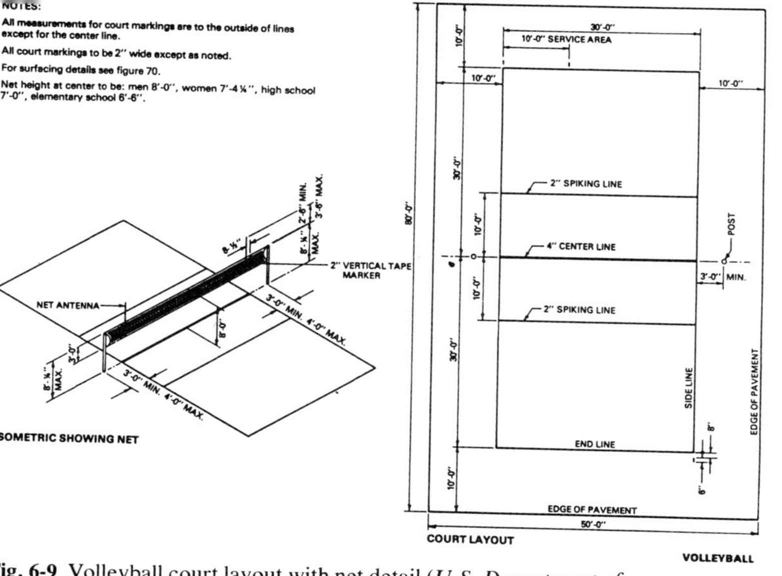

Fig. 6-9 Volleyball court layout with net detail (*U.S. Department of the Army, Office of the Chief of Engineers*).

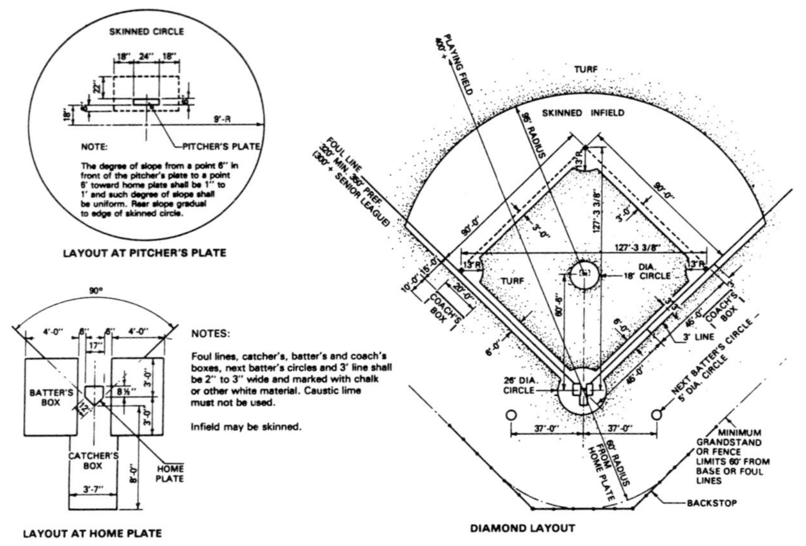

Fig. 6-10 Baseball diamond layout with home plate and pitcher's mound details (*U.S. Department of the Army, Office of the Chief of Engineers*).

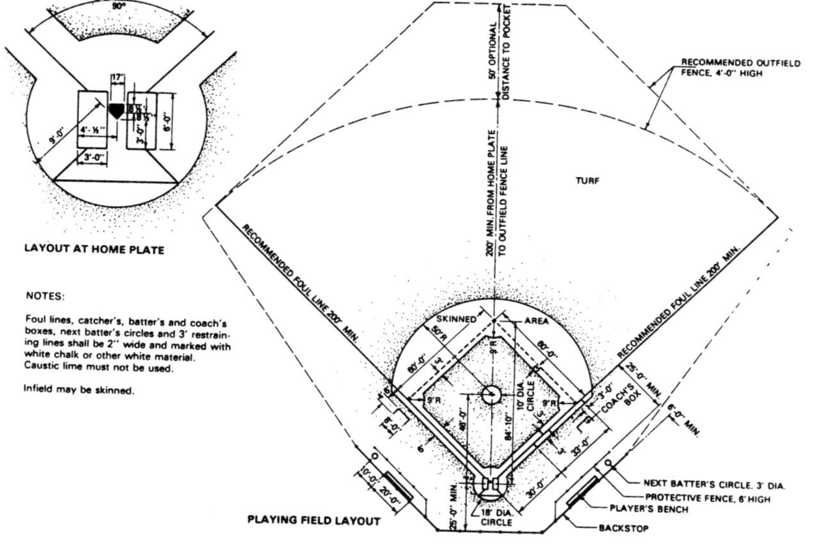

Fig. 6-11 Little league baseball playing field layout with home plate detail (*U.S. Department of the Army, Office of the Chief of Engineers*).

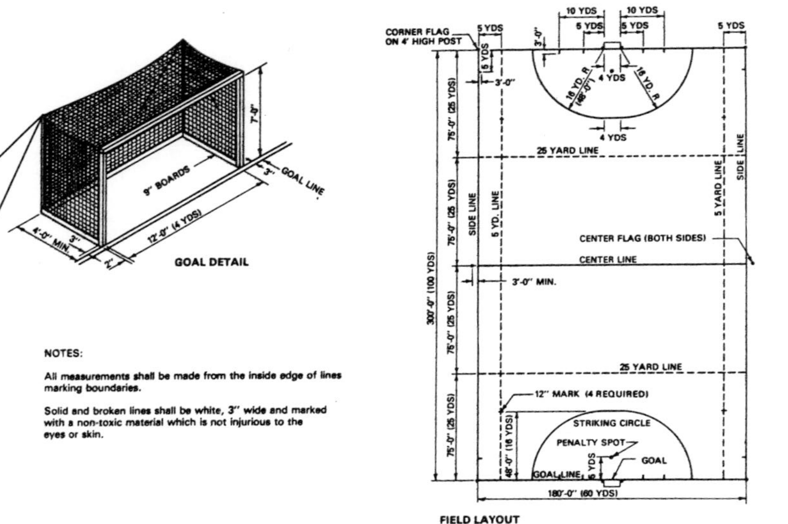

Fig. 6-12 Field hockey layout with goal detail (*U.S. Department of the Army, Office of the Chief of Engineers*).

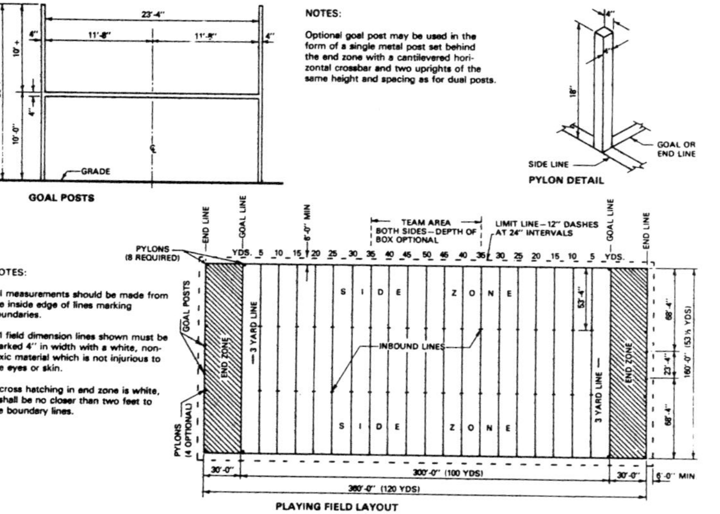

Fig. 6-13 Eleven-man football playing field layout with goal post detail (*U.S. Department of the Army, Office of the Chief of Engineers*).

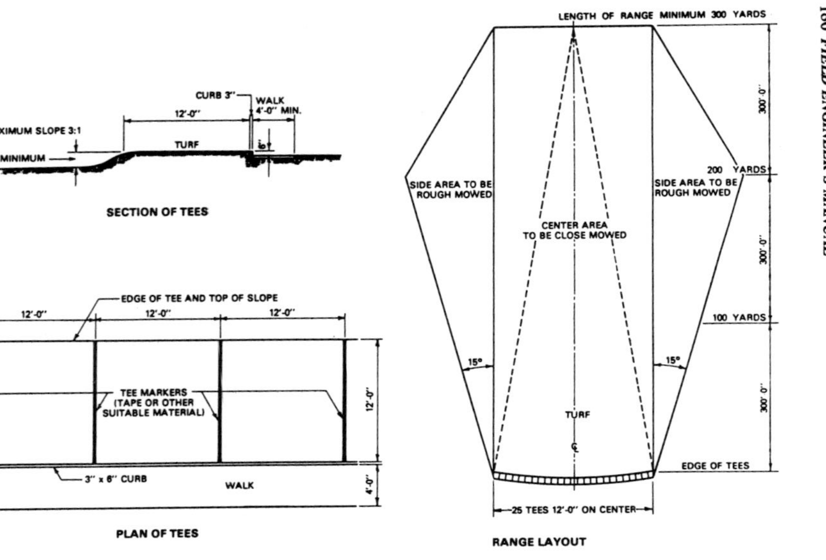

Fig. 6-14 Golf driving range with tee details (*U.S. Department of the Army, Office of the Chief of Engineers*).

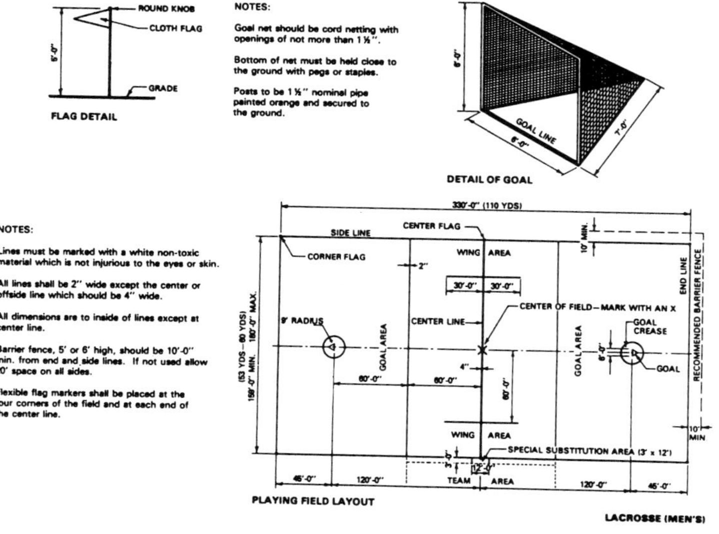

Fig. 6-15 Men's lacrosse playing field layout with goal and flag details (*U.S. Department of the Army, Office of the Chief of Engineers*).

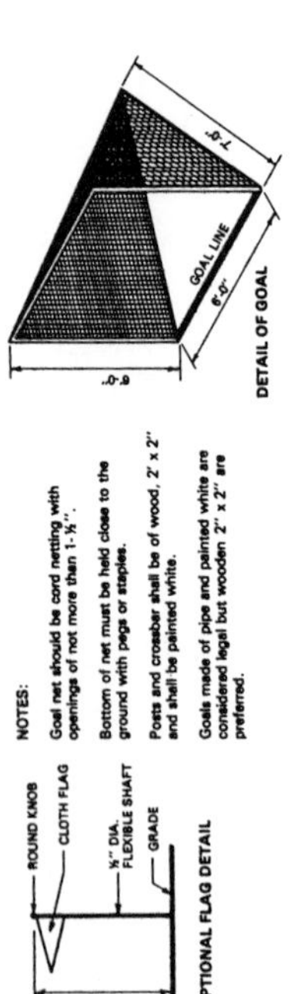

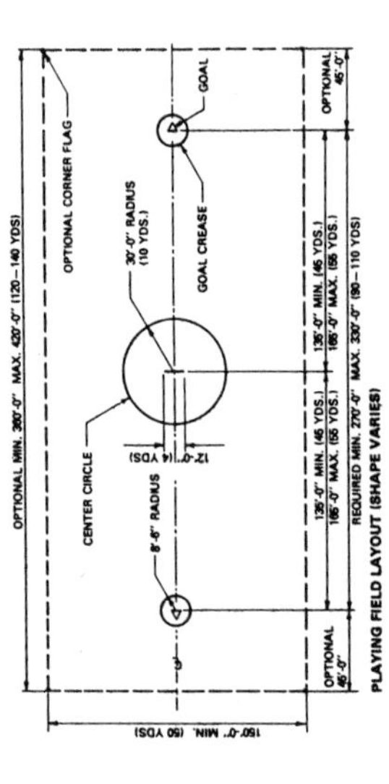

LACROSSE (WOMEN'S)

NOTES:

All marking lines shall be 2" wide and marked with a white non-toxic material which is not injurious to the eyes or skin.

Boundary lines are optional but should be 2" wide if marked.

Optional flag may be placed at the four corners or selected boundary points.

Fig. 6-16 Women's lacrosse playing field layout with flag and goal details (*U.S. Department of the Army, Office of the Chief of Engineers*).

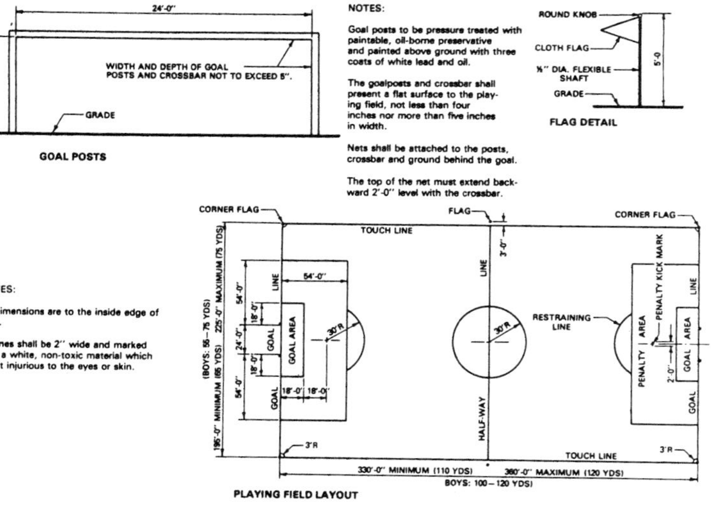
24'-0"
8'-0"
WIDTH AND DEPTH OF GOAL POSTS AND CROSSBAR NOT TO EXCEED 5".
GRADE
GOAL POSTS
NOTES:
Goal posts to be pressure treated with paintable, oil-borne preservative and painted above ground with three coats of white lead and oil.
The goalposts and crossbar shall present a flat surface to the playing field, not less than four inches nor more than five inches in width.
Nets shall be attached to the posts, crossbar and ground behind the goal.
The top of the net must extend backward 2'-0" level with the crossbar.
ROUND KNOB
CLOTH FLAG
¾" DIA. FLEXIBLE SHAFT
GRADE
5'-0
FLAG DETAIL
NOTES:
All dimensions are to the inside edge of lines.
All lines shall be 2" wide and marked with a white, non-toxic material which is not injurious to the eyes or skin.
CORNER FLAG
FLAG
CORNER FLAG
TOUCH LINE
3'-0"
LINE
HALF-WAY
54'-0"
18'-0"
24'-0"
18'-0"
54'-0"
GOAL LINE
GOAL
GOAL AREA
30'R
18'-0"
18'-0"
RESTRAINING LINE
PENALTY AREA
PENALTY KICK MARK
GOAL AREA
2'-0"
GOAL LINE
3'R
3'R
TOUCH LINE
195'-0" MINIMUM (65 YDS) 225'-0" MAXIMUM (75 YDS)
(BOYS: 55—75 YDS)
330'-0" MINIMUM (110 YDS) 360'-0" MAXIMUM (120 YDS)
BOYS: 100—120 YDS)
PLAYING FIELD LAYOUT
SOCCER (MEN'S AND BOYS')

Fig. 6-17 Men's and boys' soccer field layout with flag and goal post details (*U.S. Department of the Army, Office of the Chief of Engineers*).

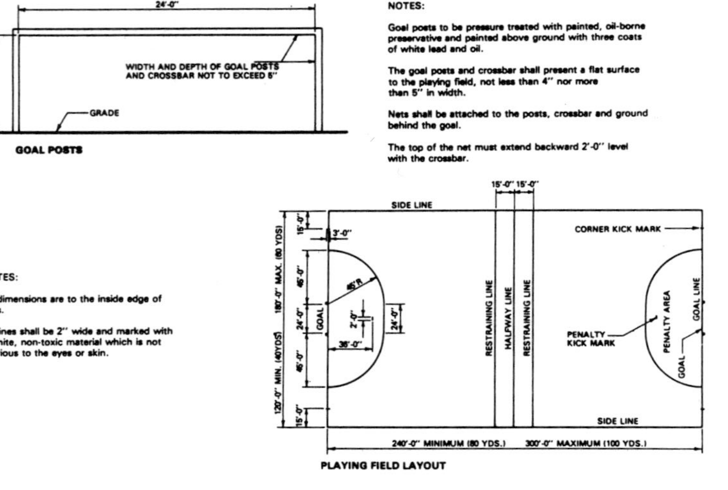

Fig. 6-18 Women's and girls' soccer field layout with goal post detail (*U.S. Department of the Army, Office of the Chief of Engineers*).

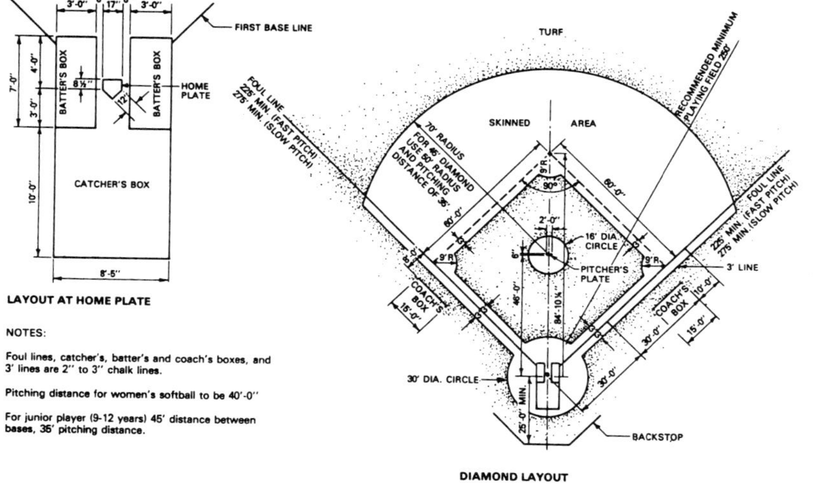

Fig. 6-19 Softball diamond (12-in) layout with home plate detail (*U.S. Department of the Army, Office of the Chief of Engineers*).

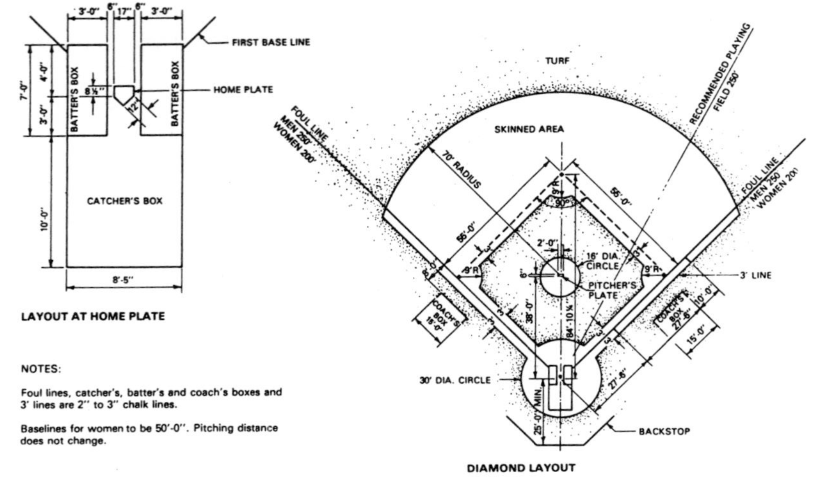

Fig. 6-20 Softball diamond (16-in) layout with home plate detail (*U.S. Department of the Army, Office of the Chief of Engineers*).

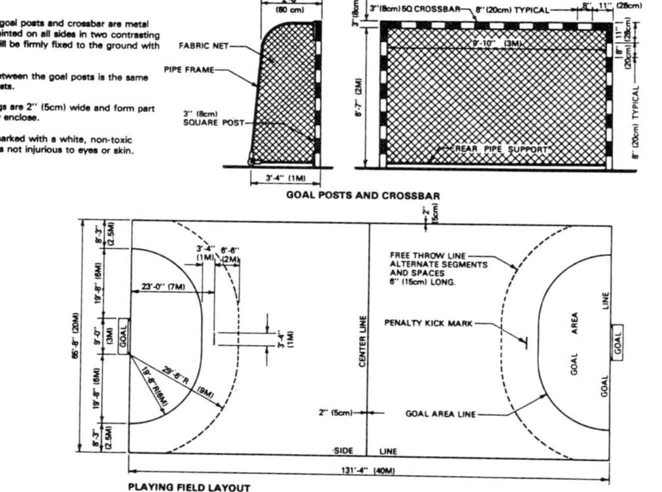

Fig. 6-21 Team-handball playing field layout with goal posts and crossbar (*U.S. Department of the Army, Office of the Chief of Engineers*).

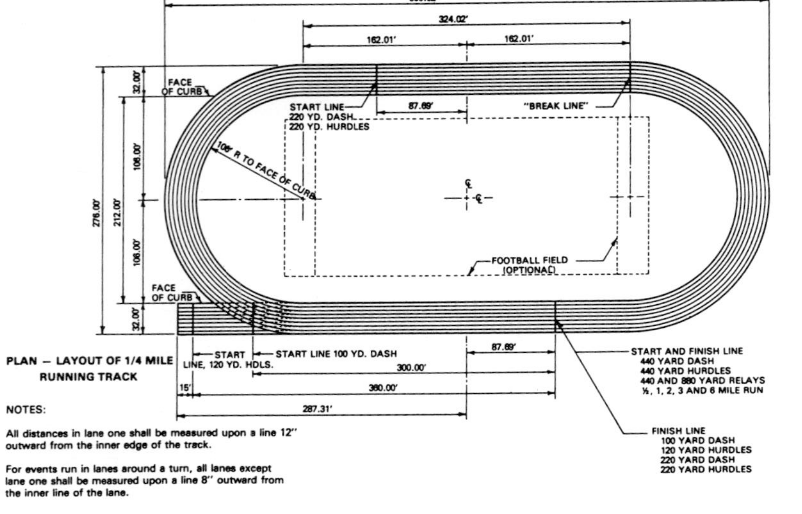

Fig. 6-22 One-quarter-mile running track (*U.S. Department of the Army, Office of the Chief of Engineers*).

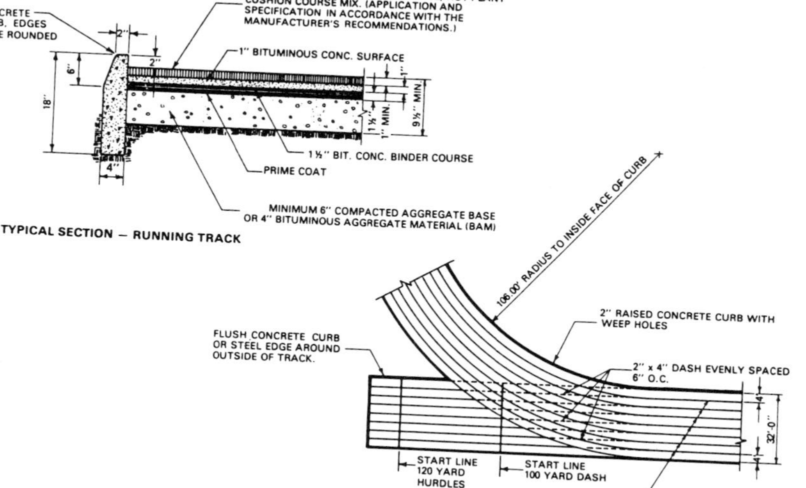

Fig. 6-23 Lane marking details (¼-mile running track) with typical track section (*U.S. Department of the Army, Office of the Chief of Engineers*).

NOTES:

All distances in lane one shall be measured upon a line 12" outward from the inner edge of the track.

For events run in lanes around a turn, all lanes except lane one shall be measured upon a line 8" outward from the inner line of the lane.

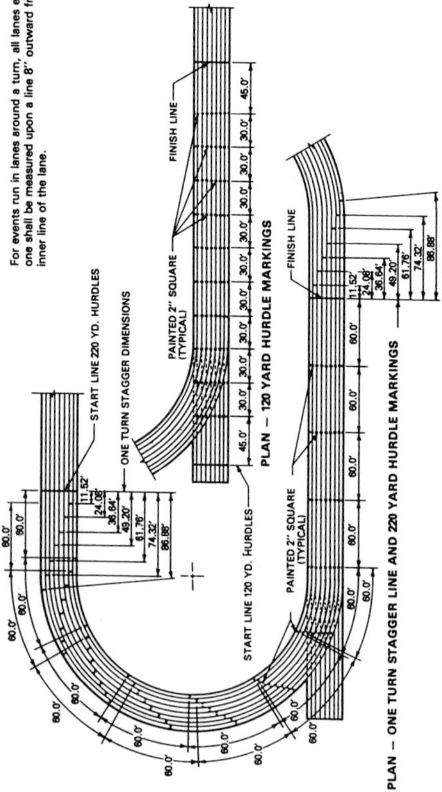

Fig. 6-24 Markings for 120- and 220-yard hurdles (*U.S. Department of the Army, Office of the Chief of Engineers*).

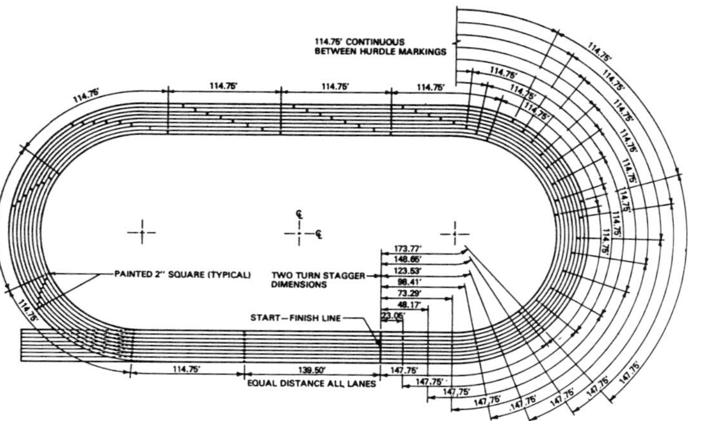

NOTES:

All distances in lane one shall be measured upon a line 12" outward from the inner edge of the track.

For events run in lanes around a turn, all lanes except lane one shall be measured upon a line 8" outward from the inner line of the lane.

Fig. 6-25 Two-turn stagger lines and 440-yard hurdle markings (¼-mile running track) (*U.S. Department of the Army, Office of the Chief of Engineers*).

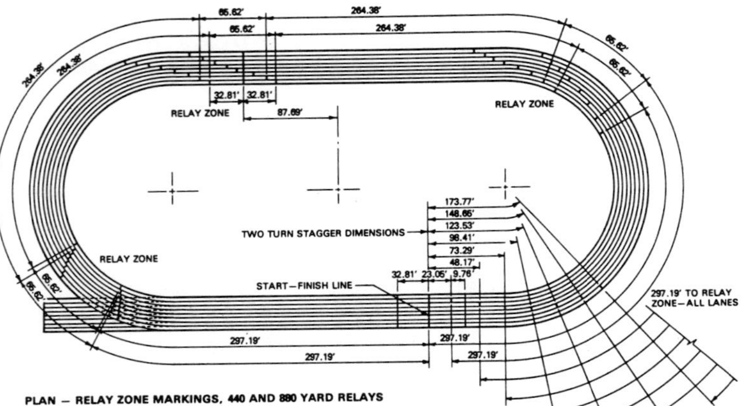

NOTES:

All distances in lane one shall be measured upon a line 12" outward from the inner edge of the track.

For events run in lanes around a turn, all lanes except lane one shall be measured upon a line 8" outward from the inner line of the lane.

Fig. 6-26 Relay zone markings—440- and 880-yard relays (¼-mile running track) (*U.S. Department of the Army, Office of the Chief of Engineers*).

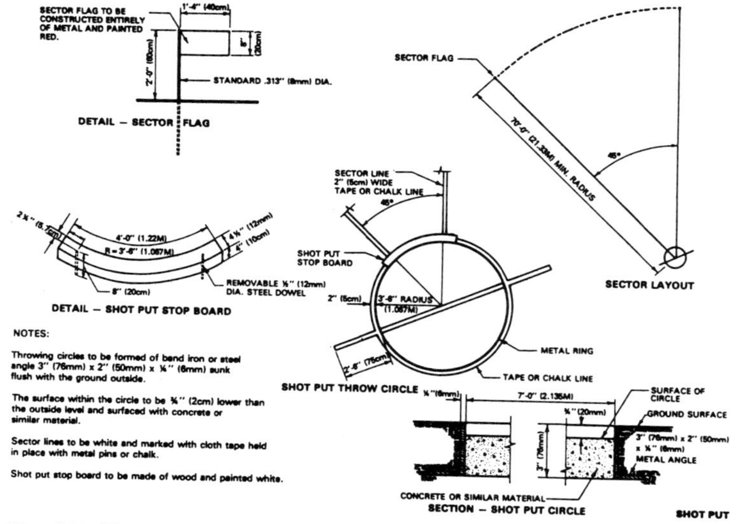

Fig. 6-27 Shot put sector layout with throw circle, stop board, and flag detail (*U.S. Department of the Army, Office of the Chief of Engineers*).

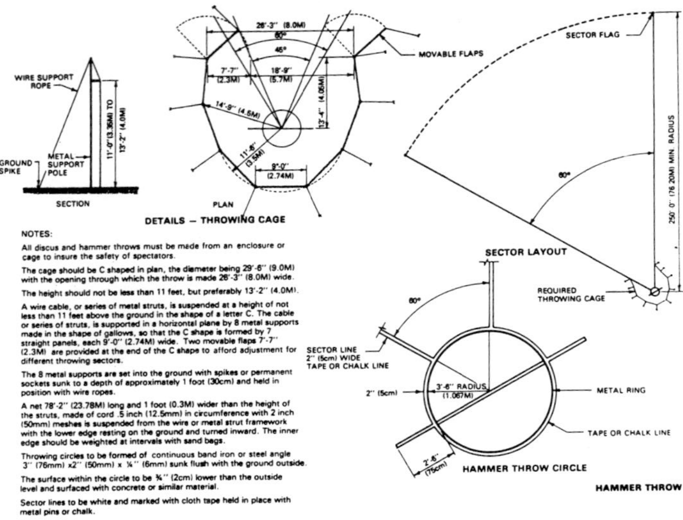

Fig. 6-28 Hammer throw sector layout with throw circle, throwing cage, and notes (*U.S. Department of the Army, Office of the Chief of Engineers*).

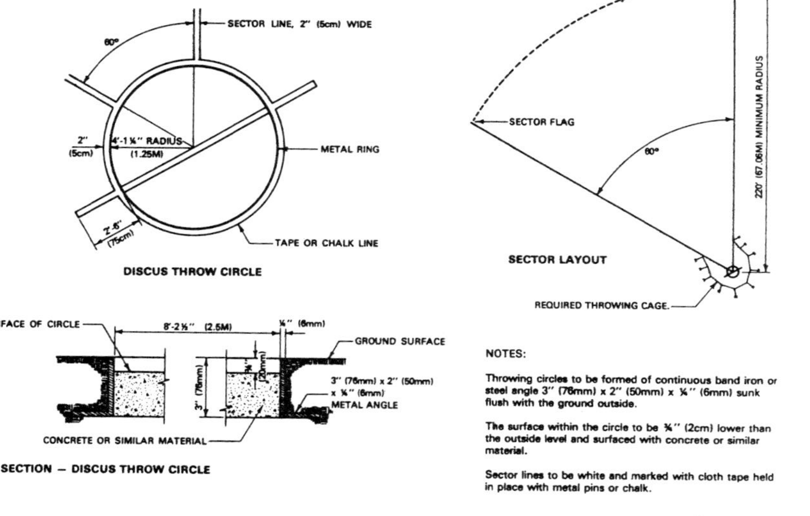

Fig. 6-29 Discus throw sector layout with throw circle and details (*U.S. Department of the Army, Office of the Chief of Engineers*).

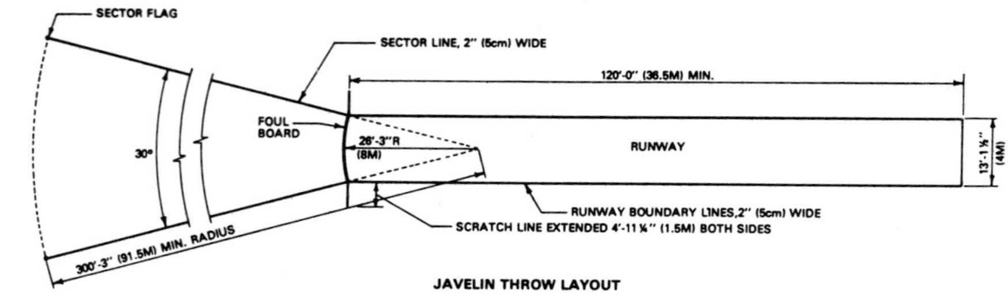

NOTES:

Sector lines to be white, 2" (5cm) wide and marked with cloth tape held in place with metal pins or chalk.

Runway may be either turf or bituminous material.

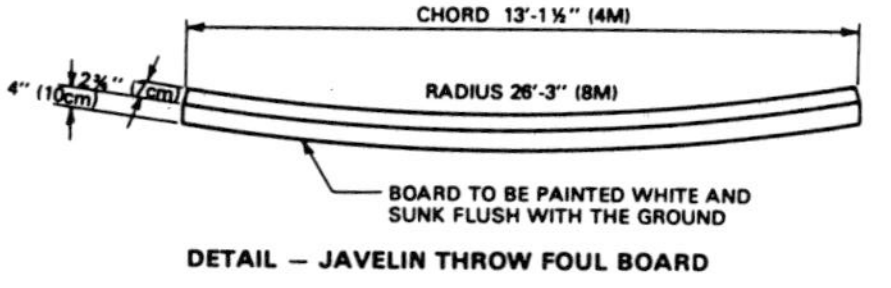

JAVELIN THROW

Fig. 6-30 Javelin throw layout (*U.S. Department of the Army, Office of the Chief of Engineers*).

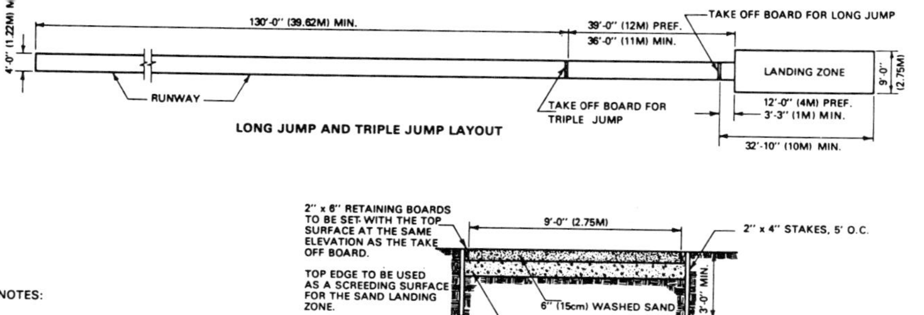

NOTES:

The edge of the takeoff board nearest the landing pit shall be the scratch, or foul line.

The construction and material of the runway shall be extended beyond the takeoff board to the nearer edge of the landing pit.

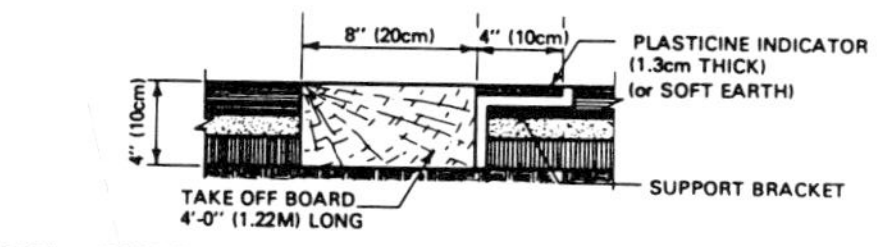

LONG JUMP AND TRIPLE JUMP

Fig. 6-31 Long jump and triple jump layout (*U.S. Department of the Army, Office of the Chief of Engineers*).

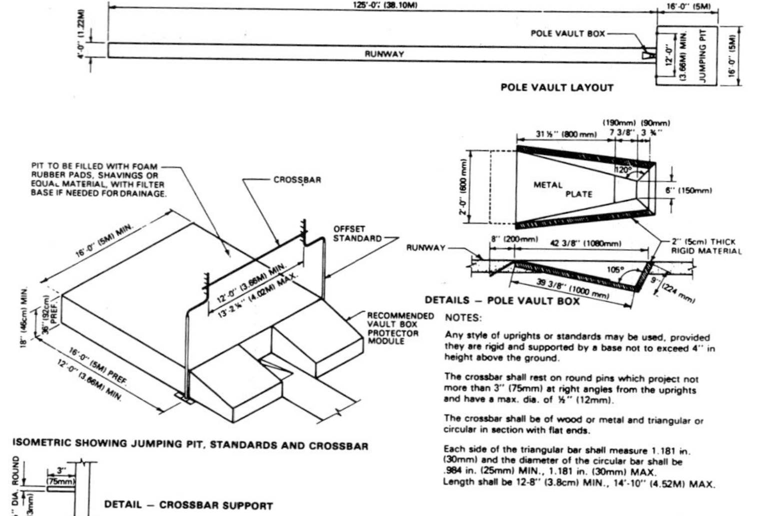

NOTES:

Any style of uprights or standards may be used, provided they are rigid and supported by a base not to exceed 4" in height above the ground.

The crossbar shall rest on round pins which project not more than 3" (75mm) at right angles from the uprights and have a max. dia. of ½" (12mm).

The crossbar shall be of wood or metal and triangular or circular in section with flat ends.

Each side of the triangular bar shall measure 1.181 in. (30mm) and the diameter of the circular bar shall be .984 in. (25mm) MIN., 1.181 in. (30mm) MAX. Length shall be 12-8" (3.8cm) MIN., 14'-10" (4.52M) MAX.

POLE VAULT

Fig. 6-32 Pole vault layout with details (*U.S. Department of the Army, Office of the Chief of Engineers*).

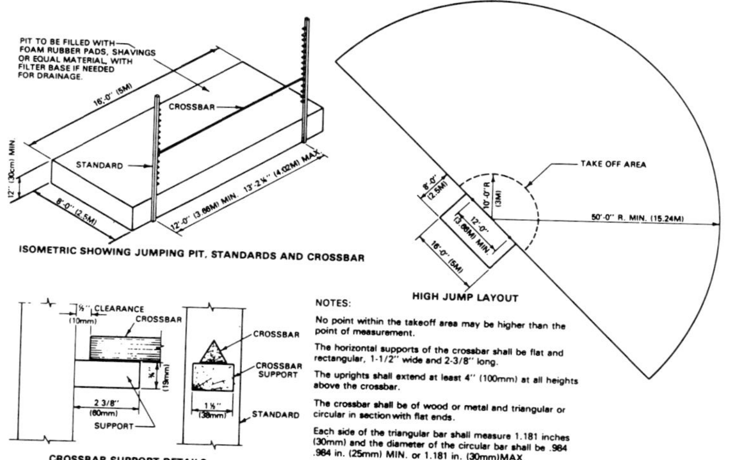
PIT TO BE FILLED WITH FOAM RUBBER PADS, SHAVINGS OR EQUAL MATERIAL, WITH FILTER BASE IF NEEDED FOR DRAINAGE.
16'-0" (5M)
CROSSBAR
STANDARD
12" (30cm) MIN.
8'-0" (2.5M)
12'-0" (3.66M) MIN. 13'-2¼" (4.02M) MAX.
ISOMETRIC SHOWING JUMPING PIT, STANDARDS AND CROSSBAR
TAKE OFF AREA
8'-0" (2.5M)
10'-0" R (3M)
12'-0" (3.66M) MIN.
16'-0" (5M)
50'-0" R. MIN. (15.24M)
HIGH JUMP LAYOUT
½" CLEARANCE (10mm)
CROSSBAR
¾" (19mm)
2 3/8" (60mm)
SUPPORT
CROSSBAR
CROSSBAR SUPPORT
1½" (38mm)
STANDARD
CROSSBAR SUPPORT DETAILS
NOTES:
No point within the takeoff area may be higher than the point of measurement.
The horizontal supports of the crossbar shall be flat and rectangular, 1-1/2" wide and 2-3/8" long.
The uprights shall extend at least 4" (100mm) at all heights above the crossbar.
The crossbar shall be of wood or metal and triangular or circular in section with flat ends.
Each side of the triangular bar shall measure 1.181 inches (30mm) and the diameter of the circular bar shall be .984 .984 in. (25mm) MIN. or 1.181 in. (30mm)MAX.
HIGH JUMP

Fig. 6-33 High jump layout with details (*U.S. Department of the Army, Office of the Chief of Engineers*).

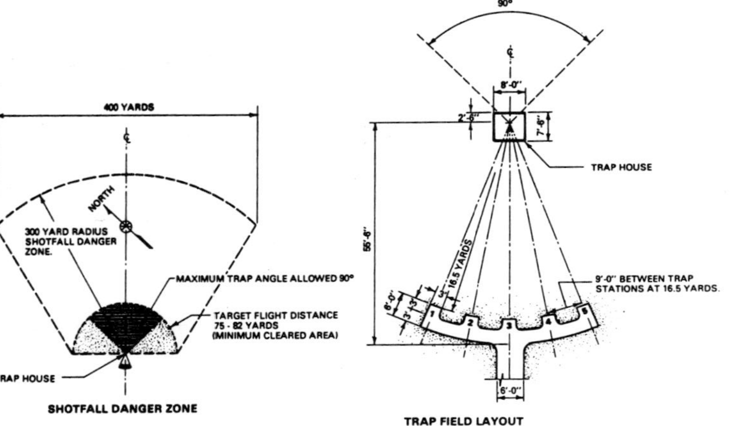

Fig. 6-34 I.S.U. automatic trap field layout with shotfall danger zone (*U.S. Department of the Army, Office of the Chief of Engineers*).

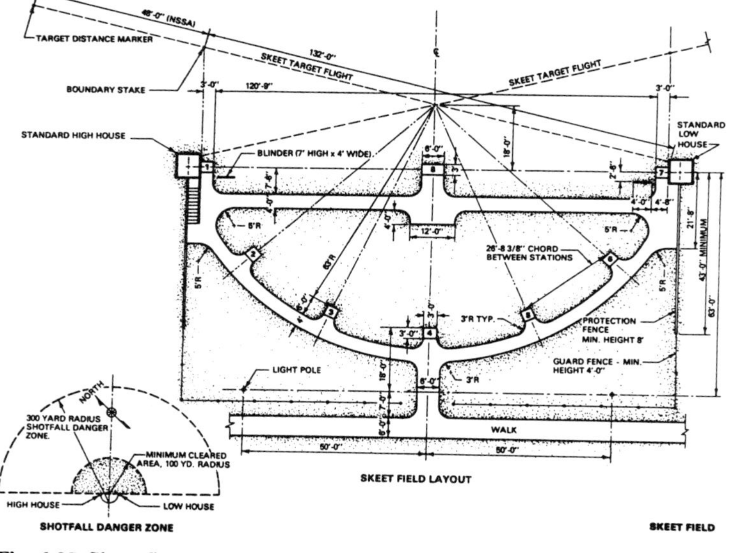

Fig. 6-35 Skeet field layout with shotfall danger zone (*U.S. Department of the Army, Office of the Chief of Engineers*).

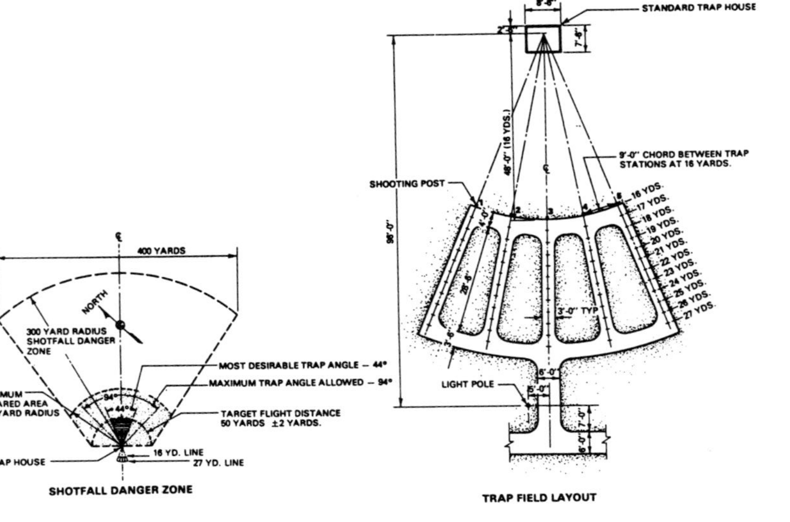

Fig. 6-36 Trap field layout with shotfall danger zone (*U.S. Department of the Army, Office of the Chief of Engineers*).

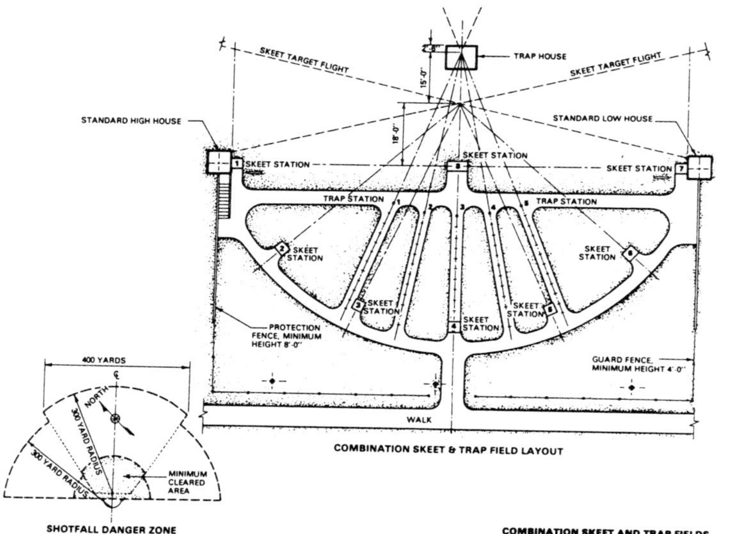

Fig. 6-37 Combination skeet and trap field layout with shotfall danger zone (*U.S. Department of the Army, Office of the Chief of Engineers*).

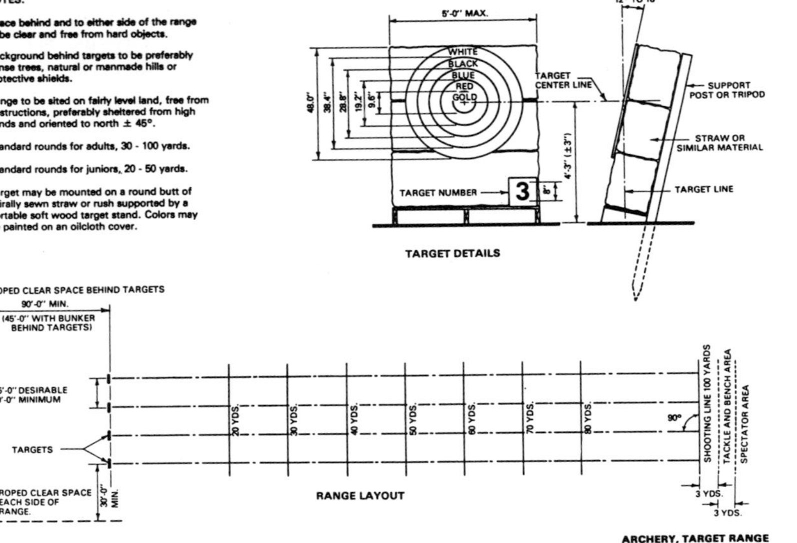
NOTES:
Space behind and to either side of the range to be clear and free from hard objects.
Background behind targets to be preferably dense trees, natural or manmade hills or protective shields.
Range to be sited on fairly level land, free from obstructions, preferably sheltered from high winds and oriented to north ± 45°.
Standard rounds for adults, 30 - 100 yards.
Standard rounds for juniors, 20 - 50 yards.
Target may be mounted on a round butt of spirally sewn straw or rush supported by a portable soft wood target stand. Colors may be painted on an oilcloth cover.
5'-0" MAX.
WHITE
BLACK
BLUE
RED
GOLD
48.0"
38.4"
28.8"
19.2"
9.6"
TARGET CENTER LINE
4'-3" (±3")
8"
TARGET NUMBER
3
12° TO 18°
SUPPORT POST OR TRIPOD
STRAW OR SIMILAR MATERIAL
TARGET LINE
TARGET DETAILS
ROPED CLEAR SPACE BEHIND TARGETS
90'-0" MIN.
(45'-0" WITH BUNKER BEHIND TARGETS)
15'-0" DESIRABLE
10'-0" MINIMUM
TARGETS
ROPED CLEAR SPACE EACH SIDE OF RANGE.
30'-0" MIN.
20 YDS.
30 YDS.
40 YDS.
50 YDS.
60 YDS.
70 YDS.
80 YDS.
90°
SHOOTING LINE 100 YARDS
TACKLE AND BENCH AREA
SPECTATOR AREA
3 YDS.
3 YDS.
RANGE LAYOUT
ARCHERY, TARGET RANGE

Fig. 6-38 Archery target range with details (*U.S. Department of the Army, Office of the Chief of Engineers*).

NOTES

NOTES

Section Seven

■

ELECTRICITY

E. J. POLASEK, P.E.

Consulting Engineer

7-1 ELECTRICAL UNITS

The commonest electrical units are listed in Table 7-1.

Table 7-1 Electrical Units

Units	*Measure of*	*Symbol*	*Common Formulas, Notes*
Volt (V)	Electrical force	E, e	$E = IR \quad E = L\frac{di}{dt} - jIX_L$
		V, v	$E = 1/C \int Idt = -jIX_c$
Ampere (A)	Electric current	I, i	$I = E/R$ (etc.) $= I(Q/T)$
Ohm (Ω)	Resistance	R, r	
	Reactance	X_L	$X_L = 2\pi fL$
	Reactance	X_C	$X_C = 1/2\pi fC$
	Impedance	Z	$Z = \sqrt{R^2 + j(X_L - X_C)^2}$
Farad (F)	Capacitance	C	$C = \frac{\text{Dielectric } K \times \text{area}}{0.113 \times 10^{14}}$ spacing for parallel plates, dimensions in cm
Henry, (H)	Inductance	L	$L = -N(\text{turns}) \frac{d\phi}{dt}$
Watt (W) Kilowatt (kW	Power	P, p	$P = EI \times \cos\theta = I^2R$ (watts) kW = W × 10^3
Wattsecond (Ws) Watthour (Wh) Kilowatt-hour (kWh)	Work, energy	W	$W = \frac{1}{2} I^2L = \frac{1}{2} E^2C$ (wattseconds) Wh = W (watts) × t(hours) kW = kW × t(hours)
Hertz (Hz)	Frequency	f	$f = \omega \div 2\pi$
Power factor*	Ratio, real to apparent power	PF	PF $= I^2R/EI$ $= \cos\theta$

NOTE: International practical units differ from U.S. practical units by about 0.03%.

*Not a unit, but convenient to list with units.

7-2 ELECTRIC CIRCUITS

Ohm's Law

$$E = IZ = I[R + (jX_L - jX_C)]$$

In the dc case, this reduces to $E = IR$, hence $I = E/R$ and $R = E/I$.
Kirchhoff's Laws (1) The algebraic sum of the voltages around a closed loop is zero. (2) The algebraic sum of the currents entering and leaving a point is zero. (See Fig. 7-1.)

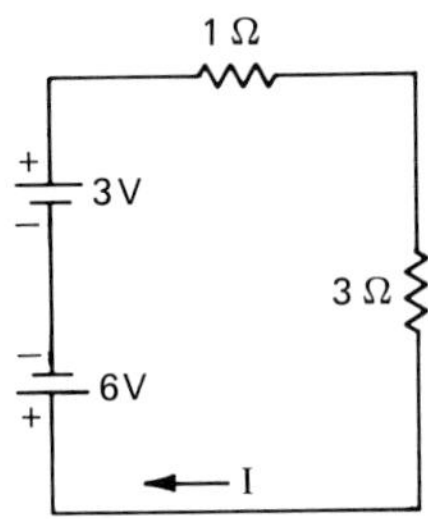

Fig. 7-1 DC series circuit.
$E = 6V - 3V = 3V$
$R = 1\Omega + 3\Omega = 4\Omega$
$I = 3A \div 4A = 0.75A$
$P = I^2R = 2.25W$

In Fig. 7-1, note that I is constant throughout the circuit while E is divided among the elements.
Series Circuits Impedances and voltages add algebraically. (See Fig. 7-2.)

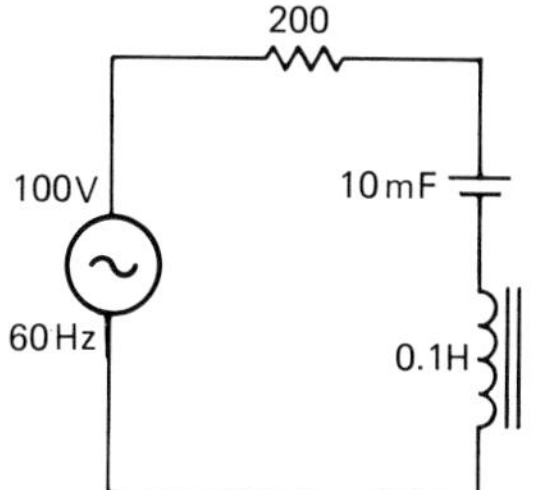

Fig. 7-2 AC series circuit.

$$
\begin{aligned}
X_c &= \frac{1}{2\pi fC} \\
&= -j265 \\
X_L &= 2\pi fL \\
&= +j377 \\
Z &= 200 + j377 - j265 \\
&= 229 \angle 29.2^\circ \\
E &= 100 \angle 0^\circ \quad \text{V} \\
I &= E \,/\, Z = 0.437 \angle -29.2^\circ \quad \text{A} \\
P &= I^2R = 38.2 \text{ W} \\
\text{PF} &= \cos \angle -29.2^\circ \\
&= 38.2/(100 \times 0.437) = 0.87
\end{aligned}
$$

Parallel Circuits Conductances, the reciprocal of impedances, add algebraically.

In Fig. 7-3, note that E is constant for all the elements while I is divided among the elements. Fig. 7-4 is more complex.

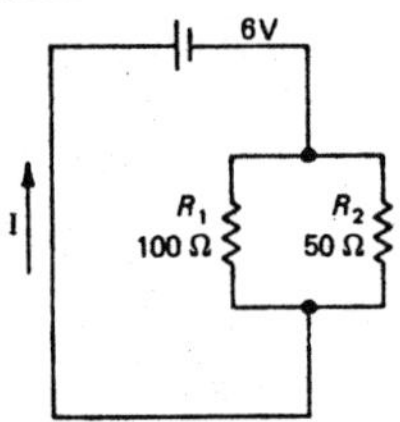

Fig. 7-3 Parallel circuit.

$$\frac{1}{R_{\text{equivalent}}} = \frac{1}{R_1} + \frac{1}{R_2} = 0.01\ \text{S} + 0.02\ \text{S} = 0.03\ \text{S}$$

$$R_{\text{equivalent}} = 33.3\ \Omega$$

$$R_e = \frac{R_1R_2}{R_1 + R_2} = \frac{R_1R_2R_3}{R_1R_2 + R_2R_3 + R_3R_1}\ \text{etc} \ldots$$

$$I = 6/33.3 = 0.18 \qquad I_{R_1} = 0.06 \qquad I_{R_2} = 0.12$$

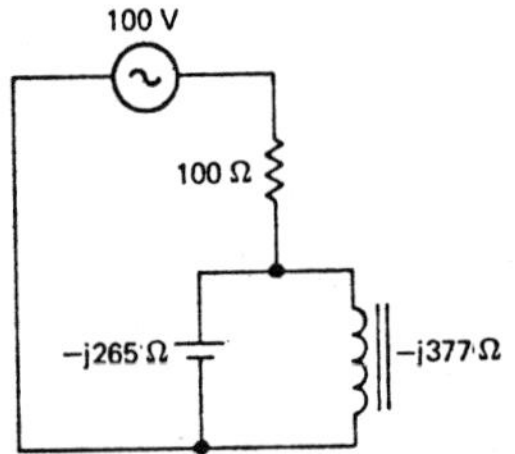

Fig. 7-4 AC series-parallel circuit.

$$Z = \sqrt{R^2 + X_e^2} \qquad X_e = \frac{(0-j377)\ (0-j265)}{j377 + j265}$$

$$= \sqrt{100^2 \pm j893^2}$$

$$= 898.6 \angle 83.6^\circ$$

$$I = 100/898.8 \angle 83.6^\circ = 0.111 \angle -83.6^\circ$$

$$E_R = 100 \times 0.111 \angle -83.6^\circ$$

$$E_L = E_C = 100 - 11.1 \angle -83.6^\circ$$

$$= 98.77 - j11.0 = 99.4\ \text{V}$$

7-3 THREE-PHASE POWER

Commercial power is normally delivered as a single-phase (1ϕ), 3 ϕ, or a combination of both. AC power is ideally of the sine-wave form (Fig. 7-5), and since power is a function of E^2, the *effective* or *root mean square* (rms) value is usually given:

$$E_{\text{eff}} = E_{\text{peak}} \div \sqrt{2} = 0.707\, E_{\text{peak}} \qquad \text{volts rms}$$

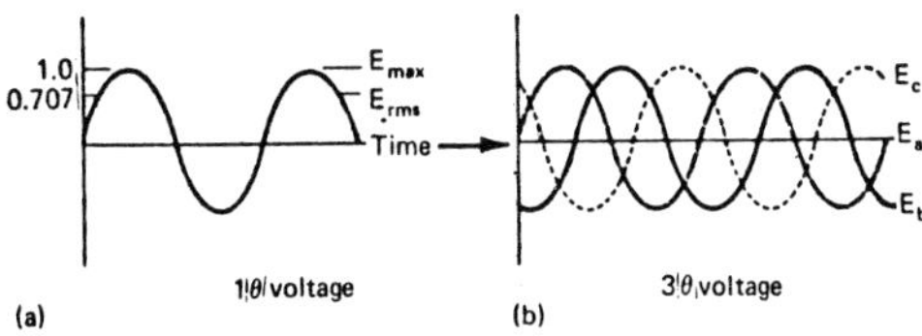

Fig. 7-5 Graphs of ac voltages vs. time; (*a*) 1ϕ voltage; (*b*) 3 ϕ voltage

A 3ϕ generator could be constructed by mounting three 1ϕ units on a common shaft, each displaced 120° electrically. In practice, three sets of windings are installed on one frame. If the three phases are connected as in Fig. 7-6, the system is said to be star- or wye-(Y)-connected

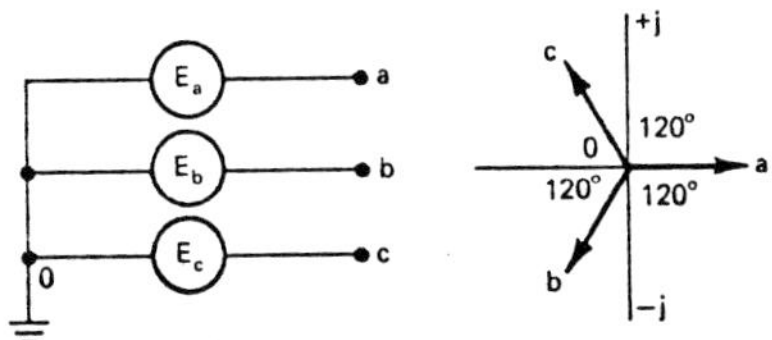

Fig. 7-6 Y-connected generator.

$$E_{a_0} = E_a \angle 0^\circ = E_a(1 + j0)$$
$$E_{b_0} = E_a \angle -120^\circ = E_a(-0.5 - j0.866)$$
$$E_{c_0} = E_a \angle -240^\circ = E_a(-0.5 + j0.866)$$

and

$$E_{ab} = -E_a + E_b$$
$$= E_a(-1 + j_0 + 0.5 - j0.866)$$
$$= \sqrt{3}E_a < -150^\circ$$

In general, line-to-line voltage $= \sqrt{3} \times$ line-to-neutral.

If the phases are connected as in Fig. 7-7, the system is said to be delta- or mesh-connected. From Fig. 7-6 or 7-7 we can get the equation for power P:

$$P = P_a + P_b + P_c$$
$$= (E_a I_a + E_b I_b + E_c I_c) \cos\theta$$

Assuming balanced voltages and balanced load, from Fig. 7-6

$P = 3EI \cos\theta$ where E is line-to-neutral

or

$P = \sqrt{3}\, E_{\phi\phi} I \cos\theta$ where $E_{\phi\phi}$ is line-to-line.

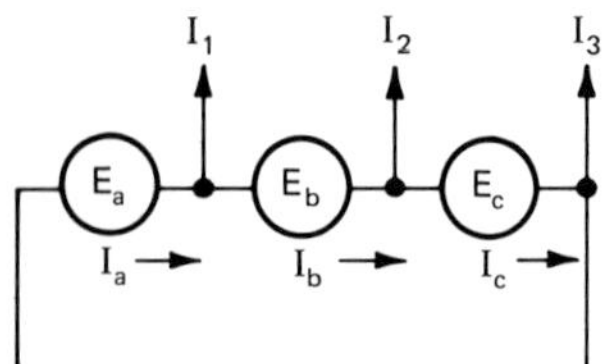

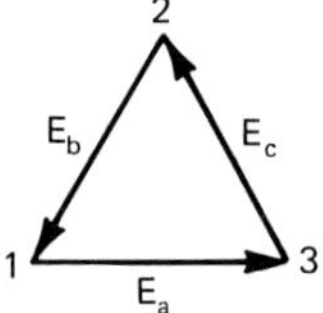

Fig. 7-7 Delta-connected generator.

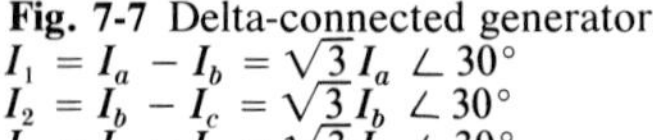

$I_1 = I_a - I_b = \sqrt{3}\, I_a \angle 30^\circ$
$I_2 = I_b - I_c = \sqrt{3}\, I_b \angle 30^\circ$
$I_3 = I_c - I_a = \sqrt{3}\, I_c \angle 30^\circ$

The same result may be obtained from Fig. 7-7 by complex algebra, and the expression is valid for either case.

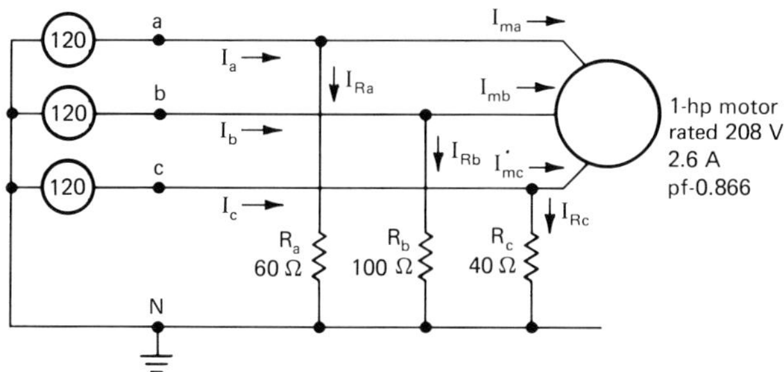

Fig. 7-8 Unbalanced 3ϕ circuit.

Fig. 7-8 shows a typical circuit with a combination of 1ϕ and 3ϕ loads. The source is Y-connected with a grounded neutral. Note the current flow in the neutral due to load unbalance.

$$I_{m_a} = 2.6\,(0.866 - j0.5) = 2.25 - j1.3$$
$$I_{m_b} = I_{m_a} \angle -120^\circ = -2.25 - j1.3$$
$$I_{m_c} = I_{m_a} \angle -240^\circ = 0 + j2.6$$
$$I_{m_a} + I_{m_b} + I_{m_c} = 0$$
$$I_{r_a} = 120 \angle 0^\circ \div 60 = 2 \angle 0^\circ = 2 + j0$$
$$I_{r_b} = 120 \angle -120^\circ \div 100 = 1.2 \angle -120^\circ = -0.6 - j1.04$$
$$I_{r_c} = 120 \angle -240^\circ \div 40 = 3 \angle -240^\circ = -1.5 + j2.60$$
$$I_N = I_{r_a} + I_{r_b} + I_{r_c} = -0.1 + j1.56 = 1.57 \angle -266^\circ$$

In the case of the motor (balanced load), it is immaterial to the external circuit whether the motor is Y-connected or delta-connected internally. In general, a Y-connected circuit may be replaced with an equivalent delta connection and vice versa if there is no neutral involved.

To replace a delta by a star (Fig. 7-9), let

$$\alpha = \frac{bc}{a + b + c} \qquad \beta = \frac{ca}{a + b + c} \qquad \gamma = \frac{ab}{a + b + c}$$

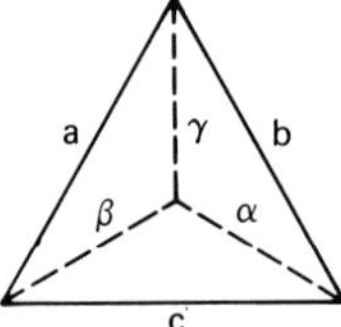

Fig. 7-9 Star-delta transformation.

To replace a star by a delta, let

$$a = \beta + \gamma + \frac{\beta\gamma}{\alpha} \qquad b = \gamma + \alpha + \frac{\gamma\alpha}{\beta} \qquad c = \alpha + \beta + \frac{\alpha\beta}{\gamma}$$

7.4 ELECTRIC POWER TRANSMISSION

Standard transmission voltages are generally based on multiples of the nominal 115-V utilization in North America and on the nominal 220/380Y voltage system used elsewhere in the world. Thus, in the United States, such voltages are 23, 34.5, 69, 138, 161, 230, or 345 kV, etc.; elsewhere one finds 88, 110, 154, or 380 kV, etc.

All ac transmission is 60-Hz 3ϕ in the United States, and generally 50-Hz 3ϕ elsewhere. High-voltage direct current is gaining prominence with the advent of more reliable and less costly rectifiers and inverters. Since *R* (Fig. 7-10) is usually much less than

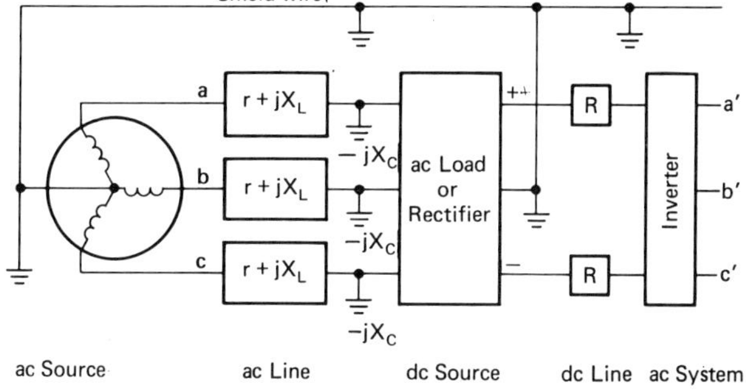

Fig. 7-10 Simplified transmission system.

$r + jX_L$, the dc line can carry more power; also, since systems *abc* and *a′b′c′* (Fig. 7-10) need not be synchronized, system stability is improved.

Table 7-2 Typical Transmission-Line Data

Voltage, kV	*Length of insulator string, ft**	*Phase spacing, ft*	*Power capability, kW × 10³*	*Cost \$1000's per mi†*	*Cost \$1000's per MW-mi*
34.5	1.5	6	18–60	34	1200
69	2.5	8	40–120	50	625
115	4.0	16	80–200	90	540
161	6.0	20	140–280	110	470
230	10	25	240–420	140	390
345	12	30	600–1800	300	250

*Based on 6 in per suspension insulator.
†1978 costs, not including right-of-way.

Underground transmission typically costs about ten times as much as overhead (See Table 7-2 and Fig. 7-11). At and below 69 kV, solid

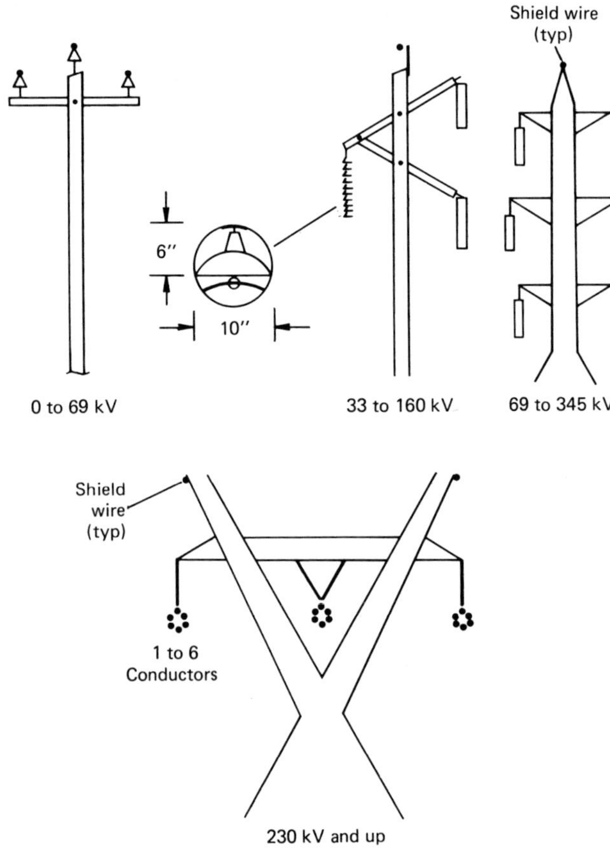

Fig. 7-11 Typical transmission structures.

polyethylene is the insulation of choice. Above 69 kV, most cables are installed within steel pipes which are filled with insulating oil under pressure. Alternatively, the bare conductor may be held centered in a large-bore steel pipe by porcelain insulators; the pipe is filled with sulfur hexafluoride gas under pressure. This last is used up to 345 kV in ultracompact substations where its extremely high cost is warranted.

7-5 OVERHEAD DISTRIBUTION

Almost worldwide, distribution is 3ϕ grounded Y with 1ϕ extensions. Delta systems are becoming so uncommon except for in-plant use that they are ignored in this book. Typical voltages are: 2.4/4.16, 4.8/8.3, and 6.6/11.4 kV (mostly in Europe); 7.2/12.47, 8.66/15 (common in Asia) and 14.4/24.9 kV. Voltages such as 19.9/34.5 and 39.9/69 kV are occasionally used in areas of sparse load, such as the American west. The legal requirements governing construction are given by the Electrical Code of the state concerned, while excellent construction standards are set out in REA Forms 803 and 804, available from the U.S. Government Printing Office.

To find sag in feet *(d)* or tension in pounds *(t)* in a dead-end span of *s* feet for a conductor weighing *w* pounds per foot,

$$t = \frac{s^2 w}{8d}$$

Conductor tensions under *fully loaded conditions* (See NEC) are usually restricted to 50 percent of ultimate. This condition normally governs for spans over 300 ft in heavy-loading districts or 500 ft in light-loading districts. Tensions under *initial stringing conditions* are usually restricted to 33.3 percent of ultimate. This condition normally governs for spans of about 200 to 300 ft.

Under *final unloaded conditions,* tensions should not exceed 18 percent of ultimate. This governs in light-loading areas.

To adjust for temperature, the following approximation may be used:

t (°F)	0	32	60	90	120
Sag (ft)	77	89	100	111	120

The clearances in Table 7-3 as well as many other unlisted provisions are subject to numerous exceptions and variations. Those in the table can be considered a general guide only.

TABLE 7-3 Basic Minimum Clearances (feet) for Power and Other Electric Lines

		Power Lines		
Surface Crossed Over	*Telephone Guys, etc.*	*0–750 V*	*750 V to 15 kV*	*15–50 kV*
Railroad tracks	27	27	28	30
Roads, streets, and other areas subject to truck traffic	18	18	20	22
Areas not subject to truck traffic	10	15	20	22
Other lands	18	18	20	22
Pedestrians only	15	15	15	17
Urban streets or alleys	18	18	20	22

TABLE 7-4 Conductor Characteristics

*Size, AWG**	*Diameter, in*	*Ultimate strength, lb*	*Weight, lb per 1000 ft*	*Resistance Ω per 1000 ft*	*Current-carrying capacity*	*X_a, Ω per 1000 ft*
COPPER CONDUCTORS (Medium-hard drawn)						
36	0.0050	1.16	0.0757	429	0.025	as magnet wire
24	0.0201	17.5	1.12	26.6	0.40	as magnet wire
18	0.0403	67.6	4.92	6.61	1.6	as magnet wire
14	0.0641	167	12.4	2.61	4.0	as magnet wire
10	0.1019	410	31.4	1.03	10	as magnet wire
6	0.1620	1010	79.4	0.409	100	0.142
2	0.2576	2450	201	0.162	175	0.135
3/0	0.464	5890	518	0.0652	320	0.117
500 MCM	0.811	17,550	1544	0.0219	630	0.100
1000 MCM	1.151	34,400	3088	0.0109	965	0.090
ALL-ALUMINUM CONDUCTORS						
6	0.184	529	24.6	0.661	80	0.119
4	0.232	826	38.2	0.416	100	0.117
2	0.292	1266	62.3	0.261	135	0.109
1/0	0.372	1538	99.1	0.207	185	0.103
4/0	0.528	3560	199	0.0820	290	0.095
336.4 MCM	0.666	5940	316	0.0520	390	0.088
477 MCM	0.795	8090	479	0.0364	480	0.085
795 MCM	1.026	13,770	746	0.0218	670	0.079

TABLE 7-4 Conductor Characteristics *(cont.)*

*Size, AWG**	*Diameter, in*	*Ultimate strength, lb*	*Weight, lb per 1000 ft*	*Resistance Ω per 1000 ft*	*Current-carrying capacity*	*X_a, Ω per 1000 ft†*
1192 MCM	1.258	21,000	1119	0.0145	900	0.074
1590 MCM	1.454	28,100	1493	0.0109	1150	0.070
ALUMINUM CONDUCTOR, STEEL REINFORCED (ACSR)						
6	0.198	1170	36.1	0.658	100	0.127
4	0.250	2288	67.0	0.413	140	0.126
2	0.325	2790	91.3	0.260	180	0.126
1/0	0.398	4280	145	0.163	230	0.124
4/0	0.556	8420	291	0.0816	340	0.110
336.4 MCM	0.721	14,050	463	0.0518	530	0.085
477 MCM	0.846	23,300	747	0.0366	670	0.081
795 MCM	1.093	31,200	1094	0.0219	910	0.076
1192 MCM	1.338	43,10	1533	0.0147	1160	0.071
1590 MCM	1.545	56,000	2046	0.0110	1380	0.068

*The AWG system is based on a ratio of 1.123 between adjacent diameters, or 1.261 between adjacent areas. Thus the area doubles for each three sizes ($1.261^3 = 2.005$). Values are accurate for one stranding only, and must be considered approximations for other strandings.

†$X_L = X_a + X_d$, where X_a is reactance at 1-ft spacing.

For other spacings:	1 ft	2 ft	4 ft	8 ft	16 ft	32 ft
X_d =	0	0.016	0.032	0.048	0.064	0.080

7-6 UNDERGROUND DISTRIBUTION

Conduit and cable systems are used within buildings and in urban areas. The conduit is often plastic or fiber, encased in concrete. Unless at least 2 in of concrete cover a duct, it must be of rigid steel for protection. Splices and terminations are done in manholes.

In pulling, tensions should not exceed 0.008 lb per cm^2 of area for copper conductors, or 0.006 lb per cm^2 for aluminum. Coefficient of friction is often taken as 0.5, but allowance must be made for bends.

TABLE 7-5 Conductor Characteristics For Conduit and Cable Systems

Allowable Ampacities of Insulated Triplexed or Three Single Conductor Aluminum Cables in Underground Raceways

Based on Conductor Temperature of 90°C, Ambient Earth Temperature of 20°C, 100% Load Factor and Thermal Resistance (RHO) of 90

One Circuit Size AWG-MCM	2001-5000 Volts Ampacity	5001-35,000 Volts Ampacity
8	50	–
6	66	70
4	86	91
2	115	120
1	130	135
1/0	150	155
2/0	170	175
3/0	195	200
4/0	225	230
250	250	250
350	305	305
500	370	370
750	470	455
1000	545	525
Three Circuit Size		
8	44	–
6	57	60
4	74	77
2	96	100
1	110	110
1/0	125	125
2/0	145	145
3/0	160	165
4/0	185	185
250	205	200
350	245	245
500	295	290
750	370	355
1000	425	405
Six Circuit Size		
8	38	–
6	48	50
4	62	64
2	80	80
1	91	90
1/0	105	105
2/0	115	115
3/0	135	130
4/0	150	150
250	165	165
350	195	195
500	240	230
750	290	280
1000	335	320

SOURCE: Reproduced with permission from 70–1978, *National Electrical Code*®, Copyright © 1977, National Fire Protection Association, Boston, MA 02210.

Direct-buried cable is used in suburban or rural areas where installation may be accomplished by trenching or plowing. Protection is provided by a concentric neutral consisting of tinned copper conductors wrapped spirally. This neutral may corrode in soils of low or variable resistivity, and cathodic protection should be supplied in these areas. Splices are normally direct-buried, but terminations are above ground, either in apparatus designed for the purpose, or on a pole or structure.

TABLE 7-6 Conductor Characteristics for Direct-Buried Systems

Allowable Ampacities of an Insulated Single Aluminum Conductor, Direct Buried

Based on Conductor Temperature of 90°C, Ambient Earth Temperature of 20°C, 100% Load Factor, Thermal Resistance (RHO) of 90, and 7½ Inch Spacing Between Conductor Center Lines, and 24 Inch Spacing Between Circuits

Conductor Size AWG-MCM	2001-5000 Volts Ampacity	5001-35,000 Volts Ampacity
One Circuit-3 Conductors		
8	85	—
6	110	100
4	140	130
2	180	165
1	205	185
1/0	230	215
2/0	265	245
3/0	300	275
4/0	340	315
250	370	345
350	445	415
500	540	510
750	665	635
1000	780	740
Two Circuits-6 Conductors		
8	80	—
6	100	95
4	130	125
2	165	155
1	190	175
1/0	215	200
2/0	245	225
3/0	275	255
4/0	310	290
250	340	320
350	410	385
500	495	470
750	610	580
1000	710	680

SOURCE: Reproduced with permission from 70–1978, *National Electrical Code®*, Copyright © 1977, National Fire Protection Association, Boston, MA 02210.

7-7 TRANSFORMERS

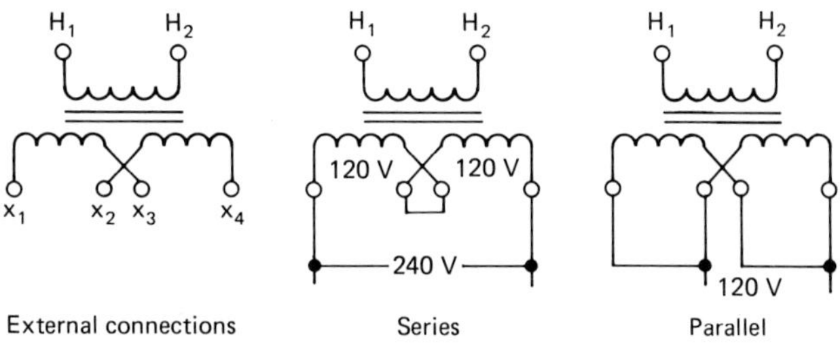

Fig. 7-12 Single-phase transformer connections.

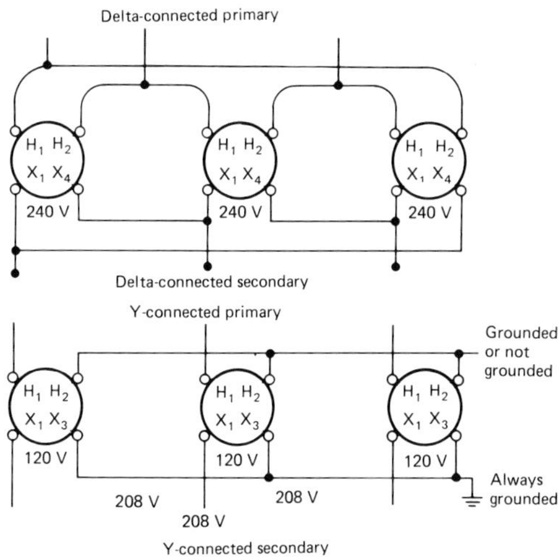

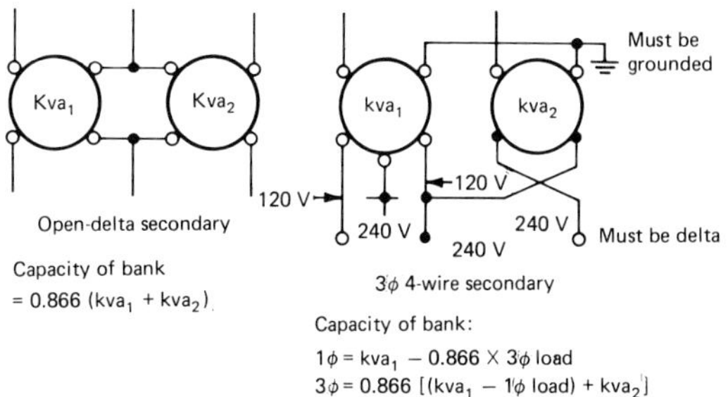

Fig. 7-13 Three-phase transformer banks.

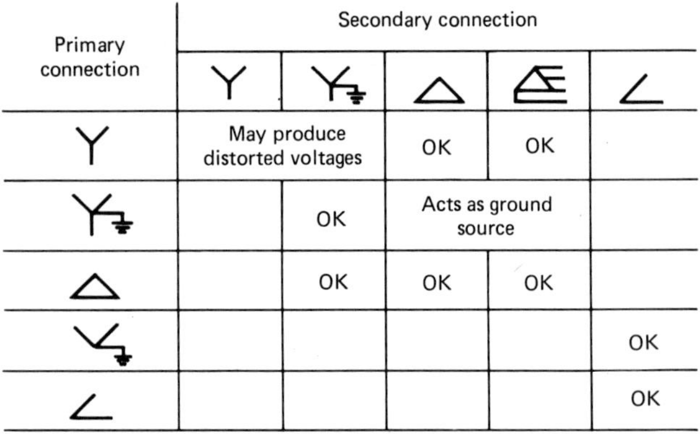

Primary connection	Secondary connection				
	Y	Y⏚	Δ	Δ⏚	∠
Y	May produce distorted voltages		OK	OK	
Y⏚		OK	Acts as ground source		
Δ		OK	OK	OK	
∠⏚ (open Y grounded)					OK
∠					OK

Fig. 7-14 Permitted connections.

TABLE 7-7 Typical Transformer Data

Specifications	*Nameplate kVA*						
	3	*5*	*10*	*25*	*50*	*100*	*250*
Weight, lb	150	175	225	350	700	1150	1900
No-load loss, %	1.0	0.8	0.6	0.5	0.4	0.4	0.4
Full-load loss, %	3.0	2.7	2.3	2.0	1.7	1.6	1.4

7-8 MOTORS

TABLE 7-8 Locked-rotor indicating code letters

Code Letter	Kilovolt-Amperes per Horsepower with Locked Rotor
A	0 – 3.14
B	3.15 – 3.54
C	3.55 – 3.99
D	4.0 – 4.49
E	4.5 – 4.99
F	5.0 – 5.59
G	5.6 – 6.29
H	6.3 – 7.09
J	7.1 – 7.99
K	8.0 – 8.99
L	9.0 – 9.99
M	10.0 – 11.19
N	11.2 – 12.49
P	12.5 – 13.99
R	14.0 – 15.99
S	16.0 – 17.99
T	18.0 – 19.99
U	20.0 – 22.39
V	22.4 – and up

SOURCE: Reproduced with permission from *National Electrical Code 1978* (NFPA 70), National Fire Protection Association, 470 Atlantic Avenue, Boston, MA 02210.

TABLE 7-9 Full-Load Current, Three-Phase ac Motors

	Induction Type Squirrel-Cage and Wound-Rotor Amperes					Synchronous Type †Unity Power Factor Amperes			
HP	**115V**	**230V**	**460V**	**575V**	**2300V**	**230V**	**460V**	**575V**	**2300V**
½	4	2	1	.8					
¾	5.6	2.8	1.4	1.1					
1	7.2	3.6	1.8	1.4					
1½	10.4	5.2	2.6	2.1					
2	13.6	6.8	3.4	2.7					
3		9.6	4.8	3.9					
5		15.2	7.6	6.1					
7½		22	11	9					
10		28	14	11					
15		42	21	17					
20		54	27	22					
25		68	34	27		53	26	21	
30		80	40	32		63	32	26	
40		104	52	41		83	41	33	
50		130	65	52		104	52	42	
60		154	77	62	16	123	61	49	12
75		192	96	77	20	155	78	62	15
100		248	124	99	26	202	101	81	20
125		312	156	125	31	253	126	101	25
150		360	180	144	37	302	151	121	30
200		480	240	192	49	400	201	161	40

For full-load currents of 208- and 200-volt motors, increase the corresponding 230-volt motor full-load current by 10 and 15 percent, respectively.

* These values of full-load current are for motors running at speeds usual for belted motors and motors with normal torque characteristics. Motors built for especially low speeds or high torques may require more running current, and multispeed motors will have full-load current varying with speed, in which case the nameplate current rating shall be used.

† For 90 and 80 percent power factor the above figures shall be multipled by 1.1 and 1.25 respectively.

The voltages listed are rated motor voltages. The currents listed shall be permitted for system voltage ranges of 110 to 120, 220 to 240, 440 to 480, and 550 to 600 volts.

SOURCE: Reproduced with permission from *National Electrical Code 1978* (NFPA 70), National Fire Protection Association, 470 Atlantic Avenue, Boston, MA 02210.

TABLE 7-10 Maximum Rating or Setting of Motor Branch-Circuit Protective Devices

Type of Motor	Percent of Full-Load Current			
	Nontime Delay Fuse	Dual-Element (Time-Delay) Fuse	Instantaneous Trip Breaker	* Inverse Time Breaker
Single-phase, all types				
No code letter	300	175	700	250
All AC single-phase and polyphase squirrel-cage and synchronous motors† with full-voltage, resistor or reactor starting:				
No code letter	300	175	700	250
Code letter F to V	300	175	700	250
Code letter B to E	250	175	700	200
Code letter A	150	150	700	150
All AC squirrel-cage and synchronous motors† with autotransformer starting:				
Not more than 30 amps				
No code letter	250	175	700	200
More than 30 amps				
No code letter	200	175	700	200
Code letter F to V	250	175	700	200
Code letter B to E	200	175	700	200
Code letter A	150	150	700	150
High-reactance squirrel-cage				
Not more than 30 amps				
No code letter	250	175	700	250
More than 30 amps				
No code letter	200	175	700	200
Wound-rotor – No code letter	150	150	700	150
Direct-current (constant voltage)				
No more than 50 hp				
No code letter	150	150	250	150
More than 50 hp				
No code letter	150	150	175	150

SOURCE: Reproduced with permission from *National Electrical Code 1978* (NFPA 70), National Fire Protection Association, 470 Atlantic Avenue, Boston, MA 02210.

7.9 INSIDE WIRING

The definitive code is the *National Electrical Code* (NFPA 70), latest edition. This document does not have the force of law except where it is incorporated by adoption into the enforceable code of the state or municipality involved. Even where it is so incorporated, great latitude is given to the local inspecting agency in the matter of interpretation. Unfortunately, many governing bodies have chosen to add further local requirements. Data given in Tables 7-11 and 7-12 are taken directly from NFPA 70, 1978 edition.

TABLE 7-11 Allowable Ampacities of Insulated Conductors Rated 0–2000 V, 60 to 90°C

Not More Than Three Conductors in Raceway or Cable or Earth (Directly Buried), Based on Ambient Temperature of 30°C (86°F)

Size	Temperature Rating of Conductor. See Table 310-13								Size
	60°C (140°F)	75°C (167°F)	85°C (185°F)	90°C (194°F)	60°C (140°F)	75°C (167°F)	85°C (185°F)	90°C (194°F)	
AWG MCM	TYPES RUW, T, TW, UF	TYPES FEPW, RH, RHW, RUH, THW, THWN, XHHW, USE, ZW	TYPES V, MI	TYPES TA, TBS, SA, AVB, SIS, †FEP, †FEPB, †RHH, †THHN, †XHHW*	TYPES RUW, T, TW, UF	TYPES RH, RHW, RUH, THW, THWN, XHHW, USE	TYPES V, MI	TYPES TA, TBS, SA, AVB, SIS, †RHH, †THHN, †XHHW*	AWG MCM
	COPPER				ALUMINUM OR COPPER-CLAD ALUMINUM				
18	....	....	....	21	....	....	....	....	....
16	....	....	22	22	....	....	....	....	....
14	15	15	25	25	....	....	....	....	....
12	20	20	30	30	15	15	25	25	12
10	30	30	40	40	25	25	30	30	10
8	40	45	50	50	30	40	40	40	8
6	55	65	70	70	40	50	55	55	6
4	70	85	90	90	55	65	70	70	4
3	80	100	105	105	65	75	80	80	3
2	95	115	120	120	75	90	95	95	2
1	110	130	140	140	85	100	110	110	1
0	125	150	155	155	100	120	125	125	0
00	145	175	185	185	115	135	145	145	00
000	165	200	210	210	130	155	165	165	000
0000	195	230	235	235	155	180	185	185	0000
250	215	255	270	270	170	205	215	215	250
300	240	285	300	300	190	230	240	240	300
350	260	310	325	325	210	250	260	260	350
400	280	335	360	360	225	270	290	290	400
500	320	380	405	405	260	310	330	330	500
600	355	420	455	455	285	340	370	370	600
700	385	460	490	490	310	375	395	395	700
750	400	475	500	500	320	385	405	405	750
800	410	490	515	515	330	395	415	415	800
900	435	520	555	555	355	425	455	455	900
1000	455	545	585	585	375	445	480	480	1000
1250	495	590	645	645	405	485	530	530	1250
1500	520	625	700	700	435	520	580	580	1500
1750	545	650	735	735	455	545	615	615	1750
2000	560	665	775	775	470	560	650	650	2000
	CORRECTION FACTORS								
Ambient Temp. °C	For ambient temperatures over 30°C, multiply the ampacities shown above by the appropriate correction factor to determine the maximum allowable load current.								Ambient Temp. °F
31-40	.82	.88	.90	.91	.82	.88	.90	.91	86-104
41-50	.58	.75	.80	.82	.58	.75	.80	.82	105-122
51-60	....	.58	.67	.71	....	.58	.67	.71	123-141
61-70	....	.35	.52	.58	....	.35	.52	.58	142-158
71-80	....	....	.30	.41	....	....	.30	.41	159-176

† The load current rating and the overcurrent protection for these conductors shall not exceed 15 amperes for 14 AWG, 20 amperes for 12 AWG, and 30 amperes for 10 AWG copper; or 15 amperes for 12 AWG and 25 amperes for 10 AWG aluminum and copper-clad aluminum.

* For dry locations only. See 75°C column for wet locations.

SOURCE: Reproduced with permission from *National Electrical Code 1978* (NFPA 70), National Fire Protection Association, 470 Atlantic Avenue, Boston MA, 02210

TABLE 7-12 Maximum Number of Conductors in Trade Sizes of Conduit or Tubing

Conduit Trade Size (Inches)		½	¾	1	1¼	1½	2	2½	3	3½	4	4½	5	6
Type Letters	Conductor Size AWG, MCM													
TW, T, RUH, RUW, XHHW (14 thru 8)	14	9	15	25	44	60	99	142						
	12	7	12	19	35	47	78	111	171					
	10	5	9	15	26	36	60	85	131	176				
	8	2	4	7	12	17	28	40	62	84	108			
RHW and RHH (without outer covering), THW	14	6	10	16	29	40	65	93	143	192				
	12	4	8	13	24	32	53	76	117	157				
	10	4	6	11	19	26	43	61	95	127	163			
	8	1	3	5	10	13	22	32	49	66	85	106	133	
TW, T, THW, RUH (6 thru 2), RUW (6 thru 2), FEPB (6 thru 2), RHW and RHH (without outer covering)	6	1	2	4	7	10	16	23	36	48	62	78	97	141
	4	1	1	3	5	7	12	17	27	36	47	58	73	106
	3	1	1	2	4	6	10	15	23	31	40	50	63	91
	2	1	1	2	4	5	9	13	20	27	34	43	54	78
	1		1	1	3	4	6	9	14	19	25	31	39	57
	0		1	1	2	3	5	8	12	16	21	27	33	49
	00		1	1	1	3	5	7	10	14	18	23	29	41
	000		1	1	1	2	4	6	9	12	15	19	24	35
	0000			1	1	1	3	5	7	10	13	16	20	29
	250			1	1	1	2	4	6	8	10	13	16	23
	300			1	1	1	2	3	5	7	9	11	14	20
	350				1	1	1	3	4	6	8	10	12	18
	400				1	1	1	2	4	5	7	9	11	16
	500				1	1	1	1	3	4	6	7	9	14
	600					1	1	1	3	4	5	6	7	11
	700					1	1	1	2	3	4	5	7	10
	750					1	1	1	2	3	4	5	6	9

...Trade Sizes of Conduit or ...ubing (cont.)

Type Letters	Conductor Size AWG, MCM	½	¾	1	1¼	1½	2	2½	3	3½	4	4½	5	6
THWN,	14	13	24	39	69	94	154							
	12	10	18	29	51	70	114	164						
	10	6	11	18	32	44	73	104	160					
	8	3	5	9	16	22	36	51	79	106	136			
THHN, FEP (14 thru 2), FEPB (14 thru 8), PFA (14 thru 4/0) PFAH (14 thru 4/0) Z (14 thru 4/0) XHHW (4 thru 500MCM)	6	1	4	6	11	15	26	37	57	76	98	125	154	
	4	1	2	4	7	9	16	22	35	47	60	75	94	137
	3	1	1	3	6	8	13	19	29	39	51	64	80	116
	2	1	1	3	5	7	11	16	25	33	43	54	67	97
	1		1	1	3	5	8	12	18	25	32	40	50	72
	0		1	1	3	4	7	10	15	21	27	33	42	61
	00		1	1	2	3	6	8	13	17	22	28	35	51
	000		1	1	1	3	5	7	11	14	18	23	29	42
	0000		1	1	1	2	4	6	9	12	15	19	24	35
	250			1	1	1	3	4	7	10	12	16	20	28
	300			1	1	1	3	4	6	8	11	13	17	24
	350			1	1	1	2	3	5	7	9	12	15	21
	400				1	1	1	3	5	6	8	10	13	19
	500				1	1	1	2	4	5	7	9	11	16
	600				1	1	1	1	3	4	5	7	9	13
	700					1	1	1	3	4	5	6	8	11
	750					1	1	1	2	3	4	6	7	11
XHHW	6	1	3	5	9	13	21	30	47	63	81	102	128	185
	600				1	1	1	1	3	4	5	7	9	13
	700					1	1	1	3	4	5	6	7	11
	750					1	1	1	2	3	4	6	7	10

SOURCE: Reproduced with permission from *National Electrical Code 1978* (NFPA 70), National Fire Protection Association, 470 Atlantic Avenue, Boston MA, 02210

7-10 ILLUMINATION

TABLE 7-13 Typical Illumination Levels

Category	*Lighting Level, fc*	*Category*	*Lighting Level, fc*	*Category*	*Lighting Level, fc*
Tennis courts	20	Offices	70–150	Fine machining	300*
Football	20–50	Reception	50	Medium machining	100
Sports, televised	100	Corridors	20	Rough machining	50
Parking, casual	0.1–1	Toilets	30	General work areas	30–100
Parking, malls	1–5	Conference	50	Sewing, low-contrast	500*
Streets, low use	0.2	Showrooms	100	Sewing, high-contrast	200*
Streets, high use	2	Warehouse	5–30		
Moonlight	0.01	Classrooms	70		
Summer shade	1000	Lecture halls	70		
Summer sun	10,000				

*Provided by local spot lighting.

To achieve a perceptible increase in lighting level, the footcandle (fc) level must be doubled, as the eye responds in a complex, nonlinear way. Contrast and glare strongly affect visibility.

$$\text{fc} = \text{lumens/ft}^2 = \frac{\text{No. of lamps} \times \text{lumens/lamp} \times \text{eff.}}{\text{area}}$$

eff. = coefficient of utilization (CU) × maintenance factor (MF)

Typical MF's: Poor, 0.5; average, 0.65; good, 0.8.

TABLE 7-14 Typical Fixture CU's

Fixture	CU
Incandescent:	
Recessed, poor	0.1–0.2
Recessed, excellent	0.5–0.8
Fluorescent:	
Surface with plastic diffuser	0.4–0.7
Recessed troffer	0.3–0.7
Metallic Vapor:	
Recessed, high-quality	0.3–0.5
Industrial high bay	0.4–0.8

In general, lower values of CU are for small rooms, higher for large rooms. If reflectivity of walls or ceiling is less than average, CU will be lower than given in Table 7-14.

TABLE 7-15 Lamp Characteristics

Lamp Type	*Wattages Available*	*Initial Output, lumens*	*Efficiency,* lumens/watt*
INCANDESCENT			
Frosted or clear	7½–1500	40–35,000	6–23
Reflector	50–500	440–6500	9–13
Quartz	200–1500	3100–35,000	15–23
FLUORESCENT†			
430 mA (conventional)	10 W/ft	500 to 800/ft	50–80
800 mA (HO	15 W/ft	700 to 1500/ft	40–80
1200 mA (VHO)	27 W/ft	1200 to 2000/ft	60–72
METALLIC VAPOR:			
Mercury, self-ballasted	160–750	2700–14,500	16–19
Mercury, external ballast	40–3000	1500–135,000	38–48
Metal halide	400–1500	34,000–155,000	85–100
Sodium, low-pressure	40–200	7000–37,000	183
Sodium, high-pressure	100–1000	13,000–140,000	130–140

*Efficiency is generally higher for larger lamps.
†Depends largely on color.

7-11 LIGHTNING PROTECTION

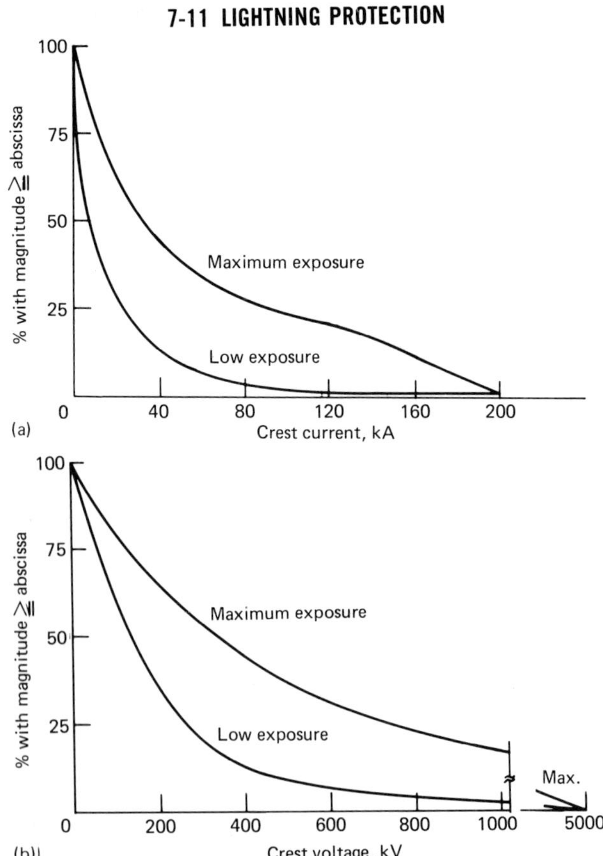

Fig. 7-15 Characteristics of strokes.

Arresters are insulators which break down at lightning levels, then recover their properties after the surge has passed. Sparkover level and *IR* drops must be less than the corresponding values for the device protected.

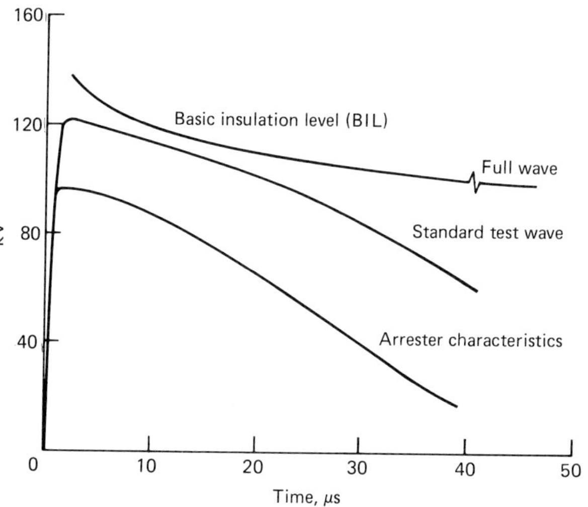

g. 7-16 Typical insulation coordination.

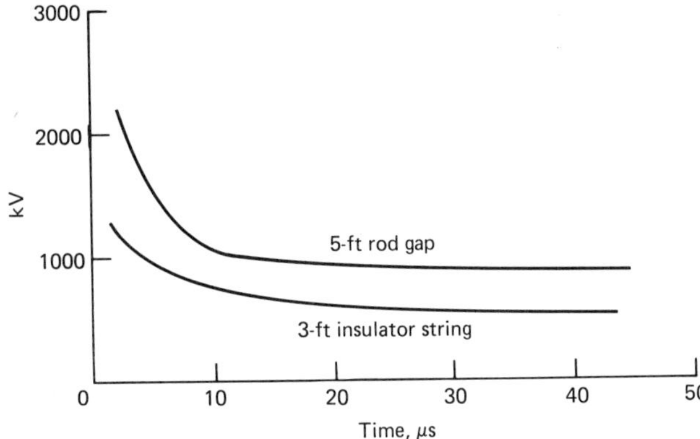

Fig. 7-17 Withstand voltage of air gaps.

Shielding is produced by placing a grounded intercept conducto[r] higher than the protected object. Figure 7-18 gives dimensions tha[t] will achieve 99.9% protection.

Figure 7-19 shows the conditions leading to development of sidestroke and also points out the need for bonding of variou[s] grounds. If the water system and the telephone ground were bonde[d] in Fig. 7-19, R_1 would be effectively short-circuited and E woul[d] be 0.

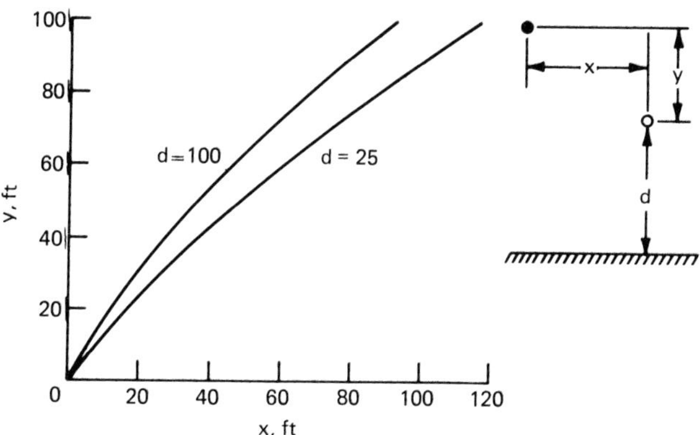

Fig. 7-18 Shielding dimensions.

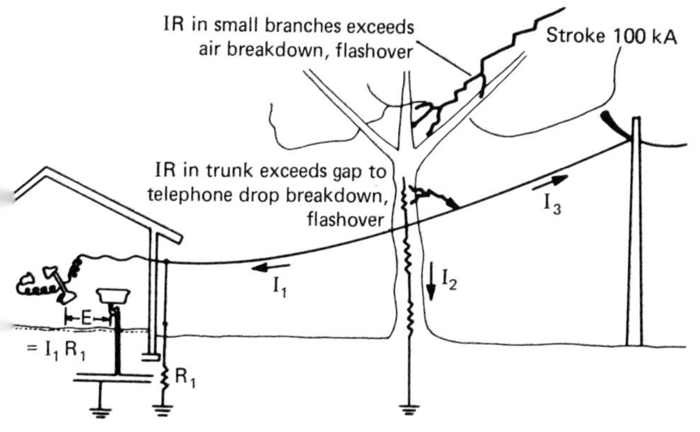

ig. 7-19 Sidestroke mechanism.

7-12 EMERGENCY POWER SOURCES

Separate emergency service is only rarely satisfactory for critical applications. Everything depends on the structure of the electric utility's system. Consult with the utility.

Connection ahead of the main switch satisfies the codes in some cases, but is of little real value.

Batteries, except in central stations and special applications, are useful for small loads such as individual exit lights. They are satisfactory and reliable only if properly maintained and expensive if the number of units is large. Much progress is being made in extending the life and reducing the cost of these devices.

Engine generators are satisfactory for virtually all applications. Proper maintenance and periodic (usually weekly) test and warm-up runs are vital. Multiple units provide redundancy. See Table 7-16.

TABLE 7-16 Engine Generator Characteristics

Prime mover	*Size Range (kW)*	*First Cost*	*Reliability*	*Maintenance*	*Weight*	*Fuel Economy*
Gasoline engine*	0.5–150	good	fair	fair	fair	fair
Diesel engine	3–5000	poor	good	good	poor	good
LPG engine†	1–150	good	fair	fair	fair	fair
Natural-gas engine‡	1–150	good	fair	fair	fair	fair
Gas turbine‡	300–	fair	good	good	good	poor

*Fuel storage is usually a problem.
†Requires a vaporizer or oversized storage in cold climates.
‡Danger of loss of fuel supply in earthquake zones.

7-13 TELEVISION

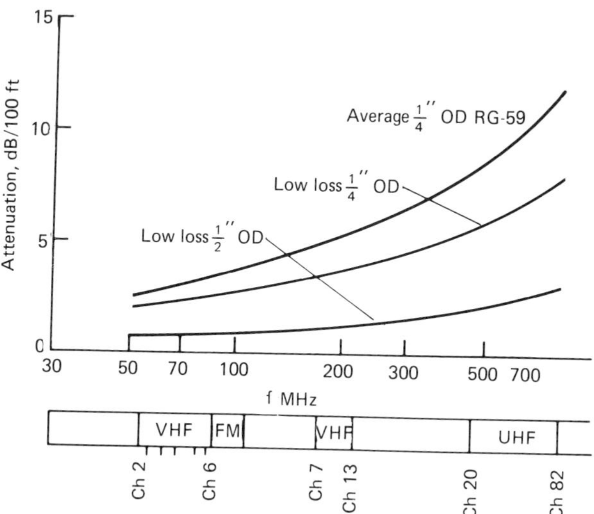

Fig. 7-20 Television frequencies and cable losses.

TABLE 7-17 Typical Antenna Gains (Over dipole, dB*)

Antenna-Type Band	*All-Channel 5-ft Boom*	*All-Channel 12-ft Boom*	*10-Element Yagi*	*Corner Reflector*	*Parabolic Reflector*
VHF lo band	4	7	9		
VHF hi band	8	12	11		
UHF	8	10	14	11	14–20

Required signal-to-noise ratios:

Excellent picture	44 dB
Fair picture	34 dB
Usable picture	28 dB

NOTES

Section Eight

■

MECHANICAL FORMULAS AND DATA

TABLE 8-1 Basic Mechanical Formulas

TORQUE AND HORSEPOWER

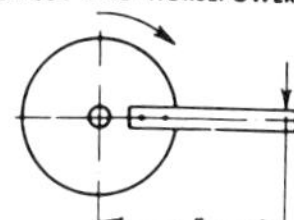

$$Q = F \times r = \frac{63025\ hp}{rpm}$$

$$hp = \frac{Q \times rpm}{63025}$$

$$rpm = \frac{63025\ hp}{Q}$$

Q = Torque (inch-pounds)
F = Force (pounds)
r = Radius (inches)
hp = Horsepower
rpm = Revolutions per minute

CENTRIFUGAL AND CENTRIPETAL FORCE

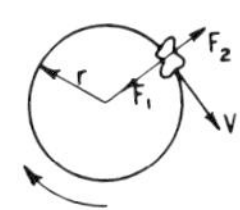

$$F_1 = F_2 = \frac{WV^2}{32.16R}$$

$$= 0.000341\ WR\ (rpm)^2$$

$$= 0.1023\ Wr\ (rps)^2$$

$$= 1.2276\ WR\ (rps)^2$$

$$= 0.669\ W\ \frac{(mph)^2}{R}$$

F_1 = Centripetal force (pounds)
F_2 = Centrifugal force (pounds)
W = Weight of rotating body (pounds)
V = Linear velocity (feet per second)
mph = Linear velocity (miles per hour)
rpm = Angular velocity (revolutions per minute)
rps = Angular velocity (revolutions per second)
r = Radius of rotation (inches)
R = Radius of rotation (feet)

FIBER STRESS IN BENDING

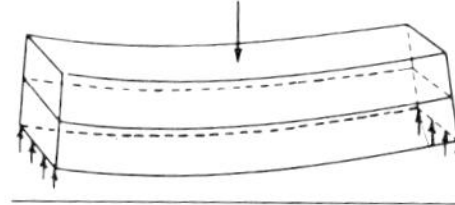

$$S_T = \frac{Mc}{I}$$

S_T = Unit tensile or compressive stress in outer fibers.
M = Bending moment at section.
c = Distance from neutral axis to extreme fiber.
I = Moment of inertia of cross section.

FIBER STRESS IN TORSION

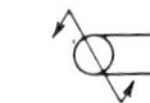

$$S_s = \frac{16Q}{\pi d^3}$$

S_s = Unit shear stress in outer fiber.
Q = Torsional moment.
d = Shaft diameter.

COMBINED STRESS

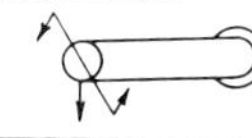

$$S'_T = \frac{S_T}{2} + \sqrt{S_s^2 + \left(\frac{S_T}{2}\right)^2}$$

$$S'_s = \sqrt{S_s^2 + \left(\frac{S_T}{2}\right)^2}$$

S'_T = Maximum unit fiber stress in tension or compression due to combined bending and torsion.
S'_s = Maximum fiber stress in shear due to combined bending and torsion.

BEARING PROJECTED AREA

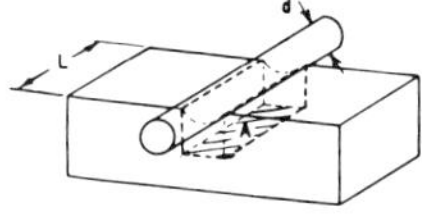

$$A = L \times d$$

A = Bearing projected area.
L = Length of bearing.
d = Diameter of shaft.

SOURCE: Clark Equipment Co., Buchanan, Mich.

SOURCE: "Tables and Formulas for the Automotive Engineer," Clark Equipment Company, Buchanan, Mich.

TABLE 8-2 Safe Torque (lb-ft) for Bolts, Cap Screws, Studs, and Nuts

SAFE TORQUE—(LBS.-FT.)
For Bolts, Cap Screws, Studs and Nuts

Material					
Material	Steel	S.A.E.—1010 S.A.E.—1020 S.A.E.—1112	S.A.E.—1038	S.A.E.—1050	S.A.E.—3135
	Brass	S.A.E.—73 S.A.E.—72	......	......	......
Heat Treat		None	R.C. 20—28	R.C. 28—39	R.C. 32—37
Tensile Strength		55,000 P.S.I.	100,000 P.S.I.	130,000 P.S.I.	150,000 P.S.I.
Yield Point—80%		22,000 P.S.I.	56,000 P.S.I.	70,400 P.S.I.	108,800 P.S.I.

THE FOLLOWING TORQUE IS APPLIED WITH DRY THREADS
For oiled bolts and nuts the Torque required is approximately 30—40% less than that indicated in this table.

		S.A.E.—1010	S.A.E.—1038	S.A.E.—1050	S.A.E.—3135
Coarse Threads	1/4—20	2	5	6	9
	5/16—18	4	10	13	20
	3/8—16	7	18	23	36
	7/16—14	12	33	38	59
	1/2—13	19	47	59	91
	9/16—12	27	68	85	132
	5/8—11	37	95	119	181
	3/4—10	69	175	219	340
	7/8— 9	113	286	359	555
	1 — 8	169	428	541	836
	1 1/8— 7	238	605	760	1175
	1 1/4— 7	345	878	1106	1705
	1 3/8— 6	497	1265	1575	2455
	1 1/2— 6	610	1551	1950	3011
Fine Threads	1/4—28	2	6	8	12
	5/16—24	5	13	16	25
	3/8—24	9	24	30	47
	7/16—20	15	37	48	74
	1/2—20	23	60	76	117
	9/16—18	34	86	108	166
	5/8—18	49	124	155	240
	3/4—16	86	218	274	424
	7/8—14	137	344	439	679
	1 —14	214	545	686	1058
	1 1/8—12	303	768	967	1494
	1 1/4—12	427	1080	1368	2115
	1 3/8—12	643	1635	2055	3180
	1 1/2—12	775	1972	2475	3830

Theoretical torque in inch pounds required on the nut of a bolt to produce either 1000# tension or 60,000# per sq. inch stress in the bolt. Based on coefficient of friction, betwen bolt and nut, of .15.

Bolt Dia.	Number of Threads	Required 1000 Lb. Tension	Torque for 60,000 Lb. Stress	Number of Threads	Required 1000 Lb. Tension	Torque for 60,000 Lb. Stress
1/4	28	53	98	20	55	87
5/16	24	67	199	18	68	178
3/8	24	79	356	16	81	319
7/16	20	93	565	14	93	501
1/2	20	105	864	13	107	771
9/16	18	118	1234	12	121	1124
5/8	18	128	1688	11	134	1538
3/4	16	156	3015	10	160	2730
7/8	14	182	4768	9	186	4371
1	14	206	7276	8	212	6554
1 1/8	12	233	10325	7	239	9295
1 1/4	12	257	14202	7	264	13086
1 1/2	12	307	25102	6	315	22686

SOURCE: "Tables and Formulas for the Automotive Engineer," Clark Equipment Company, Buchanan, Mich.

TABLE 8-3 Rules for Calculating Spur Gears

Having	To Get	Rule	Formula
Diametral Pitch	Circular Pitch	Divide 3.1416 by the Diametral Pitch	$P' = \frac{3.1416}{P}$
Outside Diameter and the Number of Teeth	Circular Pitch	Divide Outside Diameter by the Product of .3183 and Number of Teeth plus 2	$P' \frac{D}{.3183\ N + 2}$
Number of Teeth and the Circular Pitch	Pitch Diameter	Continued Product of the Number of Teeth, the Circular Pitch and .3183	$D' = NP'\ .3183$
Number of Teeth and the Outside Diameter	Pitch Diameter	Divide the Product of Number of Teeth and Outside Diameter by Number of Teeth plus 2	$D' = \frac{ND}{N + 2}$
Number of Teeth and the Circular Pitch	Outside Diameter	Continued Product of the Number of Teeth plus 2, the Circular Pitch and .3183	$D = (N + 2)\ P'\ .3183$
Circular Pitch	Thickness of Tooth	One-half the Circular Pitch	$t = \frac{P'}{2}$
Circular Pitch	Addendum	Multiply the Circular Pitch by .3183 or $s = \frac{D'}{N}$	$s = P'\ .3183$
Circular Pitch	Working Depth	Multiply the Circular Pitch by .6366	$D'' = P'\ .6366$
Thickness of Tooth	Clearance	One-tenth the Thickness of Tooth at Pitch Line	$f = \frac{t}{10}$
Circular Pitch	Diametral Pitch	Divide 3.1416 by the Circular Pitch	$P = \frac{3.1416}{P'}$
Outside Diameter and the Number of Teeth	Diametral Pitch	Divide Number of Teeth plus 2 by Outside Diameter	$P = \frac{N + 2}{D}$
Number of Teeth and the Outside Diameter	Pitch Diameter	Divide the Product of Outside Diameter and Number of Teeth by Number of Teeth plus 2	$D' = \frac{DN}{N + 2}$
Outside Diameter and the Diametral Pitch	Pitch Diameter	Subtract from the Outside Diameter the Quotient of 2 Divided by the Diametral Pitch	$D' = D - \frac{2}{P}$
Number of Teeth and the Diametral Pitch	Outside Diameter	Divide Number of Teeth plus 2 by the Diametral Pitch	$D = \frac{N + 2}{P}$
Pitch Diameter and the Number of Teeth	Outside Diameter	Divide the Number of Teeth plus 2 by the Quotient of Number of Teeth divided by the Pitch Diameter	$D = \frac{N + 2}{\frac{N}{D'}}$
Pitch Diameter and the Diametral Pitch	Number of Teeth	Multiply Pitch Diameter by the Diametral Pitch	$N = D'P$
Outside Diameter and the Diametral Pitch	Number of Teeth	Multiply Outside Diameter by the Diametral Pitch and subtract 2	$N = DP - 2$
Diametral Pitch	Thickness of Tooth	Divide 1.5708 by the Diametral Pitch	$t = \frac{1.5708}{P}$
Diametral Pitch	Working Depth	Divide 2 by the Diametral Pitch	$D'' = \frac{2}{P}$
Diametral Pitch	Whole Depth	Divide 2.157 by the Diametral Pitch	$D'' + f = \frac{2.157}{P}$
Diametral Pitch	Clearance	Divide .157 by the Diametral Pitch	$f = \frac{.157}{P}$

The stub form of tooth is designated by two figures, such as 4/5, which represents that the tooth is 4 pitch but cut to a depth of a 5-pitch tooth. These gear calculations are figured accordingly.

SOURCE: "Mechanical Details for Product Design," *Product Engineering*, Morgan-Grampian Publishing Company, New York.

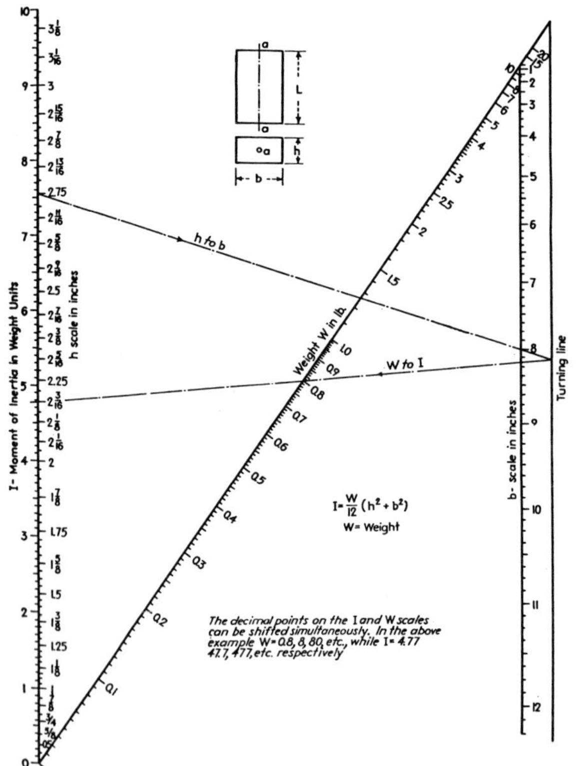

Fig. 8-1 Moment of inertia of a prism about the axis *aa* (*From* Product Engineering, *Morgan-Grampian Publishing Company, New York*).

$$\frac{2S}{T^2} = \frac{V}{2S} = \frac{V}{T} = \frac{32.16F}{W} = G$$

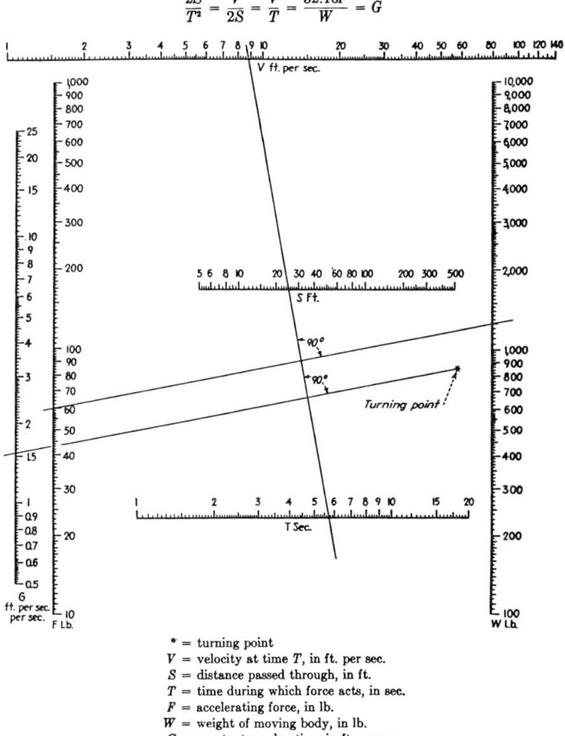

Fig. 8-2 Accelerated linear motion (*From* Product Engineering, *Morgan-Grampian Publishing Company, New York*).

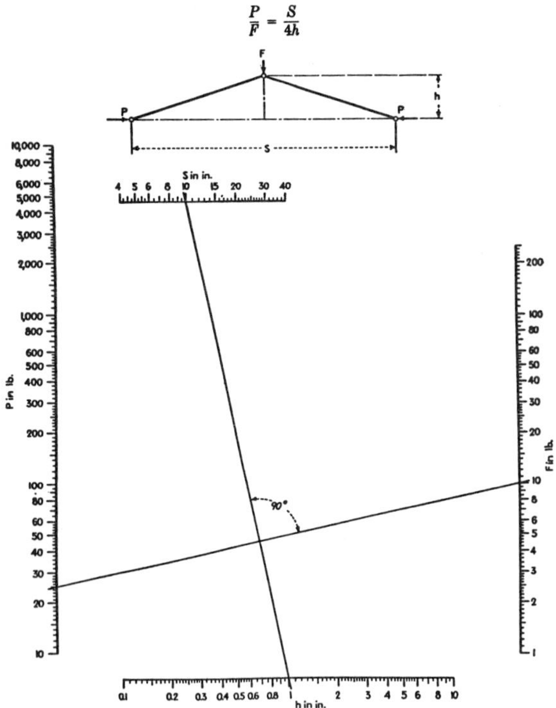

Example: Use mutually perpendicular lines drawn on tracing cloth or celluloid. In the example given for $S = 10$ in. and $h = 1$ in., a force F of 10 lb. exerts pressures P of 25 lb. each.

Fig. 8-3 Forces in toggle joint with equal arms (*From* Product Engineering, *Morgan-Grampian Publishing Company, New York*).

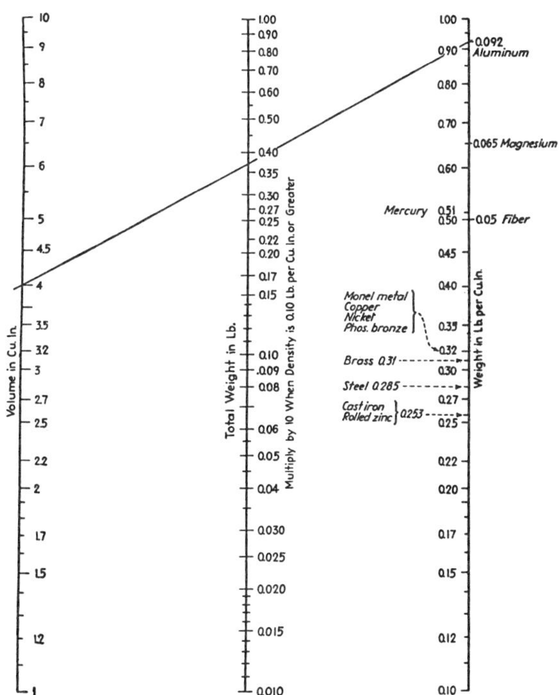

Draw a straight line through the two known points. The answer will be found at the intersection of this line with the third scale.

Example: 4 cu. in. of aluminum weighs 0.37 lb.

Fig. 8-4 Weight and volume of various materials (*From* Product Engineering, *Morgan-Grampian Publishing Company, New York*).

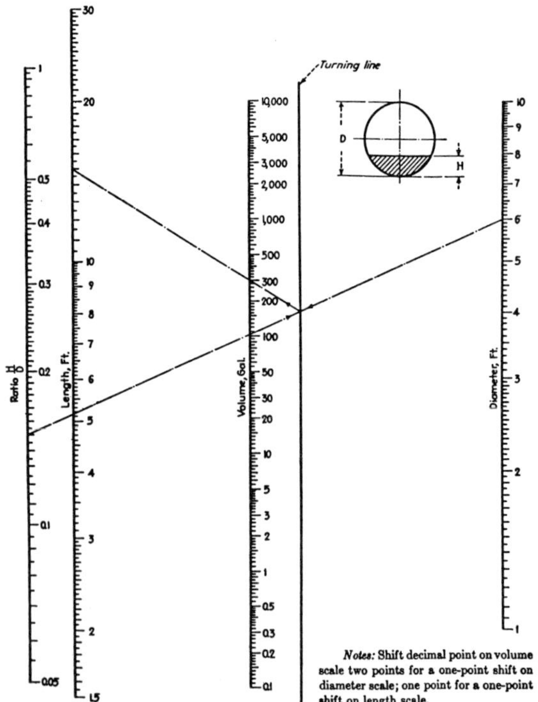

Example: Tank is 6 ft. in diameter and 15 ft. long. $H = 0.9$ ft. $H/D = 0.15$. Join 0.15 on H/D scale with 6 on diameter scale. From point of intersection with turning line, draw line to 15 ft. on the length scale. The volume scale shows 300 gal. If D had been 0.6 ft., H 0.09 ft., and length the same, the answer would be 3.00 gal.

Fig. 8-5 Volumes in horizontal round tanks with flat ends (*From* Product Engineering, *Morgan-Grampian Publishing Company, New York*).

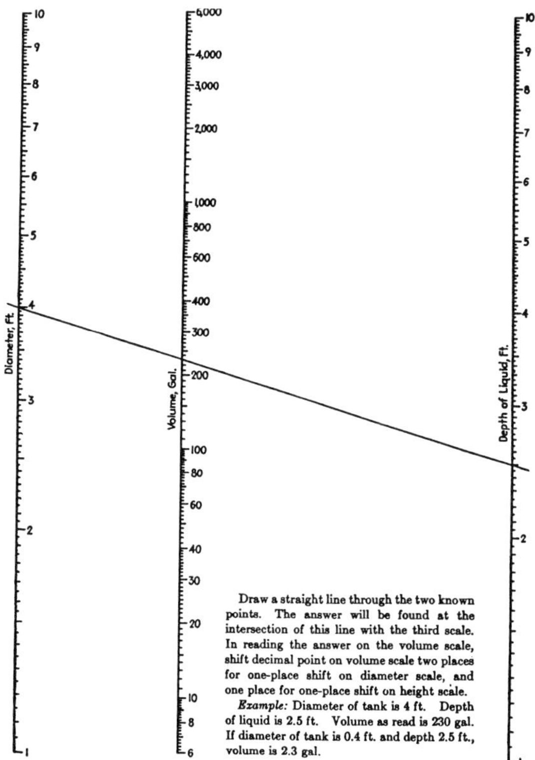
Diameter, Ft.
10
9
8
7
6
5
4
3
2
1
Volume, Gal.
6,000
4,000
3,000
2,000
1,000
800
600
400
300
200
100
80
60
40
30
20
10
8
6
Depth of Liquid, Ft.
10
9
8
7
6
5
4
3
2
1
Draw a straight line through the two known points. The answer will be found at the intersection of this line with the third scale. In reading the answer on the volume scale, shift decimal point on volume scale two places for one-place shift on diameter scale, and one place for one-place shift on height scale.
Example: Diameter of tank is 4 ft. Depth of liquid is 2.5 ft. Volume as read is 230 gal. If diameter of tank is 0.4 ft. and depth 2.5 ft., volume is 2.3 gal.

Fig. 8-6 Volumes in vertical round tanks with flat bottoms (*From* Product Engineering, *Morgan-Grampian Publishing Company, New York*).

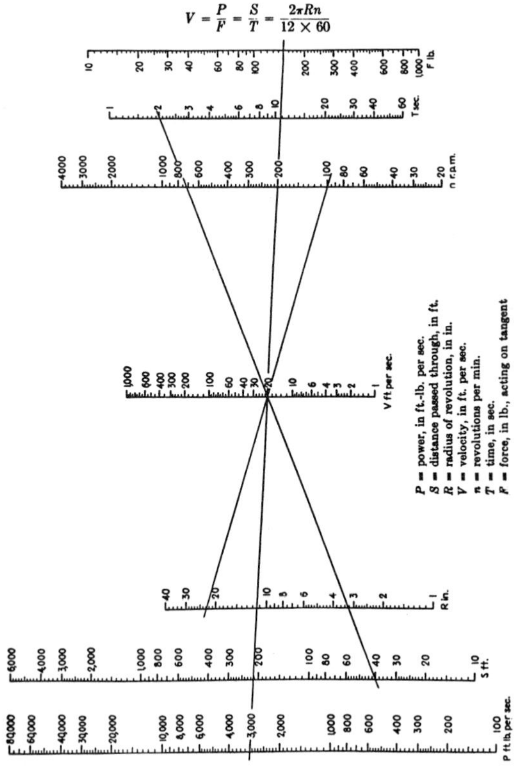

Fig. 8-7 Rotary motion (*From* Product Engineering, *Morgan-Grampian Publishing Company, New York*).

$$F = 0.000341WRn^2$$

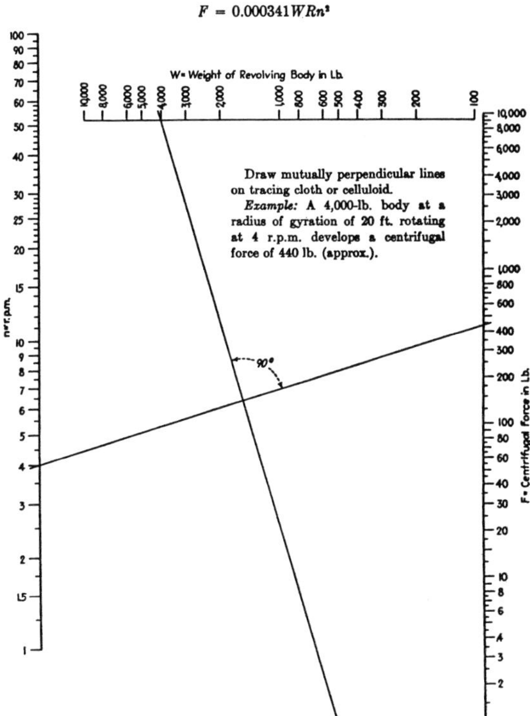

Fig. 8-8 Chart for determining centrifugal force (*From* Product Engineering, *Morgan-Grampian Publishing Company, New York*).

TABLE 8-4 Radii of Gyration for Rotating Bodies

Body	Radius of gyration
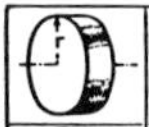 Solid cylinder about its own axis	$R^2 = \frac{r^2}{2}$
 Hollow cylinder about its own axis	$R^2 = \frac{r_1^2 + r_2^2}{2}$
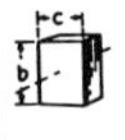 Rectangular prism about axis through center	$R^2 = \frac{b^2 + c^2}{12}$
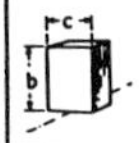 Rectangular prism about axis at one end	$R^2 = \frac{4b^2 + c^2}{12}$
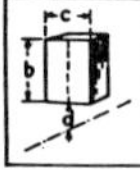 Rectangular prism about outside axis	$R^2 = \frac{4b^2 + c^2 + 12bd + 12d^2}{12}$

Body	Radius of gyration
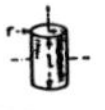 Cylinder about axis through center	$R^2 = \frac{l + 3r^2}{12}$
 Cylinder about axis at one end	$R^2 = \frac{4l^2 + 3r^2}{12}$
 Cylinder about outside axis	$R^2 = \frac{4l^2 + 3r^2 + 12dl + 12d^2}{12}$

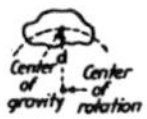

Center of gravity; Center of rotation

Any body about axis outside its center of gravity

$R_1^2 = R_s^2 + d^2$

where R_s = radius of gyration about axis through center of gravity

R_1 = radius of gyration about any other parallel axis

d = distance between center of gravity and axis of rotation

APPROXIMATIONS FOR CALCULATING MOMENTS OF INERTIA

NAME OF PART	MOMENT OF INERTIA
Flywheels (not applicable to belt pulleys)	Moment of inertia equal to 1.08 to 1.15 times that of rim alone
Flywheel (based on total weight and outside diameter)	Moment of inertia equal to two-thirds of that of total weight concentrated at the outer circumference
Spur or helical gears (teeth alone)	Moment of inertia of teeth equal to 40 per cent of that of a hollow cylinder of the limiting dimensions
Spur or helical gears (rim alone)	Figured as a hollow cylinder of same limiting dimensions
Spur or helical gears (total moment of inertia)	Equal to 1.25 times the sum of that of teeth plus rim
Spur or helical gears (with only weight and pitch diameter known)	Moment of inertia considered equal to 0.60 times the moment of inertia of the total weight concentrated at the pitch circle
Motor armature (based on total weight and outside diameter)	Multiply outer radius of armature by following factors to obtain radius of gyration: Large slow-speed motor.................. 0.75–0.85 Medium speed d-c or induction motor..... 0.70–0.80 Mill-type motor........................ 0.60–0.65

SOURCE: "Mechanical Details for Product Design," *Product Engineering*, Morgan-Grampian Publishing Company, New York.

TABLE 8-5 WR^2 of Symmetrical Bodies

For computing WR^2 of rotating masses of weight per unit volume ρ, by resolving the body into elemental shapes.

Note: ρ in pounds per cubic inch and dimensions in inches give WR^2 in lb.-in. squared.

1. Weights per Unit Volume of Materials.

Material	Weight, Lb. per Cu. In.
Cast iron	0.260
Cast-iron castings of heavy section *i.e.*, flywheel rims	0.250
Steel	0.283
Bronze	0.319
Lead	0.410
Copper	0.318

2. Cylinder, about Axis Lengthwise through the Center of Gravity.

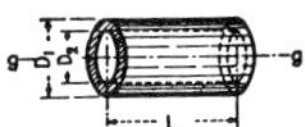

$$\text{Volume} = \frac{\pi}{4}L(D_1^2 - D_2^2)$$

(*a*) For any material:

$$WR^2 = \frac{\pi}{32}\rho L(D_1^4 - D_2^4)$$

where ρ is the weight per unit volume.

(*b*) For cast iron:

$$WR^2 = \frac{L(D_1^4 - D_2^4)}{39.2}$$

(*c*) For cast iron (heavy sections):

$$WR^2 = \frac{L(D_1^4 - D_2^4)}{40.75}$$

(*d*) For steel:

$$WR^2 = \frac{L(D_1^4 - D_2^4)}{36.0}$$

3. Cylinder, about an Axis Parallel to the Axis through Center of Gravity.

$$\text{Volume} = \frac{\pi}{4}L(D_1^2 - D_2^2)$$

(*a*) For any material:

$$WR^2_{x-x} = \frac{\pi}{4}\rho L(D_1^2 - D_2^2)\left(\frac{D_1^2 + D_2^2}{8} + y^2\right)$$

(*b*) For steel:

$$WR^2_{x-x} = \frac{(D_1^2 - D_2^2)L}{4.50}\left(\frac{D_1^2 + D_2^2}{8} + y^2\right)$$

4. Solid Cylinder, Rotated about an Axis Parallel to a Line that Passes through the Center of Gravity and Is Perpendicular to the Center Line.

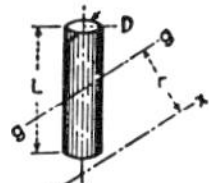

$$\text{Volume} = \frac{\pi}{4}D^2L$$

(*a*) For any material:

$$WR^2_{x-x} = \frac{\pi}{4}D^2L\rho\left(\frac{L^2}{12} + \frac{D^2}{16} + r^2\right)$$

(*b*) For steel:

$$WR^2_{x-x} = \frac{D^2L}{4.50}\left(\frac{L^2}{12} + \frac{D^2}{16} + r^2\right)$$

SOURCE: "Mechanical Details for Product Design," *Product Engineering*, Morgan-Grampian Publishing Company, New York.

TABLE 8-5 WR^2 of Symetrical Bodies *(cont.)*

5. Rod of Rectangular or Elliptical Section, Rotated about an Axis Perpendicular to and Passing through the Center Line.

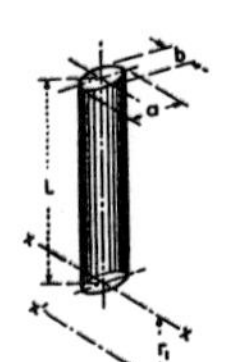

For rectangular cross sections:

$$K_1 = \tfrac{1}{12}; \qquad K_2 = 1$$

For elliptical cross sections:

$$K_1 = \frac{\pi}{64}; \qquad K_2 = \frac{\pi}{4}$$

$$\text{Volume} = K_2 abL$$

(*a*) For any material:

$$WR^2_{x'-x'} = \rho abL\left\{K_2\left[\frac{L^2}{3} + r_1(r_1 + L)\right] + K_1 a^2\right\}$$

(*b*) For a cast-iron rod of elliptical section ($\rho = 0.260$):

$$WR^2_{x'-x'} = \frac{abL}{4.90}\left[\frac{L^2}{3} + r_1(r_1 + L) + \frac{a^2}{16}\right]$$

6. Elliptical Cylinder, about an Axis Parallel to the Axis through the Center of Gravity.

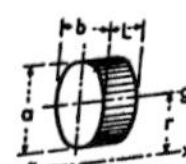

$$\text{Volume} = \frac{\pi}{4} abL$$

(*a*) For any material:

$$WR^2_{x-x} = \rho\frac{\pi}{4} abL\left(\frac{a^2 + b^2}{16} + r^2\right)$$

(*b*) For steel:

$$WR^2_{x-x} = \frac{abL}{4.50}\left(\frac{a^2 + b^2}{16} + r^2\right)$$

7. Cylinder with Frustum of a Cone Removed.

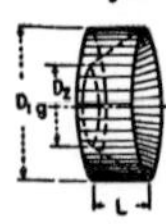

$$\text{Volume} = \frac{\pi L}{2(D_1 - D_2)}\left[\frac{1}{3}(D_1^3 - D_2^3) - \frac{D_2}{2}(D_1^2 - D_2^2)\right]$$

$$WR^2_{g-g} = \frac{\pi\rho L}{8(D_1 - D_2)}\left[\frac{1}{5}(D_1^5 - D_2^5) - \frac{D_2}{4}(D_1^4 - D_2^4)\right]$$

8. Frustum of a Cone with a Cylinder Removed.

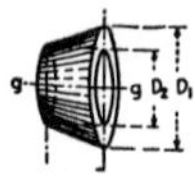

$$\text{Volume} = \frac{\pi L}{2(D_1 - D_2)}\left[\frac{D_1}{2}(D_1^2 - D_2^2) - \frac{1}{3}(D_1^3 - D_2^3)\right]$$

$$WR^2_{g-g} = \frac{\pi\rho L}{8(D_1 - D_2)}\left[\frac{D_1}{4}(D_1^4 - D_2^4) - \frac{1}{5}(D_1^5 - D_2^5)\right]$$

TABLE 8-5 WR^2 of Symetrical Bodies *(cont.)*

9. Solid Frustum of a Cone.

$$\text{Volume} = \frac{\pi L}{12}\frac{(D_1^3 - D_2^3)}{(D_1 - D_2)}$$

$$WR^2_{g-g} = \frac{\pi\rho L}{160}\frac{(D_1^5 - D_2^5)}{(D_1 - D_2)}$$

10. Chamfer Cut from Rectangular Prism Having One End Turned about a Center.

Distance to center of gravity, where $A = R_2/R_1$ and $B = C/2R_1$

$$r_x = \frac{jR_1^3B}{\text{volume} \times (1 - A)}\left[\frac{1}{3}(A^3 - 3A + 2) + \frac{B^2}{3}\left(1 - A - A\log_e\frac{1}{A}\right) + \frac{3}{40}\frac{B^4}{A}(A^2 - 2A + 1) + \frac{5}{672}\frac{B^6}{A^3}(3A_1^4 - 4A^3 + 1) \cdots\right]$$

$$\text{Volume} = \frac{jR_1^2B}{(1 - A)}\left\{(A^2 - 2A + 1) + \frac{B^2}{3}\left[\log_e\frac{1}{A} - (1 - A)\right] + \frac{1}{40}\frac{B^4}{A^2}(2A^3 - 3A + 1) + \frac{1}{224}\frac{B^6}{A^4}(4A^5 - 5A^4 + 1) + \cdots\right\}$$

$$WR^2_{x-x} = -\frac{\rho jR_1^4B}{6(1 - A)}\left\{(A^4 - 4A + 3) + B^2(A^2 - 2A + 1) + \frac{9}{10}B^4\left[\log_e\frac{1}{A} - (1 - A)\right] + \frac{5}{56}\frac{B^6}{A^2}(2A^3 - 3A^2 + 1) + \cdots\right\}$$

11. Complete Torus.

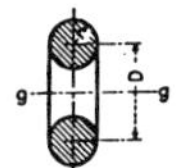

$$\text{Volume} = \pi^2Dr^2$$

$$WR^2_{g-g} = \frac{\pi^2\rho Dr^2}{4}(D^2 + 3r^2)$$

12. Outside Part of a Torus.

$$\text{Volume} = 2\pi r^2\left(\frac{\pi D}{4} + \frac{2}{3}r\right)$$

$$WR^2_{g-g} = \pi\rho r^2\left[\frac{D^2}{4}\left(\frac{\pi D}{2} + 4r\right) + r^2\left(\frac{3\pi}{8}D + \frac{8}{15}r\right)\right]$$

TABLE 8-5 WR^2 of Symetrical Bodies *(cont.)*

13. Inside Part of a Torus.

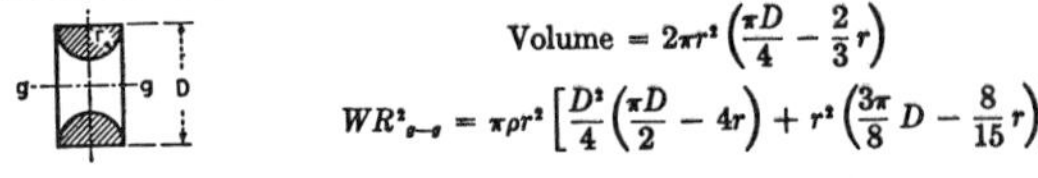

$$\text{Volume} = 2\pi r^2\left(\frac{\pi D}{4} - \frac{2}{3}r\right)$$

$$WR^2_{g-g} = \pi\rho r^2\left[\frac{D^2}{4}\left(\frac{\pi D}{2} - 4r\right) + r^2\left(\frac{3\pi}{8}D - \frac{8}{15}r\right)\right]$$

14. Circular Segment about an Axis through Center of Circle.

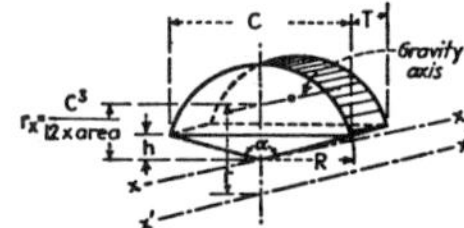

$$\alpha = 2\sin^{-1}\frac{C}{2R}\text{ deg.}$$

$$\text{Area} = \frac{R^2\alpha}{114.59} - \frac{C}{2}\sqrt{R^2 - \frac{C^2}{4}}$$

(*a*) Any material:

$$WR^2_{x-x} = \rho T\left[\frac{R^4\alpha}{229.2} - \frac{1}{6}\left(3R^2 - \frac{C^2}{2}\right)\frac{C}{2}\sqrt{R^2 - \frac{C^2}{4}}\right]$$

(*b*) For steel:

$$WR^2_{x-x} = \frac{T}{3.534}\left[\frac{R^4\alpha}{229.2} - \frac{1}{6}\left(3R^2 - \frac{C^2}{2}\right)\frac{C}{2}\sqrt{R^2 - \frac{C^2}{4}}\right]$$

15. Circular Segment about Any Axis Parallel to an Axis through the Center of the Circles. (Refer to 14 for Figure.)

$$WR^2_{x'-x'} = WR^2_{x-x} + \text{weight}\ (r^2 - r^2_x)$$

16. Rectangular Prism about an Axis Parallel to the Axis through the Center of Gravity.

$$\text{Volume} = WLT$$

(*a*) For any material:

$$WR^2_{x-x} = \rho WLT\left(\frac{W^2 + L^2}{12} + y^2\right)$$

(*b*) For steel:

$$WR^2_{x-x} = \frac{WLT}{3.534}\left(\frac{W^2 + L^2}{12} + y^2\right)$$

TABLE 8-5 WR^2 of Symetrical Bodies *(cont.)*

17. Isosceles Triangular Prism, Rotated about an Axis through Its Vertex.

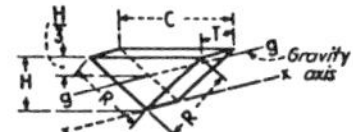

$$\text{Volume} = \frac{CHT}{2}$$

$$WR^2_{x-x} = \frac{\rho CHT}{2}\left(\frac{R^2}{2} - \frac{C^2}{12}\right)$$

18. Isosceles Triangular Prism, Rotated about Any Axis Parallel to an Axis through the Vertex.

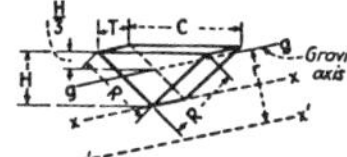

$$\text{Volume} = \frac{CHT}{2}$$

$$WR^2_{x'-x'} = \frac{\rho CHT}{2}\left(\frac{R^2}{2} - \frac{C^2}{12} - \frac{4}{9}H^2 + r^2\right)$$

19. Prism with Square Cross Section and Cylinder Removed, along Axis through Center of Gravity of Square.

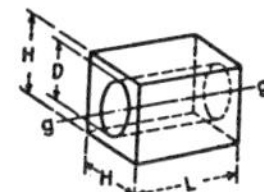

$$\text{Volume} = L\left(H^2 - \frac{\pi D^2}{4}\right)$$

$$WR^2_{g-g} = \frac{\pi \rho L}{32}(1.697H^4 - D^4)$$

20. Any Body about an Axis Parallel to the Gravity Axis, When WR^2 about the Gravity Axis Is Known.

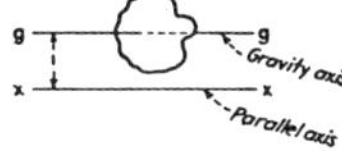

$$WR^2_{x-x} = WR^2_{g-g} + \text{weight} \times r^2$$

21. WR^2 of a Piston, Effective at the Cylinder Center Line, about the Crankshaft Center Line.

$$WR^2 = r^2 W_p\left(\frac{1}{2} + \frac{r^2}{8L^2}\right)$$

where r = crank radius
L = center-to-center length of connecting rod
W_p = weight of complete piston, rings, and pin

TABLE 8-5 WR^2 of Symetrical Bodies *(cont.)*

22. WR^2 of a Connecting Rod, Effective at the Cylinder Center Line, about the Crankshaft Center Line.

$$WR^2 = r^2\left[W_1 + W_2\left(\frac{1}{2} + \frac{r^2}{8L^2}\right)\right]$$

where r = crank radius
L = center-to-center length of connecting rod
W_1 = weight of the lower or rotating part of the rod = $[W_s(L - L_1)]/L$
W_2 = weight of the upper or reciprocating part of the rod = W_sL_1/L
W_s = $W_1 + W_2$, the weight of the complete rod
L_1 = distance from the center line of the crankpin to the center of gravity of the connecting rod

23. Mass Geared to a Shaft.—The equivalent flywheel effect at the shaft in question is

$$WR^2 = h^2(WR^2)'$$

where h = gear ratio $= \dfrac{\text{r.p.m. of mass geared to shaft}}{\text{r.p.m. of shaft}}$
$(WR^2)'$ = flywheel effect of the body in question about its own axis of rotation

24. Mass Geared to Main Shaft and Connected by a Flexible Shaft.—The effect of the mass $(WR^2)'$ at the position of the driving gear on the main shaft is

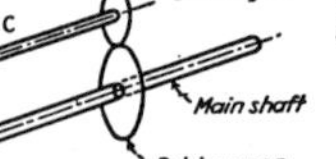

$$WR^2 = \frac{h^2(WR^2)'}{1 - \dfrac{(WR^2)'f^2}{9.775C}}$$

where h = gear ratio $= \dfrac{\text{r.p.m. of driven gear}}{\text{r.p.m. of driving gear}}$
$(WR^2)'$ = flywheel effect of geared-on mass
f = natural torsional frequency of the shafting system, in vibrations per sec.
C = torsional rigidity of flexible connecting shaft, in pound-inches per radian

25. Belted Drives.—The equivalent flywheel effect of the driven mass at the driving shaft is

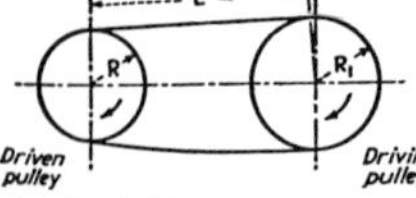

$$WR^2 = \frac{h^2(WR^2)'}{1 - \dfrac{(WR^2)'f^2}{9.775C}}$$

where h = R_1/R $= \dfrac{\text{r.p.m. of pulley belted to shaft}}{\text{r.p.m. of shaft}}$
$(WR^2)'$ = flywheel effect of the driven body about its own axis of rotation
f = natural torsional frequency of the system, in vibrations per sec.
C = R^2AE/L
A = cross-sectional area of belt, in sq. in.
E = modulus of elasticity of belt material in tension, in lb. per sq. in.
R = radius of driven pulley, in in.
L = length of tight part of belt which is clear of the pulley, in in.

TABLE 8-5 WR^2 of Symetrical Bodies *(cont.)*

26. Effect of the Flexibility of Flywheel Spokes on WR^2 of Rim.—The effective WR^2 of the rim is

$$WR^2 = \frac{(WR^2)'}{1 - \dfrac{(WR^2)'f^2}{9.775C}}$$

where $(WR^2)'$ = flywheel effect of the rim

f = natural torsional frequency of the system of which the flywheel is a member, in vibrations per sec.

C = torque required to move the rim through one radian relative to the hub

$$C = \frac{12gEka^3bR}{L^3}\left(\frac{L}{3R} + \frac{R}{L} - 1\right)$$

where g = number of spokes

E = bending modulus of elasticity of the spoke material

$k = \pi/64$ for elliptical, and $k = 1/12$ for rectangular section spokes

All dimensions are in inches.

For cast-iron spokes of elliptical section:

$$E = 15 \times 10^6 \text{ lb. per sq. in.}$$

$$C = \frac{ga^3bR \times 10^6}{0.1132L^3}\left(\frac{L}{3R} + \frac{R}{L} - 1\right)\frac{\text{lb.-in.}}{\text{radians}}.$$

Note: It is found by comparative calculations that with spokes of moderate taper very little error is involved in assuming the spoke to be straight and using cross section at mid-point for area calculation.

TYPICAL EXAMPLE

The flywheel shown below is used in a Diesel engine installation. It is required to determine effective WR^2 for calculation of one of the natural frequencies of torsional vibration. The anticipated natural frequency of the system is 56.4 vibrations per sec.

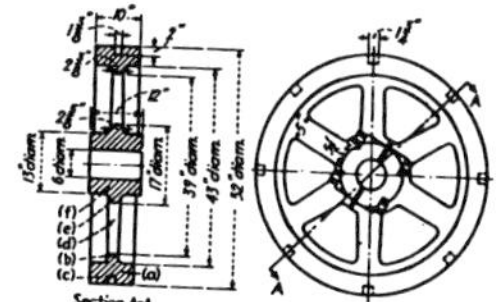

Note: Since the beads at the ends of the spokes comprise but a small part of the flywheel WR^2, very little error will result in assuming them to be of rectangular cross section. Also, because of the effect of the clamping bolts, the outer hub will be considered a square equal to the diameter. The spokes will be assumed straight and of mid-point cross section.

Part of fly wheel	Formula	WR^2
(a)	2c	$\frac{10[(52)^4 - (43)^4]}{40.75} = 955{,}300$
(b)	2b	$\frac{2.375[(43)^4 - (39)^4]}{39.2} = 67{,}000$
(c)	16a neglecting $\left(\frac{W^2 + L^2}{12}\right)$	$-0.250 \times 1.75 \times 2 \times 1.375(25)^2 \times 8 = -6{,}000$
		Total for rim = 1,016,300 lb.-in. squared
(d)	5b	$6 \times \frac{5.25 \times 2.5 \times 11}{4.90}\left[\frac{(11)^2}{3} + 8.5(8.5+11) + \frac{(5.25)^2}{16}\right] = 36{,}800$
(e)	2b	$\frac{2.625[(17)^4 - (13)^4]}{39.2} = 3{,}700$
(f)	19	$\frac{\pi \times 0.250 \times 12}{32}[1.697 \times (13)^4 - (6)^4] = 13{,}900$
		Total for remainder of flywheel = 54,400 lb.-in. squared

From formula (26)

$$C = \frac{6 \times (5.25)^3 \times 2.5 \times 19.5 \times 10^6}{0.1132 \times (11)^3}\left(\frac{11}{3 \times 19.5} + \frac{19.5}{11} - 1\right) = 2{,}970 \times 10^6 \frac{\text{lb.-in.}}{\text{radians}}$$

and

$$WR^2 = \frac{1{,}016{,}300}{1 - \dfrac{1{,}016{,}300 \times (56.4)^2}{9.775 \times 2{,}970 \times 10^6}} + 54{,}400$$

$$= \underline{1{,}197{,}000} \text{ lb.-in. squared}$$

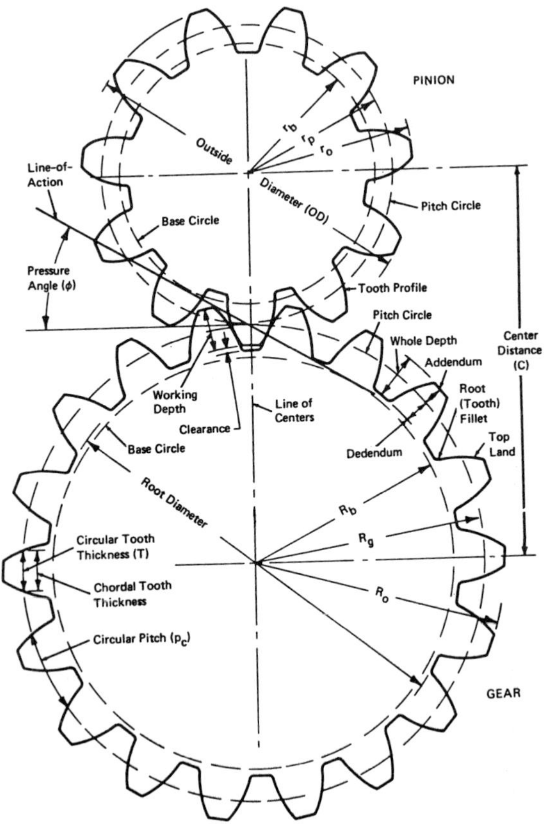

Fig. 8-9 Basic gear geometry *(Sterling Instrument, New Hyde Park, N.Y.).*

TABLE 8-6 Nomenclature and Gear Symbols

The terminology, nomenclature and symbols used in this writing are summarized below. This is consistent with most gear literature and the publications of AGMA and ASA.

Symbol	Nomenclature and Definition
B	backlash, linear amount along pitch circle
B_{LA}	backlash, linear amount along line-of-action
$_aB$	backlash in arc minutes
C	center distance
ΔC	change in center distance
C_o	operating center distance
C_{std}	standard center distance
D	pitch diameter
D_b	base circle diameter
D_o	outside diameter
F	face width
K	factor, general
L	length, general; also lead of worm
M	measurement over-pins
N	number teeth, usually gear
N_c	critical number teeth for no undercutting
N_v	virtual number of teeth for helical gear
P_d	diametral pitch
P_{dn}	normal diametral pitch
P_t	horsepower, transmitted
R	pitch radius, gear or general use
R_b	base circle radius, gear
R_o	outside radius, gear
R_T	testing radius
S	radial clearance
T	tooth thickness, gear
W_b	beam tooth strength
Y	Lewis factor, diametral pitch
Z	mesh velocity ratio
a	addendum
b	dedendum
c	clearance
d	pitch diameter, pinion
d_w	pin diameter, for over-pins measurement
e	eccentricity
m_p	contact ratio
n	number teeth, pinion
n_w	number of threads in worm
p_a	axial pitch
p_b	base pitch
p_c	circular pitch
p_{cn}	normal circular pitch
r	pitch radius, pinion
r_b	base circle radius, pinion
r_o	outside radius, pinion
t	tooth thickness, and for general use for tolerance
y_c	Lewis factor, circular pitch
γ	pitch angle, bevel gear
θ	rotation angle, general
λ	lead angle, worm gearing
μ	mean value
ν	gear stage velocity ratio
ϕ	pressure angle
ϕ_o	operating pressure angle
ψ	helix angle
ω	angular velocity

SOURCE: Sterling Instrument, New Hyde Park, N.Y.

TABLE 8-7 Summary of Gear Fundamentals

To Obtain	From Known	Formula
SPUR GEARS		
Pitch diameter	no. teeth and pitch	$D = \frac{N}{P_d} = \frac{N \cdot P_c}{\pi}$
Circular pitch	diametral pitch or no. teeth and pitch diameter	$P_c = \frac{\pi}{P_d} = \frac{\pi D}{N}$
Diametral pitch	circular pitch or no. teeth and pitch diameter	$P_d = \frac{\pi}{P_c} = \frac{N}{D}$
No. teeth	pitch and pitch diameter	$N = DP_d = \frac{\pi D}{P_c}$
Outside diameter	pitch and pitch diameter or pitch and number of teeth	$D_o = D + \frac{2}{P_d} = \frac{N + 2}{P_d}$
Root diameter	pitch diameter and dedendum	$D_R = D - 2b$
Base circle diameter	pitch diameter and pressure angle	$D_b = D \cos\phi$
Base pitch	circular pitch and pressure angle	$P_b = P_c \cos\phi$
Tooth thickness at standard pitch diameter	circular pitch	$T_{std} = \frac{P_c}{2} = \frac{\pi D}{2N}$
Addendum	diametral pitch	$A = \frac{1}{P_d}$
Center distance	pitch diameters or no. teeth and pitch	$C = \frac{D_1 + D_2}{2} = \frac{N_1 + N_2}{2P_d} = \frac{P_c (N_1 + N_2)}{2\pi}$
Contact ratio	outside radii, base radii, center distance and pressure angle	$m_p = \frac{\sqrt{R_o^2 - R_b^2} + \sqrt{r_o^2 - r_b^2} - C \sin\phi}{P_c \cos\phi}$
Backlash (linear)	from change in center distance	$B = 2(\overline{\Delta C}) \tan\phi$
Backlash (linear)	from change in tooth thickness	$B = \Delta T$
Backlash, linear along line-of-action	linear backlash along pitch circle	$B_{LA} = B \cos\phi$
Backlash, angular	linear backlash	$_aB = 6880 \frac{B}{D}$ (arc minutes)
Min. no. teeth for no undercutting	pressure angle	$N = \frac{2}{\sin^2\phi}$
HELICAL GEARING		
Normal circular pitch	transverse circular pitch	$P_{cn} = P_c \cos\psi$
Normal diametral pitch	transverse diametral pitch	$P_{dn} = \frac{P_d}{\cos\psi}$
Axial pitch	circular pitches	$P_a = P_c \cot\psi = \frac{P_{cn}}{\sin\psi}$
Normal pressure angle	transverse pressure angle	$\tan\phi_n = \tan\phi \cos\psi$
Pitch diameter	number teeth and pitch	$D = \frac{N}{P_d} = \frac{N}{P_{dn} \cos\psi}$
Center distance (parallel-shafts)	number teeth and pitch	$C = \frac{N_1 + N_2}{2P_{dn} \cos\psi}$
Center distance (crossed-shafts)	number of teeth and pitch	$C = \frac{1}{2P_{dn}}\left(\frac{N_1}{\cos\psi_1} + \frac{N_2}{\cos\psi_2}\right)$
Shaft angle (crossed-shafts)	helix angles of 2 mated gears	$\theta = \psi_1 + \psi_2$

TABLE 8-7 Summary of Gear Fundamentals *(cont.)*

To Obtain	From Known	Formula
WORM MESHES		
Pitch diameter of worm	number of teeth and pitch	$d_w = \frac{n_w P_{cn}}{\pi \sin \lambda}$
Pitch diameter of worm gear	number of teeth and pitch	$D_g = \frac{N_g P_{cn}}{\pi \cos \lambda}$
Lead angle	pitch, diameter, teeth	$\lambda = \tan^{-1} \frac{n_w}{P_d d_w} = \sin^{-1} \frac{n_w P_{cn}}{\pi d_w}$
Lead of worm	no teeth and pitch	$L = n_w P_c = \frac{n_w P_{cn}}{\cos \lambda}$
Normal circular pitch	transverse pitch and lead angle	$P_{cn} = p_c \cos \lambda$
Center distance	pitch diameters	$C = \frac{d_w + D_g}{2}$
Center distance	pitch, lead angle, teeth	$C = \frac{P_{cn}}{2\pi} \left[\frac{N_g}{\cos \lambda} + \frac{n_w}{\sin \lambda} \right]$
Velocity ratio	number teeth	$Z = \frac{N_g}{n_w}$
BEVEL GEARING		
Velocity ratio	number teeth	$Z = \frac{N_1}{N_2}$
Velocity ratio	pitch diameters	$Z = \frac{D_1}{D_2}$
Velocity ratio	pitch angles	$Z = \frac{\sin \gamma_1}{\sin \gamma_2}$
Shaft angle	pitch angles	$\Sigma = \gamma_1 + \gamma_2$

SOURCE: Sterling Instrument, New Hyde Park, N.Y.

TABLE 8-8 Definitions of Symbols in AGMA Rating Formulas

Term	Strength	Durability
Load:		
Transmitted Load	W_t	W_t
Dynamic Factor	K_v	C_v
Overload Factor	K_o	C_o
Size:		
Pinion Pitch Diameter	-	d
Net Face Width	F	F
Transverse Diametral Pitch	P_d	-
Size Factor	K_s	C_s
Stress Distribution:		
Load Distribution Factor	K_m	C_m
Geometry Factor	J	I
Surface Condition Factor	-	C_f
Stress:		
Calculated Stress	s_t	s_c
Allowable Stress	s_{at}	s_{ac}
Elastic Coefficient	-	C_p
Hardness-Ratio Factor	-	C_H
Life Factor	K_L	C_L
Temperature Factor	K_T	C_T
Factor of Safety	K_R	C_R

SOURCE: Sterling Instrument, New Hyde Park, N.Y.

TABLE 8-9 Lewis Y Factors

Number of Teeth	14½° Full Depth Involute	20° Full Depth Involute
10	0.176	0.201
11	0.192	0.226
12	0.210	0.245
13	0.223	0.264
14	0.236	0.276
15	0.245	0.289
16	0.255	0.295
17	0.264	0.302
18	0.270	0.308
19	0.277	0.314
20	0.283	0.320
22	0.292	0.330
24	0.302	0.337
26	0.308	0.344
28	0.314	0.352
30	0.318	0.358
32	0.322	0.364
34	0.325	0.370
36	0.329	0.377
38	0.332	0.383
40	0.336	0.389
45	0.340	0.399
50	0.346	0.408
55	0.352	0.415
60	0.355	0.421
65	0.358	0.425
70	0.360	0.429
75	0.361	0.433
80	0.363	0.436
90	0.366	0.442
100	0.368	0.446
150	0.375	0.458
200	0.378	0.463
300	0.382	0.471
Rack	0.390	0.484

SOURCE: Sterling Instrument, New Hyde Park, N.Y.

TABLE 8-10 Properties of Nylon Gear Materials

(Data at 70° F)

Property	ASTM Type 66		ASTM Type 6	
	0.2% Moisture	2.5% Moisture	0.2% Moisture	2.8% Moisture
Tensile strength	11,800	11,200	9,000	8,700
Yield strength	13,800	8,500	7,400	5,900
Compressive strength-psi	13,000	7,200	8,000	---
Water absorption, % (24 hrs.)	1.5	---	1.6 to 2.0	---
Saturation, %	8	---	9.5	---
Density-Lbs/In^3	.041	---	.041	---
Modulus of elasticity-psi (Flexural)	4.1 x10^5	1.75 x10^5	2.5 x10^5	1.1 x10^5
Coefficient of linear thermal expansion/°F	5 x 10^{-5}	---	4.6 to 5.4 x 10^{-5}	---

SOURCE: Sterling Instrument, New Hyde Park, N.Y.

TABLE 8-11 Properties of Phenolic Laminates Used for Gears

Properties	NEMA		GRADE	
	X	XXX	C	L
Base	Kraft Paper	Paper	Cotton Canvas Fabric	Fine Weave Cotton Linen Fabric
Tensile strength-psi				
Lengthwise	21,000	16,000	11,500	14,500
Crosswise	17,000	13,000	9,000	11,000
Flexural strength-psi				
Lengthwise	26,000	14,000	22,000	23,000
Crosswise	24,000	12,000	18,000	18,000
Compressive strength-psi				
Flatwise	36,000	32,000	37,000	35,000
Modules of elasticity in flexure, psi				
Lengthwise	1.8×10^6	1.3×10^6	1.0×10^6	1.1×10^6
Crosswise	1.3×10^6	1.0×10^6	0.9×10^6	0.8×10^6
Water absorption % (24 hr)	0.9	0.3	1.0	0.5
Coefficient of thermal expansion per °F				
Lengthwise	1.1×10^{-5}	0.94×10^{-5}	1.04×10^{-5}	0.77×10^{-5}
Crosswise	1.4×10^{-5}	1.28×10^{-5}	1.22×10^{-5}	1.04×10^{-5}

SOURCE: Sterling Instrument, New Hyde Park, N.Y.

TABLE 8-12 Properties of Gear Steels

AISI #	Tensile Strength (Psi)	Yield Strength (Psi)	Hardness Brinell #	Elongation % in 2"	Condition
1040	70,000	40,000	137	30	Hot rolled
	99,000	64,000	200	27	Cold worked
	123,000	93,000	255	20	Heat treated
1060	98,000	54,000	200	25	Hot rolled
	122,000	90,000	255	19	Heat treated
	175,000	165,000	350	17	Heat treated
1117	62,000	34,000	120	23	Hot rolled
	71,000	45,000	135	20	Hot rolled
	80,000	68,000	163	19	Cold worked
3140	105,000	90,000	280	.27	Heat treated
	140,000	128,000	350	18	Heat treated
	228,000	209,000	450	11	Heat treated
4140	95,000	60,000	197	26	Annealed
	145,000	120,000	290	17	Normalized
	185,000	165,000	360	12	Heat treated
	215,000	190,000	440	2	Heat treated
4340	108,000	65,500	220	25	Annealed
	122,000	105,000	248	15	Cold drawn
	135,000	120,000	300	16	Heat treated
	242,000	222,000	690	12	Heat treated
4820	150,000	125,000	325	18	Heat treated
	206,000	166,000	415	14	Heat treated
8620	76,000	51,000	155	32	Annealed
	91,000	64,000	185	29	Hot rolled
	142,000	112,000	295	16	Heat treated
	173,000	142,000	375	14	Heat treated
9310	152,000	120,000	350	18	Heat treated
	180,000	140,000	375	15	Heat treated
302, 303	85,000	35,000	150	50	Annealed
& 304	110,000	75,000	240	35	Cold worked
416	75,000	40,000	155	30	Annealed
	125,000	105,000	250	21	Heat treated
	160,000	140,000	350	19	Heat treated
420	95,000	50,000	190	25	Annealed
	230,000	135,000	500	8	Heat treated
440C	110,000	65,000	230	14	Annealed
	285,000	275,000	580	2	Heat treated

SOURCE: Sterling Instrument, New Hyde Park, N.Y.

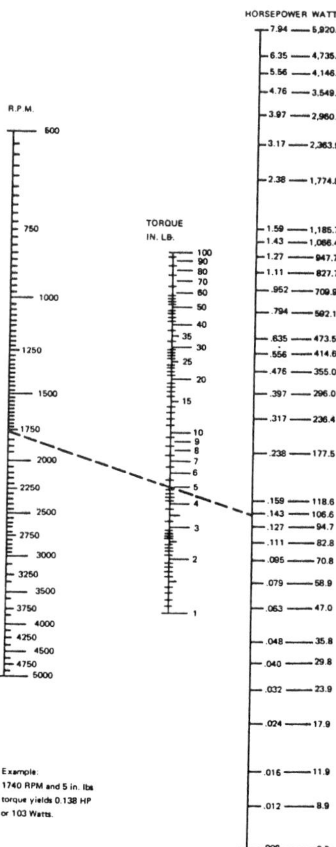

Fig. 8-10 Power nomogram *(Sterling Instrument, New Hyde Park, N.Y.)*.

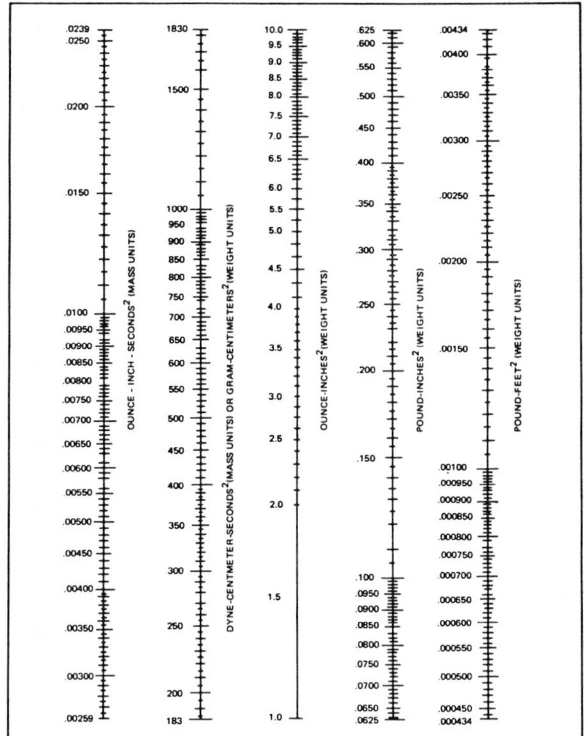

This chart can be used to convert units of moment of inertia as applicable. Values given in the scales above are the equivalent of each other. Care should be exercised to move the decimal point the same number of places and in the proper direction for values of both the input scale and the output scale. Example: .045 oz-in^2 = 8.23 gm-cm^2

Fig. 8-11 Nomograph for conversion of units of inertia *(Sterling Instrument, New Hyde Park, N.Y.).*

TABLE 8-13 Torque Table for Anti-Backlash and Mini-Lash Gears

Pitch Diameter (Linear Scale)	Example No. of Teeth for Each Below Listed Diametral Pitch: 48	64	72	80	96	120	Pitch Diameter (Linear Scale)
3.0000	144	192	216	240	288	360	3.0000
2.5000	120	160	180	200	240	300	2.5000
2.0000	96	128	144	160	192	240	2.0000
1.5000	72	96	108	120	144	180	1.5000
1.0000		64	72	80	96	120	1.0000
.5000		32	36	40	48	60	.5000

Spring Torque (oz. in. Approx) @ maximum & @ Zero Tooth Loading Positions
39 oz. in. @ Max 6 oz. in. @ Zero
29 oz. in. @ Max 4 oz. in. @ Zero
20 oz. in. @ Max 2½ oz. in. @ Zero
15 oz. in @ Max 2 oz. in. @ Zero
4 oz. in. @ Max 0 oz. in @ Zero (No Preload)

Number of Teeth Loaded at Maximum Spring Torque Value for Each Diametral Pitch: 48	64	72	80	96	120
5	7	8	8	10	12
4	6	7	7	9	11
3	5	6	6	8	10
5	6	5	8	7	9
4	5	4	7	10	13
5	6	8	6	9	12
4	5	7	8	8	11
3	6	6	7	10	10
	5	7	6	9	13
	4	6	7	8	12
	3	5	6	7	11
	2	4	5	6	10
		3	4	5	9
			3	4	8
					7
					6
					5

SOURCE: Sterling Instrument, New Hyde Park, N.Y.

TABLE 8-14 Military and Federal Specifications

Item No.	Designation	MIL or FED Specification	Corresp. Finish
		Materials	
1	Aluminum alloy 2024-T4	QQ-A-225/6 (Bar Material)	A, D, E
2	Aluminum alloy 2024-T4	QQ-A-250/4 (Sheet Material)	A, D, E
3	Beryllium copper	QQ-C-530 Alloy 172 cond. H	H
4	Bronze	QQ-B-637 Alloy 464	–
5	No. 1020 Carbon steel	QQ-S-633 comp. FS 1020	C, H
6	No. 1095 Carbon steel	QQ-S-633 comp. FS 1095	C, H
7	Cast aluminum alloy 356-T4	QQ-A-601	A, E
8	Laminated brass	MIL-S-22499 comp 2, CL 1	–
9	Linen phenolic	MIL-P-15035 FBI	G
10	Nylon	MIL-P-20693 Type I	–
11	Oil - less bronze	MIL-B-5687 Type I, comp. A	–
12	Phenolic (molded)	MIL-P-79 FBE	G
13	Phosphor bronze	QQ-B-750 Alloy 510 or 544	H
14	Plexiglass	MIL-P-5425	–
15	Spring wire	MIL-W-6713 Type 302	C
16	No. 302 stainless steel	QQ-S-766	F
17	No. 303 stainless steel	QQ-S-764	B, F
18	No. 416 stainless steel	QQ-S-764	B, F
19	Steel spring	QQ-W-428 Type III, cond A	C
		Finishes	**Corresp. Material**
A	Black anodized	MIL-A-8625 Type II	1, 2, 7
B	Black oxide finish	MIL-C-13924 Class 2	17, 18
C	Cadium plated	QQ-P-416 Type II, CL 2	15, 19
D	Chemical film	MIL-C-5541	1, 2
E	Chromic acid anodized	MIL-A-8625 Type I	1, 2, 7
F	Clear passivated	QQ-P-35	16, 17, 18
G	Fungus proofed	MIL-T-152	9, 12
H	Nickel plated	QQ-N-290 CL 1, Grade G	3, 13

SOURCE: Sterling Instrument, New Hyde Park, N.Y.

TABLE 8-15 Noncircular Gears

Noncircular gears have been known and used for a great many years. However, they usually have been considered mathematical curiosities, rather than practical machine elements.

Actually, these gears can be designed to generate almost any mathematical function that does not reverse its slope, have discontinuities or excessive ratios, or cause mechanical interference.

These gears may be used for either position or velocity function generation. The characteristics of certain stock noncircular gears are summarized in the following table.

	Basic Equations	Displacement	Velocity Ratio $\frac{\omega_B}{\omega_A}$
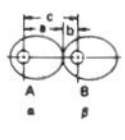 ELLIPTICAL (1-st ORDER)	$\tan\frac{\beta}{2} = \frac{a}{b}\tan\frac{\alpha}{2}$ $\frac{\omega_B}{\omega_A} = \frac{2ab}{c^2 - c(a-b)\cos\alpha - 2ab}$ $\ddot{\beta} \approx .7\left(\frac{a}{b} - \frac{b}{a}\right)\omega_A^2$ for ω_A = const.	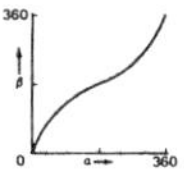	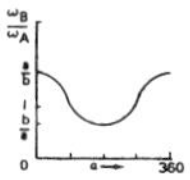
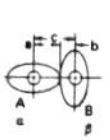 ELLIPTICAL (2-nd ORDER)	$\tan\beta = \frac{a}{b}\tan\alpha$ $\frac{\omega_B}{\omega_A} = \frac{2ab}{(c^2 - 2ab - (a-b)c\cos 2\alpha}$ $\ddot{\beta} \approx 1.4\left(\frac{a}{b} - \frac{b}{a}\right)\omega_A^2$ for ω_A = const.	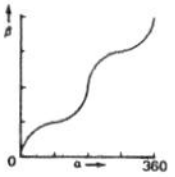	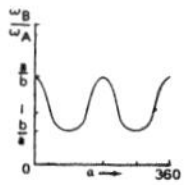
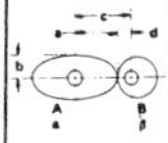 ELLIPTICAL (COMBINATION)	$\tan\frac{\beta}{2} = \frac{a}{2d}\tan\alpha$ $\frac{\omega_B}{\omega_A} = \frac{2ab}{c(a+b) - c(a-b)\cos 2\alpha - 2ab}$ $\ddot{\beta} \approx 1.4\left(\frac{a}{d} - \frac{b}{c-b}\right)\omega_A^2$ for ω_A = const.	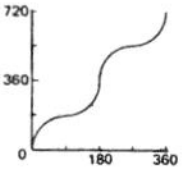	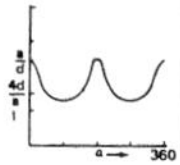
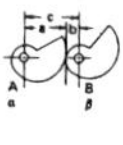 LOGARITHMIC	$\beta = G\log_{10}\alpha$ $\frac{a}{b} = \frac{.43429G}{\alpha}$ G . . . . . degrees/decade	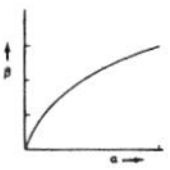	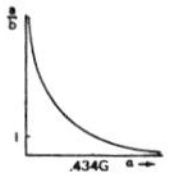
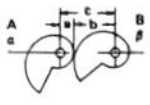 RECIPROCAL	$\beta = \frac{H}{\alpha}$ $\frac{a}{b} = \frac{H}{\alpha^2}$	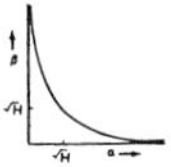	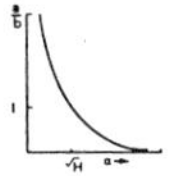

Note: $\ddot{\beta}$ indicates second derivative of β, i.e., angular acceleration.

SOURCE: Sterling Instrument, New Hyde Park, N.Y.

TABLE 8-16 Typical Gear Lubricants

Lubricant Type	Military Specification	Useful Temp Range (deg F)	Remarks	Applications
Oils:				
Petroleum oil	MIL-L-644B	-10 to 250	Good general purpose lubricant	All quality gears having a narrow range of operating temperature.
Diester oil	MIL-L-6085A	-67 to 350	General purpose, low starting torque, stable over a wide temperature range.	Precision instrument gears and small machinery gears.
Diester oil	MIL-L-7808C	-67 to 400	Suitable for oil spray or mist system, at high temperature	High speed gears.
Silicone oil		-100 to 600	Best load carrier of silicone oils, widest temperature range.	Power gears requiring wide temperature ranges.
Greases:				
Diester oil-lithium soap	MIL-G7421A	-100 to 200	Particularly suited for low starting torques, low temperatures.	Moderately loaded gears.
Diester oil-lithium soap	MIL-G-3278A	-67 to 250	General purpose light grease	Precision instrument gears, and generally lightly loaded gears.
Petroleum oil-sodium soap	MIL-L-3545	-20 to 300	High temperature only	High speed and high loads
Silicone oil-non-soap	MIL-G-25013B	-65 to 400	Good high temperature features	High temperature, moderately loaded gear trains
Silicone oil-lithium soap	MIL-G-15719A	0 to 350	High temperature use only	Light to moderately loaded gears, low speeds
Solid Lubricants:				
Molybdenum disulfide (MoS_2) powder	MIL-M-7866A	-350 to 750 above 2000 in vacuum	Highly stable, radiation resistant, useable in vacuum over wide temperature range.	Light duty precision gears.
Graphite in resin binder		-100 to 450	Application by spray and baking up to 350°F. Film thickness .0003 to .001 in.	Low precision and commercial quality gears. Light loads.
MoS_2 in resin binder		-100 to 450	Application by spray and baking up to 350°F. Film thickness .0003 to .001 in. Stable in vacuum.	Space gear trains and vacuum

SOURCE: Sterling Instrument, New Hyde Park, N.Y.

TABLE 8-17 Screw Head Dimensions

SCREW SIZE		2	4	6	8	10	1/4"	5/16"	3/8"
FILLISTER HEAD	A	.140 .124	.183 .166	.226 .208	.270 .250	.313 .292	.414 .389	.518 .490	.622 .590
	B	.083 .070	.107 .094	.132 .118	.156 .141	.180 .165	.237 .219	.295 .276	.355 .333
	C	.062 .055	.079 .072	.096 .089	.113 .106	.130 .123	.170 .161	.211 .201	.253 .242
ROUND HEAD	A	.162 .146	.211 .193	.260 .240	.309 .287	.359 .334	.472 .443	.590 .557	.708 .670
	B	.069 .059	.086 .075	.103 .091	.120 .107	.137 .123	.175 .160	.216 .198	.256 .237
	C	.031 .023	.039 .031	.048 .039	.054 .045	.060 .050	.075 .064	.084 .072	.094 .081
PAN HEAD (Binding)	A	.167 .155	.219 .205	.270 .256	.322 .306	.373 .357	.492 .473	.615 .594	.740 .716
	B	.053 .045	.068 .058	.082 .072	.096 .085	.110 .099	.144 .130	.178 .162	.212 .195
	C	.031 .023	.039 .031	.048 .039	.054 .045	.060 .050	.075 .064	.084 .072	.094 .081
TRUSS HEAD	A	.194 .180	.257 .241	.321 .303	.384 .364	.448 .425	.573 .546	.698 .668	.823 .787
	B	.053 .044	.069 .059	.086 .074	.102 .088	.118 .103	.150 .133	.183 .162	.215 .191
	C	.031 .023	.039 .031	.048 .039	.054 .045	.060 .050	.075 .064	.084 .072	.094 .081
FLAT HEAD 80°	A	.172 .156	.225 .207	.279 .257	.332 .308	.385 .359	.507 .477	.635 .600	.762 .722
	B	.051 .040	.067 .055	.083 .069	.100 .084	.116 .098	.153 .131	.191 .165	.230 .200
	C	.031 .023	.039 .031	.048 .039	.054 .045	.060 .050	.075 .064	.084 .072	.094 .081
FLAT HEAD 100°	A	.172 .156	.225 .207	.279 .257	.332 .308	.385 .359	.507 .477	.635 .600	.762 .722
	B	.036 .029	.048 .040	.060 .051	.072 .062	.083 .072	.110 .097	.138 .123	.165 .148
	C	.031 .023	.039 .031	.048 .039	.054 .045	.060 .050	.075 .064	.084 .072	.094 .081
OVAL HEAD 80°	A	.172 .156	.225 .207	.279 .257	.332 .308	.385 .359	.507 .477	.635 .600	.762 .722
	B	.051 .040	.067 .055	.083 .069	.100 .084	.116 .098	.153 .131	.191 .165	.230 .200
	C	.080 .063	.104 .084	.128 .105	.152 .126	.176 .148	.232 .187	.290 .249	.347 .300
HEX. HEAD	A	.125 .120	.187 .181	.250 .244	.250 .244	.312 .305	.375 .367	.500 .491	.562 .552
	B	.144 .134	.217 .205	.289 .275	.289 .275	.361 .344	.433 .413	.577 .557	.650 .621
	C	.050 .040	.060 .049	.080 .067	.110 .096	.120 .105	.190 .172	.230 .208	.295 .270
HEX. SOCKET HD	A	.140 .136	.183 .178	.226 .221	.270 .265	.312 .306	.375 .367	.468 .457	.562 .553
	B	.086 .083	.112 .109	.138 .134	.164 .160	.190 .185	.250 .244	.312 .306	.375 .368

)URCE: Sterling Instrument, New Hyde Park, N.Y.

TABLE 8-18 Standards for Threads and Accessories

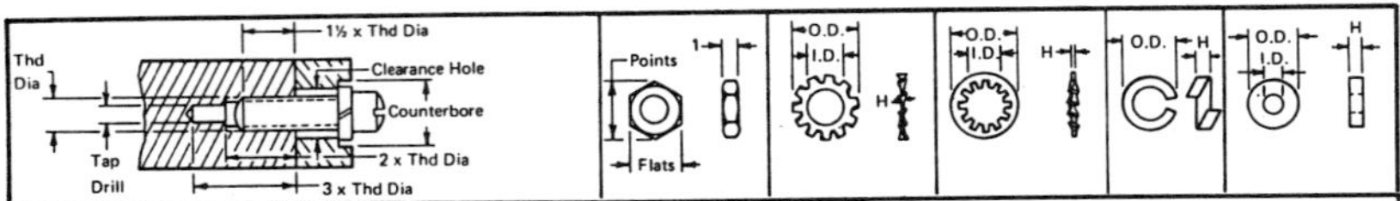

SCREW SIZE & THD DIA	THDS PER INCH	TAP DRILL OR BORE		CLEARANCE DRILL				COUNTERBORE OR SPOTFACE DIA			FASTENING ACCESSORIES														SCREW SIZE
				MACHINED PART		SHEET METAL					MACHINE NUTS			EXTERNAL WASHER			INTERNAL WASHER			SPLIT WASHER		PLAIN WASHER			
		NO.	DEC.	NO.	DEC.	NO.	DEC.	NO. WASHER	SPLIT LOCK-WASHER	PLAIN WASHER	FLTS.	PTS.	H.	O.D.	I.D.	H	O.D.	I.D.	H	O.D.	H	O.D.	I.D.	H	
0 .060	80	3/64	.046	50	.070	48	.076	1/8			.156 .150	.180 .171	.050 .043							.137	.020	.187	.078	.025	0
1 .073	64 72	53	.059	43	.089	3/32	.093	5/32	3/16	9/32	.156 .150	.180 .171	.050 .043							.150	.020	.250	.099	.032	1
2 .086	56 64	50	.070	37	.104	7/64	.109	3/16	3/16	9/32	.187 .180	.217 .205	.066 .057				.185 .175	.095 .089	.013	.175	.020	.250	.099	.032	2
3 .099	48 56	5/64 45	.078 .082	31	.120	1/8	.125	3/16	7/32	9/32	.187 .180	.217 .205	.066 .057	.230 .220	.109 .102	.013	.225 .215	.109 .102	.015	.198	.025	.250	.105	.032	3
4 .112	40 48	43 42	.089 .093	29	.136	9/64	.140	7/32	1/4	3/8	.250 .241	.289 .275	.098 .087	.255 .245	.123 .116	.015	.265 .255	.123 .116	.015	.212	.025	.312	.125	.032	4
6 .138	32 40	36 33	.106 .113	20	.161	11/64	.171	9/32	9/32	3/8	.312 .302	.361 .344	.114 .102	.317 .306	.150 .142	.020	.288 .278	.150 .142	.018	.253	.031	.375	.149	.032	6
8 .164	32 36	29	.136	12	.189	13/64	.203	5/16	5/16	1/2	.343 .332	.397 .378	.130 .117	.381 .370	.176 .168	.020	.336 .325	.176 .168	.020	296	.040	.375	.174	.032	8
10 .190	24 32	25 21	.149 .159	3	.213	7/32	.218	11/32	3/8	17/32	.375 .362	.433 .413	.130 .117	.406 .395	.204 .195	.022	.381 .370	.204 .195	.022	337	.047	.437	.203	.062	10
1/4" .250	20 28	7 3	.201 .213	9/32	.281	19/64	.296	7/16	1/2	11/16	.437 .423	.505 .482	.193 .178	.506 .494	.267 .256	.025	.478 .466	.267 .256	.025	.493	.062	.500	.265	.062	1/4

SOURCE: Sterling Instrument, New Hyde Park, N.Y.

NOTES

NOTES

Section Nine

▪

HEATING AND VENTILATING

TABLE 9-1 Basic Heating Formulas

Formula	Legend
$\text{Btu} = \dfrac{TC}{55}$	Btu = British thermal unit
$\text{Btu/h} = \dfrac{60TC}{55}$	T = Temperature rise, degrees Fahrenheit C = Cubic feet of air delivery
$DS = \dfrac{144C}{60v}$	DS = Cross section of duct, inches v = Flow velocity, feet per minute ΔT = Final temperature of air − temperature of entering air
$\text{Btu}/C = 1.085 \times C \times \Delta T$	Btu/C = Heating capacity of coil, Btu/
$SHL = \dfrac{24hd\,(t_i - t_a)}{t_i - t_e}$	SHL = Seasonal heat loss, Btu h = Hourly building heat loss, Btu d = Total days of heating season t_i = Interior design temperature, °F t_e = Exterior design temperature, °F t_a = Average outside temperature for heating season, °F 24 = Hours per day
$U = \dfrac{1}{R}$ $R = \dfrac{1}{U}$ $R = 1/c$	U = Overall conductance (equal to number of Btu that will flow through 1 ft^2 of structure from air to air due to a temperature difference of 1° F, in 1 h) R = Thermal resistance (reciprocal of conductance) c = Thermal conductance (heat flow through a given thickness of 1 ft^2 of material with 1° F temperature differential)

TABLE 9-2 Air Conditioning Data and Formulas*

Formula	*Legend*
$T_L = \dfrac{T_{CL}}{12{,}000}$	T_L = Load expressed in tons of refrigeration
	T_{CL} = Total cooling load, expressed in Btu/h
Heating Gain Formulas	
$q_s = 1.08\,Q(T_o - T_i)$	q_s = Sensible load due to outside ventilation air, expressed in Btu/h
	Q = Outside air entering conditioned room (make-up air), expressed in cfm (ft³/min)
	T_o = Design dry-bulb temperature of outside air
	T_i = Design dry-bulb temperature of conditioned air
$q_L - 0.67Q(G_o - G_i)$	q_L = Latent load due to outside ventilation air, expressed in Btu/h
	G_o = Moisture content of outside air, expressed in gr/lb of air
	G_i = Moisture content of inside air, expressed in gr/lb of air
q_e = watts × 3.42	q_e = Gain from electric light fixtures, expressed in Btu
q_{ef} = watts × 3.42 × 1.25	q_{ef} = Gain from fluorescent lighting (25 percent increase for ballast), expressed in Btu
$q_{fan} = R \times Q_s$	q_{fan} = Gain from fan motor, expressed in Btu
	R = 0.05 for small systems 0.35 for large systems
	Q_s = Total sensible load, expressed in Btu/h
q_{os} = no. of persons × H_s	q_{os} = Sensible heat load from occupants, expressed in Btu/h
	H_s = Sensible heat per person—see ASHRAE tables
q_{oL} = no. of persons × H_L	q_{oL} = Latent heat load from occupants, expressed in Btu/h
	H_L = Latent heat per person—see ASHRAE tables

*1 ton of refrigeration = amount of cooling that can be done by 1 ton of ice melting in a 24-h period.

TABLE 9-3 Basic Heating Data

1 Btu (British thermal unit) = Heat required to raise 1lb of water 1° F
Btu (British thermal units) = 3.413 × W
Specific heat of water = 1 Btu
Specific heat of air = 0.24 Btu
1 ft^3 of air @ 32° F = 0.0807 lb
1 Btu will heat 1 ft^3 of air 55° F
1 gal (U.S.) of water = 8.33 lb
1 gal (U.S.) of water = 231 in^3 @ 39.2° F
1 ft^3 of water = 7.48 gal
1 ft^3 of water = 62.418 lb @ 39.2° F
1 ft^2 of steam radiation = 240 Btu
1 ft^2 of hot water radiation = 150 Btu
No. 2 fuel oil = 20,571 Btu/lb or 144,000 Btu/gal
Natural gas = 18,000 Btu/lb or 1,030 Btu/ft^3
Propane = 21,564 Btu/lb or 2,572 Btu/ft^3
Butane = 21,440 Btu/lb or 3,200 Btu/ft^3
1 Degree day = 1° F, in mean outdoor temperature, below 65° F (24-h avg.) per day
1 watt (W) = 1 joule (J) per second (s)
1 kWh = 1.341 hph
1 Btu = 778 ft-lb = 1,055 J = 252 cal
1 hph = 2545 Btu = 0.7457 kWh
1 gal of LP gas = 36.6 ft^3

TABLE 9-4 Wind Chill Chart

WIND SPEED M.P.H.	ACTUAL THERMOMETER READING (°F)										
	50	40	30	20	10	0	-10	-20	-30	-40	-50
	WIND CHILL TEMPERATURE (°F)										
CALM	50	40	30	20	10	0	-10	-20	-30	-40	-50
5	48	37	27	16	6	-5	-15	-26	-36	-47	-57
10	40	29	16	-4	-9	-21	-33	-46	-58	-70	-83
15	36	22	9	-5	-18	-36	-46	-58	-72	-85	-99
20	32	18	4	-10	-25	-39	-53	-67	-82	-96	-110
25	30	16	0	-15	-29	-44	-59	-74	-88	-104	-113
30	28	13	-2	-18	-33	-48	-63	-79	-94	-109	-123
35	27	11	-4	-20	-35	-49	-67	-82	-98	-113	-129
40	26	10	-6	-21	-37	-53	-69	-85	-100	-116	-132
	VERY COLD				BITTER COLD			EXTREME COLD			

TABLE 9-5 Beaufort Scale for Wind Velocity

Beaufort scale No.	Description	Indicators of velocity	Velocity, mph
0	Calm	Calm air; smoke rises vertically	Less than 1
1	Light air	Direction of wind shown by smoke drift but not by wind vanes	1-3
2	Slight breeze	Wind felt on face; leaves rustle; ordinary vane moved by wind	4-7
3	Gentle breeze	Leaves and small twigs in constant motion; wind extends light flag	8-12
4	Moderate breeze	Raises dust and loose paper; small branches are moved	13-18
5	Fresh breeze	Small trees in leaf sway; crested wavelets form on inland waters	19-24
6	Strong breeze	Large branches in motion; whistling heard in telegraph wires; umbrellas used with difficulty	25-31
7	High wind	Whole trees in motion; inconvenience when walking against wind	32-38
8	Gale	Breaks twigs off trees; wind generally impedes progress	39-46
9	Strong gale	Slight structural damage occurs to signs; branches broken	47-54
10	Whole gale	Trees uprooted or broken; considerable structural damage occurs	55-63
11	Storm	Very rarely experienced; accompanied by widespread damage	64-75
12	Hurricane		Above 75

SOURCE: N. Foster, *Practical Tables for Building Construction,* copyright © 1963 by McGraw-Hill, Inc., New York. (Used with permission of McGraw-Hill Book Company.)

NOTES

Section Ten

▪

STRUCTURAL

TABLE 10-1 Beam Diagrams and Formulas: Nomenclature

E	Modulus of Elasticity of steel at 29,000 ksi.
I	Moment of Inertia of beam (in.^{4}).
M_{max}	Maximum moment (kip in.).
M_1	Maximum moment in left section of beam (kip in.).
M_2	Maximum moment in right section of beam (kip in.).
M_3	Maximum positive moment in beam with combined end moment conditions (kip in.).
M_x	Moment at distance x from end of beam (kip in.).
P	Concentrated load (kips).
P_1	Concentrated load nearest left reaction (kips).
P_2	Concentrated load nearest right reaction, and of different magnitude than P_1 (kips).
R	End beam reaction for any condition of symmetrical loading (kips).
R_1	Left end beam reaction (kips).
R_2	Right end or intermediate beam reaction (kips).
R_3	Right end beam reaction (kips).
V	Maximum vertical shear for any condition of symmetrical loading (kips).
V_1	Maximum vertical shear in left section of beam (kips).
V_2	Vertical shear at right reaction point, or to left of intermediate reaction point of beam (kips).
V_3	Vertical shear at right reaction point, or to right of intermediate reaction point of beam (kips).
V_x	Vertical shear at distance x from end of beam (kips).
W	Total load on beam (kips).
a	Measured distance along beam (in.).
b	Measured distance along beam which may be greater or less than "a" (in.).
l	Total length of beam between reaction points (in.).
w	Uniformly distributed load per unit of length (kips per in.).
w_1	Uniformly distributed load per unit of length nearest left reaction (kips per in.).
w_2	Uniformly distributed load per unit of length nearest right reaction, and of different magnitude than w_1 (kips per in.).
x	Any distance measured along beam from left reaction (in.).
x_1	Any distance measured along overhang section of beam from nearest reaction point (in.).
Δ_{max}	Maximum deflection (in.).
Δ_a	Deflection at point of load (in.).
Δ_x	Deflection at any point x distance from left reaction (in.).
Δ_{x_1}	Deflection of overhang section of beam at any distance from nearest reaction point (in.).

SOURCE: *Manual of Steel Construction,* 7th ed., American Institute of Steel Construction, Inc. Chicago, Ill, 1973.

TABLE 10-2 Frequently Used Formulas for Beams

The formulas given below are frequently required in structural designing. They are included herein for the convenience of those engineers who have infrequent use for such formulas and hence may find reference necessary.

BEAMS

Flexural stress at extreme fiber:

$$f = Mc/I = M/S$$

Flexural stress at any fiber:

$f = My/I$ y = distance from neutral axis to fiber.

Average vertical shear (for maximum see below):

$v = V/A = V/dt$ (for beams and girders)

Horizontal shearing stress at any section A-A:

$v = VQ/I\,b$ Q = statical moment about the neutral axis of the entire section of that portion of the cross-section lying outside of section A-A,

b = width at section A-A

(Intensity of vertical shear is equal to that of horizontal shear acting normal to it at the same point and both are usually a maximum at mid-height of beam.)

Slope and deflection at any point:

$$EI\frac{d^2y}{dx^2} = M$$

x and y are abscissa and ordinate respectively of a point on the neutral axis, referred to axes of rectangular co-ordinates through a selected point of support.

(First integration gives slopes; second integration gives deflections. Constants of integration must be determined.)

CONTINUOUS BEAMS (THE THEOREM OF THREE MOMENTS)

Uniform load:

$$M_a\frac{l_1}{I_1} + 2M_b\left(\frac{l_1}{I_1} + \frac{l_2}{I_2}\right) + M_c\frac{l_2}{I_2} = -\frac{1}{4}\left(\frac{w_1 l_1^3}{I_1} + \frac{w_2 l_2^3}{I_2}\right)$$

Concentrated loads:

$$M_a\frac{l_1}{I_1} + 2M_b\left(\frac{l_1}{I_1} + \frac{l_2}{I_2}\right) + M_c\frac{l_2}{I_2} = -\frac{P_1 a_1 b_1}{I_1}\left(1 + \frac{a_1}{l_1}\right) - \frac{P_2 a_2 b_2}{I_2}\left(1 + \frac{b_2}{l_2}\right)$$

Considering any two consecutive spans in any continuous structure:

- M_a, M_b, M_c = moments at left, center, and right supports respectively, of any pair of adjacent spans.
- l_1 and l_2 = length of left and right spans respectively, of the pair.
- I_1 and I_2 = moment of inertia of left and right spans respectively.
- w_1 and w_2 = load per unit of length on left and right spans respectively.
- P_1 and P_2 = concentrated loads on left and right spans respectively.
- a_1 and a_2 = distance of concentrated loads from left support in left and right spans respectively.
- b_1 and b_2 = distance of concentrated loads from right support in left and right spans respectively.

The above equations are for beams with moment of inertia constant in each span but differing in different spans, continuous over three or more supports. By writing such an equation for each successive pair of spans and introducing the known values (usually zero) of end moments, all other moments can be found.

SOURCE: *Manual of Steel Construction,* 7th ed., American Institute of Steel Construction, Inc. Chicago, Ill, 1973.

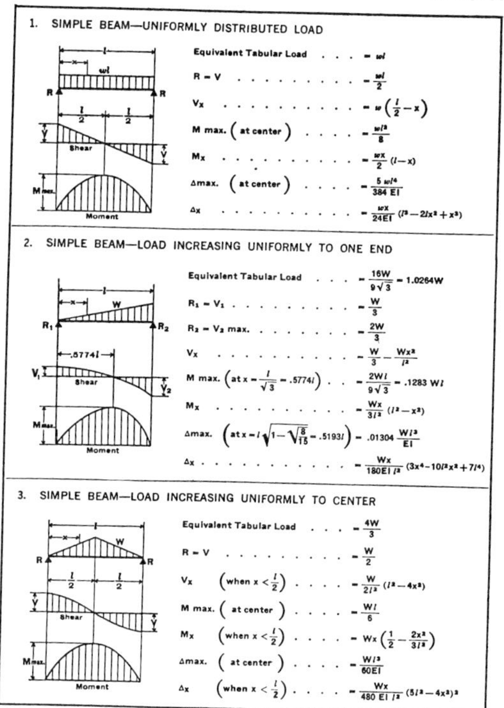

1. SIMPLE BEAM—UNIFORMLY DISTRIBUTED LOAD
Equivalent Tabular Load . . . . $= wl$
R = V . . . . $= \frac{wl}{2}$
Vx . . . . $= w\left(\frac{l}{2} - x\right)$
M max. (at center) . . . . . $= \frac{wl^2}{8}$
Mx . . . . $= \frac{wx}{2}(l-x)$
Δmax. (at center) . . . . . $= \frac{5wl^4}{384EI}$
Δx . . . . $= \frac{wx}{24EI}(l^3 - 2lx^2 + x^3)$
Shear
Moment
2. SIMPLE BEAM—LOAD INCREASING UNIFORMLY TO ONE END
Equivalent Tabular Load . . . . $= \frac{16W}{9\sqrt{3}} = 1.0264W$
R₁ = V₁ . . . . $= \frac{W}{3}$
R₂ = V₂ max. . . . . $= \frac{2W}{3}$
Vx . . . . $= \frac{W}{3} - \frac{Wx^2}{l^2}$
M max. $\left(\text{at } x = \frac{l}{\sqrt{3}} = .5774l\right)$ . . . $= \frac{2Wl}{9\sqrt{3}} = .1283\ Wl$
Mx . . . . $= \frac{Wx}{3l^2}(l^2 - x^2)$
Δmax. $\left(\text{at } x = l\sqrt{1 - \sqrt{\frac{8}{15}}} = .5193l\right)$ $= .01304 \frac{Wl^3}{EI}$
Δx . . . . $= \frac{Wx}{180EIl^2}(3x^4 - 10l^2x^2 + 7l^4)$
.5774l
Shear
Moment
3. SIMPLE BEAM—LOAD INCREASING UNIFORMLY TO CENTER
Equivalent Tabular Load . . . . $= \frac{4W}{3}$
R = V . . . . $= \frac{W}{2}$
Vx $\left(\text{when } x < \frac{l}{2}\right)$ . . . . $= \frac{W}{2l^2}(l^2 - 4x^2)$
M max. (at center) . . . . . $= \frac{Wl}{6}$
Mx $\left(\text{when } x < \frac{l}{2}\right)$ . . . . $= Wx\left(\frac{1}{2} - \frac{2x^2}{3l^2}\right)$
Δmax. (at center) . . . . . $= \frac{Wl^3}{60EI}$
Δx $\left(\text{when } x < \frac{l}{2}\right)$ . . . . $= \frac{Wx}{480EIl^2}(5l^2 - 4x^2)^2$
Shear
Moment

Fig. 10-1 Beam diagrams and formulas for various static loading conditions (equivalent tabular load is the uniformly distributed load given in the beam tables): 1. simple beam—uniformly distributed load; 2. simple beam—load increasing uniformly to one end; 3. simple beam—load increasing uniformly to center. For meaning of symbols see Table 10-1 *(American Institute of Steel Construction, Inc., Chicago, Ill.)*

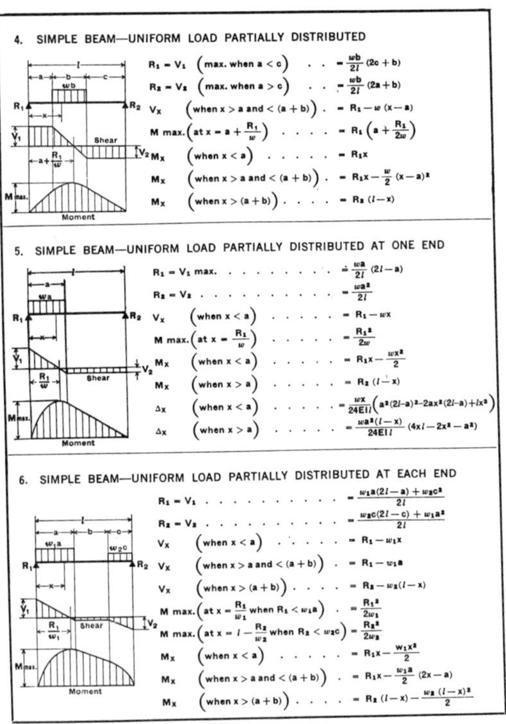

4. SIMPLE BEAM—UNIFORM LOAD PARTIALLY DISTRIBUTED

$R_1 = V_1 \left(\text{max. when } a < c\right) \ldots = \frac{wb}{2l}(2c + b)$

$R_2 = V_2 \left(\text{max. when } a > c\right) \ldots = \frac{wb}{2l}(2a + b)$

$V_x \left(\text{when } x > a \text{ and } < (a + b)\right) = R_1 - w(x - a)$

$M \text{ max.} \left(\text{at } x = a + \frac{R_1}{w}\right) \ldots = R_1\left(a + \frac{R_1}{2w}\right)$

$M_x \left(\text{when } x < a\right) \ldots = R_1 x$

$M_x \left(\text{when } x > a \text{ and } < (a + b)\right) = R_1 x - \frac{w}{2}(x - a)^2$

$M_x \left(\text{when } x > (a + b)\right) \ldots = R_2 (l - x)$

5. SIMPLE BEAM—UNIFORM LOAD PARTIALLY DISTRIBUTED AT ONE END

$R_1 = V_1 \text{ max.} \ldots = \frac{wa}{2l}(2l - a)$

$R_2 = V_2 \ldots = \frac{wa^2}{2l}$

$V_x \left(\text{when } x < a\right) \ldots = R_1 - wx$

$M \text{ max.} \left(\text{at } x = \frac{R_1}{w}\right) \ldots = \frac{R_1^2}{2w}$

$M_x \left(\text{when } x < a\right) \ldots = R_1 x - \frac{wx^2}{2}$

$M_x \left(\text{when } x > a\right) \ldots = R_2 (l - x)$

$\Delta x \left(\text{when } x < a\right) \ldots = \frac{wx}{24EIl}\left(a^2(2l-a)^2 - 2ax^2(2l-a) + lx^3\right)$

$\Delta x \left(\text{when } x > a\right) \ldots = \frac{wa^2(l - x)}{24EIl}(4xl - 2x^2 - a^2)$

6. SIMPLE BEAM—UNIFORM LOAD PARTIALLY DISTRIBUTED AT EACH END

$R_1 = V_1 \ldots = \frac{w_1 a(2l - a) + w_2 c^2}{2l}$

$R_2 = V_2 \ldots = \frac{w_2 c(2l - c) + w_1 a^2}{2l}$

$V_x \left(\text{when } x < a\right) \ldots = R_1 - w_1 x$

$V_x \left(\text{when } x > a \text{ and } < (a + b)\right) = R_1 - w_1 a$

$V_x \left(\text{when } x > (a + b)\right) \ldots = R_2 - w_2(l - x)$

$M \text{ max.} \left(\text{at } x = \frac{R_1}{w_1} \text{ when } R_1 < w_1 a\right) = \frac{R_1^2}{2w_1}$

$M \text{ max.} \left(\text{at } x = l - \frac{R_2}{w_2} \text{ when } R_2 < w_2 c\right) = \frac{R_2^2}{2w_2}$

$M_x \left(\text{when } x < a\right) \ldots = R_1 x - \frac{w_1 x^2}{2}$

$M_x \left(\text{when } x > a \text{ and } < (a + b)\right) = R_1 x - \frac{w_1 a}{2}(2x - a)$

$M_x \left(\text{when } x > (a + b)\right) \ldots = R_2 (l - x) - \frac{w_2 (l - x)^2}{2}$

Fig. 10-1 Beam diagrams and formulas for various static loading conditions *(cont.)*: 4. simple beam—uniform load partially distributed; 5. simple beam—uniform load partially distributed at one end; 6. simple beam—uniform load partially distributed at each end. For meaning of symbols, see Table 10-1 *(American Institute of Steel Construction, Inc., Chicago, Ill.)*

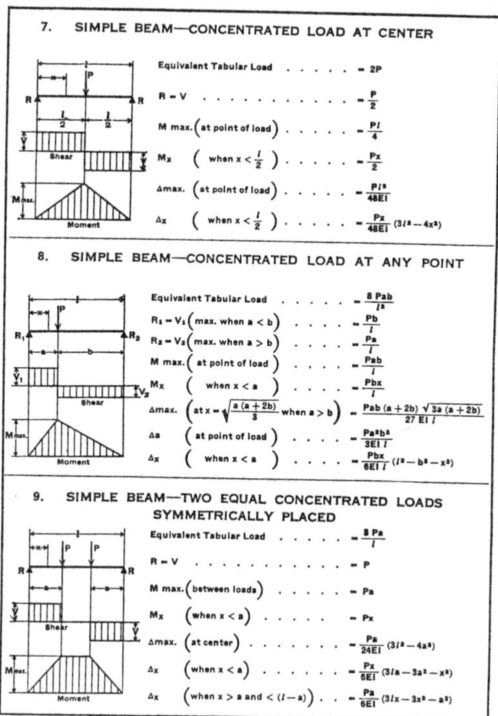

7. SIMPLE BEAM—CONCENTRATED LOAD AT CENTER
Equivalent Tabular Load . . . . . . = 2P
R = V . . . . = P/2
M max. (at point of load) . . . . . . = Pl/4
Mx (when x < l/2) . . . . . . = Px/2
Δmax. (at point of load) . . . . . . = Pl³/48EI
Δx (when x < l/2) . . . . . . = Px/48EI (3l² − 4x²)
Shear
Moment
M max.
8. SIMPLE BEAM—CONCENTRATED LOAD AT ANY POINT
Equivalent Tabular Load . . . . . . = 8 Pab/l²
R₁ = V₁ (max. when a < b) . . . . = Pb/l
R₂ = V₂ (max. when a > b) . . . . = Pa/l
M max. (at point of load) . . . . = Pab/l
Mx (when x < a) . . . . = Pbx/l
Δmax. (at x = √(a (a + 2b)/3) when a > b) = Pab (a + 2b) √(3a (a + 2b)) / 27 EI l
Δa (at point of load) . . . . = Pa²b²/3EI l
Δx (when x < a) . . . . = Pbx/6EI l (l² − b² − x²)
Shear
Moment
M max.
9. SIMPLE BEAM—TWO EQUAL CONCENTRATED LOADS SYMMETRICALLY PLACED
Equivalent Tabular Load . . . . . . = 8 Pa/l
R = V . . . . = P
M max. (between loads) . . . . . . = Pa
Mx (when x < a) . . . . . . = Px
Δmax. (at center) . . . . = Pa/24EI (3l² − 4a²)
Δx (when x < a) . . . . = Px/6EI (3la − 3a² − x²)
Δx (when x > a and < (l − a)) . . = Pa/6EI (3lx − 3x² − a²)
Shear
Moment
M max.

ig. 10-1 Beam diagrams and formulas for various static loading onditions *(cont.)*: 7. simple beam—concentrated load at center; 8. imple beam—concentrated load at any point; 9. simple beam—wo equally concentrated loads symmetrically placed. Equivalent abular load is the uniformly distributed load given in beam tables. 'or meaning of symbols, see Table 10-1. *(American Institute of teel Construction, Inc., Chicago, Ill.)*

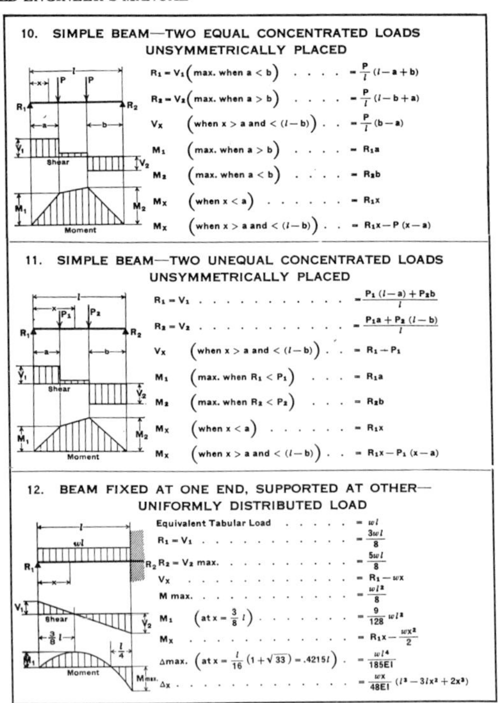

10. SIMPLE BEAM—TWO EQUAL CONCENTRATED LOADS UNSYMMETRICALLY PLACED
R1 = V1 (max. when a < b) . . . . = P/l (l − a + b)
R2 = V2 (max. when a > b) . . . . = P/l (l − b + a)
Vx (when x > a and < (l − b)) . . = P/l (b − a)
M1 (max. when a > b) . . . . = R1a
M2 (max. when a < b) . . . . = R2b
Mx (when x < a) . . . . . . = R1x
Mx (when x > a and < (l − b)) . . = R1x − P (x − a)
Shear
Moment
11. SIMPLE BEAM—TWO UNEQUAL CONCENTRATED LOADS UNSYMMETRICALLY PLACED
R1 = V1 . . . . = (P1 (l − a) + P2b) / l
R2 = V2 . . . . = (P1a + P2 (l − b)) / l
Vx (when x > a and < (l − b)) . . = R1 − P1
M1 (max. when R1 < P1) . . . = R1a
M2 (max. when R2 < P2) . . . = R2b
Mx (when x < a) . . . . . . = R1x
Mx (when x > a and < (l − b)) . . = R1x − P1 (x − a)
Shear
Moment
12. BEAM FIXED AT ONE END, SUPPORTED AT OTHER—UNIFORMLY DISTRIBUTED LOAD
Equivalent Tabular Load . . . . . = wl
R1 = V1 . . . . = 3wl/8
R2 = V2 max. . . . . = 5wl/8
Vx . . . . = R1 − wx
M max. . . . . = wl²/8
M1 (at x = 3/8 l) . . . . = 9/128 wl²
Mx . . . . = R1x − wx²/2
Δmax. (at x = l/16 (1 + √33) = .4215l) . = wl⁴/185EI
Δx . . . . = wx/48EI (l³ − 3lx² + 2x³)
Shear
Moment

Fig. 10-1 Beam diagrams and formulas for various static loading conditions *(cont.)*: 10. simple beam—two equally concentrated loads unsymmetrically placed; 11. simple beam—two unequal concentrated loads unsymmetrically placed; 12. beam fixed at one end supported at other—uniformly distributed load. Equivalent tabular load is the uniformly distributed load given in beam tables. For meaning of symbols, see Table 10-1. *(American Institute of Steel Construction, Inc., Chicago, Ill.)*

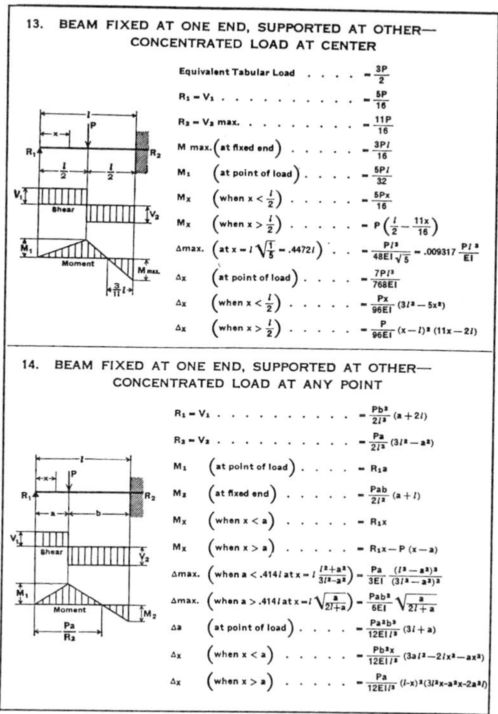

13. BEAM FIXED AT ONE END, SUPPORTED AT OTHER—CONCENTRATED LOAD AT CENTER

Equivalent Tabular Load		$= \frac{3P}{2}$
$R_1 = V_1$		$= \frac{5P}{16}$
$R_2 = V_2$ max.		$= \frac{11P}{16}$
M max.	(at fixed end)	$= \frac{3Pl}{16}$
M_1	(at point of load)	$= \frac{5Pl}{32}$
M_x	$\left(\text{when } x < \frac{l}{2}\right)$	$= \frac{5Px}{16}$
M_x	$\left(\text{when } x > \frac{l}{2}\right)$	$= P\left(\frac{l}{2} - \frac{11x}{16}\right)$
Δmax.	$\left(\text{at } x = l\sqrt{\frac{1}{5}} = .4472l\right)$	$= \frac{Pl^3}{48EI\sqrt{5}} = .009317\frac{Pl^3}{EI}$
Δ_x	(at point of load)	$= \frac{7Pl^3}{768EI}$
Δ_x	$\left(\text{when } x < \frac{l}{2}\right)$	$= \frac{Px}{96EI}(3l^2 - 5x^2)$
Δ_x	$\left(\text{when } x > \frac{l}{2}\right)$	$= \frac{P}{96EI}(x - l)^2(11x - 2l)$

14. BEAM FIXED AT ONE END, SUPPORTED AT OTHER—CONCENTRATED LOAD AT ANY POINT

$R_1 = V_1$		$= \frac{Pb^2}{2l^3}(a + 2l)$
$R_2 = V_2$		$= \frac{Pa}{2l^3}(3l^2 - a^2)$
M_1	(at point of load)	$= R_1a$
M_2	(at fixed end)	$= \frac{Pab}{2l^2}(a + l)$
M_x	(when $x < a$)	$= R_1x$
M_x	(when $x > a$)	$= R_1x - P(x - a)$
Δmax.	$\left(\text{when } a < .414l \text{ at } x = l\frac{l^2+a^2}{3l^2-a^2}\right)$	$= \frac{Pa}{3EI}\frac{(l^2 - a^2)^3}{(3l^2 - a^2)^2}$
Δmax.	$\left(\text{when } a > .414l \text{ at } x = l\sqrt{\frac{a}{2l+a}}\right)$	$= \frac{Pab^2}{6EI}\sqrt{\frac{a}{2l+a}}$
Δ_a	(at point of load)	$= \frac{Pa^2b^3}{12EIl^3}(3l + a)$
Δ_x	(when $x < a$)	$= \frac{Pb^2x}{12EIl^3}(3al^2 - 2lx^2 - ax^2)$
Δ_x	(when $x > a$)	$= \frac{Pa}{12EIl^3}(l-x)^2(3l^2x - a^2x - 2a^2l)$

Fig. 10-1 Beam diagrams and formulas for various static loading conditions *(cont.)*: 13. beam fixed at one end, supported at other—concentrated load at center; 14. beam fixed at one end, supported at other—concentrated load at any point. Equivalent tabular load is the uniformly distributed load given in beam tables. For meaning of symbols, see Table 10-1. *(American Institute of Steel Construction, Inc., Chicago, Ill.)*

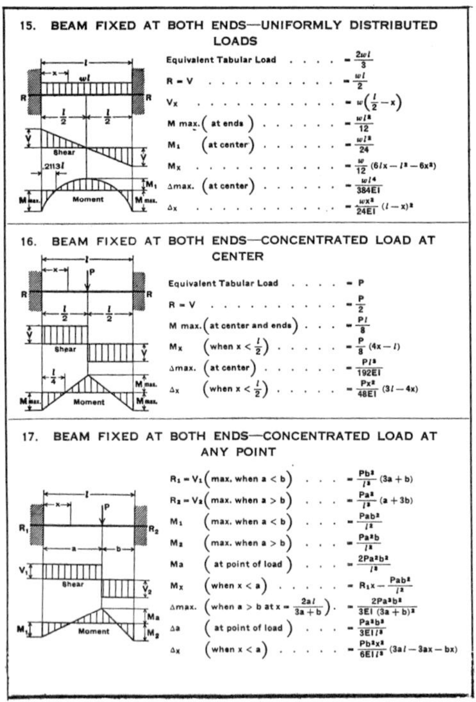

Fig. 10-1 Beam diagrams and formulas for various static loading conditions *(cont.)*: 15. beam fixed at both ends—uniformly distributed loads; 16. beam fixed at both ends—concentrated load at center; 17. beam fixed at both ends—concentrated load at any point. Equivalent tabular load is the uniformly distributed load given in beam tables. For meaning of symbols, see Table 10-1. *(American Institute of Steel Construction, Inc., Chicago, Ill.)*

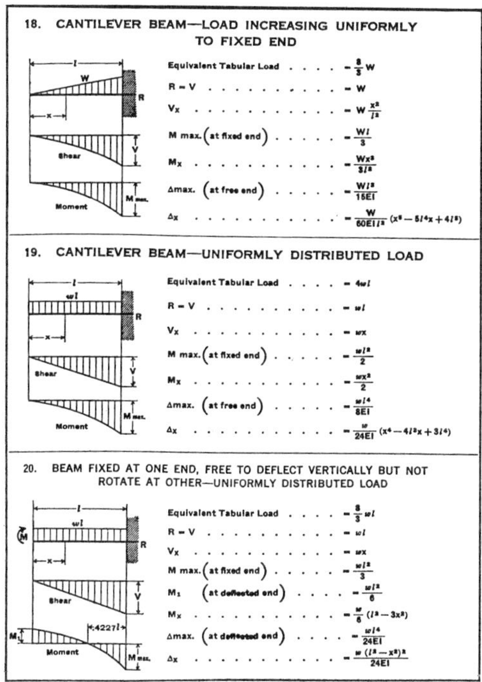

18. CANTILEVER BEAM—LOAD INCREASING UNIFORMLY TO FIXED END
Equivalent Tabular Load . . . . . $= \frac{8}{3}W$
R = V . . . . $= W$
V_x . . . . $= W\frac{x^2}{l^2}$
M max. (at fixed end) . . . . . $= \frac{Wl}{3}$
M_x . . . . $= \frac{Wx^3}{3l^2}$
Δmax. (at free end) . . . . . $= \frac{Wl^3}{15EI}$
Δx . . . . $= \frac{W}{60EIl^2}(x^5 - 5l^4x + 4l^5)$
W
R
x
Shear
V
Moment
M max.
19. CANTILEVER BEAM—UNIFORMLY DISTRIBUTED LOAD
Equivalent Tabular Load . . . . . $= 4wl$
R = V . . . . $= wl$
V_x . . . . $= wx$
M max. (at fixed end) . . . . . $= \frac{wl^2}{2}$
M_x . . . . $= \frac{wx^2}{2}$
Δmax. (at free end) . . . . . $= \frac{wl^4}{8EI}$
Δx . . . . $= \frac{w}{24EI}(x^4 - 4l^3x + 3l^4)$
wl
R
x
Shear
V
Moment
M max.
20. BEAM FIXED AT ONE END, FREE TO DEFLECT VERTICALLY BUT NOT ROTATE AT OTHER—UNIFORMLY DISTRIBUTED LOAD
Equivalent Tabular Load . . . . . $= \frac{8}{3}wl$
R = V . . . . $= wl$
V_x . . . . $= wx$
M max. (at fixed end) . . . . . $= \frac{wl^2}{3}$
M_1 (at deflected end) . . . . $= \frac{wl^2}{6}$
M_x . . . . $= \frac{w}{6}(l^2 - 3x^2)$
Δmax. (at deflected end) . . . . $= \frac{wl^4}{24EI}$
Δx . . . . $= \frac{w(l^2 - x^2)^2}{24EI}$
wl
M
R
x
Shear
V
.4227l
M_1
Moment
M max.

'ig. **10-1** Beam diagrams and formulas for various static loading onditions *(cont.)*: 18. cantilever beam—load increasing uniformly ɔ fixed end; 19. cantilever beam—uniformly distributed load; 20. eam fixed at one end, free to deflect vertically but not rotate at ther—uniformly distributed load. Equivalent tabular load is the niformly distributed load given in beam tables. For meaning of ymbols, see Table 10-1. *(American Institute of Steel Construction, ıc., Chicago, Ill.)*

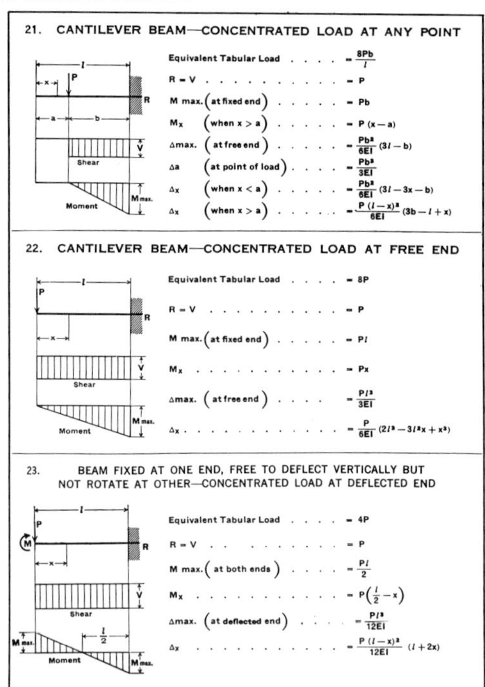

21. CANTILEVER BEAM—CONCENTRATED LOAD AT ANY POINT
Equivalent Tabular Load . . . . = 8Pb/l
R = V . . . . = P
M max. (at fixed end) . . . . . = Pb
Mx (when x > a) . . . . . = P (x − a)
Δmax. (at free end) . . . . . = Pb²/6EI (3l − b)
Δa (at point of load) . . . . = Pb³/3EI
Δx (when x < a) . . . . . = Pb²/6EI (3l − 3x − b)
Δx (when x > a) . . . . . = P (l − x)²/6EI (3b − l + x)
Shear
Moment
M max.
22. CANTILEVER BEAM—CONCENTRATED LOAD AT FREE END
Equivalent Tabular Load . . . . = 8P
R = V . . . . = P
M max. (at fixed end) . . . . . = Pl
Mx . . . . = Px
Δmax. (at free end) . . . . = Pl³/3EI
Δx . . . . = P/6EI (2l³ − 3l²x + x³)
Shear
Moment
M max.
23. BEAM FIXED AT ONE END, FREE TO DEFLECT VERTICALLY BUT NOT ROTATE AT OTHER—CONCENTRATED LOAD AT DEFLECTED END
Equivalent Tabular Load . . . . = 4P
R = V . . . . = P
M max. (at both ends) . . . . = Pl/2
Mx . . . . = P(l/2 − x)
Δmax. (at deflected end) . . . . = Pl³/12EI
Δx . . . . = P (l − x)²/12EI (l + 2x)
Shear
Moment
M max.

Fig. 10-1 Beam diagrams and formulas for various static loading conditions *(cont.)*: 21. cantilever beam—concentrated load at any point; 22. cantilever beam—concentrated load at free end; 23. beam fixed at one end, free to deflect vertically but not rotate at other—concentrated load at deflected end. Equivalent tabular load is the uniformly distributed load given in beam tables. For meaning of symbols, see Table 10-1. *(American Institute of Steel Construction Inc., Chicago, Ill.)*

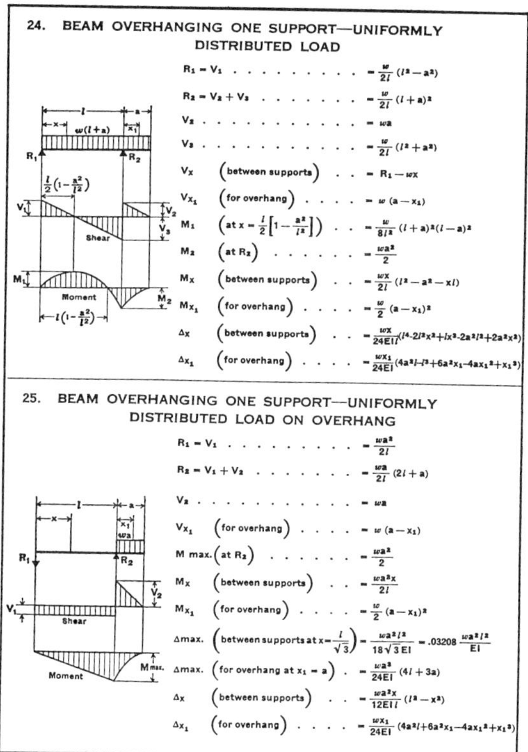

24. BEAM OVERHANGING ONE SUPPORT—UNIFORMLY DISTRIBUTED LOAD

$R_1 = V_1$ . . . . $= \frac{w}{2l}(l^2 - a^2)$

$R_2 = V_2 + V_3$ . . . . $= \frac{w}{2l}(l + a)^2$

V_2 . . . . $= wa$

V_3 . . . . $= \frac{w}{2l}(l^2 + a^2)$

V_x (between supports) . . $= R_1 - wx$

V_{x_1} (for overhang) . . . . $= w(a - x_1)$

M_1 $\left(\text{at } x = \frac{l}{2}\left[1 - \frac{a^2}{l^2}\right]\right)$ . . $= \frac{w}{8l^2}(l + a)^2(l - a)^2$

M_2 (at R_2) . . . . . . $= \frac{wa^2}{2}$

M_x (between supports) . . $= \frac{wx}{2l}(l^2 - a^2 - xl)$

M_{x_1} (for overhang) . . . . $= \frac{w}{2}(a - x_1)^2$

Δx (between supports) . . $= \frac{wx}{24EIl}(l^4 - 2l^2x^2 + lx^3 - 2a^2l^2 + 2a^2x^2)$

Δx_1 (for overhang) . . . . $= \frac{wx_1}{24EI}(4a^2l - l^3 + 6a^2x_1 - 4ax_1^2 + x_1^3)$

25. BEAM OVERHANGING ONE SUPPORT—UNIFORMLY DISTRIBUTED LOAD ON OVERHANG

$R_1 = V_1$ . . . . $= \frac{wa^2}{2l}$

$R_2 = V_1 + V_2$ . . . . $= \frac{wa}{2l}(2l + a)$

V_2 . . . . $= wa$

V_{x_1} (for overhang) . . . . $= w(a - x_1)$

M max. (at R_2) . . . . . . $= \frac{wa^2}{2}$

M_x (between supports) . . $= \frac{wa^2x}{2l}$

M_{x_1} (for overhang) . . . . $= \frac{w}{2}(a - x_1)^2$

Δmax. $\left(\text{between supports at } x = \frac{l}{\sqrt{3}}\right) = \frac{wa^2l^2}{18\sqrt{3}EI} = .03208\frac{wa^2l^2}{EI}$

Δmax. (for overhang at $x_1 = a$) . $= \frac{wa^3}{24EI}(4l + 3a)$

Δx (between supports) . . $= \frac{wa^2x}{12EIl}(l^2 - x^2)$

Δx_1 (for overhang) . . . . $= \frac{wx_1}{24EI}(4a^2l + 6a^2x_1 - 4ax_1^2 + x_1^3)$

ig. 10-1 Beam diagrams and formulas for various static loading onditions *(cont.)*: 24. beam overhanging one support—uniformly istributed load; 25. beam overhanging one support—uniformly dis-ributed load on overhang. For meaning of symbols, see Table 10-1. *merican Institute of Steel Construction, Inc., Chicago, Ill.)*

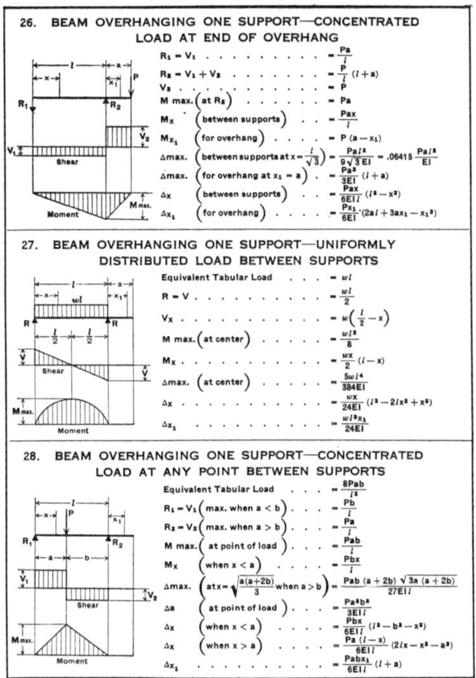

Fig. 10-1 Beam diagrams and formulas for various static loading conditions *(cont.)*: 26. beam overhanging one support—concentrated load at end of overhang; 27. beam overhanging one support—uniformly distributed load between supports; 28. beam overhanging one support—concentrated load at any point between supports. Equivalent tabular load is the uniformly distributed load given in beam tables. For meaning of symbols, see Table 10-1. *(American Institute of Steel Construction, Inc., Chicago, Ill.)*

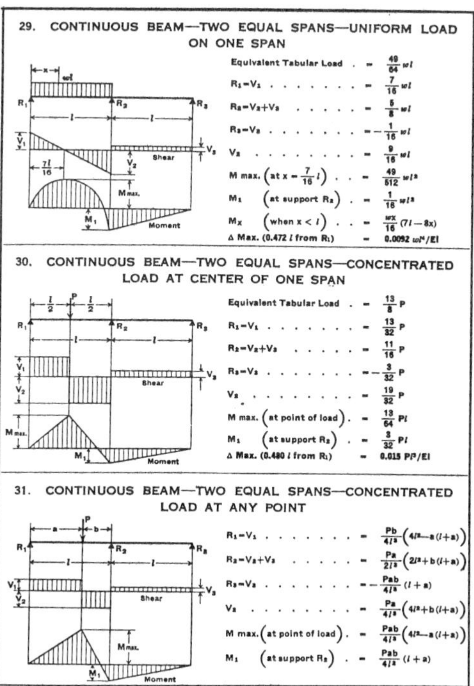

29. CONTINUOUS BEAM—TWO EQUAL SPANS—UNIFORM LOAD ON ONE SPAN
Equivalent Tabular Load . = 49/64 wl
R1=V1 . . . . = 7/16 wl
R2=V2+V3 . . . . . = 5/8 wl
R3=V3 . . . . =−1/16 wl
V2 . . . . = 9/16 wl
M max. (at x = 7/16 l) . . = 49/512 wl²
M1 (at support R2) . = 1/16 wl²
Mx (when x < l) . . = wx/16 (7l − 8x)
Δ Max. (0.472 l from R1) = 0.0092 wl⁴/EI
Shear
Moment
30. CONTINUOUS BEAM—TWO EQUAL SPANS—CONCENTRATED LOAD AT CENTER OF ONE SPAN
Equivalent Tabular Load . = 13/8 P
R1=V1 . . . . = 13/32 P
R2=V2+V3 . . . . . = 11/16 P
R3=V3 . . . . =−3/32 P
V2 . . . . = 19/32 P
M max. (at point of load) . = 13/64 Pl
M1 (at support R2) . = 3/32 Pl
Δ Max. (0.480 l from R1) = 0.015 Pl³/EI
Shear
Moment
31. CONTINUOUS BEAM—TWO EQUAL SPANS—CONCENTRATED LOAD AT ANY POINT
R1=V1 . . . . = Pb/4l³ (4l² − a(l+a))
R2=V2+V3 . . . . . = Pa/2l³ (2l² + b(l+a))
R3=V3 . . . . =− Pab/4l³ (l + a)
V2 . . . . = Pa/4l³ (4l² + b(l+a))
M max. (at point of load) . = Pab/4l³ (4l² − a(l+a))
M1 (at support R2) . = Pab/4l³ (l + a)
Shear
Moment

Fig. 10-1 Beam diagrams and formulas for various static loading conditions *(cont.)*: 29. continuous beam, two equal spans—uniform load on one span; 30. continuous beam, two equal spans—concentrated load at center of one span; 31. continuous beam, two equal spans—concentrated load at any point. Equivalent tabular load is the uniformly distributed load given in beam tables. For meaning of symbols, see Table 10-1. *(American Institute of Steel Construction, Inc., Chicago, Ill.)*

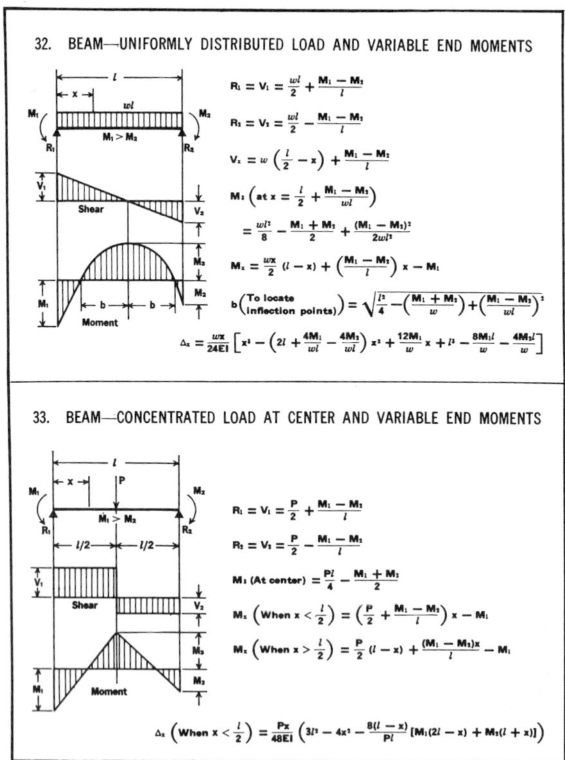

32. BEAM—UNIFORMLY DISTRIBUTED LOAD AND VARIABLE END MOMENTS
l
x
wl
M₁ > M₂
M₁
M₂
R₁
R₂
V₁
V₂
Shear
M₃
Moment
b
$R_1 = V_1 = \frac{wl}{2} + \frac{M_1 - M_2}{l}$
$R_2 = V_2 = \frac{wl}{2} - \frac{M_1 - M_2}{l}$
$V_x = w\left(\frac{l}{2} - x\right) + \frac{M_1 - M_2}{l}$
$M_3 \left(\text{at } x = \frac{l}{2} + \frac{M_1 - M_2}{wl}\right)$
$= \frac{wl^2}{8} - \frac{M_1 + M_2}{2} + \frac{(M_1 - M_2)^2}{2wl^2}$
$M_x = \frac{wx}{2}(l - x) + \left(\frac{M_1 - M_2}{l}\right)x - M_1$
$b \left(\text{To locate inflection points}\right) = \sqrt{\frac{l^2}{4} - \left(\frac{M_1 + M_2}{w}\right) + \left(\frac{M_1 - M_2}{wl}\right)^2}$
$\Delta_x = \frac{wx}{24EI}\left[x^3 - \left(2l + \frac{4M_1}{wl} - \frac{4M_2}{wl}\right)x^2 + \frac{12M_1}{w}x + l^3 - \frac{8M_1 l}{w} - \frac{4M_2 l}{w}\right]$
33. BEAM—CONCENTRATED LOAD AT CENTER AND VARIABLE END MOMENTS
l
x
P
M₁ > M₂
M₁
M₂
R₁
R₂
l/2
l/2
V₁
V₂
Shear
M₃
Moment
$R_1 = V_1 = \frac{P}{2} + \frac{M_1 - M_2}{l}$
$R_2 = V_2 = \frac{P}{2} - \frac{M_1 - M_2}{l}$
$M_3 \text{ (At center)} = \frac{Pl}{4} - \frac{M_1 + M_2}{2}$
$M_x \left(\text{When } x < \frac{l}{2}\right) = \left(\frac{P}{2} + \frac{M_1 - M_2}{l}\right)x - M_1$
$M_x \left(\text{When } x > \frac{l}{2}\right) = \frac{P}{2}(l - x) + \frac{(M_1 - M_2)x}{l} - M_1$
$\Delta_x \left(\text{When } x < \frac{l}{2}\right) = \frac{Px}{48EI}\left(3l^2 - 4x^2 - \frac{8(l - x)}{Pl}[M_1(2l - x) + M_2(l + x)]\right)$

Fig. 10-1 Beam diagrams and formulas for various static loading conditions *(cont.)*: 32. beam—uniformly distributed load and variable end moments; 33. beam—concentrated load at center and variable end moments. For meaning of symbols, see Table 10-1. *(American Institute of Steel Construction, Inc., Chicago, Ill.)*

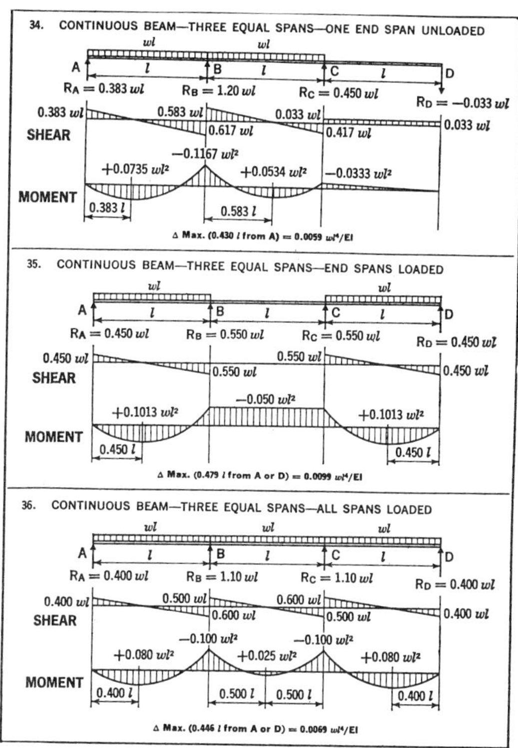

Fig. 10-2 Beam diagrams and deflections for various static loading conditions 34. continuous beam, three equal spans—one end unloaded; 35. continuous beam, three equal spans—end spans loaded; 36. continuous beam, three equal spans—all spans loaded. For meaning of symbols, see Table 10-1. *(American Institute of Steel Construction, Inc., Chicago, Ill.)*

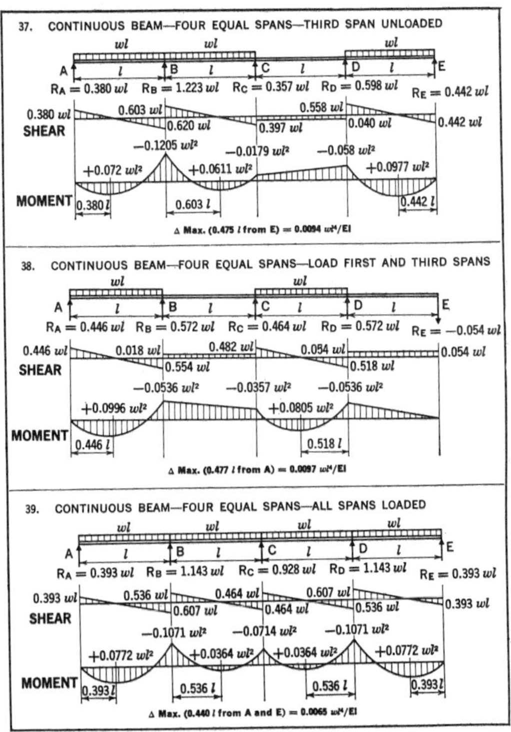

Fig. 10-2 Beam diagrams and deflections for various static loading conditions *(cont.)*: 37. continuous beam, four equal spans—third span unloaded; 38. continuous beam, four equal spans—load on first and third spans; 39. continuous beam, four equal spans—all spans loaded. For meaning of symbols see Table 10-1. *(American Institute of Steel Construction, Inc., Chicago, Ill.)*

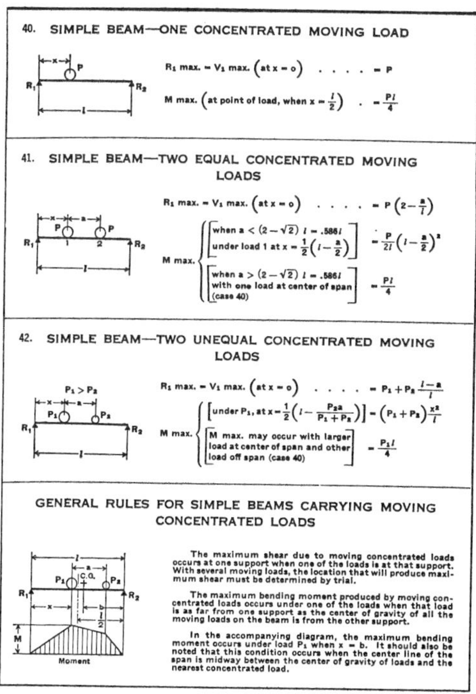

40. SIMPLE BEAM—ONE CONCENTRATED MOVING LOAD
R_1 max. = V_1 max. (at x = o) . . . . . = P
M max. (at point of load, when x = $\frac{l}{2}$) . = $\frac{Pl}{4}$
41. SIMPLE BEAM—TWO EQUAL CONCENTRATED MOVING LOADS
R_1 max. = V_1 max. (at x = o) . . . . . = $P\left(2-\frac{a}{l}\right)$
M max.
when a < $(2-\sqrt{2})\,l$ = .586l under load 1 at x = $\frac{1}{2}\left(l-\frac{a}{2}\right)$ = $\frac{P}{2l}\left(l-\frac{a}{2}\right)^2$
when a > $(2-\sqrt{2})\,l$ = .586l with one load at center of span (case 40) = $\frac{Pl}{4}$
42. SIMPLE BEAM—TWO UNEQUAL CONCENTRATED MOVING LOADS
$P_1 > P_2$
R_1 max. = V_1 max. (at x = o) . . . . = $P_1 + P_2\frac{l-a}{l}$
M max.
[under P_1, at x = $\frac{1}{2}\left(l-\frac{P_2a}{P_1+P_2}\right)$] = $(P_1+P_2)\frac{x^2}{l}$
M max. may occur with larger load at center of span and other load off span (case 40) = $\frac{P_1l}{4}$
GENERAL RULES FOR SIMPLE BEAMS CARRYING MOVING CONCENTRATED LOADS
The maximum shear due to moving concentrated loads occurs at one support when one of the loads is at that support. With several moving loads, the location that will produce maximum shear must be determined by trial.
The maximum bending moment produced by moving concentrated loads occurs under one of the loads when that load is as far from one support as the center of gravity of all the moving loads on the beam is from the other support.
In the accompanying diagram, the maximum bending moment occurs under load P_1 when x = b. It should also be noted that this condition occurs when the center line of the span is midway between the center of gravity of loads and the nearest concentrated load.
C.G.
Moment

Fig. 10-3 Beam diagrams and formulas for various concentrated moving loads (The values given in these formulas do not include impact which varies according to the requirements of each case. For meaning of symbols see Table 10-1.): 40. simple beam—one concentrated moving load; 41. simple beam—two equal concentrated moving loads; 42. simple beam—two unequal concentrated moving loads. *American Institute of Steel Construction, Inc., Chicago, Ill.)*

GENERAL RULES FOR SIMPLE BEAMS CARRYING MOVING CONCENTRATED LOADS*

TABLE 10-3 Camber and Deflection Coefficients

Given the simple span length, the depth of a beam, girder or truss, and the design unit bending stress, the center deflection in inches may be found by multiplying the span length in feet by the tabulated coefficients given in the following table.

For the unit stress values not tabulated, the deflection can be found by the equation 0.00103448 $(L^2 f_b/d)$ where L is the span in feet, f_b is the fiber stress in kips per square inch, and d is the depth in inches.

The maximum fiber stresses listed in this table correspond to the allowable unit stresses as provided in Sections 1.5.1.4.1 and 1.5.1.4.5 of the AISC Specification for steels having yield points ranging between 36 ksi and 65 ksi when $F_b = 0.66F_y$, and between 36 ksi and 100 ksi when $F_b = 0.60F_y$.

The table values, as given, assume a uniformly distributed load. For a single load at center span multiply these factors by 0.80; for two equal concentrated loads at third points, multiply by 1.02. Likewise, for three equal concentrated loads at quarter points multiply by 0.95.

The tabulated factors are correct for beams of constant cross section; reasonably accurate for cover plated beams and girders; and approximate for trusses.

Ratio of Depth/Span	Maximum Fiber Stress in Kips Per Sq. Inch												
	10.0	22.0	24.0	25.2	27.0	28.0	30.0	33.0	36.0	39.0	42.9	54.0	60.0
1/8	.0069	.0152	.0166	.0174	.0186	.0193	.0207	.0228	.0248	.0269	.0296	.0372	.0414
1/9	.0078	.0171	.0186	.0196	.0209	.0217	.0233	.0256	.0279	.0303	.0333	.0419	.0466
1/10	.0086	.0190	.0207	.0217	.0233	.0241	.0259	.0284	.0310	.0336	.0370	.0466	.0517
1/11	.0095	.0209	.0228	.0239	.0256	.0266	.0284	.0313	.0341	.0370	.0407	.0512	.0569
1/12	.0103	.0228	.0248	.0261	.0279	.0290	.0310	.0341	.0372	.0403	.0444	.0559	.0621
1/13	.0112	.0247	.0269	.0282	.0303	.0314	.0336	.0370	.0403	.0437	.0481	.0605	.0672
1/14	.0121	.0266	.0290	.0304	.0326	.0338	.0362	.0398	.0434	.0471	.0518	.0652	.0724
1/15	.0129	.0284	.0310	.0326	.0349	.0362	.0388	.0427	.0466	.0504	.0555	.0698	.0776
1/16	.0138	.0303	.0331	.0348	.0372	.0386	.0414	.0455	.0497	.0538	.0592	.0745	.0828
1/17	.0147	.0322	.0352	.0369	.0396	.0410	.0440	.0484	.0528	.0572	.0629	.0791	.0879
1/18	.0155	.0341	.0372	.0391	.0419	.0434	.0446	.0512	.0559	.0605	.0666	.0838	.0931
1/19	.0164	.0360	.0393	.0413	.0442	.0459	.0491	.0541	.0590	.0639	.0703	.0885	.0983
1/20	.0172	.0379	.0414	.0434	.0466	.0483	.0517	.0569	.0621	.0672	.0740	.0931	.1035
1/21	.0181	.0398	.0434	.0456	.0489	.0507	.0543	.0597	.0652	.0706	.0777	.0978	.1086
1/22	.0190	.0417	.0455	.0478	.0512	.0531	.0569	.0626	.0683	.0740	.0814	.1024	.1138
1/23	.0198	.0436	.0476	.0500	.0535	.0555	.0595	.0654	.0714	.0773	.0851	.1071	.1190
1/24	.0207	.0455	.0497	.0521	.0559	.0579	.0621	.0683	.0745	.0807	.0888	.1117	.1241
1/25	.0216	.0474	.0517	.0543	.0582	.0603	.0647	.0711	.0776	.0841	.0925	.1164	.1293
1/26	.0224	.0493	.0538	.0565	.0605	.0628	.0672	.0740	.0807	.0874	.0962	.1210	.1345
1/27	.0233	.0512	.0559	.0587	.0628	.0652	.0698	.0768	.0838	.0908	.0999	.1257	.1397
1/28	.0241	.0531	.0579	.0608	.0652	.0676	.0724	.0797	.0869	.0941	.1036	.1303	.1448
1/29	.0250	.0550	.0600	.0630	.0675	.0700	.0750	.0825	.0900	.0975	.1073	.1350	.1500
1/30	.0259	.0569	.0621	.0652	.0698	.0724	.0776	.0853	.0931	.1009	.1110	.1397	.1552

SOURCE: *Manual of Steel Construction,* 7th ed., American Institute of Steel Construction, Inc., Chicago, Ill., 1973.

*Reprinted by permission of American Institute of Steel Construction, Inc., Chicago, Ill.

TABLE 10-4 Rivets and Threaded Fasteners (⅝ to 1½ in)—Tension (Allowable loads in kips)

Unfinished Bolts and Threaded Parts

Tension on tensile stress area

ASTM Designation or Yield Stress		Allowable Tensile Stress F_t ksi	Nominal Diameter, in.							
			⅝	¾	⅞	1	1⅛	1¼	1⅜	1½
			Tensile Stress Area, sq. in.							
			0.2260	0.3345	0.4617	0.6057	0.7633	0.9691	1.1549	1.4053
A307 Bolts		20.0	4.52	6.69	9.23	12.11	15.27	19.38	23.10	28.11
Threaded Parts F_y, ksi	36	22.0	4.97	7.36	10.16	13.33	16.79	21.32	25.41	30.92
	42	25.2	5.70	8.41	11.64	15.27	19.23	24.42	29.23	35.53
	45	27.0	6.10	9.03	12.47	16.35	20.61	26.17	31.18	37.94
	50	30.0	6.78	10.04	13.85	18.17	22.90	29.07	34.65	42.16
	55	33.0	7.46	11.04	15.24	19.99	25.19	31.98	38.11	46.37
	60	36.0	8.14	12.04	16.62	21.81	—	—	—	—

The definition of tensile stress area is given in the AISC Specification, Section 1.5.2.1. Values are based on UNC thread dimensions.

Nuts must meet specifications compatible with threaded parts.

For Upset Rods see AISC Specification, Section 1.5.2.1.

Rivets and High Strength Bolts

Tension on gross (nominal) area

ASTM Designation		Allowable Tensile Stress F_t ksi	Nominal Diameter, in.							
			⅝	¾	⅞	1	1⅛	1¼	1⅜	1½
			Gross (Nominal) Area, sq. in.							
			0.3068	0.4418	0.6013	0.7854	0.9940	1.2272	1.4849	1.7671
Rivets	A502-1	20.0	6.14	8.84	12.03	15.71	19.88	24.54	29.70	35.34
	A502-2	27.0	8.28	11.93	16.24	21.21	26.84	33.13	40.09	47.71
Bolts	A325	40.0	12.27	17.67	24.05	31.42	39.76	49.09	59.40	70.68
	A490	54.0[a]	16.57[a]	23.86[a]	32.47[a]	42.41[a]	53.68[a]	66.27[a]	80.18[a]	95.42[a]

[a] For static loading only.

For allowable combined shear and tension loads, see AISC Specification, Section 1.6.3.

SOURCE: *Manual of Steel Construction,* 7th ed., American Institute of Steel Construction, Inc., Chicago, Ill., 1973.

TABLE 10-5 Rivets and Threaded Fasteners (⅝ to ¾ in)—Shear (Allowable loads in kips)

Power Driven Shop and Field Rivets

Diam. — Area	⅝ in. — 0.3068 sq. in.		¾ in. — 0.4418 sq. in.	
ASTM Designation	A502-1	A502-2	A502-1	A502-2
Shear F_v, ksi	15.0	20.0	15.0	20.0
Single Shear, kips	4.60	6.14	6.63	8.84
Double Shear, kips	9.20	12.27	13.25	17.67

Unfinished Bolts, ASTM A307, and Threaded Parts, $F_y = 36$ ksi[a]

Diam. — Area	⅝ in. — 0.3068 sq. in.		¾ in. — 0.4418 sq. in.	
ASTM Designation or Yield Stress, ksi	A307	$F_y = 36$	A307	$F_y = 36$
Shear F_v, ksi	10.0	10.8[a]	10.0	10.8[a]
Single Shear, kips	3.07	3.31	4.42	4.77
Double Shear, kips	6.14	6.63	8.84	9.54

High Strength Bolts in Friction Type Connections and in Bearing Type Connections with Threads in Shear Planes

Diam. — Area	⅝ in. — 0.3068 sq. in.			¾ in. — 0.4418 sq. in.		
[b]ASTM Designation	A325-F A325-N	A490-F	A490-N	A325-F A325-N	A490-F	A490-N
Shear F_v, ksi	15.0	20.0	22.5	15.0	20.0	22.5
Single Shear, kips	4.60	6.14	6.90	6.63	8.84	9.94
Double Shear, kips	9.20	12.27	13.81	13.25	17.67	19.88

High Strength Bolts in Bearing Type Connections with Threads Excluded from Shear Planes

Diam. — Area	⅝ in. — 0.3068 sq. in.		¾ in. — 0.4418 sq. in.	
[b]ASTM Designation	A325-X	A490-X	A325-X	A490-X
Shear F_v, ksi	22.0	32.0	22.0	32.0
Single Shear, kips	6.75	9.82	9.72	14.14
Double Shear, kips	13.50	19.64	19.44	28.28

[a] For threaded parts of material other than $F_y = 36$ ksi steel, use $F_v = 0.30\ F_y$.
[b] The letter suffixes following the ASTM Designations A325 and A490 represent the following:
F: Friction type connection
N: Bearing type connection with threads **included** in shear plane
X: Bearing type connection with threads **excluded** from shear plane

SOURCE: *Manual of Steel Construction,* 7th ed., American Institute of Steel Construction, Inc., Chicago, Ill., 1973.

TABLE 10-6 Rivets and Threaded Fasteners (⅝ to ¾ in) — Bearing (Allowable loads in kips. All rivets and bolts in bearing-type connections)

Diam., in.	⅝								¾							
F_y, ksi	36	42	45	50	55	60	65	100	36	42	45	50	55	60	65	100
Bearing F_p, ksi	48.6	56.7	60.8	67.5	74.3	81.0	87.8	135	48.6	56.7	60.8	67.5	74.3	81.0	87.8	135
MATERIAL THICKNESS ⅛	3.80	4.43	4.75	5.27	5.80	6.33	6.86	10.5	4.56	5.31	5.70	6.33	6.97	7.59	8.23	12.7
3/16	5.70	6.64	7.13	7.91	8.71	9.49	10.3	15.8	6.83	7.97	8.55	9.49	10.4	11.4	12.3	19.0
¼	7.59	8.86	9.50	10.6	11.6	12.7	13.7	21.1	9.11	10.6	11.4	12.7	13.9	15.2	16.5	25.3
5/16	9.49	11.1	11.9	13.2	14.5	15.8	17.1	26.4	11.4	13.3	14.3	15.8	17.4	19.0	20.6	31.6
⅜	11.4	13.3	14.3	15.8	17.4	19.0	20.6	31.6	13.7	15.9	17.1	19.0	20.9	22.8	24.7	38.0
7/16	13.3	15.5	16.6	18.5	20.3	22.1		36.9	15.9	18.6	19.9	22.1	24.4	26.6	28.8	44.3
½	15.2	17.7	19.0	21.1				42.2	18.2	21.3	22.8	25.3	27.9	30.4		50.6
9/16	17.1	19.9	21.4					47.5	20.5	23.9	25.7	28.5	31.3			57.0
⅝	19.0							52.7	22.8	26.6	28.5					63.3
11/16	20.9							58.0	25.1	29.2						69.6
¾								63.3	27.3							75.9
13/16								68.6	29.6							82.3
⅞								73.8								88.6
15/16								79.1								94.9
1	30.4	35.4	38.0	42.2	46.4	50.6	54.9	84.4	36.5	42.5	45.6	50.6	55.7	60.8	65.9	101

This table not applicable to fasteners in friction type connections.

F_y is the yield stress of the connected material; see AISC Specification, Sect. 1.5.2.2.

F_p, the unit bearing stress, applies equally to conditions of single shear and enclosed bearing.

Values for thicknesses not listed may be obtained by multiplying the unlisted thickness by the value given for a 1″ thickness in the appropriate F_y column.

Values for F_y's not listed may be obtained by multiplying the value given for $F_y = 100$ ksi by the unlisted F_y and dividing by 100.

SOURCE: *Manual of Steel Construction,* 7th ed., American Institute of Steel Construction, Inc., Chicago, Ill., 1973.

TABLE 10-7 Rivets and Threaded Fasteners (⅞ to 1 in)—Shear (Allowable loads in kips)

Power Driven Shop and Field Rivets

Diam.—Area	⅞ in.—0.6013 sq. in.		1 in.—0.7854 sq. in.	
ASTM Designation	A502-1	A502-2	A502-1	A502-2
Shear F_v, ksi	15.0	20.0	15.0	20.0
Single Shear, kips	9.02	12.03	11.78	15.71
Double Shear, kips	18.04	24.05	23.56	31.42

Unfinished Bolts, ASTM A307, and Threaded Parts, F_y = 36 ksi[a]

Diam.—Area	⅞ in.—0.6013 sq. in.		1 in.—0.7854 sq. in.	
ASTM Designation or Yield Stress, ksi	A307	$F_y = 36$	A307	$F_y = 36$
Shear F_v, ksi	10.0	10.8[a]	10.0	10.8[a]
Single Shear, kips	6.01	6.49	7.85	8.48
Double Shear, kips	12.03	12.99	15.71	16.96

High Strength Bolts in Friction Type Connections and in Bearing Type Connections with Threads in Shear Planes

Diam.—Area	⅞ in.—0.6013 sq. in.			1 in.—0.7854 sq. in.		
[b]ASTM Designation	A325-F A325-N	A490-F	A490-N	A325-F A325-N	A490-F	A490-N
Shear F_v ksi	15.0	20.0	22.5	15.0	20.0	22.5
Single Shear, kips	9.02	12.03	13.53	11.78	15.71	17.67
Double Shear, kips	18.04	24.05	27.06	23.56	31.42	35.34

High Strength Bolts in Bearing Type Connections with Threads Excluded from Shear Planes

Diam.—Area	⅞ in.—0.6013 sq. in.		1 in.—0.7854 sq. in.	
[b]ASTM Designation	A325-X	A490-X	A325-X	A490-X
Shear F_v, ksi	22.0	32.0	22.0	32.0
Single Shear, kips	13.23	19.24	17.28	25.13
Double Shear, kips	26.46	38.48	34.56	50.27

[a] For threaded parts of material other than $F_y = 36$ ksi steel, use $F_v = 0.30\ F_y$.

[b] The letter suffixes following the ASTM Designations A325 and A490 represent the following:
F: Friction type connection
N: Bearing type connection with threads **included** in shear plane
X: Bearing type connection with threads **excluded** from shear plane

SOURCE: *Manual of Steel Construction,* 7th ed., American Institute of Steel Construction, Inc., Chicago, Ill., 1973.

TABLE 10-8 Rivets and Threaded Fasteners (⅞ to 1 in)—Bearing (Allowable loads in kips. All rivets and bolts in bearing-type connections.)

Diam., in.	⅞								1							
F_y, ksi	36	42	45	50	55	60	65	100	36	42	45	50	55	60	65	100
Bearing F_p, ksi	48.6	56.7	60.8	67.5	74.3	81.0	87.8	135	48.6	56.7	60.8	67.5	74.3	81.0	87.8	135
MATERIAL THICKNESS																
⅛	5.32	6.20	6.65	7.38	8.13	8.86	9.60	14.8	6.08	7.09	7.60	8.44	9.29	10.1	11.0	16.9
3/16	7.97	9.30	9.98	11.1	12.2	13.3	14.4	22.1	9.11	10.6	11.4	12.7	13.9	15.2	16.5	25.3
¼	10.6	12.4	13.3	14.8	16.3	17.7	19.2	29.5	12.2	14.2	15.2	16.9	18.6	20.3	22.0	33.8
5/16	13.3	15.5	16.6	18.5	20.3	22.1	24.0	36.9	15.2	17.7	19.0	21.1	23.2	25.3	27.4	42.2
⅜	15.9	18.6	20.0	22.1	24.4	26.6	28.8	44.3	18.2	21.3	22.8	25.3	27.9	30.4	32.9	50.6
7/16	18.6	21.7	23.3	25.8	28.4	31.0	33.6	51.7	21.3	24.8	26.6	29.5	32.5	35.4	38.4	59.1
½	21.3	24.8	26.6	29.5	32.5	35.4	38.4	59.1	24.3	28.4	30.4	33.8	37.2	40.5	43.9	67.5
9/16	23.9	27.9	29.9	33.2	36.6	39.9	43.2	66.4	27.3	31.9	34.2	38.0	41.8	45.6	49.4	75.9
⅝	26.6	31.0	33.3	36.9	40.6			73.8	30.4	35.4	38.0	42.2	46.4	50.6	54.9	84.4
11/16	29.2	34.1	36.6	40.6				81.2	33.4	39.0	41.8	46.4	51.1			92.8
¾	31.9	37.2	39.9					88.6	36.5	42.5	45.6	50.6				101
13/16	34.6	40.3						96.0	39.5	46.1	49.4					110
⅞	37.2							103	42.5	49.6	53.2					118
15/16	39.9							111	45.6	53.2						127
1	42.5	49.6	53.2	59.1	65.0	70.9	76.8	118	48.6	56.7	60.8	67.5	74.3	81.0	87.8	135
1 1/16								126	51.6							143

This table not applicable to fasteners in friction type connections.

F_y is the yield stress of the connected material; see AISC Specification, Sect. 1.5.2.2.

F_p, the unit bearing stress, applies equally to conditions of single shear and enclosed bearing.

Values for thicknesses not listed may be obtained by multiplying the unlisted thickness by the value given for a 1″ thickness in the appropriate F_y column.

Values for F_y's not listed may be obtained by multiplying the value given for $F_y = 100$ ksi by the unlisted F_y and dividing by 100.

SOURCE: *Manual of Steel Construction,* 7th ed., American Institute of Steel Construction, Inc., Chicago, Ill., 1973.

TABLE 10-9 Rivets and Threaded Fasteners (1⅛ to 1¼ in)—Shear (Allowable loads in kips)

Power Driven Shop and Field Rivets				
Diam.—Area	1⅛ in.—0.9940 sq. in.		1¼ in.—1.2272 sq. in.	
ASTM Designation	A502-1	A502-2	A502-1	A502-2
Shear F_v, ksi	15.0	20.0	15.0	20.0
Single Shear, kips Double Shear, kips	14.91 29.82	19.88 39.76	18.41 36.82	24.54 49.09

Unfinished Bolts, ASTM A307, and Threaded Parts, F_y = 36 ksi[a]				
Diam.—Area	1⅛ in.—0.9940 sq. in.		1¼ in.—1.2272 sq. in.	
ASTM Designation or Yield Stress, ksi	A307	F_y = 36	A307	F_y = 36
Shear F_v, ksi	10.0	10.8[a]	10.0	10.8[a]
Single Shear, kips Double Shear, kips	9.94 19.88	10.74 21.47	12.27 24.54	13.25 26.51

High Strength Bolts in Friction Type Connections and in Bearing Type Connections with Threads in Shear Planes						
Diam.—Area	1⅛ in.—0.9940 sq. in.			1¼ in.—1.2272 sq. in.		
[b]ASTM Designation	A325-F A325-N	A490-F	A490-N	A325-F A325-N	A490-F	A490-N
Shear F_v, ksi	15.0	20.0	22.5	15.0	20.0	22.5
Single Shear, kips Double Shear, kips	14.91 29.82	19.88 39.76	22.37 44.73	18.41 36.82	24.54 49.09	27.61 55.22

High Strength Bolts in Bearing Type Connections with Threads Excluded from Shear Planes				
Diam.—Area	1⅛ in.—0.9940 sq. in.		1¼ in.—1.2272 sq. in.	
[b]ASTM Designation	A325-X	A490-X	A325-X	A490-X
Shear F_v, ksi	22.0	32.0	22.0	32.0
Single Shear, kips Double Shear, kips	21.87 43.74	31.81 63.62	27.00 54.00	39.27 78.54

[a] For threaded parts of material other than F_y = 36 ksi steel, use $F_v = 0.30\ F_y$.
[b] The letter suffixes following the ASTM Designations A325 and A490 represent the following:
F: Friction type connection
N: Bearing type connection with threads **included** in shear plane
X: Bearing type connection with threads **excluded** from shear plane

SOURCE: *Manual of Steel Construction,* 7th ed., American Institute of Steel Construction, Inc., Chicago, Ill., 1973.

TABLE 10-10 Rivets and Threaded Fasteners (1⅛ to 1¼ in)—Bearing (Allowable loads in kips. All rivets and bolts in bearing-type connection.)

Diam., in.	1⅛								1¼							
F_y, ksi	36	42	45	50	55	60	65	100	36	42	45	50	55	60	65	100
Bearing F_p, ksi	48.6	56.7	60.8	67.5	74.3	81.0	87.8	135	48.6	56.7	60.8	67.5	74.3	81.0	87.8	135
MATERIAL THICKNESS																
⅛	6.83	7.97	8.55	9.49	10.4	11.4	12.3	19.0	7.59	8.86	9.50	10.6	11.6	12.7	13.7	21.1
3/16	10.2	12.0	12.8	14.2	15.7	17.1	18.5	28.5	11.4	13.3	14.3	15.8	17.4	19.0	20.6	31.6
¼	13.7	15.9	17.1	19.0	20.9	22.8	24.7	38.0	15.2	17.7	19.0	21.1	23.2	25.3	27.4	42.2
5/16	17.1	19.9	21.4	23.7	26.1	28.5	30.9	47.5	19.0	22.1	23.7	26.4	29.0	31.6	34.3	52.7
⅜	20.5	23.9	25.7	28.5	31.3	34.2	37.0	57.0	22.8	26.6	28.5	31.6	34.8	38.0	41.2	63.3
7/16	23.9	27.9	29.9	33.2	36.6	39.9	43.2	66.4	26.6	31.0	33.3	36.9	40.6	44.3	48.0	73.8
½	27.3	31.9	34.2	38.0	41.8	45.6	49.4	75.9	30.4	35.4	38.0	42.2	46.4	50.6	54.9	84.4
9/16	30.8	35.9	38.5	42.7	47.0	51.3	55.6	85.4	34.2	39.9	42.7	47.5	52.2	57.0	61.7	94.9
⅝	34.2	39.9	42.7	47.5	52.2	57.0	61.7	94.9	38.0	44.3	47.5	52.7	58.1	63.3	68.6	105
11/16	37.6	43.9	47.0	52.2	57.5	62.6	67.9	104	41.8	48.7	52.5	58.0	63.9	69.6	75.5	116
¾	41.0	47.8	51.3	57.0	62.7	68.3		114	45.6	53.2	57.0	63.3	69.7	75.9	82.3	127
13/16	44.4	51.8	55.6	61.7	67.9			123	49.4	57.6	61.7	68.6	75.5	82.3		137
⅞	47.8	55.8	59.9	66.4				133	53.2	62.0	66.5	73.8	81.3			148
15/16	51.3	59.8	64.1					142	57.0	66.4	71.3	79.1				158
1	54.7	63.8	68.4	75.9	83.6	91.1	98.8	152	60.8	70.9	76.0	84.4	92.9	101	110	169
1 1/16	58.1							161	64.5	75.3	80.7					179
1⅛	61.5							171	68.3	79.7						190
1 3/16	64.9							180	72.1							200
1¼									75.9							211
1 5/16									79.7							221

This table not applicable to fasteners in friction type connections.

F_y is the yield stress of the connected material; see AISC Specification, Sect. 1.5.2.2.

F_p, the unit bearing stress, applies equally to conditions of single shear and enclosed bearing.

Values for thicknesses not listed may be obtained by multiplying the unlisted thickness by the value given for a 1" thickness in the appropriate F_y column.

Values for F_y's not listed may be obtained by multiplying the value given for $F_y = 100$ ksi by the unlisted F_y and dividing by 100.

SOURCE: *Manual of Steel Construction,* 7th ed., American Institute of Steel Construction, Inc., Chicago, Ill., 1973.

TABLE 10-11 Rivets and Threaded Fasteners (1⅜ to 1½ in)—Shear (Allowable load in kips)

Power Driven Shop and Field Rivets

Diam. — Area	1⅜ in. — 1.4849 sq. in.		1½ in. — 1.7671 sq. in.	
ASTM Designation	A502-1	A502-2	A502-1	A502-2
Shear F_v, ksi	15.0	20.0	15.0	20.0
Single Shear, kips Double Shear, kips	22.27 44.55	29.70 59.40	26.51 53.01	35.34 70.68

Unfinished Bolts, ASTM A307, and Threaded Parts, F_y = 36 ksi[a]

Diam. — Area	1⅜ in. — 1.4849 sq. in.		1½ in. — 1.7621 sq. in.	
ASTM Designation or Yield Stress, ksi	A307	$F_y = 36$	A307	$F_y = 36$
Shear F_v, ksi	10.0	10.8[a]	10.0	10.8[a]
Single Shear, kips Double Shear, kips	14.85 29.70	16.04 32.07	17.67 35.34	19.08 38.17

High Strength Bolts in Friction Type Connections and in Bearing Type Connections with Threads in Shear Planes

Diam. — Area	1⅜ in. — 1.4849 sq. in.			1½ in. — 1.7671 sq. in.		
[b]ASTM Designation	A325-F A325-N	A490-F	A490-N	A325-F A325-N	A490-F	A490-N
Shear F_v, ksi	15.0	20.0	22.5	15.0	20.0	22.5
Single Shear, kips Double Shear, kips	22.27 44.55	29.70 59.40	33.41 66.82	26.51 53.01	35.34 70.68	39.76 79.52

High Strength Bolts in Bearing Type Connections with Threads Excluded from Shear Planes

Diam. — Area	1⅜ in. — 1.4849 sq. in.		1½ in. — 1.7671 sq. in.	
[b]ASTM Designation	A325-X	A490-X	A325-X	A490-X
Shear F_v, ksi	22.0	32.0	22.0	32.0
Single Shear, kips Double Shear, kips	32.67 65.34	47.52 95.03	38.88 77.75	56.55 113.09

[a] For threaded parts of material other than F_y = 36 ksi steel, use $F_v = 0.30\ F_y$.

[b] The letter suffixes following the ASTM Designations A325 and A490 represent the following:
F: Friction type connection
N: Bearing type connection with threads **included** in shear plane
X: Bearing type connection with threads **excluded** from shear plane

SOURCE: *Manual of Steel Construction,* 7th ed., American Institute of Steel Construction, Inc., Chicago, Ill., 1973.

TABLE 10-12 Rivets and Threaded Fasteners (1⅜ to 1½ in)—Bearing (Allowable loads in kips. All rivets and bolts in bearing-type connections.)

Diam., in.	1⅜								1½							
F_y, ksi	36	42	45	50	55	60	65	100	36	42	45	50	55	60	65	100
Bearing F_p, ksi	48.6	56.7	60.8	67.5	74.3	81.0	87.8	135	48.6	56.7	60.8	67.5	74.3	81.0	87.8	135
MATERIAL THICKNESS																
⅛	8.35	9.75	10.5	11.6	12.8	13.9	15.1	23.2	9.11	10.6	11.4	12.7	13.9	15.2	16.5	25.3
3/16	12.5	14.6	15.7	17.4	19.2	20.9	22.6	34.8	13.7	15.9	17.1	19.0	20.9	22.8	24.7	38.0
¼	16.7	19.5	20.9	23.2	25.5	27.8	30.2	46.4	18.2	21.3	22.8	25.3	27.9	30.4	32.9	50.6
5/16	20.9	24.4	26.1	29.0	31.9	34.8	37.7	58.0	22.8	26.6	28.5	31.6	34.8	38.0	41.2	63.3
⅜	25.1	29.2	31.3	34.8	38.3	41.8	45.3	69.6	27.3	31.9	34.2	38.0	41.8	45.6	49.4	75.9
7/16	29.2	34.1	36.6	40.6	44.7	48.7	52.8	81.2	31.9	37.2	39.9	44.3	48.8	53.2	57.6	88.6
½	33.4	39.0	41.8	46.4	51.1	55.7	60.4	92.8	36.5	42.5	45.6	50.6	55.7	60.8	65.9	101
9/16	37.6	43.9	47.0	52.2	57.5	62.6	67.9	104	41.0	47.8	51.3	57.0	62.7	68.3	74.1	114
⅝	41.8	48.7	52.3	58.0	63.9	69.6	75.5	116	45.6	53.2	57.0	63.3	69.7	75.9	82.3	127
11/16	45.9	53.6	57.5	63.8	70.2	76.6	83.0	128	50.1	58.5	62.7	69.6	76.6	83.5	90.5	139
¾	50.1	58.5	62.7	69.6	76.6	83.5	90.5	139	54.7	63.8	68.4	75.9	83.6	91.1	98.8	152
13/16	54.3	63.3	67.9	75.4	83.0	90.5	98.1	151	59.2	69.1	74.1	82.3	90.6	98.7	107	165
⅞	58.5	68.2	73.1	81.2	89.4	97.5		162	63.8	74.4	79.8	88.6	97.5	106	115	177
15/16	62.7	73.1	78.4	87.0	95.8			174	68.3	79.7	85.5	94.9	104	114		190
1	66.8	78.0	83.6	92.8	102	111	121	186	72.9	85.1	91.2	101	111	122	132	203
1 1/16	71.0	82.8	88.8	98.6				197	77.5	90.4	96.9	108	118			215
1⅛	75.2	87.7	94.1					209	82.0	95.7	103	114				228
1 3/16	79.4	92.6	99.3					220	86.6	101	108					240
1¼	83.5	97.5						232	91.1	106	114					253
1 5/16	87.7							244	95.7	112						266
1⅜	91.9							255	100	117						278
1 7/16	96.1							267	105							291
1½									109							304
1 9/16									114							316

This table not applicable to fasteners in friction type connections.

F_y is the yield stress of the connected material; see AISC Specification, Sect. 1.5.2.2.

F_p, the unit bearing stress, applies equally to conditions of single shear and enclosed bearing.

Values for thicknesses not listed may be obtained by multiplying the unlisted thickness by the value given for a 1" thickness in the appropriate F_y column.

Values for F_y's not listed may be obtained by multiplying the value given for $F_y = 100$ ksi by the unlisted F_y and dividing by 100.

SOURCE: *Manual of Steel Construction,* 7th ed., American Institute of Steel Construction, Inc., Chicago, Ill., 1973.

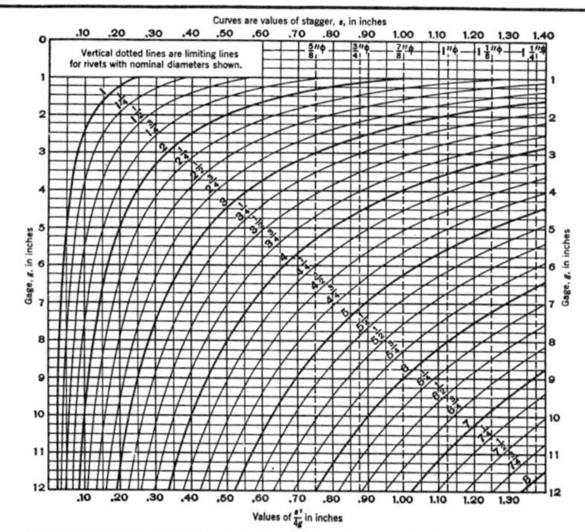

The above chart will simplify the application of the rule for net width, Sections 1.14.3 and 1.14.4 of the AISC Specification. Entering the chart at left or right with the gage g and proceeding horizontally to intersection with the curve for the pitch s, thence vertically to top or bottom, the value of $s^2/4g$ may be read directly.

Step 1 of the example below illustrates the application of the rule and the use of the chart. Step 2 illustrates the application of the 85% of gross area limitation.

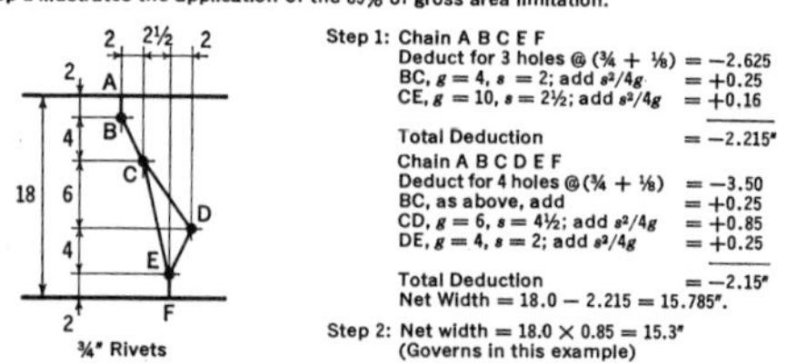

Step 1: Chain A B C E F
Deduct for 3 holes @ (¾ + ⅛) = −2.625
BC, $g = 4$, $s = 2$; add $s^2/4g$ = +0.25
CE, $g = 10$, $s = 2½$; add $s^2/4g$ = +0.16

Total Deduction = −2.215"

Chain A B C D E F
Deduct for 4 holes @ (¾ + ⅛) = −3.50
BC, as above, add = +0.25
CD, $g = 6$, $s = 4½$; add $s^2/4g$ = +0.85
DE, $g = 4$, $s = 2$; add $s^2/4g$ = +0.25

Total Deduction = −2.15"
Net Width = 18.0 − 2.215 = 15.785".

Step 2: Net width = 18.0 × 0.85 = 15.3"
(Governs in this example)

In comparing the path CDE with the path CE, it is seen that if the sum of the two values of $s^2/4g$ for CD and DE exceed the single value of $s^2/4g$ for CE, by more than the deduction for one hole, then the path CDE is not critical as compared with CE.

Evidently if the value of $s^2/4g$ for one leg CD of the path CDE is greater than the deduction for one hole, the path CDE cannot be critical as compared with CE. The vertical dotted lines in the chart serve to indicate, for the respective rivet diameters noted at the top thereof, that any value of $s^2/4g$ to the right of such line is derived from a non-critical chain which need not be further considered.

Fig. 10-4 Net section of tension members *(American Institute of Steel Construction, Inc., Chicago, Ill.)*

TABLE 10-13 Rivets—Spacing

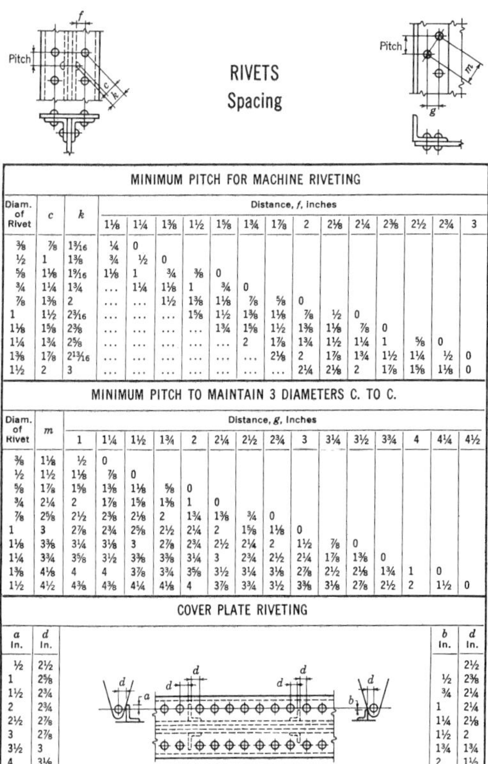

MINIMUM PITCH FOR MACHINE RIVETING

Diam. of Rivet	c	k	Distance, f, inches													
			1⅛	1¼	1⅜	1½	1⅝	1¾	1⅞	2	2⅛	2¼	2⅜	2½	2¾	3
⅜	⅞	1³⁄₁₆	¼	0												
½	1	1⅜	¾	½	0											
⅝	1⅛	1⁹⁄₁₆	1⅛	1	¾	⅜	0									
¾	1¼	1¾	...	1¼	1⅛	1	¾	0								
⅞	1⅜	2	...	...	1½	1⅜	1⅛	⅞	⅝	0						
1	1½	2³⁄₁₆	...	...	...	1⅝	1½	1⅜	1⅛	⅞	½	0				
1⅛	1⅝	2⅜	...	...	...	...	1¾	1⅝	1½	1⅜	1⅛	⅞	0			
1¼	1¾	2⅝	...	...	...	...	...	2	1⅞	1¾	1½	1¼	1	⅝	0	
1⅜	1⅞	2¹³⁄₁₆	...	...	...	...	...	...	2⅛	2	1⅞	1¾	1½	1¼	½	0
1½	2	3	...	...	...	...	...	...	...	2¼	2⅛	2	1⅞	1⅝	1⅛	0

MINIMUM PITCH TO MAINTAIN 3 DIAMETERS C. TO C.

Diam. of Rivet	m	Distance, g, Inches														
		1	1¼	1½	1¾	2	2¼	2½	2¾	3	3¼	3½	3¾	4	4¼	4½
⅜	1⅛	½	0													
½	1½	1⅛	⅞	0												
⅝	1⅞	1⅝	1⅜	1⅛	⅝	0										
¾	2¼	2	1⅞	1⅝	1⅜	1	0									
⅞	2⅝	2½	2⅜	2⅛	2	1¾	1⅜	¾	0							
1	3	2⅞	2¾	2⅝	2½	2¼	2	1⅝	1⅛	0						
1⅛	3⅜	3¼	3⅛	3	2⅞	2¾	2½	2¼	2	1½	⅞	0				
1¼	3¾	3⅝	3½	3⅜	3⅜	3¼	3	2¾	2½	2¼	1⅞	1⅜	0			
1⅜	4⅛	4	4	3⅞	3¾	3⅝	3½	3¼	3⅛	2⅞	2½	2⅛	1¾	1	0	
1½	4½	4⅜	4⅜	4¼	4⅛	4	3⅞	3¾	3½	3⅜	3⅛	2⅞	2½	2	1½	0

COVER PLATE RIVETING

a In.	d In.	b In.	d In.
½	2½		2½
1	2⅝	½	2⅜
1½	2¾	¾	2¼
2	2¾	1	2¼
2½	2⅞	1¼	2⅛
3	2⅞	1½	2
3½	3	1¾	1¾
4	3⅛	2	1½
5	3¼	2¼	1
6	3⅜	2½	...

SOURCE: *Manual of Steel Construction,* 7th ed., American Institute of Steel Construction, Inc., Chicago, Ill., 1973.

TABLE 10-14 Threaded Fasteners: Bolt Heads

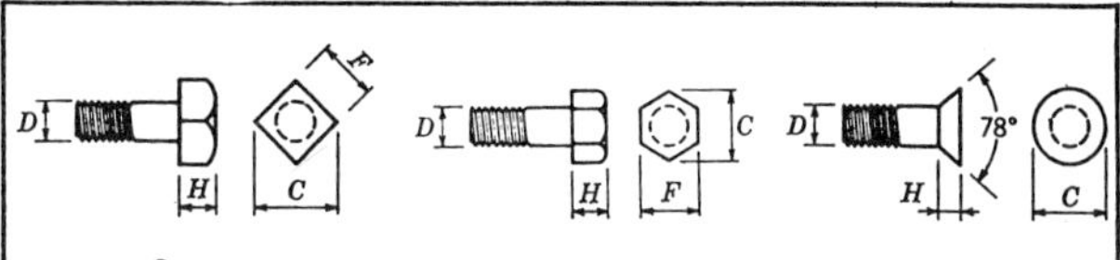

Square Hex Countersunk

Bolt head dimensions, rounded to nearest 1/16 inch, are in accordance with ANSI B18.2.1—1965 (Square and Hex) and ANSI 18.5—1959 (Countersunk)

Standard Dimensions for Bolt Heads

Diam. of Bolt D	Square			Hex			Heavy Hex			Countersunk	
	Width F	Width C	Height H	Width F	Width C	Height H	Width F	Width C	Height H	Diam. C	Height H
In.	In.	In.	In.	In.	In.	In.	In.	In.	In.	In.	In.
1/4	3/8	1/2	3/16	7/16	1/2	3/16	...	...	...	1/2	1/8
3/8	9/16	13/16	1/4	9/16	5/8	1/4	...	...	...	11/16	3/16
1/2	3/4	1 1/16	3/8	3/4	7/8	3/8	7/8	1	3/8	7/8	1/4
5/8	15/16	1 5/16	7/16	15/16	1 1/16	7/16	1 1/16	1 1/4	7/16	1 1/8	5/16
3/4	1 1/8	1 9/16	1/2	1 1/8	1 5/16	1/2	1 1/4	1 7/16	1/2	1 3/8	3/8
7/8	1 5/16	1 7/8	5/8	1 5/16	1 1/2	5/8	1 7/16	1 11/16	5/8	1 9/16	7/16
1	1 1/2	2 1/8	11/16	1 1/2	1 3/4	11/16	1 5/8	1 7/8	11/16	1 13/16	1/2
1 1/8	1 11/16	2 3/8	3/4	1 11/16	1 15/16	3/4	1 13/16	2 1/16	3/4	2 1/16	9/16
1 1/4	1 7/8	2 5/8	7/8	1 7/8	2 3/16	7/8	2	2 5/16	7/8	2 1/4	5/8
1 3/8	2 1/16	2 15/16	15/16	2 1/16	2 3/8	15/16	2 3/16	2 1/2	15/16	2 1/2	11/16
1 1/2	2 1/4	3 3/16	1 1/16	2 1/4	2 5/8	1 1/16	2 3/8	2 3/4	1 1/16	2 11/16	3/4
1 3/4	...	...	...	2 5/8	3	1 3/16	2 3/4	3 3/16	1 3/16	...	...
2	...	...	...	3	3 7/16	1 3/8	3 1/8	3 5/8	1 3/8	...	...
2 1/4	...	...	...	3 3/8	3 7/8	1 9/16	3 1/2	4 1/16	1 9/16	...	...
2 1/2	...	...	...	3 3/4	4 5/16	1 11/16	3 7/8	4 1/2	1 11/16	...	...
2 3/4	...	...	...	4 1/8	4 3/4	1 7/8	4 1/4	4 15/16	1 7/8	...	...
3	...	...	...	4 1/2	5 3/16	2 1/16	4 5/8	5 5/16	2 1/16	...	...
3 1/4	...	...	...	4 7/8	5 5/8	2 1/4	...	...	...	...	...
3 1/2	...	...	...	5 1/4	6 1/16	2 3/8	...	...	...	...	...
3 3/4	...	...	...	5 5/8	6 1/2	2 9/16	...	...	...	...	...
4	...	...	...	6	6 15/16	2 3/4	...	...	...	...	...

SOURCE: *Manual of Steel Construction,* 7th ed., American Institute of Steel Construction, Inc., Chicago, Ill., 1973.

TABLE 10-15 Threaded Fasteners: Nuts

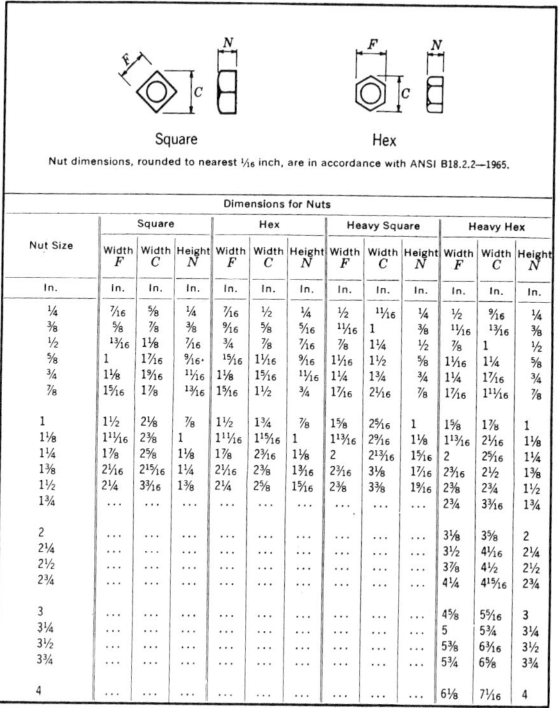

Nut dimensions, rounded to nearest 1/16 inch, are in accordance with ANSI B18.2.2—1965.

	Dimensions for Nuts											
Nut Size	Square			Hex			Heavy Square			Heavy Hex		
	Width F	Width C	Height N	Width F	Width C	Height N	Width F	Width C	Height N	Width F	Width C	Height N
In.	In.	In.	In.	In.	In.	In.	In.	In.	In.	In.	In.	In.
1/4	7/16	5/8	1/4	7/16	1/2	1/4	1/2	11/16	1/4	1/2	9/16	1/4
3/8	5/8	7/8	3/8	9/16	5/8	5/16	11/16	1	3/8	11/16	13/16	3/8
1/2	13/16	1 1/8	7/16	3/4	7/8	7/16	7/8	1 1/4	1/2	7/8	1	1/2
5/8	1	1 7/16	9/16	15/16	1 1/16	9/16	1 1/16	1 1/2	5/8	1 1/16	1 1/4	5/8
3/4	1 1/8	1 9/16	11/16	1 1/8	1 5/16	11/16	1 1/4	1 3/4	3/4	1 1/4	1 7/16	3/4
7/8	1 5/16	1 7/8	13/16	1 5/16	1 1/2	3/4	1 7/16	2 1/16	7/8	1 7/16	1 11/16	7/8
1	1 1/2	2 1/8	7/8	1 1/2	1 3/4	7/8	1 5/8	2 5/16	1	1 5/8	1 7/8	1
1 1/8	1 11/16	2 3/8	1	1 11/16	1 15/16	1	1 13/16	2 9/16	1 1/8	1 13/16	2 1/16	1 1/8
1 1/4	1 7/8	2 5/8	1 1/8	1 7/8	2 3/16	1 1/8	2	2 13/16	1 5/16	2	2 5/16	1 1/4
1 3/8	2 1/16	2 15/16	1 1/4	2 1/16	2 3/8	1 3/16	2 3/16	3 1/8	1 7/16	2 3/16	2 1/2	1 3/8
1 1/2	2 1/4	3 3/16	1 3/8	2 1/4	2 5/8	1 5/16	2 3/8	3 3/8	1 9/16	2 3/8	2 3/4	1 1/2
1 3/4	...	...	...	...	...	...	...	...	...	2 3/4	3 3/16	1 3/4
2	...	...	...	...	...	...	...	...	...	3 1/8	3 5/8	2
2 1/4	...	...	...	...	...	...	...	...	...	3 1/2	4 1/16	2 1/4
2 1/2	...	...	...	...	...	...	...	...	...	3 7/8	4 1/2	2 1/2
2 3/4	...	...	...	...	...	...	...	...	...	4 1/4	4 15/16	2 3/4
3	...	...	...	...	...	...	...	...	...	4 5/8	5 5/16	3
3 1/4	...	...	...	...	...	...	...	...	...	5	5 3/4	3 1/4
3 1/2	...	...	...	...	...	...	...	...	...	5 3/8	6 3/16	3 1/2
3 3/4	...	...	...	...	...	...	...	...	...	5 3/4	6 5/8	3 3/4
4	...	...	...	...	...	...	...	...	...	6 1/8	7 1/16	4

SOURCE: *Manual of Steel Construction,* 7th ed., American Institute of Steel Construction, Inc., Chicago, Ill., 1973.

TABLE 10-16 Threaded Fasteners: Weight of Bolts (With square heads and hexagon nuts, in lb/100)

Length Under Head Inches	Diameter of Bolt in Inches								
	¼	⅜	½	⅝	¾	⅞	1	1⅛	1¼
1	2.38	6.11	13.0	24.1	38.9	...	...	...	...
1¼	2.71	6.71	14.0	25.8	41.5	...	...	...	...
1½	3.05	7.47	15.1	27.6	44.0	67.3	95.1	...	...
1¾	3.39	8.23	16.5	29.3	46.5	70.8	99.7	...	...
2	3.73	8.99	17.8	31.4	49.1	74.4	104	143	...
2¼	4.06	9.75	19.1	33.5	52.1	77.9	109	149	...
2½	4.40	10.5	20.5	35.6	55.1	82.0	114	155	206
2¾	4.74	11.3	21.8	37.7	58.2	86.1	119	161	213
3	5.07	12.0	23.2	39.8	61.2	90.2	124	168	221
3¼	5.41	12.8	24.5	41.9	64.2	94.4	129	174	229
3½	5.75	13.5	25.9	44.0	67.2	98.5	135	181	237
3¾	6.09	14.3	27.2	46.1	70.2	103	140	188	246
4	6.42	15.1	28.6	48.2	73.3	107	145	195	254
4¼	6.76	15.8	29.9	50.3	76.3	111	151	202	262
4½	7.10	16.6	31.3	52.3	79.3	115	156	208	271
4¾	7.43	17.3	32.6	54.4	82.3	119	162	215	279
5	7.77	18.1	33.9	56.5	85.3	123	167	222	288
5¼	8.11	18.9	35.3	58.6	88.4	127	172	229	296
5½	8.44	19.6	36.6	60.7	91.4	131	178	236	304
5¾	8.78	20.4	38.0	62.8	94.4	136	183	242	313
6	9.12	21.1	39.3	64.9	97.4	140	188	249	321
6¼	9.37	21.7	40.4	66.7	100	143	193	255	329
6½	9.71	22.5	41.8	68.7	103	147	198	262	337
6¾	10.1	23.3	43.1	70.8	106	151	204	269	345
7	10.4	24.0	44.4	72.9	109	156	209	275	354
7¼	10.7	24.8	45.8	75.0	112	160	214	282	362
7½	11.0	25.5	47.1	77.1	115	164	220	289	371
7¾	11.4	26.3	48.5	79.2	118	168	225	296	379
8	11.7	27.0	49.8	81.3	121	172	231	303	387
8½	...	28.6	52.5	85.5	127	180	241	316	404
9	...	30.1	55.2	89.7	133	189	252	330	421
9½	...	31.6	57.9	93.9	139	197	263	343	438
10	...	33.1	60.6	98.1	145	205	274	357	454
10½	...	34.6	63.3	102	151	213	284	371	471
11	...	36.2	66.0	106	157	221	295	384	488
11½	...	37.7	68.7	110	163	230	306	398	505
12	...	39.2	71.3	115	170	238	316	411	522
12½	...	...	74.0	119	176	246	327	425	538
13	...	...	76.7	123	182	254	338	439	556
13½	...	...	79.4	127	188	263	349	452	572
14	...	...	82.1	131	194	271	359	466	589
14½	...	...	84.8	135	200	279	370	479	605
15	...	...	87.5	140	206	287	381	493	622
15½	...	...	90.2	144	212	296	392	507	639
16	...	...	92.9	148	218	304	402	520	656
Per Inch Additional	1.3	3.0	5.4	8.4	12.1	16.5	21.4	27.2	33.6

Note: Bolt is Square Bolt, ANSI B18.2.1—65 and nut is Hex Nut, ANSI B18.2.2—65. This table conforms to weight standards adopted by the Industrial Fasteners Institute.

SOURCE: *Manual of Steel Construction,* 7th ed., American Institute of Steel Construction, Inc., Chicago, Ill., 1973.

TABLE 10-17 Threaded Fasteners: Weight of Bolts (Special cases, in lb/100)

VARIATIONS IN BOLT AND NUT TYPES

Weights for combinations of bolt heads and nuts, other than square heads and hex nuts, may be determined by making the appropriate additions and deductions tabulated below from the weight per 100 shown on the previous page.

Combination	Add or Subtract	Diameter of Bolt in Inches								
		¼	⅜	½	⅝	¾	⅞	1	1⅛	1¼
Square bolt with square nut	+	0.1	1.0	2.0	3.4	3.5	5.5	8.0	12.2	16.3
Square bolt with heavy square nut	+	0.6	2.1	4.1	7.0	11.6	17.2	23.2	32.1	41.2
Square bolt with heavy hex nut	+	0.4	1.5	2.8	4.6	7.6	10.7	14.2	18.9	24.3
Hex bolt with square nut	+	0.1	0.6	1.1	1.4	0.2	0.5	−0.2	−0.1	−1.7
Hex bolt with hex nut	−	0.0	0.4	0.9	2.0	3.3	5.0	8.2	12.3	18.0
Hex bolt with heavy square nut	+	0.6	1.7	3.2	5.0	8.3	12.2	15.0	19.8	23.2
Hex bolt with heavy hex nut	+	0.4	1.1	1.9	2.6	4.3	5.7	6.0	6.6	6.3
Heavy hex bolt with heavy square nut	+	...	...	4.7	7.3	11.3	16.5	20.7	27.0	33.6
Heavy hex bolt with heavy hex nut	+	...	...	3.4	4.9	7.3	10.0	11.7	13.8	16.7

LARGE DIAMETER BOLTS

Weights of bolts over 1¼ inches in diameter may be calculated from the following data. Standard practice is hex head bolts with heavy hex nut. Square head bolts and square nuts are not standard in sizes over 1½ inches.

Weight of 100 Each	Diameter of Bolt in Inches											
	1⅜	1½	1¾	2	2¼	2½	2¾	3	3¼	3½	3¾	4
Square heads	105	130	...	...	...	...	...	...	...	...	...	...
Hex heads	84	112	178	259	369	508	680	900	1120	1390	1730	2130
Heavy hex heads	95	124	195	280	397	541	720	950	...	...	...	...
Square nuts	94.5	122	...	...	...	...	...	...	...	...	...	...
Heavy square nuts	125	161	...	...	...	...	...	...	...	...	...	...
Heavy hex nuts	102	131	204	299	419	564	738	950	1190	1530	1810	2180
Linear inch of threaded shank	35.0	42.5	57.4	75.5	97.4	120	147	178	210	246	284	325
Linear inch of unthreaded shank	42.0	50.0	68.2	89.0	113	139	168	200	235	272	313	356

SOURCE: *Manual of Steel Construction,* 7th ed., American Institute of Steel Construction, Inc., Chicago, Ill., 1973.

TABLE 10-18 Threaded Fasteners: Weight of ASTM A325 or A490 High-Strength Bolts
(Heavy hex structural bolts with heavy hex nuts, in lb/100)

Length Under Head Inches	Diameter of Bolt in Inches								
	½	⅝	¾	⅞	1	1⅛	1¼	1⅜	1½
1	16.5	29.4	47.0	...	...	...	...	...	...
1¼	17.8	31.1	49.6	74.4	104	...	...	...	...
1½	19.2	33.1	52.2	78.0	109	148	197	...	...
1¾	20.5	35.3	55.3	81.9	114	154	205	261	333
2	21.9	37.4	58.4	86.1	119	160	212	270	344
2¼	23.3	39.8	61.6	90.3	124	167	220	279	355
2½	24.7	41.7	64.7	94.6	130	174	229	290	366
2¾	26.1	43.9	67.8	98.8	135	181	237	300	379
3	27.4	46.1	70.9	103	141	188	246	310	391
3¼	28.8	48.2	74.0	107	146	195	255	321	403
3½	30.2	50.4	77.1	111	151	202	263	332	416
3¾	31.6	52.5	80.2	116	157	209	272	342	428
4	33.0	54.7	83.3	120	162	216	280	353	441
4¼	34.3	56.9	86.4	124	168	223	289	363	453
4½	35.7	59.0	89.5	128	173	230	298	374	465
4¾	37.1	61.2	92.7	133	179	237	306	384	478
5	38.5	63.3	95.8	137	184	244	315	395	490
5¼	39.9	65.5	98.9	141	190	251	324	405	503
5½	41.2	67.7	102	146	196	258	332	416	515
5¾	42.6	69.8	105	150	201	265	341	426	527
6	44.0	71.9	108	154	207	272	349	437	540
6¼	...	74.1	111	158	212	279	358	447	552
6½	...	76.3	114	163	218	286	367	458	565
6¾	...	78.5	118	167	223	293	375	468	577
7	...	80.6	121	171	229	300	384	479	589
7¼	...	82.8	124	175	234	307	392	489	602
7½	...	84.9	127	179	240	314	401	500	614
7¾	...	87.1	130	183	246	321	410	510	626
8	...	89.2	133	187	251	328	418	521	639
8¼	...	...	...	192	257	335	427	531	651
8½	...	...	...	196	262	342	435	542	664
8¾	...	...	...	...	...	...	444	552	676
9	...	...	...	...	...	...	453	563	689
Per inch additional and	5.5	8.6	12.4	16.9	22.1	28.0	34.4	42.5	49.7
For each 100 plain round washers add	2.1	3.6	4.8	7.0	9.4	11.3	13.8	16.8	20.0
For each 100 beveled square washers add	23.1	22.4	21.0	20.2	19.2	34.0	31.6	...	...

Note: This table conforms to weight standards adopted by the Industrial Fasteners Institute, 1965, updated for washer weights.

SOURCE: *Manual of Steel Construction,* 7th ed., American Institute of Steel Construction, Inc., Chicago, Ill., 1973.

TABLE 10-19 Threaded Fasteners: Screw Threads

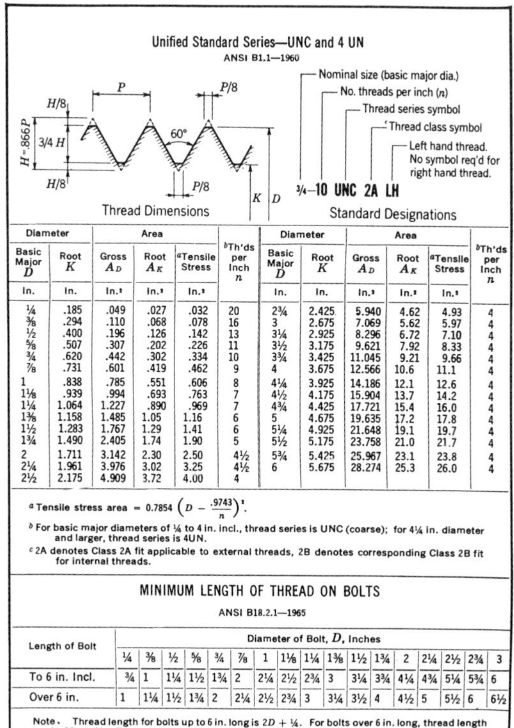

Unified Standard Series—UNC and 4 UN

ANSI B1.1—1960

Diameter: Basic Major D (In.)	Diameter: Root K (In.)	Area: Gross A_D (In.²)	Area: Root A_K (In.²)	Area: [a]Tensile Stress (In.²)	[b]Th'ds per Inch n	Diameter: Basic Major D (In.)	Diameter: Root K (In.)	Area: Gross A_D (In.²)	Area: Root A_K (In.²)	Area: [a]Tensile Stress (In.²)	[b]Th'ds per Inch n
¼	.185	.049	.027	.032	20	2¾	2.425	5.940	4.62	4.93	4
⅜	.294	.110	.068	.078	16	3	2.675	7.069	5.62	5.97	4
½	.400	.196	.126	.142	13	3¼	2.925	8.296	6.72	7.10	4
⅝	.507	.307	.202	.226	11	3½	3.175	9.621	7.92	8.33	4
¾	.620	.442	.302	.334	10	3¾	3.425	11.045	9.21	9.66	4
⅞	.731	.601	.419	.462	9	4	3.675	12.566	10.6	11.1	4
1	.838	.785	.551	.606	8	4¼	3.925	14.186	12.1	12.6	4
1⅛	.939	.994	.693	.763	7	4½	4.175	15.904	13.7	14.2	4
1¼	1.064	1.227	.890	.969	7	4¾	4.425	17.721	15.4	16.0	4
1⅜	1.158	1.485	1.05	1.16	6	5	4.675	19.635	17.2	17.8	4
1½	1.283	1.767	1.29	1.41	6	5¼	4.925	21.648	19.1	19.7	4
1¾	1.490	2.405	1.74	1.90	5	5½	5.175	23.758	21.0	21.7	4
2	1.711	3.142	2.30	2.50	4½	5¾	5.425	25.967	23.1	23.8	4
2¼	1.961	3.976	3.02	3.25	4½	6	5.675	28.274	25.3	26.0	4
2½	2.175	4.909	3.72	4.00	4						

[a] Tensile stress area $= 0.7854 \left(D - \frac{.9743}{n}\right)^2$.

[b] For basic major diameters of ¼ to 4 in. incl., thread series is UNC (coarse); for 4¼ in. diameter and larger, thread series is 4UN.

[c] 2A denotes Class 2A fit applicable to external threads, 2B denotes corresponding Class 2B fit for internal threads.

MINIMUM LENGTH OF THREAD ON BOLTS

ANSI B18.2.1—1965

Length of Bolt	Diameter of Bolt, D, Inches																
	¼	⅜	½	⅝	¾	⅞	1	1⅛	1¼	1⅜	1½	1¾	2	2¼	2½	2¾	3
To 6 in. Incl.	¾	1	1¼	1½	1¾	2	2¼	2½	2¾	3	3¼	3¾	4¼	4¾	5¼	5¾	6
Over 6 in.	1	1¼	1½	1¾	2	2¼	2½	2¾	3	3¼	3½	4	4½	5	5½	6	6½

Note. Thread length for bolts up to 6 in. long is $2D + ¼$. For bolts over 6 in. long, thread length is $2D + ½$. These proportions may be used to compute thread length for diameters not shown in the table. Bolts which are too short for listed or computed thread lengths are threaded as close to the head as possible.

SOURCE: *Manual of Steel Construction,* 7th ed., American Institute of Steel Construction, Inc., Chicago, Ill., 1973.

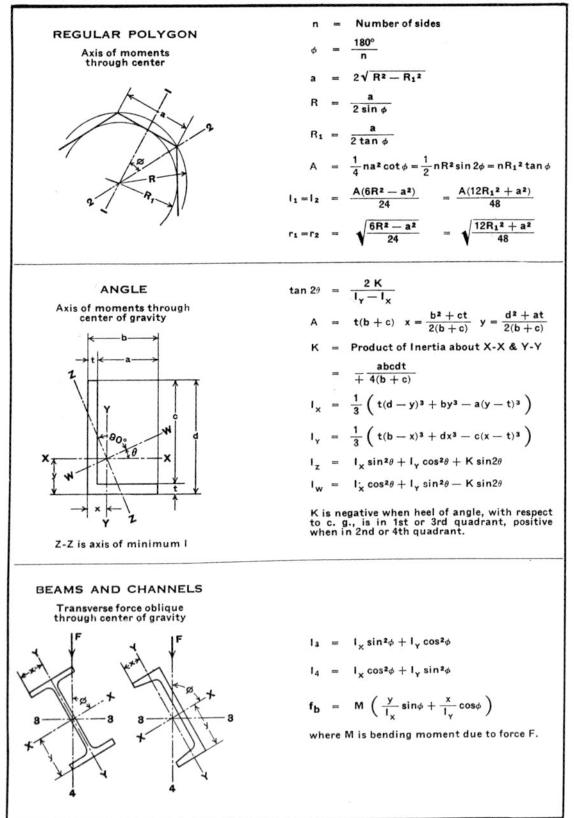

REGULAR POLYGON
Axis of moments through center
n = Number of sides
$\phi = \frac{180^\circ}{n}$
$a = 2\sqrt{R^2 - R_1^2}$
$R = \frac{a}{2 \sin \phi}$
$R_1 = \frac{a}{2 \tan \phi}$
$A = \frac{1}{4} na^2 \cot \phi = \frac{1}{2} nR^2 \sin 2\phi = nR_1^2 \tan \phi$
$I_1 = I_2 = \frac{A(6R^2 - a^2)}{24} = \frac{A(12R_1^2 + a^2)}{48}$
$r_1 = r_2 = \sqrt{\frac{6R^2 - a^2}{24}} = \sqrt{\frac{12R_1^2 + a^2}{48}}$
ANGLE
Axis of moments through center of gravity
$\tan 2\theta = \frac{2K}{I_Y - I_X}$
$A = t(b + c)$ $x = \frac{b^2 + ct}{2(b + c)}$ $y = \frac{d^2 + at}{2(b + c)}$
K = Product of Inertia about X-X & Y-Y
$= \mp \frac{abcdt}{4(b + c)}$
$I_X = \frac{1}{3}\left(t(d - y)^3 + by^3 - a(y - t)^3\right)$
$I_Y = \frac{1}{3}\left(t(b - x)^3 + dx^3 - c(x - t)^3\right)$
$I_Z = I_X \sin^2\theta + I_Y \cos^2\theta + K \sin 2\theta$
$I_W = I_X \cos^2\theta + I_Y \sin^2\theta - K \sin 2\theta$
K is negative when heel of angle, with respect to c. g., is in 1st or 3rd quadrant, positive when in 2nd or 4th quadrant.
Z-Z is axis of minimum I
BEAMS AND CHANNELS
Transverse force oblique through center of gravity
$I_3 = I_X \sin^2\phi + I_Y \cos^2\phi$
$I_4 = I_X \cos^2\phi + I_Y \sin^2\phi$
$f_b = M\left(\frac{y}{I_X} \sin\phi + \frac{x}{I_Y} \cos\phi\right)$
where M is bending moment due to force F.

Fig. 10-5 Properties of geometric sections and structural shapes *(American Institute of Steel Construction, Inc., Chicago, Ill.)*

NOTES

NOTES

Section Eleven

■

WELDING

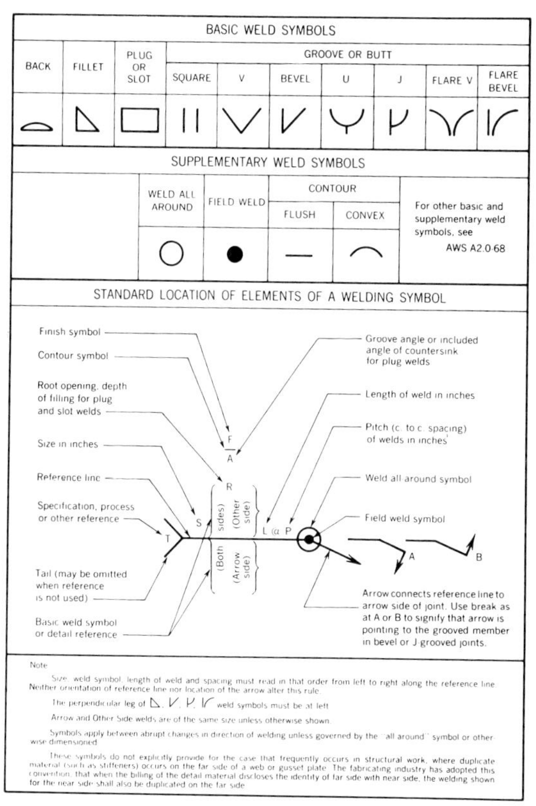

Fig. 11-1 Standard symbols for welding *(American Institute of Steel Construction, Inc., Chicago, Ill.).*

11-1 WELDING DESIGNATIONS*

The designations such as **B-L1a, B-L2, B-U2, B-P3** which are given on the following pages are those used in the AWS standards. Groove welds are classified using the following convention:

1. *Symbols for Joint Types*
   B—butt joint
   C—corner joint
   T—tee joint
   BC—butt or corner joint
   TC—tee or corner joint
   BTC—butt, tee or corner joint

2. *Symbols for Base Metal Thickness and Penetration*
   L—limited thickness, complete joint penetration
   U—unlimited thickness, complete joint penetration
   P—partial joint penetration

3. *Symbols for Weld Types*
   1—square groove
   2—single-vee groove
   3—double-vee groove
   4—single-bevel groove
   5—double-bevel groove
   6—single-U groove
   7—double-U groove
   8—single-J groove
   9—double-J groove

4. *Symbols for Welding Processes*
   If not manual shielded metal arc:
   S—submerged arc
   G—gas metal arc
   F—flux cored arc

*Reprinted by permission of American Institute of Steel Construction, Inc., Chicago, Ill.

11-2 WELDED JOINTS: COMPLETE PENETRATION

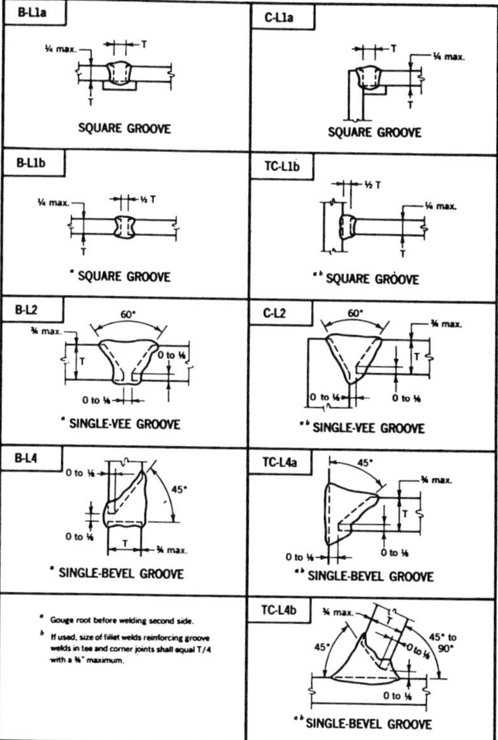

Fig. 11-2 Manual shielded metal-arc welded joints of *limited* thickness *(American Institute of Steel Construction, Inc., Chicago, Ill.)*.

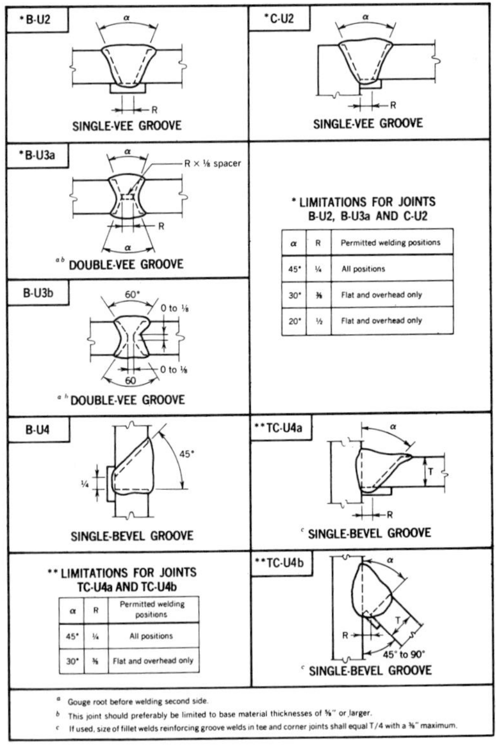

* LIMITATIONS FOR JOINTS B-U2, B-U3a AND C-U2

α	R	Permitted welding positions
45°	¼	All positions
30°	⅜	Flat and overhead only
20°	½	Flat and overhead only

** LIMITATIONS FOR JOINTS TC-U4a AND TC-U4b

α	R	Permitted welding positions
45°	¼	All positions
30°	⅜	Flat and overhead only

[a] Gouge root before welding second side.

[b] This joint should preferably be limited to base material thicknesses of ⅝" or larger.

[c] If used, size of fillet welds reinforcing groove welds in tee and corner joints shall equal T/4 with a ⅜" maximum.

Fig. 11-3 Manual shielded metal-arc welded joints of *unlimited* thickness *(American Institute of Steel Construction, Inc., Chicago, Ill.).*

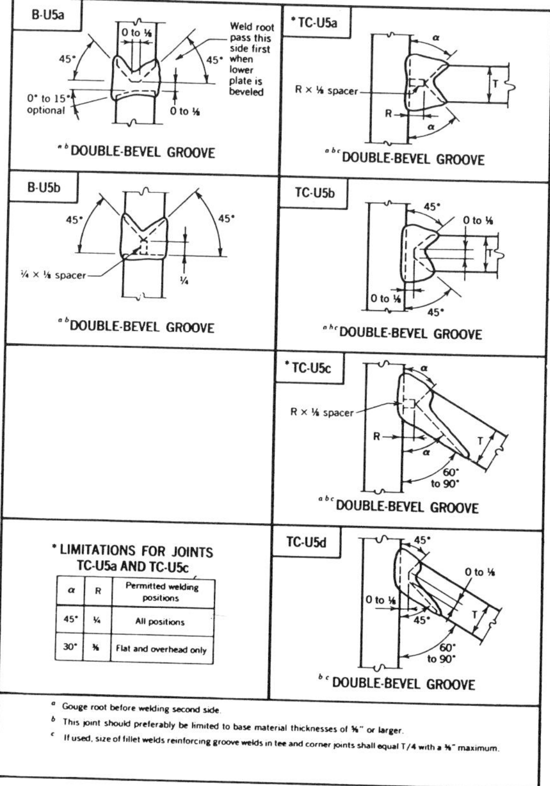

* LIMITATIONS FOR JOINTS TC-U5a AND TC-U5c

α	R	Permitted welding positions
45°	1/4	All positions
30°	3/8	Flat and overhead only

[a] Gouge root before welding second side.

[b] This joint should preferably be limited to base material thicknesses of 5/8" or larger.

[c] If used, size of fillet welds reinforcing groove welds in tee and corner joints shall equal T/4 with a 3/8" maximum.

Fig. 11-3 Manual shielded metal-arc welded joints of *unlimited* thickness *(cont.)*

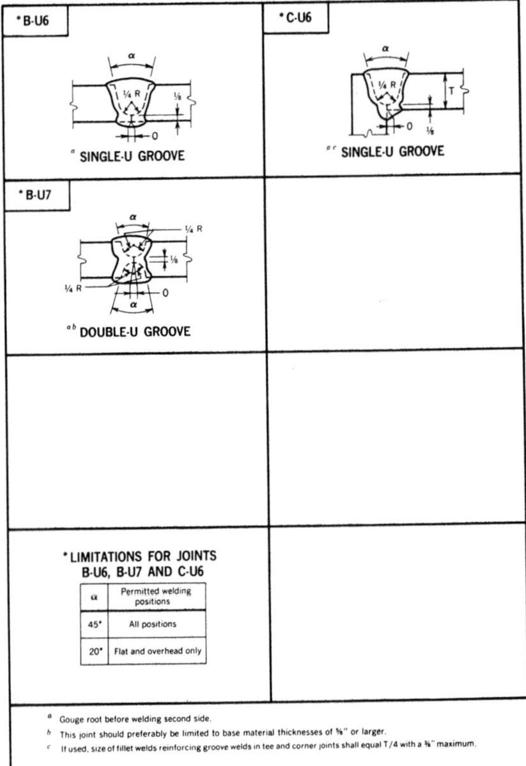

α	Permitted welding positions
45°	All positions
20°	Flat and overhead only

[a] Gouge root before welding second side.
[b] This joint should preferably be limited to base material thicknesses of ⅝" or larger.
[c] If used, size of fillet welds reinforcing groove welds in tee and corner joints shall equal T/4 with a ⅜" maximum.

Fig. 11-3 Manual shielded metal-arc welded joints of *unlimited* thickness *(cont.)*

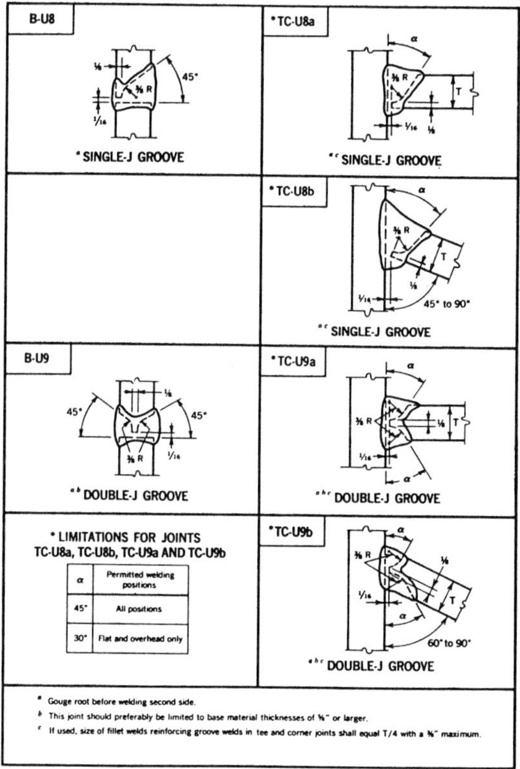

α	Permitted welding positions
45°	All positions
30°	Flat and overhead only

[a] Gouge root before welding second side.
[b] This joint should preferably be limited to base material thicknesses of 5/8" or larger.
[c] If used, size of fillet welds reinforcing groove welds in tee and corner joints shall equal T/4 with a 3/8" maximum.

Fig. 11-3 Manual shielded metal-arc welded joints of *unlimited* thickness *(cont.)*

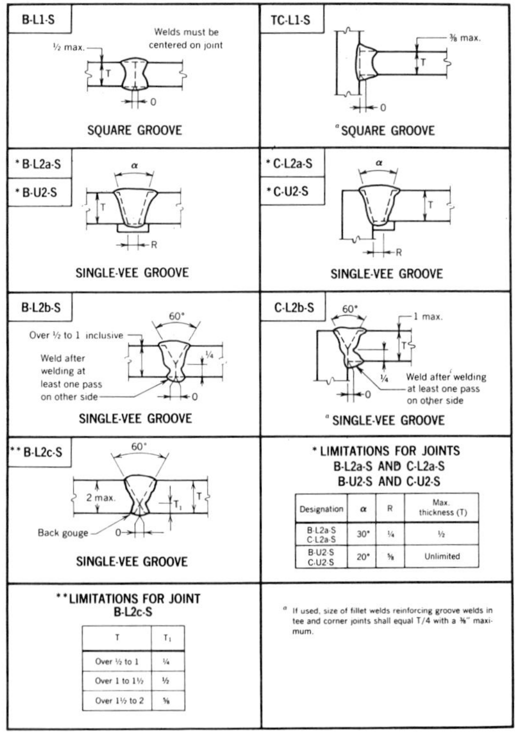

*** LIMITATIONS FOR JOINTS B-L2a-S AND C-L2a-S B-U2-S AND C-U2-S**

Designation	α	R	Max. thickness (T)
B-L2a-S C-L2a-S	30°	¼	½
B-U2-S C-U2-S	20°	⅝	Unlimited

****LIMITATIONS FOR JOINT B-L2c-S**

T	T_1
Over ½ to 1	¼
Over 1 to 1½	½
Over 1½ to 2	⅝

[a] If used, size of fillet welds reinforcing groove welds in tee and corner joints shall equal T/4 with a ⅜" maximum.

Fig. 11-4 Submerged arc welded joints of *limited* and *unlimited* thickness *(American Institute of Steel Construction, Inc., Chicago, Ill.)*.

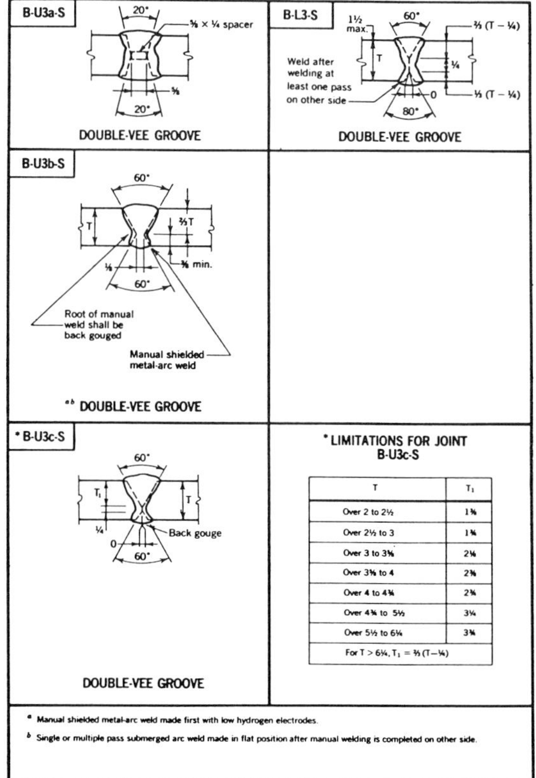

T	T_1
Over 2 to 2½	1⅜
Over 2½ to 3	1¾
Over 3 to 3⅝	2⅛
Over 3⅝ to 4	2⅜
Over 4 to 4¾	2¾
Over 4¾ to 5½	3¼
Over 5½ to 6¼	3¾
For $T > 6\frac{1}{4}$, $T_1 = \frac{2}{3}(T - \frac{1}{4})$	

[a] Manual shielded metal-arc weld made first with low hydrogen electrodes.

[b] Single or multiple pass submerged arc weld made in flat position after manual welding is completed on other side.

Fig. 11-4 Submerged arc welded joints of *limited* and *unlimited* thickness *(cont.)*

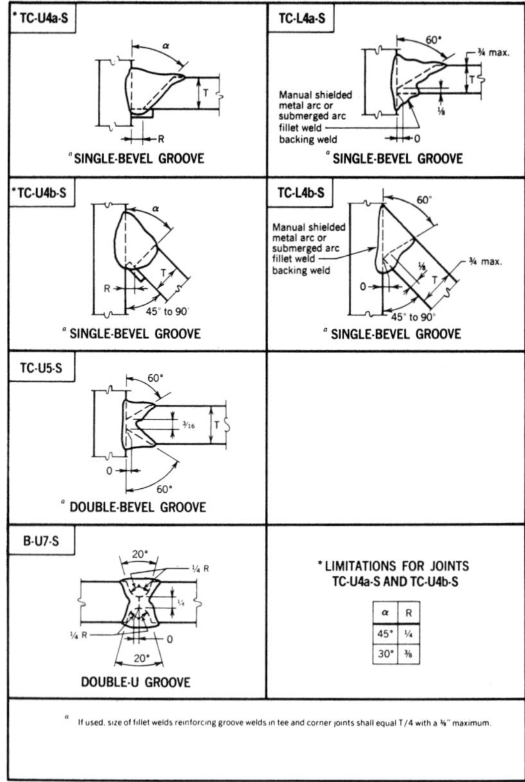

α	R
45°	¼
30°	⅜

[a] If used, size of fillet welds reinforcing groove welds in tee and corner joints shall equal T/4 with a ⅜" maximum.

Fig. 11-4 Submerged arc welded joints of *limited* and *unlimited* thickness *(cont.)*

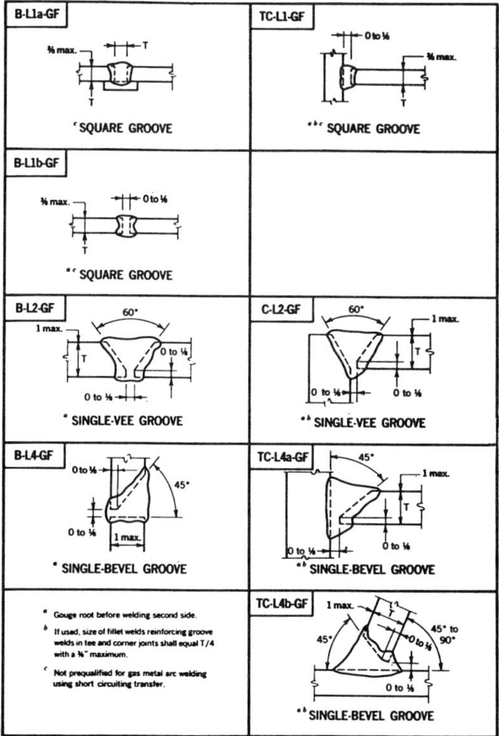

Fig. 11-5 Gas metal and flux cored arc welded joints of *limited* thickness *(American Institute of Steel Construction, Inc., Chicago, Ill.).*

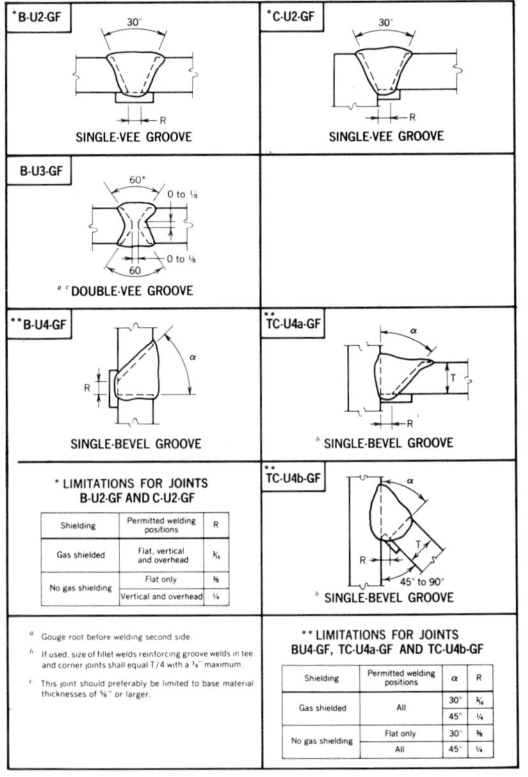

* LIMITATIONS FOR JOINTS B-U2-GF AND C-U2-GF

Shielding	Permitted welding positions	R
Gas shielded	Flat, vertical and overhead	3/16
No gas shielding	Flat only	3/8
	Vertical and overhead	1/4

** LIMITATIONS FOR JOINTS BU4-GF, TC-U4a-GF AND TC-U4b-GF

Shielding	Permitted welding positions	α	R
Gas shielded	All	30°	3/16
		45°	1/4
No gas shielding	Flat only	30°	3/8
	All	45°	1/4

[a] Gouge root before welding second side.

[b] If used, size of fillet welds reinforcing groove welds in tee and corner joints shall equal T/4 with a 3/8" maximum.

[c] This joint should preferably be limited to base material thicknesses of 5/8" or larger.

Fig. 11-6 Gas metal and flux cored arc welded joints of *unlimited* thickness *(American Institute of Steel Construction, Inc., Chicago, Ill.)*.

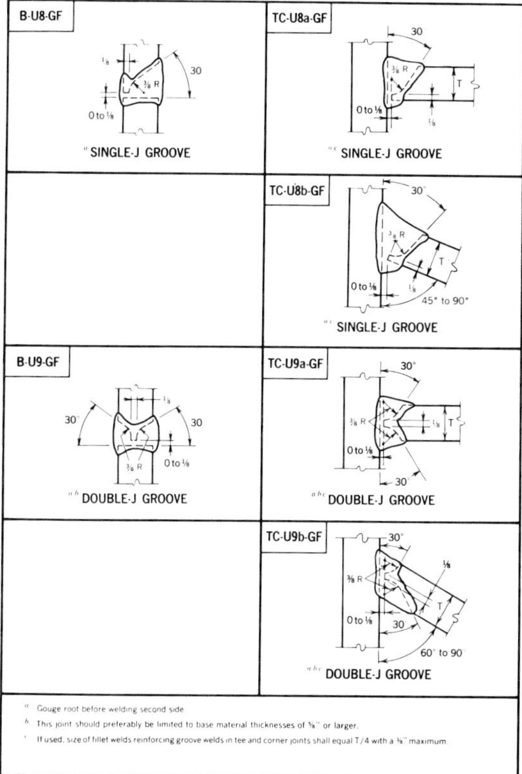

Fig. 11-6 Gas metal and flux cored arc welded joints of *unlimited* thickness *(cont.)*.

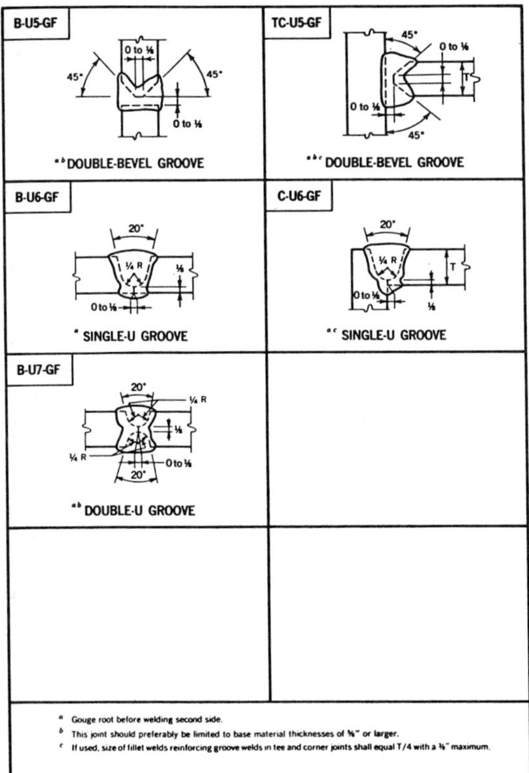

Fig. 11-6 Gas metal and flux cored arc welded joints of *unlimited* thickness *(cont.)*.

11-3 WELDED JOINTS: PARTIAL PENETRATION

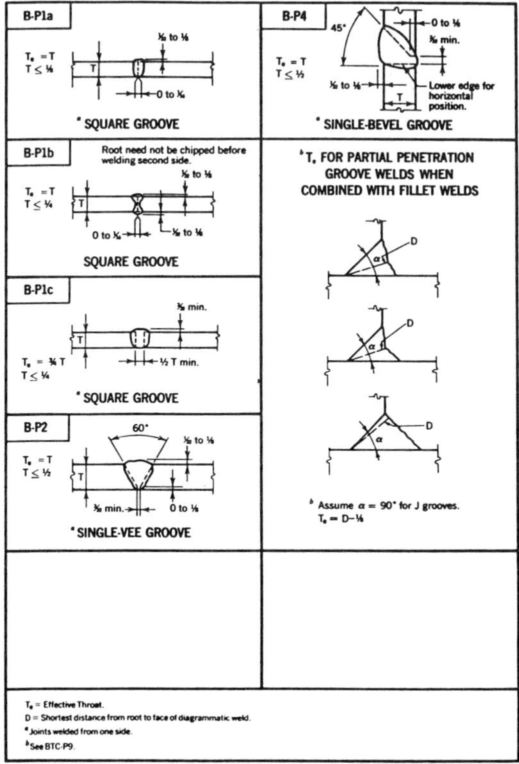

Fig. 11-7 Manual shielded metal-arc welded joints *(American Institute of Steel Construction, Inc., Chicago, Ill.).*

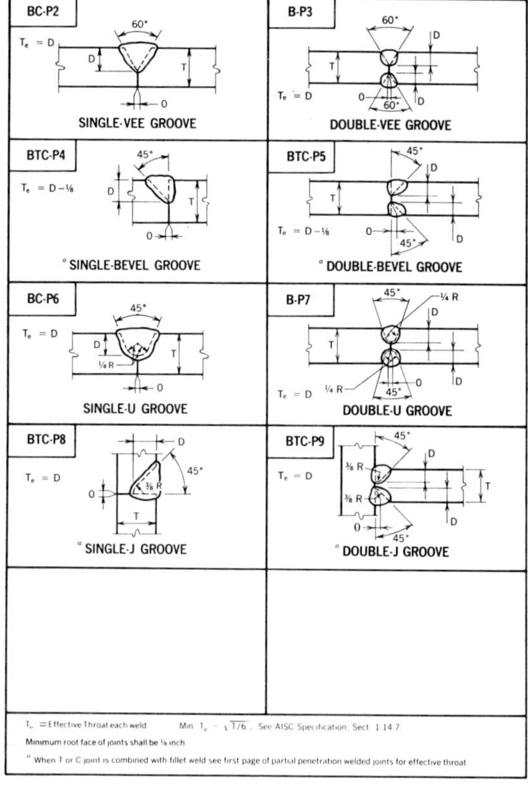

Fig. 11-7 Manual shielded metal-arc welded joints *(cont.)*.

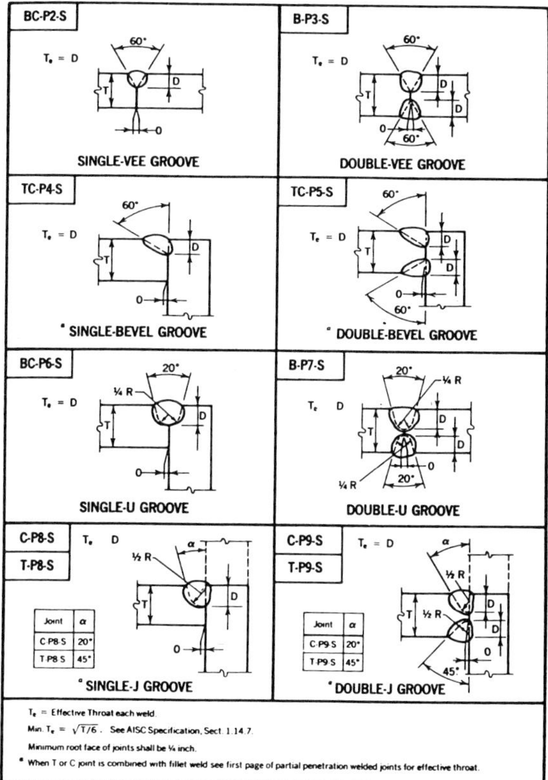

Fig. 11-8 Submerged arc welded joints *(American Institute of Steel Construction, Inc., Chicago, Ill.).*

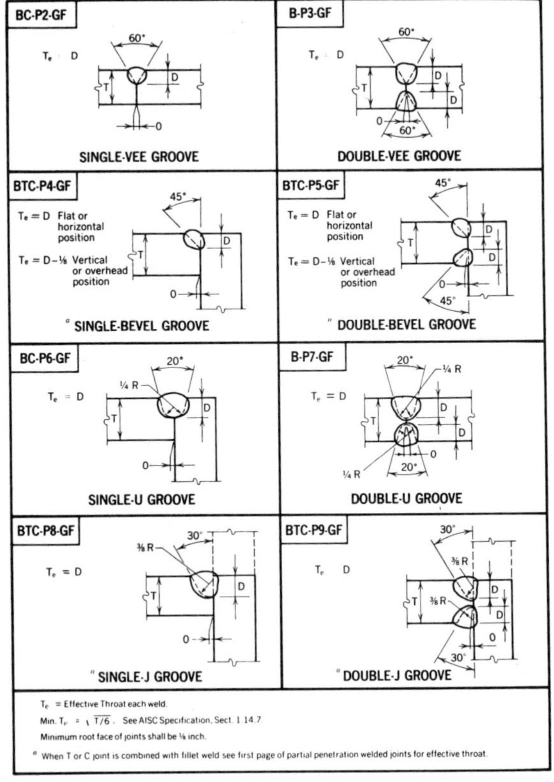

Fig. 11-9 Gas metal and flux cored arc welded joints *(American Institute of Steel Construction, Inc., Chicago, Ill.).*

11-4 WELDED JOINTS: DETAILS OF FILLET WELDS

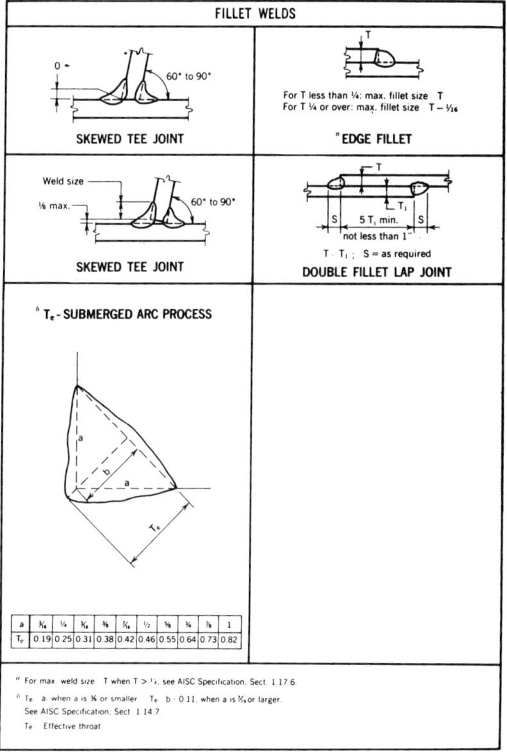

a	3/16	1/4	5/16	3/8	7/16	1/2	5/8	3/4	7/8	1
T_e	0.19	0.25	0.31	0.38	0.42	0.46	0.55	0.64	0.73	0.82

[a] For max. weld size T_1 when $T > 1/4$, see AISC Specification, Sect. 1.17.6.

[b] T_e = a, when a is 3/8 or smaller. T_e = b + 0.11, when a is 7/16 or larger. See AISC Specification, Sect. 1.14.7.

T_e Effective throat

Fig. 11-10 Fillet welds for manual shielded metal-arc, submerged arc, gas metal, and flux cored welded joints *(American Institute of Steel Construction, Inc., Chicago, Ill.).*

NOTES

Section Twelve

▪

Surveying

The following tables and illustrations have been extracted, with permission, from two manuals published by the Keuffel & Esser Company, 20 Whippany Road, Morristown, New Jersey 07960.

"K & E Surveying Instruments" and "Solar Ephemeris 1978" may be obtained by writing to the Keuffel & Esser Company. Sources of basic astronomical data may be obtained from the U.S. Naval Observatory Publications, Washington, D.C. as printed by the U.S. Government Printing Office.

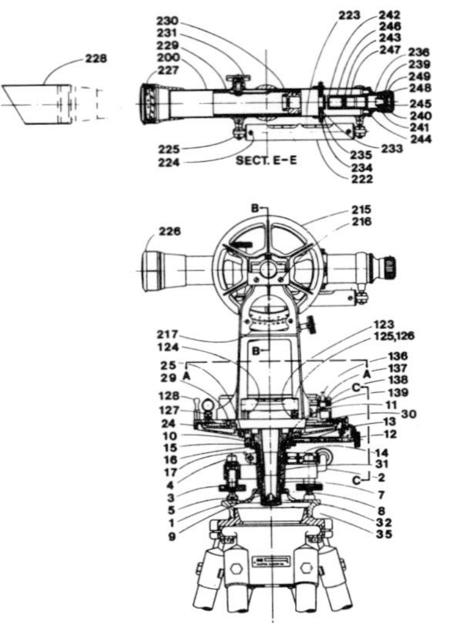

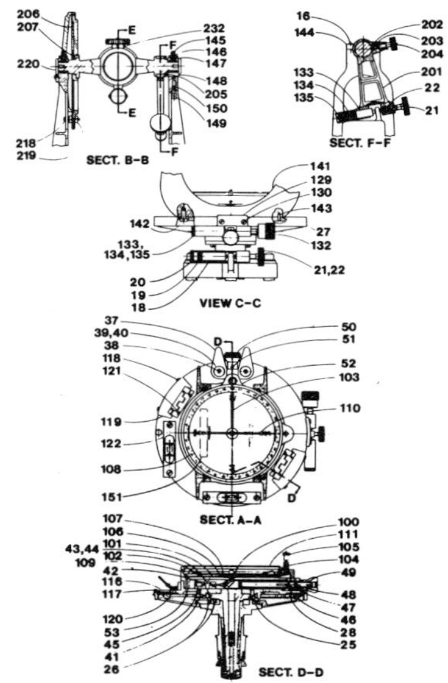

Fig. 12-1 Typical transit *(Keuffel & Esser Company).*

TABLE 12-1 Component Parts of Transit in Fig. 12-1

NO.	DESCRIPTION	NO.	DESCRIPTION
1	Tripod Plate	46	Tube, Opt Plummet
2	Leveling Head	47	Tube Guiding Screw, Opt Plummet
3	Leveling Sleeve Ass'y	48	Reticle, Opt Plummet
4	Leveling Screw Stem	49	Tube End, Opt Plummet
5	Leveling Screw Shoe	50	Eyepiece, Opt Plummet
6	Shifting Plate	51	Eyepiece Mtg Screw, Opt Plummet
7	Half Ball	52	Body, Opt Plummet
8	Half Ball Lock Screw	53	Body Mtg Screw, Opt Plummet
9	Outer Center	55	Leveling Screw Cap
10	Vernier Plate Clamp	100	Compass Needle Pivot & Bushing Ass'y
11	Vernier Plate Clamp Gib	101	Compass Dial
12	Vernier Plate Clamp Screw	102	Compass Dial Mtg Screw
13	Vernier Plate Clamp Screw Pin	103	Compass Needle Lifter
14	Lower Clamp	104	Compass Needle Lifter Bushing
15	Lower Clamp Collar	105	Compass Needle Lifter Bushing Screw
16	Mtg Screw	106	Compass Variation Index
17	Lower Clamp Screw	107	Compass Variation Index Mtg. Screw
18	Lower Clamp Plunger	108	Compass Ring
19	Lower Clamp Spring	109	Compass Ring Tension
20	Lower Clamp Spring Cap	110	Compass Needle
21	Tangent Screw	111	Compass Cover Glass & Mount
22	Tangent Screw Pivot	112	Compass Box Stud
24	Horizontal Circle	113	Compass Box Cover
25	Mtg Screw	114	Compass Box Mtg Screw
26	Adjusting Screw	115	Compass Needle Lift. Bush. Repl. Screw
27	Vernier Plate	116	Vernier Cover Glass
28	Spanned Vernier	117	Vernier Cover Glass Strap
29	Spanned Vernier Mtg Screw	118	Vernier Cover Glass Strap Mtg Screw
30	Spanned Vernier Height Adj Screw	119	Vernier Reflector & Hinge Ass'y
31	Inner Center	120	Vernier Hinge Block
32	Center Nut	121	Vernier Hinge Block Mtg Screw
33	Center Ball	122	Plate Level Vial
34	Center Spring	123	Plate Level Vial Post Adjustable
35	Center Cap	124	Plate Level Vial Post Fixed
36	Plumb Bob Chain & Hook	125	Plate Level Vial Post Bushing
37	Optical Plummet Housing	126	Plate Level Vial Post Spring
38	Stud, Opt Plummet Housing	127	Plate Level Vial Adj Nut
39	Stud Washer, Opt Plummet Housing	128	Plate Level Vial Guard
40	Stud Nut, Opt Plummet Housing	129	Vernier Plate Clamp Spring Box
41	Mount, Opt Plummet Housing	130	Vernier Plate Clamp Spring Box Mtg Screw
42	Mount Mtg Screw, Opt Plummet Housing	131	Vernier Plate Clamp Tangent Screw
43	Mtg Screw, Opt Plummet Housing	132	Vernier Plate Clamp Tangent Screw 2 Speed
44	Mtg Screw Washer, Opt Plummet Hsg	133	Tangent Screw Plunger
45	Adj Screw, Opt Plummet Hsg	134	Tangent Screw Spring

SOURCE: Keuffel & Esser Company, Whippany, N.J.

TABLE 12-1 Component Parts of Transit in Fig. 12-1 *(cont.)*

NO.	DESCRIPTION
135	Tangent Screw Spring Cap
136	Declination Adjustment Pinion
137	Declination Adjustment Pinion Washer
138	Declination Adjustment Pinion Spacer
139	Declination Adjustment Pinion Set Screw
140	Plug
141	Standard
142	Standard Mtg Screw
143	Standard Mtg Pin
144	Bearing Cap
145	Bearing Cap Bushing
146	Tension Plug
147	Tension Plug Washer
148	Bearing Block
149	Bearing Block Adj Screw
150	Bearing Block Adj Screw Pin
151	Name Plate
200	Telescope Barrel & Axle
201	Telescope Clamp
202	Telescope Clamp Gib
203	Telescope Clamp Pin
204	Telescope Clamp Screw
205	Telescope Clamp Nut
206	Vertical Circle
207	Vertical Circle Mtg Screw
208	Stadia Circle
209	Stadia Circle Holder
210	Stadia Circle Holder Mtg Screw
211	Index Ring Horizontal
212	Index Ring Vertical
213	Index Ring Bushing
214	Index Ring Bushing Screw
215	Vertical Circle Guard
216	Vertical Circle Guard Mtg Screw
217	Vertical Vernier
218	Vertical Vernier Screw
219	Vertical Vernier Nut
220	Telescope Axle End Cap
221	Telescope Axle Cover
222	Telescope Level Vial Ass'y
223	Telescope Level Vial Only
224	Telescope Level Vial End Lock Screw
225	Telescope Level Adj Nut
226	Objective Cap

TABLE 12-1 Component Parts of Transit in Fig. 12-1 ***(cont.)***

NO.	DESCRIPTION
227	Objective Lens & Mount
228	Sunshade
229	Telescope Draw Tube
230	Telescope Focusing Lens Ass'y
231	Telescope Focusing Pinion Ass'y
232	Telescope Focusing Pinion Lock Screw
233	Reticle
234	Reticle Adj Screw
235	Reticle Shutter
236	Eyepiece Body & Tube
237	Eyepiece End Ring
238	Eyepiece Tube
239	Eyepiece Tube Screw
240	Eyepiece Focusing Sleeve
241	Eyepiece Focusing Sleeve Screw
242	Eyepiece Lens I & Mount
243	Eyepiece Lens II & Mount
244	Eyepiece Lens III & Mount
245	Eyepiece Lens IV & Mount
246	Diaphragm Erector
247	Eyepiece Threaded Bushing
248	Eyepiece Cap
249	Eyepiece Cap Screw
250	Eyepiece End Ring Tension Spring
251	Telescope Focusing Lens Mount Screw
252	Telescope Level Vial Spacer

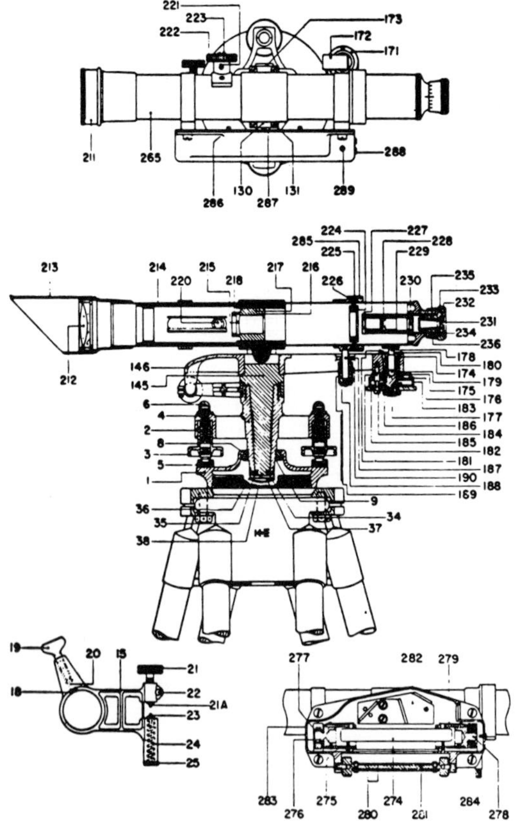

Fig. 12-2 Tilting level *(Keuffel & Esser Company)*.

TABLE 12-2 Component Parts of Tilting Level in Fig. 12-2

1 Tripod Plate
2 Leveling Head
3 Leveling Screw Head
4 Leveling Screw Stem
5 Leveling Screw Shoe
6 Leveling Screw Cap
8 Half Ball
9 Half Ball Lock Screw

15 Clamp
18 Clamp Gib
19 Clamp Screw
20 Clamp Screw Pin
21 Clamp Tangent Screw
21A Clamp Tangent Screw Pivot Pin

22 Clamp Tangent Screw Tension Screw
23 Clamp Tangent Screw Plunger
24 Clamp Tangent Screw Spring
25 Clamp Tangent Screw Cap
34 Center Nut
35 Center Nut Lock Screw
36 Center Cap
37 Center Spring
38 Center Ball
130 Trunnion Cap
131 Trunnion Cap Screw
145 Center
146 Level Bar
171 Circular Level
172 Circular Level Reflector
173 Telescope Trunnion Rosette
174 Tilting Screw Bushing
175 Tilt. Sc. Index Drum
176 Tilt. Sc. Bushing Lock Nut
177 Tilting Screw
178 Tilt. Sc. Pivot
179 Tilt. Sc. Pivot Ball
180 Tilt. Sc. Pivot Guide Plate
181 Tilt. Sc. Pivot Guide Pl. Screw
182 Tilt. Sc. Scale Drum Stop Screw (Fixed)
183 Tilt. Sc. Scale Drum
184 Tilt. Sc. Scale Dr. Stop Screw
185 Tilt. Sc. Knob
186 Tilt. Sc. Knob Lock Nut
187 Tilt. Sc. Tension Stud
188 Tilt Sc. Tension Stud Lock Nut
189 Tilt. Sc. Tension Spring
190 Tilt. Sc. Tension Spring Housing
211 Objective Cap
212 Objective Lens & Mount
213 Sunshade
214 Tele. Draw Tube
215 Telescope Focusing Lens
216 Telescope Focusing Lens Mount
217 Tele. Focusing Lens Mount Lock Screw
218 Tele. Focusing Lens Mount Lock Ring
219 Telescope Focusing Pinion, Pinion Head (222) & Screw (223), complete
220 Tele. Focusing Pinion
221 Telescope Focusing Pinion Lock Screw
222 Telescope Focusing Pinion Head
223 Telescope Focusing Pinion Head Screw
224 Reticle
225 Reticle Adj. Screw
226 Reticle Adjusting Screw Shutter
227 Eyepiece Lens I & Mount
228 Eyepiece Lens II & Mount
229 Eyepiece Tube
230 Eyepiece Lens III & Mount
231 Eyepiece Focusing Lens & Mount
232 Eyepiece Foc. Ring
233 Eyepiece Focusing Ring Set Screw
234 Eyepiece Cap
235 Eyepiece Focusing Sleeve
236 Eyepiece Focusing Sleeve Screw
265 Telescope Barrel
274 Level Vial
275 Level Vial Tube
276 Level Vial Tube Ball End
277 Level Vial Tube Ball End Tension Spring
278 Level Vial Tube Adjustable End
279 Level Vial Tension Spring
280 Level Vial Light Reflector
281 Level Vial Light Reflector Holder
282 Level Viewing Unit
283 Level Vial Housing
284 Level Vial Housing Screw
285 Reticle Adj. Sc. Cover
286 Bubble Adj. Lock Scr.
287 Bubble Adj. Screw
288 Mirror Attach. Screw
289 Screw for Bubble Adjusting Hole

SOURCE: Keuffel & Esser Company, Whippany, N.J.

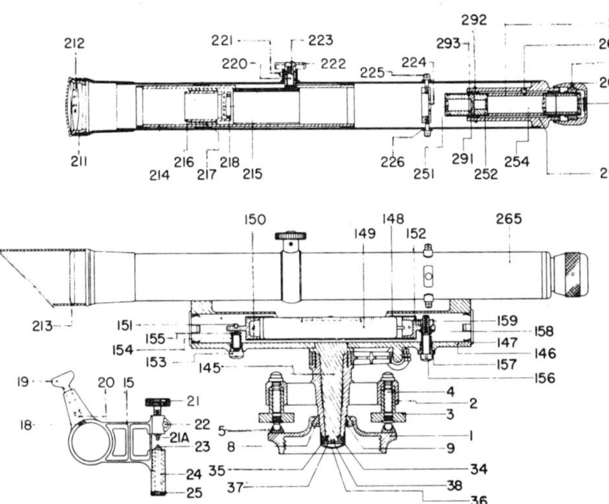

Fig. 12-3 Dumpy level *(Keuffel & Esser Company)*.

TABLE 12-3 Component Parts of Dumpy Level in Fig. 12-3

1	Tripod Plate	23	Clamp Tangent Screw Plunger
2	Leveling Head	24	Clamp Tangent Screw Spring
3	Leveling Screw Head	25	Clamp Tangent Screw Cap
4	Leveling Screw Stem	34	Center Nut
5	Leveling Screw Shoe	35	Center Nut Lock Screw
8	Half Ball	36	Center Cap
9	Half Ball Lock Screw	37	Center Spring
15	Clamp	38	Center Ball
18	Clamp Gib	145	Center
19	Clamp Screw	146	Level Bar
20	Clamp Screw Pin	147	Level Bar End Cap
21	Clamp Tangent Screw		
21A	Clamp Tangent Screw Pivot Pin		
22	Clamp Tangent Screw Tension Screw		

TABLE 12-3 Component Parts of Dumpy Level in Fig. 12-3 ***(cont.)***

No.	Part
148	Telescope Level Vial Tube & Ends Complete
149	Telescope Level Vial Only
150	Telescope Level Vial Tube End Lock Screw
151	Telescope Level Pivot
152	Telescope Level Tube End Position Pin
153	Telescope Level Tension Screw
154	Telescope Level Tension Screw Spring
155	Telescope Level Tension Screw Stud
156	Telescope Level Adjusting Screw
157	Telescope Level Adjusting Screw Bushing
158	Telescope Level Adj. Sc. Washer
159	Telescope Level Adjusting Screw Nut
211	Objective Cap
212	Objective Lens & Mount
213	Sunshade
214	Telescope Draw Tube
215	Telescope Focusing Lens
216	Telescope Focusing Lens Mount
217	Telescope Focusing Lens Mount Lock Screw
218	Tele. Focusing Lens Mount Lock Ring
219	Telescope Focusing Pinion, Pinion Head (222) and Screw (223) complete
220	Telescope Focusing Pinion
221	Telescope Focusing Pinion Lock Screw
222	Telescope Focusing Pinion Head
223	Telescope Focusing Pinion Head Screw
224	Reticle (See 74 0300)
225	Reticle Adjusting Screw
226	Reticle Adjusting Screw Shutter
251	Reticle Lens I & Mount
252	Reticle Lens II & Mount
254	Reticle Lens III & Mount
255	Reticle Lens IV & Mount
265	Telescope Barrel
266	Eyepiece Body
267	Eyepiece Focusing Ring
268	Eyepiece Focusing Ring Lock Screw
269	Eyepiece Cam Screw
270	Eyepiece Draw Tube
291	Front Aperture
292	Tension Spring
293	Ring, Eyepiece Guide

SOURCE: Keuffel & Esser Company, Whippany, N.J.

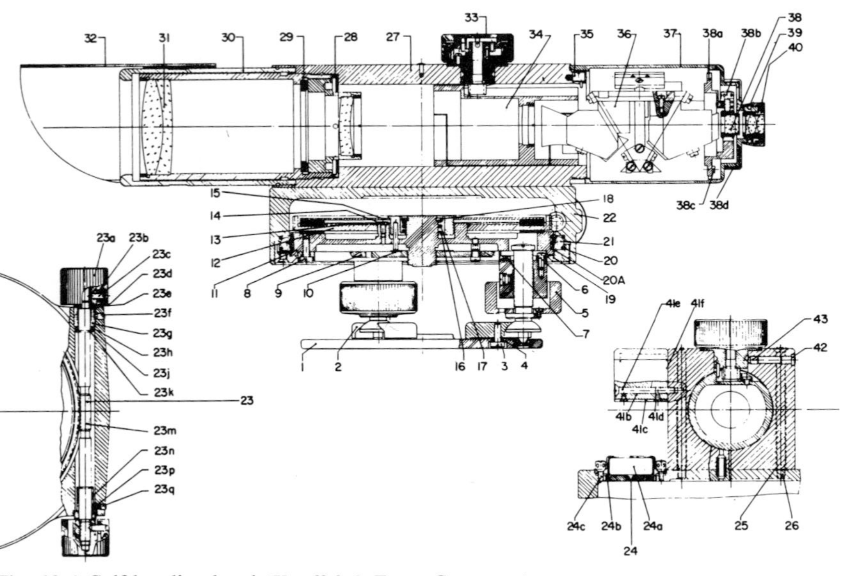

Fig. 12-4 Self-leveling level *(Keuffel & Esser Company).*

TABLE 12-4 Component Parts of Self-Leveling Level in Fig. 12-4

No.	Part
1	Lower Foot Plate
2	Foot Clamp
3	Foot Clamp Screw
4	Washer
5	Leveling Screw, Complete Assembly
6	Leveling Screw Assembly Fastening Screw
7	Leveling Screw Mounting Ring
8	Mounting Ring Screw
9	Friction Adjusting Disk with Stop Screw
10	Membrane
11	Lower Friction Plate with Cork
12	Worm Gear Plate
13	Membrane
14	Membrane Mounting Screw
15	Membrane with Cork Surface
16	Set Spring Housing
17	Set Spring
18	Set Spring Lock Ring
19	Ball Cage with Shoe & Lock Screw
20	Ball Race, Outer Ring
20a	Ball Race, Inner Ring
21	Ball (Mention Quantity Required When Ordering)
22	Box Housing for Telescope
23	Lateral Drive, Complete Assembly
23a	Knob
23b	Gib Nut
23c	Gib Screw
23d	Spring Washer
23e	Plastic Washer
23f	Right Bearing Set Screw
23g	Right Bearing
23h	Plastic Washer
23j	Metal Washer
23k	Spring Washer
23m	Worm
23n	Left Bearing
23p	Coil Spring
23q	Coil Spring Screw
23s	Left Knob Set Screw
24	Circular Vial & Housing Complete Assembly
24a	Circular Vial
24b	Mount
24c	Adjusting Screw
25	Telescope Supporting Disc
26	Telescope Housing Mounting Screw
27	Telescope Housing (furnished only as a Unit with Draw 34)
28	Stadia Ratio Adjusting Ring
29	Stadia Ratio Adjusting Ring
30	Objective Protection Sleeve
31	Objective Unit, Complete Assembly
32	Sunshade
33	Telescope Focusing Pinion Complete Assembly
34	Telescope Draw Complete with Lens (furnished only as a Unit with Telescope Housing 27)
35	Excentric Stop for Draw, Complete
36	Compensator Unit, Complete Assembly
37	Ocular, Complete Assembly
38	Reticle, Complete Assembly
38a	Reticle Centering Mount
38b	Reticle Adjusting Screw
38c	Reticle Centering Mount Screw
38d	Reticle Screw Cover
39	Eyepiece Cap with Diopter Scale
40	Eyepiece Cap Lock Screw
41a	Prism (not shown)
41b	Mount
41c	Lock Plate
41d	Prism Plate Screw
41e	Observation Prism Assembly Mounting Screw
41f	Intermediate Disk
42	Cover Screw
43	Focusing Pinion Lock Screw

SOURCE: Keuffel & Esser Company, Whippany, N.J.

12-1 THREE-WIRE LEVELING METHOD*

For fast, accurate, self-checking benchmark leveling, the three-wire method is outstanding. Both U. S. Geological Survey and the U. S. Coast and Geodetic Survey have found it to be the most economical and the most accurate method developed. In spite of this, the three-wire method is unfamiliar to many engineers.

K&E leveling equipment is designed so this method can be applied to the best advantage whenever desired.

Method. The three-wire method can be applied whenever the reticle of the level has stadia lines. The rod is read at each of the three lines and the average is used for the final result. The bubble is centered before each reading. The half-stadia intervals are compared to check for blunders. Example:

	Reading	
Upper Wire	8.798	
Middle Wire	6.563	2.235 Upper Intercept
Lower Wire	4.332	2.231 Lower Intercept
Sum	19.693	
Average	6.564	

The final rod reading is 6.564 feet. The upper and lower intercepts differ by only 0.004 foot, a normal accidental error, so it is evident that no blunder has been made.

Theory.

1. Tests have shown the levelman can read the rod more accurately than he can set a target.
2. The three readings can be made quickly as the adjustment of the bubble between readings never requires more than a slight movement.
3. The note-keeper can check the intercepts more quickly than the reading can be checked by the rodman.
4. The accuracy is as great as if three lines of levels had been run and the results averaged.
5. Since the stadia intercept is available, the length of each sight is known and the total length of run between benchmarks can be computed. This can be used to compute the accuracy of the work and to adjust a level net.
6. The unbalance between the horizontal lengths of the foresight and the backsight at each instrument position is known at once.
7. When the instrumental error per unit distance is determined by the peg method, the effect of the error can be eliminated, since the residual unbalance between the foresights and the backsights can be computed.

*Reprinted by permission of Keuffel & Esser Company, Whippany, N.J.

12-2 STADIA MEASUREMENTS*

Purpose. Stadia provides a method of measuring distances and differences in elevation by merely sighting a rod. It is very rapid, and the accuracy is better than is necessary for even large scale mapping. Stadia is always used for measurements with a plane table and it is nearly always used for locating topography with a transit. Stadia is often employed exclusively for complete traverses as well as for ties to other control.

Theory. Two supplementary horizontal lines called stadia lines are placed at equal distances above and below the central horizontal cross line of the instrument. When a graduated rod is sighted, the length of rod between the stadia lines can be observed. This is called the stadia intercept, *S*.

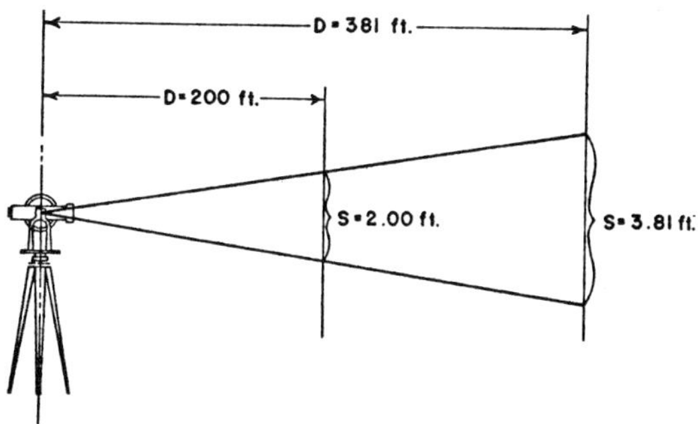

In modern K&E internal focusing instruments the distance between the stadia lines is such that, when the telescope is horizontal and the rod is vertical, the distance *D* from the center of the instrument to the rod is equal to *100 times the stadia intercept.*

*Reprinted by permission of Keuffel & Esser Company, Whippany, N.J.

TABLE 12-5 Horizontal Corrections for Stadia Intercept 1.00 ft

Vert. Angle	Hor. Cor. for 1.00 ft.	Vert. Angle	Hor. Cor. for 1.00 ft.	Vert. Angle	Hor. Cor. for 1.00 ft.
0°00′		5°36′		8°02′	
	0.0 ft		1.0 ft.		2.0 ft.
1°17′		5°53′		8°14′	
	0.1 ft.		1.1 ft.		2.1 ft.
2°13′		6°09′		8°26′	
	0.2 ft.		1.2 ft.		2.2 ft.
2°52′		6°25′		8°38′	
	0.3 ft.		1.3 ft.		2.3 ft.
3°23′		6°40′		8°49′	
	0.4 ft.		1.4 ft.		2.4 ft.
3°51′		6°55′		9°00′	
	0.5 ft.		1.5 ft.		2.5 ft.
4°15′		7°09′		9°11′	
	0.6 ft.		1.6 ft.		2.6 ft.
4°37′		7°23′		9°22′	
	0.7 ft.		1.7 ft.		2.7 ft.
4°58′		7°36′		9°33′	
	0.8 ft.		1.8 ft.		2.8 ft.
5°17′		7°49′		9°43′	
	0.9 ft.		1.9 ft.		2.9 ft.
5°36′		8°02′		9°53′	
					3.0 ft.
				10°03′	

Results from Table are correct to the nearest foot at 1000 feet and to the nearest 1/10 foot at 100 feet, etc.

Multiply the stadia intercept by the tabular value and subtract the product from the horizontal distance.

Example. Vertical angle, 4°22′; stadia intercept, 3.58 ft.

Corrected Hor. Dist. =

$$358 - (3.58 \times 0.6) = 356 \text{ ft.}$$

FORMULAS TO USE WITH VERTICAL ANGLES GREATER THAN 10°

	Internal Focusing	External Focusing
Hor. Dist.	$100\,S\cos^2 a$	$100\,(S + 0.01)\cos^2 a$
Vert. Ht.	$100\,S\sin a\cos a$	$100\,(S + 0.01)\sin a\cos a$

Where: S = stadia intercept; a = vertical angle

SOURCE: Keuffel & Esser Company, Whippany, N.J.

TABLE 12-6 Vertical Heights for Stadia Intercept 1.00 ft

Min.	0°	1°	2°	3°	4°	5°	6°	7°	8°	9°
0	0.00	1.74	3.49	5.23	6.96	8.68	10.40	12.10	13.78	15.45
2	0.06	1.80	3.55	5.28	7.02	8.74	10.45	12.15	13.84	15.51
4	0.12	1.86	3.60	5.34	7.07	8.80	10.51	12.21	13.89	15.56
6	0.17	1.92	3.66	5.40	7.13	8.85	10.57	12.27	13.95	15.62
8	0.23	1.98	3.72	5.46	7.19	8.91	10.62	12.32	14.01	15.67
10	0.29	2.04	3.78	5.52	7.25	8.97	10.68	12.38	14.06	15.73
12	0.35	2.09	3.84	5.57	7.30	9.03	10.74	12.43	14.12	15.78
14	0.41	2.15	3.89	5.63	7.36	9.08	10.79	12.49	14.17	15.84
16	0.47	2.21	3.95	5.69	7.42	9.14	10.85	12.55	14.23	15.89
18	0.52	2.27	4.01	5.75	7.48	9.20	10.91	12.60	14.28	15.95
20	0.58	2.33	4.07	5.80	7.53	9.25	10.96	12.66	14.34	16.00
22	0.64	2.38	4.13	5.86	7.59	9.31	11.02	12.72	14.40	16.06
24	0.70	2.44	4.18	5.92	7.65	9.37	11.08	12.77	14.45	16.11
26	0.76	2.50	4.24	5.98	7.71	9.43	11.13	12.83	14.51	16.17
28	0.81	2.56	4.30	6.04	7.76	9.48	11.19	12.88	14.56	16.22
30	0.87	2.62	4.36	6.09	7.82	9.54	11.25	12.94	14.62	16.28
32	0.93	2.67	4.42	6.15	7.88	9.60	11.30	13.00	14.67	16.33
34	0.99	2.73	4.47	6.21	7.94	9.65	11.36	13.05	14.73	16.39
36	1.05	2.79	4.53	6.27	7.99	9.71	11.42	13.11	14.79	16.44
38	1.11	2.85	4.59	6.32	8.05	9.77	11.47	13.17	14.84	16.50
40	1.16	2.91	4.65	6.38	8.11	9.83	11.53	13.22	14.90	16.55
42	1.22	2.97	4.71	6.44	8.17	9.88	11.59	13.28	14.95	16.61
44	1.28	3.02	4.76	6.50	8.22	9.94	11.64	13.33	15.01	16.66
46	1.34	3.08	4.82	6.56	8.28	10.00	11.70	13.39	15.06	16.72
48	1.40	3.14	4.88	6.61	8.34	10.05	11.76	13.45	15.12	16.77
50	1.45	3.20	4.94	6.67	8.40	10.11	11.81	13.50	15.17	16.83
52	1.51	3.26	4.99	6.73	8.45	10.17	11.87	13.56	15.23	16.88
54	1.57	3.31	5.05	6.79	8.51	10.22	11.93	13.61	15.28	16.94
56	1.63	3.37	5.11	6.84	8.57	10.28	11.98	13.67	15.34	16.99
58	1.69	3.43	5.17	6.90	8.63	10.34	12.04	13.73	15.40	17.05
60	1.74	3.49	5.23	6.96	8.68	10.40	12.10	13.78	15.45	17.10

Example. Vertical angle, 4°22′; stadia intercept, 3.58 ft.
Vertical Height = 3.58 × 7.59 = 27.2 ft.

SOURCE: Keuffel & Esser Company, Whippany, N.J.

12-3 SETTING OFF MAGNETIC DECLINATION*

The magnetic declination is the angle measured from the true north to the direction of the compass needle. A 12° west declination is present when the needle points 12° west of true north. The declination is different in different localities, and in any locality it is continually changing. The U.S. Coast and Geodetic Survey publishes charts from which the declination can be estimated for any time or place in the United States.

When the compass circle is set in its normal position, the needle gives the magnetic bearing of the line of sight. This is shown in Fig. 1 with the instrument pointed to the true north. The compass circle can be rotated by turning the capstan head pinion (near the south or west mark) so that the needle will give the true bearing of the line of sight.

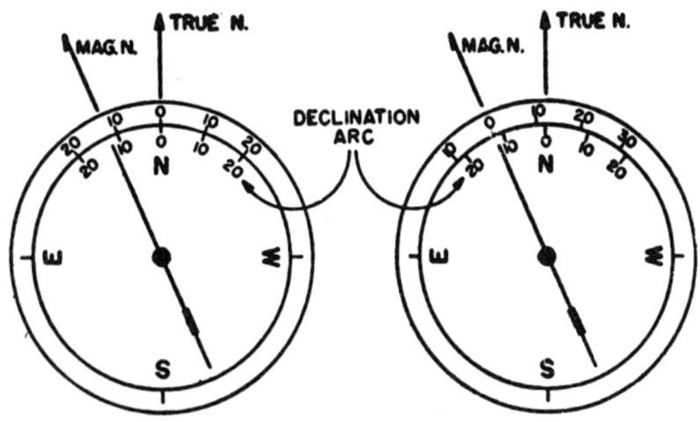

Fig. 1. **Fig. 2.**

The extent of rotation of the circle is measured *by the declination arc* in the base of the compass box next to the north part of the circle.

To set the circle for a declination of 12° west, turn the pinion clockwise and hence the circle counter-clockwise until the zero of the circle coincides with 12° to the left of N on the declination arc. The result is shown in Fig. 2. The instrument will now read true bearings for all points of the compass. The procedure is the same in the southern hemisphere.

*Reprinted by permission of Keuffel & Esser Company, Whippany, N.J.

TABLE 12-7 Conversion of Time to Arc

HOURS OF TIME INTO ARC

T.	A.	T.	A.	T.	A.	T.	A.	T.	A.	T.	A.
hrs.	°	hrs.	°	hrs.	°	hrs.	°	hrs.	°	hrs.	°
1	15	5	75	9	135	13	195	17	255	21	315
2	30	6	90	10	150	14	210	18	270	22	330
3	45	7	105	11	165	15	225	19	285	23	345
4	60	8	120	12	180	16	240	20	300	24	360

MINUTES OF TIME TO ARC
SECONDS OF TIME TO ARC

Min. / *Sec.*	° / ′	′ / ″	Min. / *Sec.*	° / ′	′ / ″	Min. / *Sec.*	° / ′	′ / ″
1	0	15	21	5	15	41	10	15
2	0	30	22	5	30	42	10	30
3	0	45	23	5	45	43	10	45
4	1	0	24	6	0	44	11	0
5	1	15	25	6	15	45	11	15
6	1	30	26	6	30	46	11	30
7	1	45	27	6	45	47	11	45
8	2	0	28	7	0	48	12	0
9	2	15	29	7	15	49	12	15
10	2	30	30	7	30	50	12	30
11	2	45	31	7	45	51	12	45
12	3	0	32	8	0	52	13	0
13	3	15	33	8	15	53	13	15
14	3	30	34	8	30	54	13	30
15	3	45	35	8	45	55	13	45
16	4	0	36	9	0	56	14	0
17	4	15	37	9	15	57	14	15
18	4	30	38	9	30	58	14	30
19	4	45	39	9	45	59	14	45
20	5	0	40	10	0	60	15	0

HUNDREDTHS OF A SECOND OF TIME TO ARC

100ths of sec. of time s.	.00 ″	.01 ″	.02 ″	.03 ″	.04 ″	.05 ″	.06 ″	.07 ″	.08 ″	.09 ″
0.00	0.00	0.15	0.30	0.45	0.60	0.75	0.90	1.05	1.20	1.35
.10	1.50	1.65	1.80	1.95	2.10	2.25	2.40	2.55	2.70	2.85
.20	3.00	3.15	3.30	3.45	3.60	3.75	3.90	4.05	4.20	4.35
.30	4.50	4.65	4.80	4.95	5.10	5.25	5.40	5.55	5.70	5.85
.40	6.00	6.15	6.30	6.45	6.60	6.75	6.90	7.05	7.20	7.35
0.50	7.50	7.65	7.80	7.95	8.10	8.25	8.40	8.55	8.70	8.85
.60	9.00	9.15	9.30	9.45	9.60	9.75	9.90	10.05	10.20	10.35
.70	10.50	10.65	10.80	10.59	11.10	11.25	11.40	11.55	11.70	11.85
.80	12.00	12.15	12.30	12.45	12.60	12.75	12.90	13.05	13.20	13.35
.90	13.50	13.65	13.80	13.95	14.10	14.25	14.40	14.55	14.70	14.85

SOURCE: Keuffel & Esser Company, Whippany, N.J.

TABLE 12-8 Temperature Correction for Steel Tape

BASED ON COEF. OF EXP. OF 0.00000645 PER DEGREE

For these temperatures subtract correction	Correction per 1000 ft.	For these temperatures add correction	For these temperatures subtract correction	Correction per 1000 ft.	For these temperatures add correction
68°F.	.00000	68°F.	28°F	.25800	108°F.
67	.00645	69	27	.26445	109
66	.01290	70	26	.27090	110
65	.01935	71	25	.27735	111
64	.02580	72	24	.28380	112
63	.03225	73	23	.29025	113
62	.03870	74	22	.29670	114
61	.04515	75	21	.30315	115
60	.05160	76	20	.30960	116
59	.05805	77	19	.31605	117
58	.06450	78	18	.32250	118
57	.07095	79	17	.32895	119
56	.07740	80	16	.33540	120
55	.08385	81	15	.34185	121
54	.09030	82	14	.34830	122
53	.09675	83	13	.35475	123
52	.10320	84	12	.36120	124
51	.10965	85	11	.36765	125
50	.11610	86	10	.37410	126
49	.12255	87	9	.38055	127
48	.12900	88	8	.38700	128
47	.13545	89	7	.39345	129
46	.14190	90	6	.39990	130
45	.14835	91	5	.40635	
44	.15480	92	4	.41280	
43	.16125	93	3	.41925	
42	.16770	94	2	.42570	
41	.17415	95	1	.43215	
40	.18060	96	0	.43860	
39	.18705	97	− 1	.44505	
38	.19350	98	− 2	.45150	
37	.19995	99	− 3	.45795	
36	.20640	100	− 4	.46440	
35	.21285	101	− 5	.47085	
34	.21930	102	− 6	.47730	
33	.22575	103	− 7	.48375	
32	.23220	104	− 8	.49020	
31	.23865	105	− 9	.49665	
30	.24510	106	−10	.50310	
29	.25155	107	−11	.50955	

Example: Measured distance at 29°F. = 782.36
Correction −.25155 x .782 = −.20
Corrected Length = 782.16

SOURCE: Keuffel & Esser Company, Whippany, N.J.

12-4 CORRECTION FOR MERIDIAN CONVERGENCE*

Apply when traverse angles are checked by celestial observation.

$C = \Delta\lambda \sin\phi$ $\qquad\qquad C'' = 52.13\ m \tan\phi$

where: C is angular convergence, $\Delta\lambda$ is long. diff., o is mean lat., m is distance east or west in miles.

12-5 CORRECTION FOR SLOPE*

Square the difference in height of the two ends and divide by twice the slope measurement. Subtract the result from the slope measurement.

Slope meas. 100 ft. Diff. in ht. 12 ft.

$100 - \frac{144}{200} = 99.28$ ft. (error 0.003 ft.)

12-6 CORRECTION FOR CURVATURE AND REFRACTION IN LEVELING*

The correction equals $-.000209\ S^2$ where S is the number of hundreds of feet in the sight.

Length of sight 220 ft.

Rod reading = 8.276
Correction: $-.000209\,(2.2)^2 = -.001$
Corrected reading = 8.275

12-7 PROBABLE ERROR*

If d_1, d_2, d_3, etc. are the discrepancies of various results from the mean, and if Σd^2 = the sum of the squares of these differences, and n = the number of observations, then the Probable Error is computed thus:

$$\text{P.E. Mean} = \pm\, 0.6745 \sqrt{\frac{\Sigma d^2}{n(n-1)}}$$

$$\text{P.E. One Obser.} = \pm\, 0.6745 \sqrt{\frac{\Sigma d^2}{n-1}}$$

*Reprinted by permission of Keuffel & Esser Company, Whippany, N.J.

TABLE 12-9 Measurement Equivalents

π = 3.1415927 –	log	= 0.49714987
1 radian = 57.295780 – deg.	log	= 1.75812263
1 radian = 3437.7468 – min.	log	= 3.53627388
1 radian = 206264.81 – sec.	log	= 5.31442513
1 degree = 0.017453293 – rad.	log	= 8.24187737 – 10
1 minute = 0.000290888 + rad.	log	= 6.46372612 – 10
1 second = 0.000004848 + rad.	log	= 4.68557487 – 10

SOURCE: Keuffel & Esser Company, Whippany, N.J.

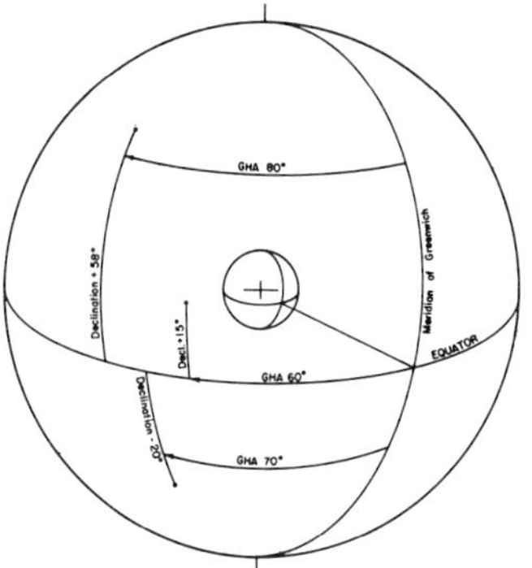

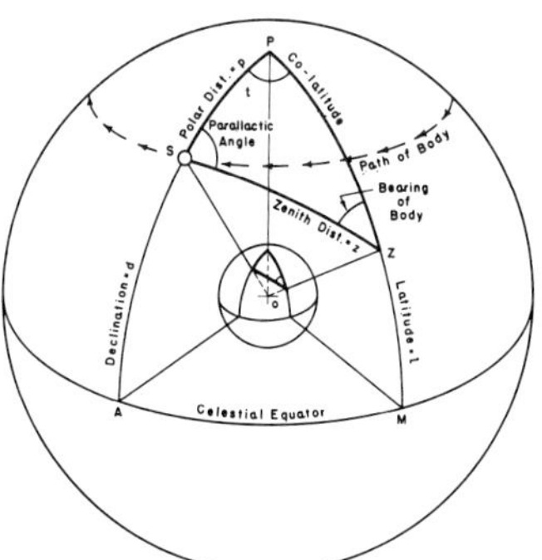

Fig. 12-5 Spherical geometry *(Keuffel & Esser Company)*.

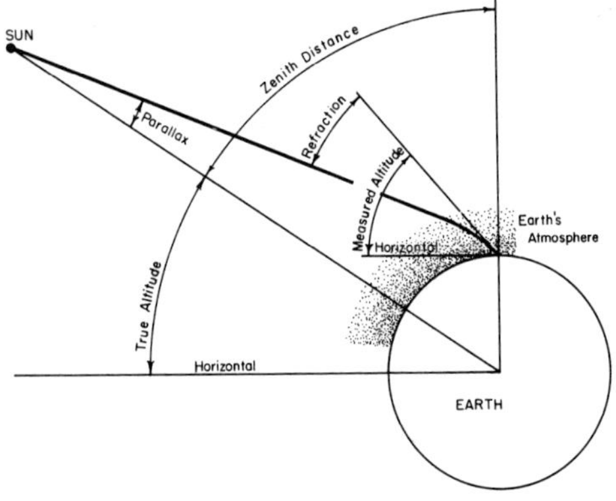

Fig. 12-6 True altitude = measured altitude − refraction + parallax *(Keuffel & Esser Company).*

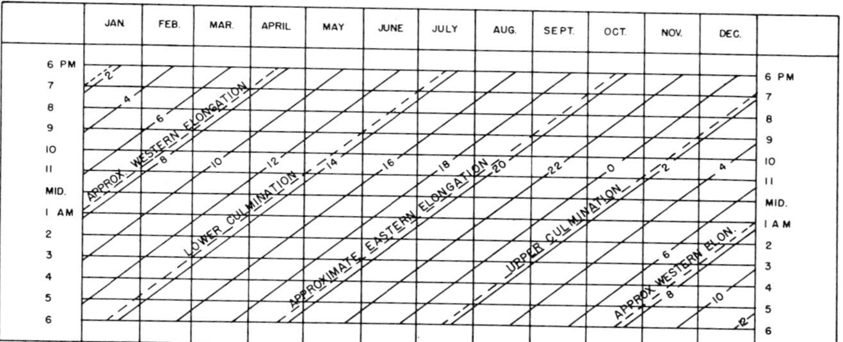

The *LCT* used in the chart is substantially correct for any year. To convert watch time to *LCT*, add to the watch time 4 minutes for every degree of longitude that the observation is east of the meridian from which the watch time is reckoned. Subtract if west. Example: watch 10:00 *CST* (90th mer.); observation 87° W. Long. 10:00+3x4 = 10:12 (*LCT*). This conversion is necessary only when watch time and *LCT* differ substantially.

Fig. 12-7 LCT Chart *(Keuffel & Esser Company)*.

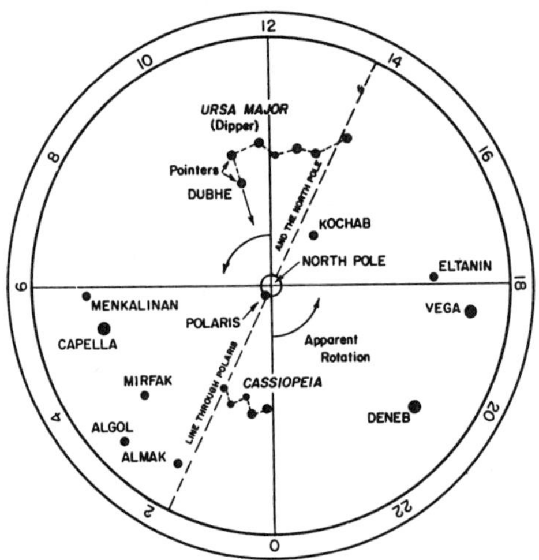

Fig. 12-8 Northern sky *(Keuffel & Esser Company).*

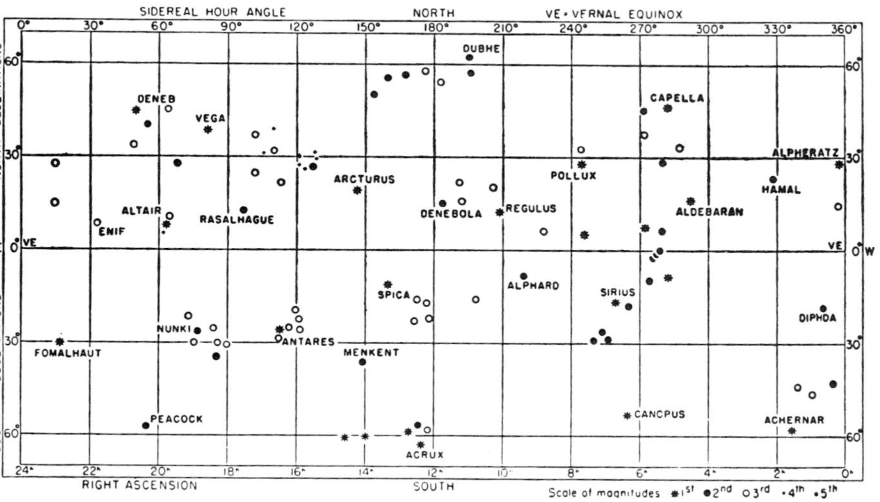

Fig. 12-9 Star chart *(Keuffel & Esser Company).*

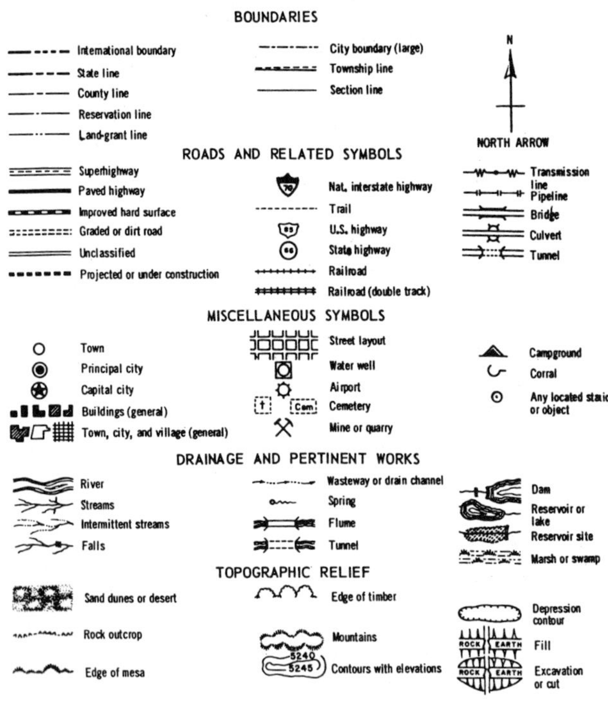

Fig. 12-10 Standard map symbols *(U.S. Dept. of the Interior, Bureau of Mines).*

NOTES

NOTES

Section Thirteen

■

DRAINAGE

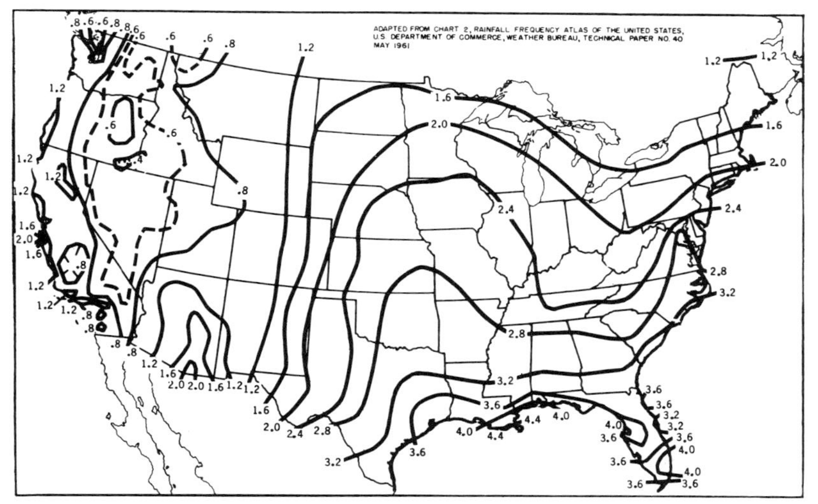

Fig. 13-1 Map of the United States showing 2-year, 30-minute rainfall intensity *(American Iron and Steel Institute)*.

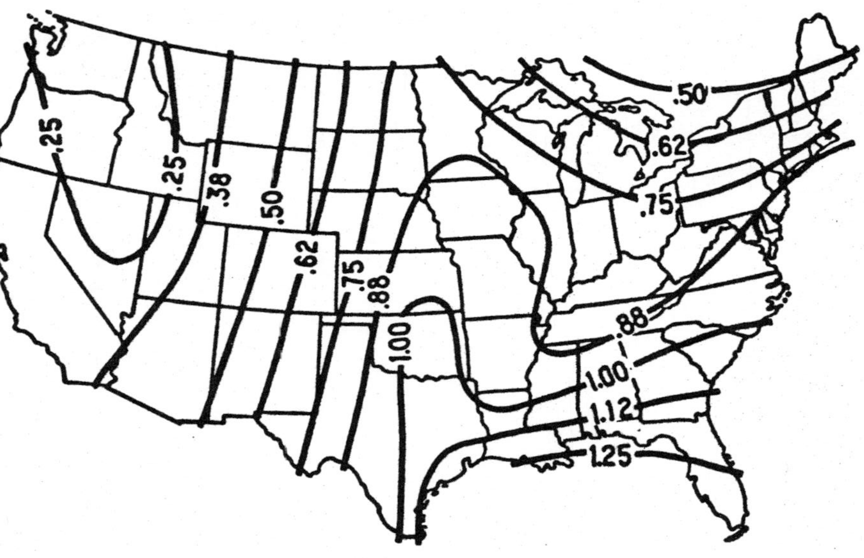

Fig. 13-2 Fifteen-minute rainfall, in inches, to be expected once in 2 years (*National Clay Pipe Institute*).

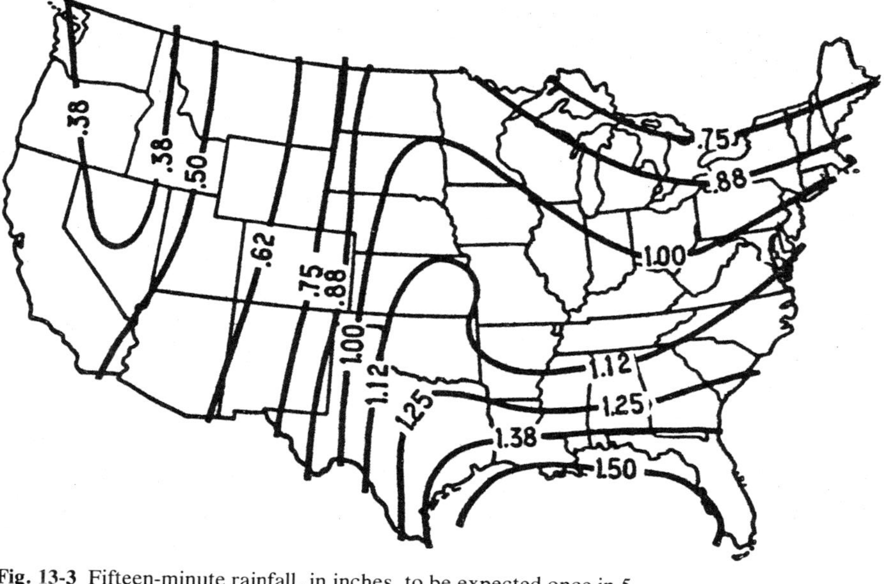

Fig. 13-3 Fifteen-minute rainfall, in inches, to be expected once in 5 years (*National Clay Pipe Institute*).

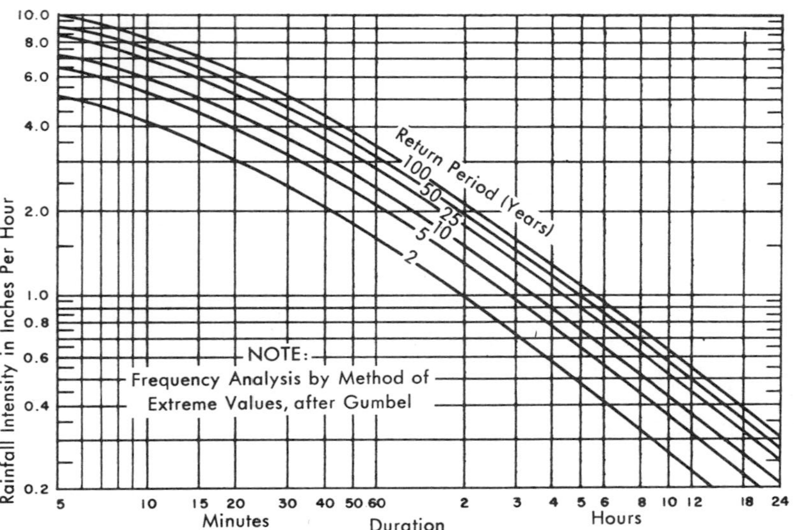

Fig. 13-4 Rainfall intensity for various durations and return periods (*American Iron and Steel Institute*).

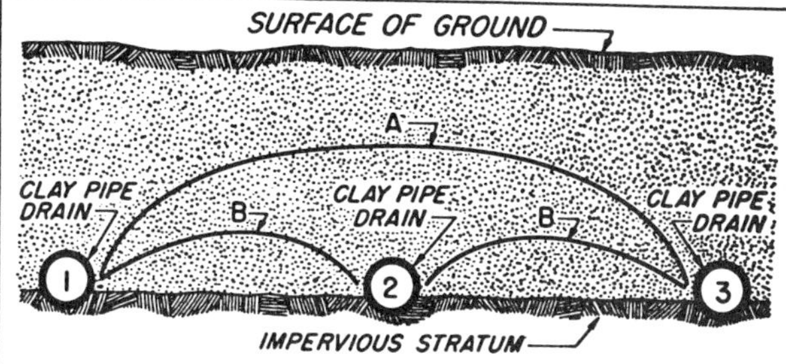

As previously shown, the amount by which ground water tables are lowered is influenced to a high degree by the spacing of the subdrains. Ground-water profiles follow the laws of hydraulic flow through porous mediums and resemble the shapes indicated in the above diagram. The ground water adjacent to an open conduit is always at the same elevation as the water in the conduit. The height to which the water rises between drains is a function of the fineness of the soil. Closer spacing between subdrains not only reduces the level of the ground water materially but also reduces, to a great degree, the amount of ground water remaining in the soil. This leaves the soil with a much greater storage capacity for subsequent percolation.

In the above diagram curve A shows the water-table for pipe lines 1 and 3 (omitting pipe 2.) Curve B shows the effect of the intermediate line. The area between curve A and B shows the increase in soil volume drained.

(a)

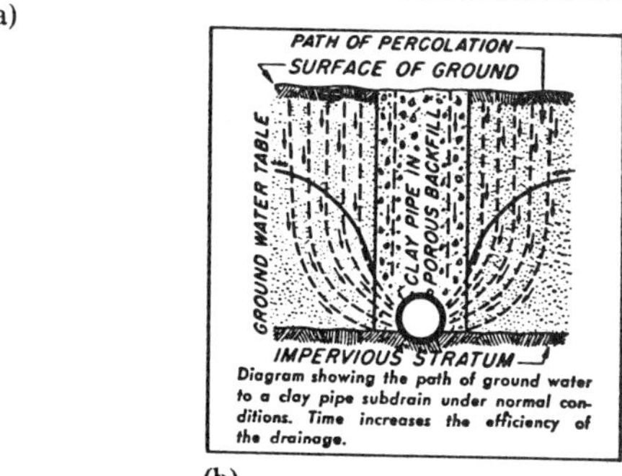

Diagram showing the path of ground water to a clay pipe subdrain under normal conditions. Time increases the efficiency of the drainage.

(b)

Fig. 13-5 Subsurface Drainage: (a) ground-water drainage curves; (b) percolation path *(National Clay Pipe Institute).*

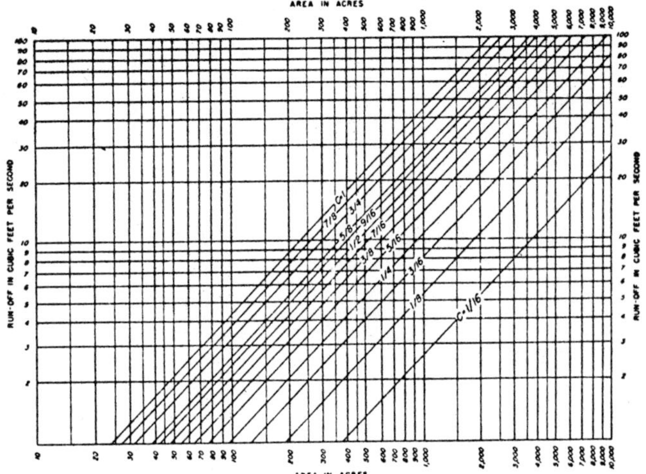

Fig. 13-6 Runoff Diagram for Subsurface Drainage. R = run-off, ft^3/s; A = area, acres; C = runoff coefficient, in inches of rainfall per 24 hours. *(National Clay Pipe Institute)*

$$R = \frac{121}{2880} AC$$

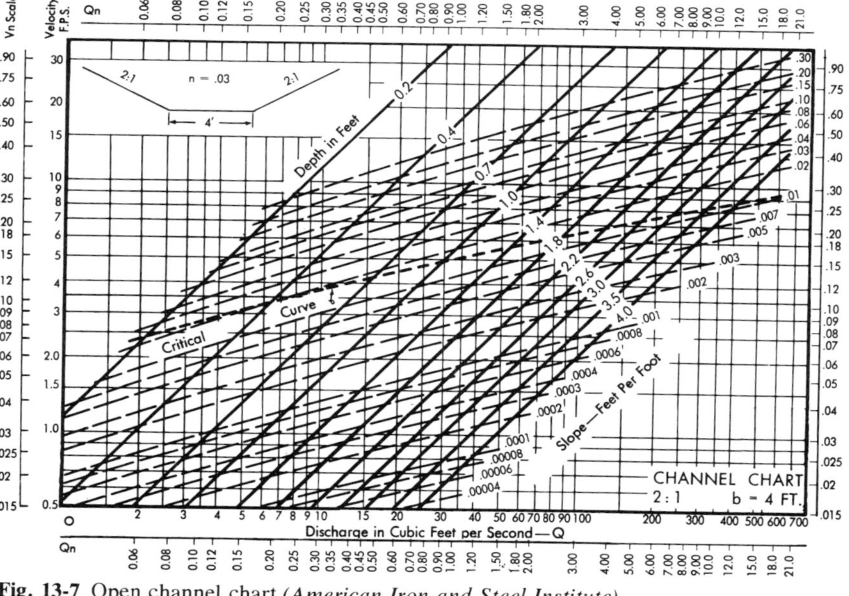

Fig. 13-7 Open channel chart (*American Iron and Steel Institute*).

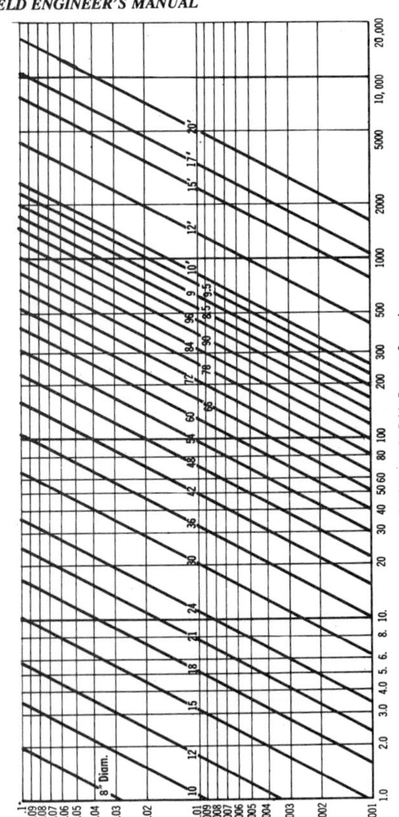

Fig. 13-8 Head loss/discharge chart (*American Iron and Steel Institute*).

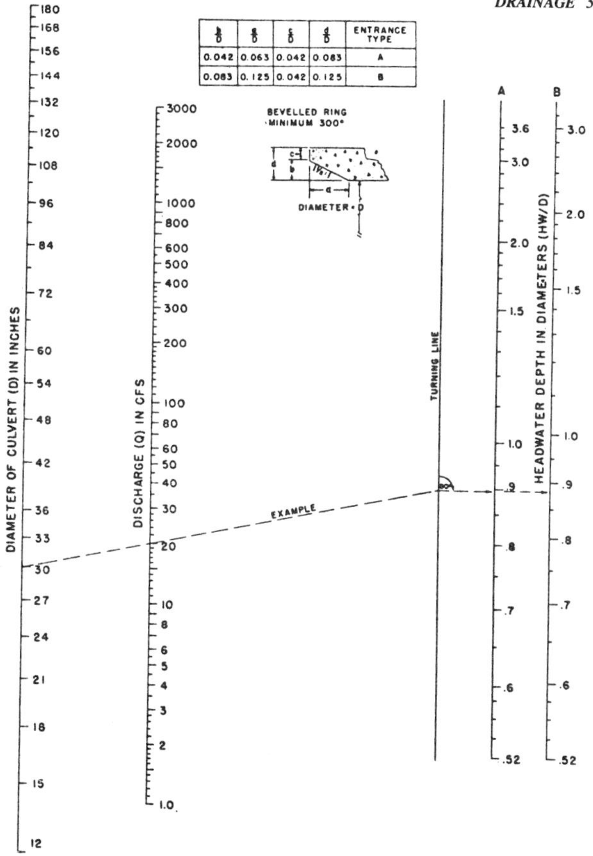

Fig. 13-9 Headwater depth for circular pipe culverts with beveled ring inlet control *(American Iron and Steel Institute)*.

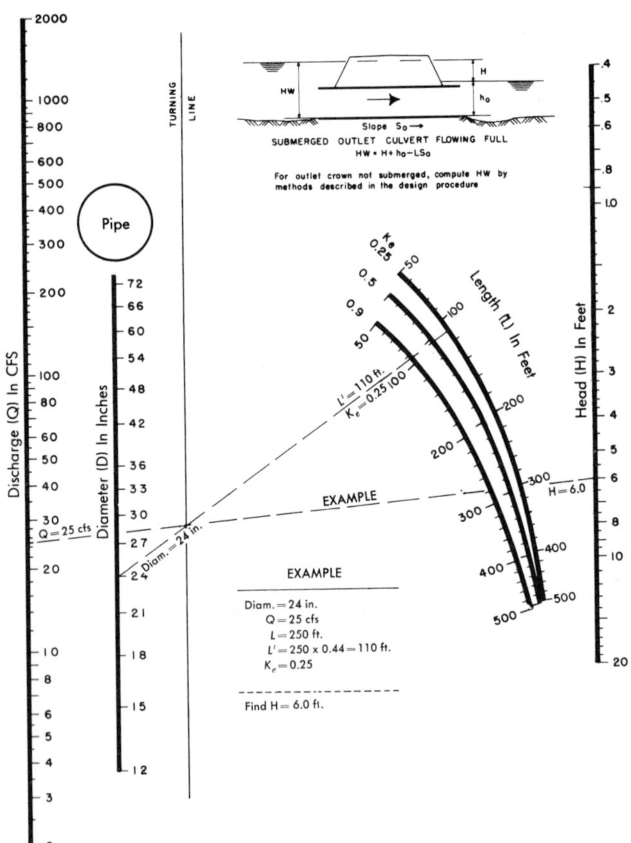

Fig. 13-10 Head for standard corrugated steel pipe culverts; flowing full—outlet control. n = 0.024. See note under sketch at top. *(American Iron and Steel Institute)*.

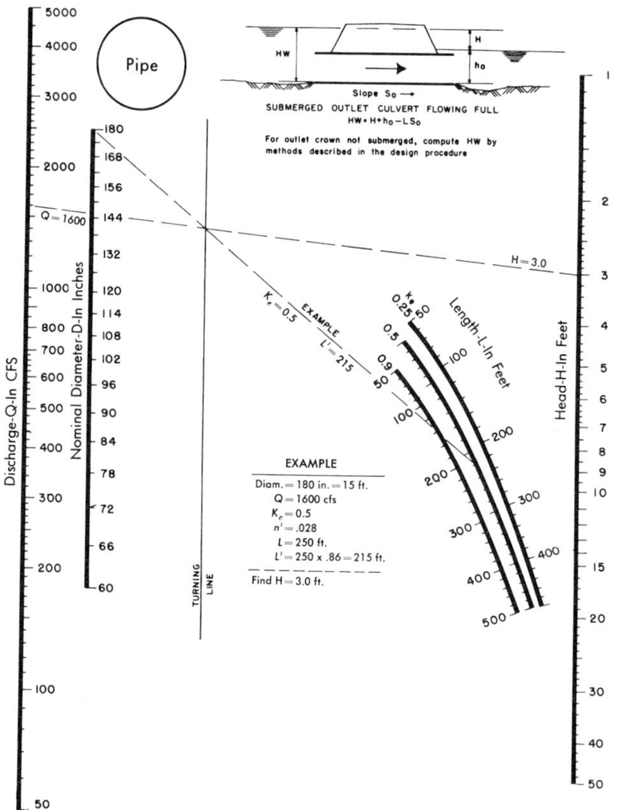

Fig. 13-11 Head for structural plate corrugated steel pipe culverts: flowing full—outlet control *(American Iron and Steel Institute)*.

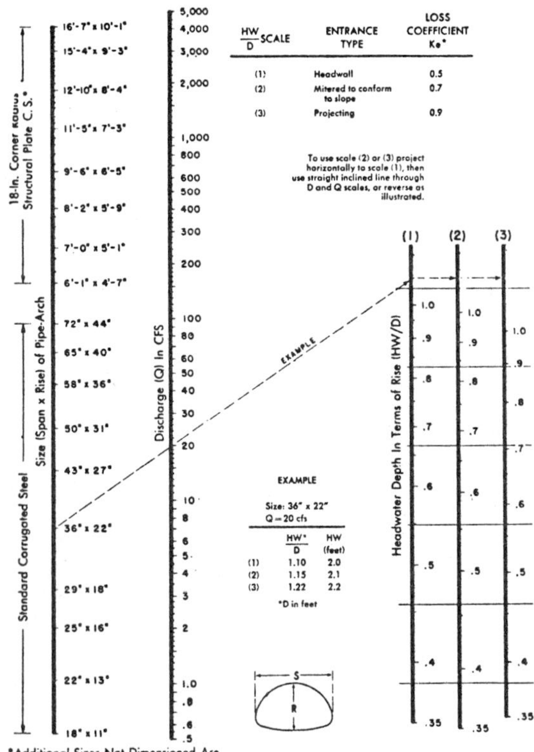

Fig. 13-12 Headwater depth for corrugated steel pipe-arch culverts with inlet control. Headwater depth should be kept low because pipe-arches are generally used where headroom is limited. *(American Iron and Steel Institute)*.

TABLE 13-1 Shapes and Uses of Corrugated Conduits*

Shape	Range of Sizes	Common Uses
Round 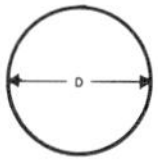	6 in. to 21 ft	Culverts, subdrains, sewers, service tunnels, etc. All plates same radius. For medium and high fills (or trenches).
Vertically-elongated (ellipse) 5% is common 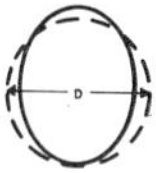	4 ft to 21 ft nominal; before elongating	Culverts, sewers, service tunnels, recovery tunnels. Plates of varying radii; shop fabrication. For appearance and where backfill compaction is only moderate.
Pipe-arch 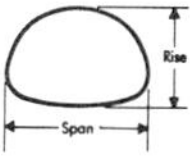	span x rise 18 in. x 11 in. to 20 ft 7 in. x 13 ft 2 in.	Where headroom is limited. Has hydraulic advantages at low flows. Corner plate radius, 18 inches or 31 inches for structural plate.
Underpass* 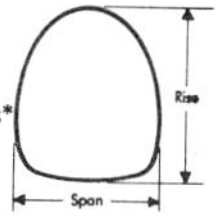	span x rise 5 ft 8 in. x 5 ft 9 in. to 20 ft 4 in. x 17 ft 10 in.	For pedestrians, livestock or vehicles (structural plate).
Arch 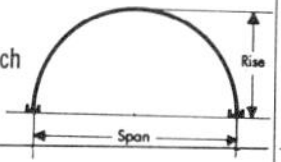	span x rise 6 ft x 1 ft 9½ in. to 25 ft x 12 ft 6 in.	For low clearance large waterway opening, and aesthetics (structural plate).
Specials	Various	For lining old structures or other special purposes. Special fabrication.

*For equal area or clearance, the round shape is generally more economical and simpler to assemble.

SOURCE: *Handbook of Steel Drainage and Highway Construction Products*, American Iron and Steel Institute, Washington, D.C. 1971.

*For equal area or clearance, the round shape is generally more economical and simpler to assemble.

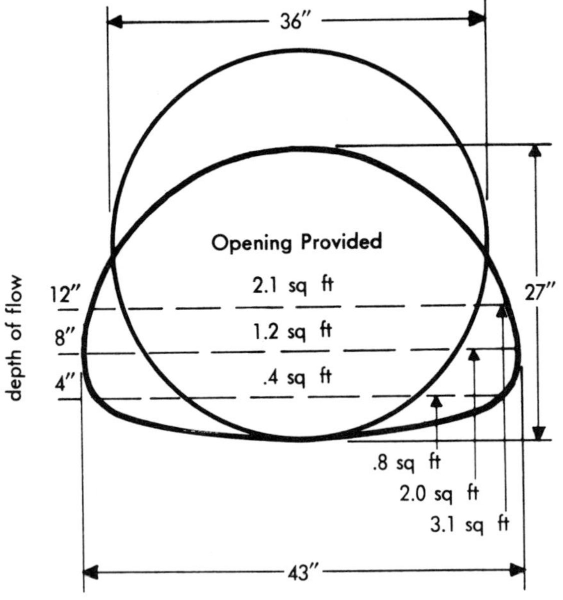

Fig. 13-13 Cross-sectional comparison of steel-pipe and pipe-arch. The pipe-arch handles a larger volume at the lower lines of flow *(American Iron and Steel Institute)*.

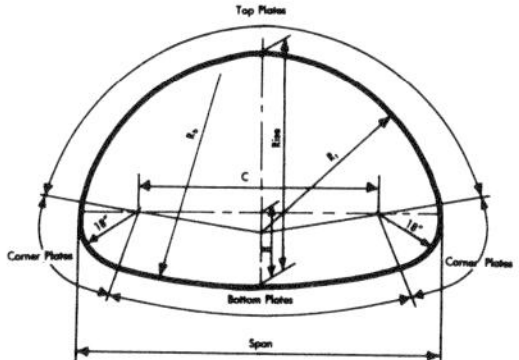

18″ R_c Corner Radius

H 20 LIVE LOAD 6″ x 2″ Corrugations

Size		Minimum Specified Thickness Required in In.	Minimum Cover	Maximum Height-of-Fill Over Pipe-Arch for the Following Corner Bearing Pressures in Tons per Sq Ft		
Span Ft-In.	Rise Ft-In.			2 Tons	3 Tons	4 Tons
6-1	4-7	0.109	12 inches	19	28	38
6-4	4-9			18	27	36
6-9	4-11			17	26	34
7-0	5-1			16	25	33
7-3	5-3			16	24	32
7-8	5-5			15	23	30
7-11	5-7			14	22	29
8-2	5-9		18 inches	14	22	28
8-7	5-11			13	21	27
8-10	6-1			13	20	26
9-4	6-3			12	18	25
9-6	6-5			12	18	24
9-9	6-7			10	17	24
10-3	6-9			9	16	22
10-8	6-11			9	16	21
10-11	7-1			9	16	21
11-5	7-3			8	15	20
11-7	7-5			7	15	20
11-10	7-7			7	14	19
12-4	7-9		24 inches	6	12	19
12-6	7-11			6	12	18
12-8	8-1			6	11	18
12-10	8-4			6	11	18
13-5	8-5			5	11	17
13-11	8-7			5	10	16
14-1	8-9			5	10	16
14-3	8-11				10	14
14-10	9-1				10	13
15-4	9-3				9	13
15-6	9-5				9	12
15-8	9-7				9	12
15-10	9-10	0.109			9	12
16-5	9-11	0.138	30 in.		9	12
16-7	10-1	0.138			9	12

Fig. 13-14 Height-of-cover limits for structural plate pipe-arch *(American Iron and Steel Institute).*

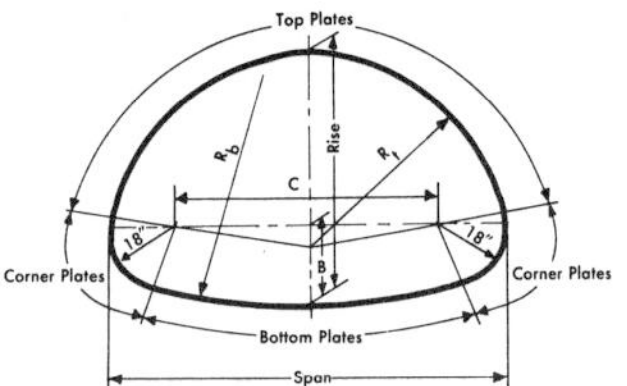

6 in. x 2 in. Corrugations—Bolted Seams

18-inch Corner Radius R_c

Dimensions			Layout Dimensions			Periphery		
Span Ft-In.	Rise Ft-In.	Waterway Area in Sq Ft	B in In.	R_t in Ft	R_b in Ft	No. of Plates	Total N	Total Pi
6-1	4-7	22	21.0	3.07	6.36	5	22	66
6-4	4-9	24	20.5	3.18	8.22	5	23	69
6-9	4-11	26	22.0	3.42	6.96	5	24	72
7-0	5-1	28	21.4	3.53	8.68	5	25	75
7-3	5-3	31	20.8	3.63	11.35	6	26	78
7-8	5-5	33	22.4	3.88	9.15	6	27	81
7-11	5-7	35	21.7	3.98	11.49	6	28	84
8-2	5-9	38	20.9	4.08	15.24	6	29	87
8-7	5-11	40	22.7	4.33	11.75	7	30	90
8-10	6-1	43	21.8	4.42	14.89	7	31	93
9-4	6-3	46	23.8	4.68	12.05	7	32	96
9-6	6-5	49	22.9	4.78	14.79	7	33	99
9-9	6-7	52	21.9	4.86	18.98	7	34	102
10-3	6-9	55	23.9	5.13	14.86	7	35	105
10-8	6-11	58	26.1	5.41	12.77	7	36	108
10-11	7-1	61	25.1	5.49	15.03	7	37	111
11-5	7-3	64	27.4	5.78	13.16	7	38	114
11-7	7-5	67	26.3	5.85	15.27	8	39	117
11-10	7-7	71	25.2	5.93	18.03	8	40	120
12-4	7-9	74	27.5	6.23	15.54	8	41	123
12-6	7-11	78	26.4	6.29	18.07	8	42	126
12-8	8-1	81	25.2	6.37	21.45	8	43	129
12-10	8-4	85	24.0	6.44	26.23	8	44	132
13-5	8-5	89	26.3	6.73	21.23	9	45	135
13-11	8-7	93	28.9	7.03	18.39	9	46	138
14-1	8-9	97	27.6	7.09	21.18	9	47	141
14-3	8-11	101	26.3	7.16	24.80	9	48	144
14-10	9-1	105	28.9	7.47	21.19	9	49	147
15-4	9-3	109	31.6	7.78	18.90	9	50	150
15-6	9-5	113	30.2	7.83	21.31	10	51	153
15-8	9-7	118	28.8	7.89	24.29	10	52	156
15-10	9-10	122	27.4	7.96	28.18	10	53	159
16-5	9-11	126	30.1	8.27	24.24	10	54	162
16-7	10-1	131	28.7	8.33	27.73	10	55	165

Dimensions are to inside crests and are subject to manufacturing tolerances.
$N = 3$ Pi $= 9.6$ in.

Fig. 13-15 Sizes and layout details: structural plate steel pipe-arches *(American Iron and Steel Institute).*

TABLE 13-2 Height-of-Cover Limits for Corrugated Steel Pipe-Arch

H 20 LIVE LOAD 2⅔″ x ½″ Corrugations

Size in Inches — Span x Rise	Minimum Specified Thickness Required in In.	Maximum Height-of-Fill Over Pipe-Arch for the Following Corner Bearing Pressures in Tons per Sq Ft		
		2 Tons	3 Tons	4 Tons
18 x 11	0.064	15	22	30
22 x 13	0.064	14	21	28
25 x 16	0.064	12	19	25
29 x 18	0.064	12	18	24
36 x 22	0.064	12	18	24
43 x 27	0.064	10	15	20
50 x 31	0.079	9	14	19
58 x 36	0.109	9	14	19
65 x 40	0.109	9	14	19
72 x 44	0.138	9	14	19
79 x 49	0.168	9	14	19
85 x 54	0.168	10	15	20

Minimum cover from top of pipe-arch to top of subgrade = 1 ft

H 20 LIVE LOAD 3″ x 1″ Corrugations

Size in Inches — Equiv. Pipe Diam.	Size in Inches — Span x Rise	Minimum Specified Thickness Required in Inches	Minimum Cover	Maximum Height-of-Fill Over Pipe-Arch for the Following Corner Bearing Pressures in Tons per Sq Ft		
				2 Tons	3 Tons	4 Tons
36	43 x 27	0.064	12 inches	14	21	28
42	50 x 31	|	|	14	21	28
48	58 x 36	|	|	14	21	28
54	65 x 40	|	|	14	21	28
60	72 x 44	|	|	14	21	28
66	73 x 55	|	|	19	28	38
72	81 x 59	|	|	17	26	34
78	87 x 63	|	|	16	24	32
84	95 x 67	|	12 inches	15	22	29
90	103 x 71	0.064	18 inches	13	20	27
96	112 x 75	0.079	18 inches	13	18	25
102	117 x 79	0.109	18 inches	12	18	24
108	128 x 83	0.109	24 inches	11	16	22
114	137 x 87	0.109	24 inches	10	15	20
120	142 x 91	0.138	24 inches	10	15	20

SOURCE: *Handbook of Steel Drainage and Highway Construction Products*, American Iron and Steel Institute, Washington, D.C. 1971.

TABLE 13-3 Height-of-Cover Limits for Structural Plate Pipe-Arch

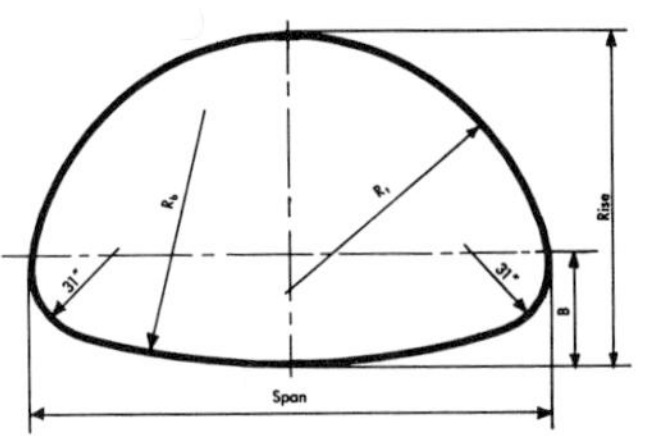

31" R_c Corner Radius

H 20 LIVE LOAD 6" x 2" Corrugations

Size		Minimum Specified Thickness Required in In.	Minimum Cover	Maximum Height-of-Fill Over Pipe-Arch for the Following Corner Bearing Pressures in Tons per Sq Ft		
Span Ft-In.	Rise Ft-In.			2 Tons	3 Tons	4 Tons
13-3	9-4	0.109	30 inches	15	22	30
13-6	9-6			14	22	29
14-0	9-8			12	21	28
14-2	9-10			12	21	28
14-5	10-0			11	20	28
14-11	10-2			11	20	27
15-4	10-4			11	19	26
15-7	10-6			11	19	25
15-10	10-8	0.109		10	19	25
16-3	10-10	0.138		10	18	24
16-6	11-0			10	18	24
17-0	11-2			10	17	23
17-2	11-4			10	17	23
17-5	11-6			9	17	22
17-11	11-8	0.138		9	16	22
18-1	11-10	0.168		9	16	22
18-7	12-0			9	16	21
18-9	12-2			9	16	21
19-3	12-4			8	15	20
19-6	12-6			8	15	20
19-8	12-8			7	15	20
19-11	12-10	0.168		7	15	20
20-5	13-0	0.188	36 in.	7	14	19
20-7	13-2	0.188		6	14	19

SOURCE: *Handbook of Steel Drainage and Highway Construction Products*, American Iron and Steel Institute, Washington, D.C. 1971.

TABLE 13-4 Design Details of Corrugated Steel Pipe-Arches*
(A: 2⅔ × ½ in; B: 3 × 1 in)

(A)

Diam of Pipe of Equal Periphery, Inches	Span, Inches	Rise, Inches	Waterway Area, Sq Ft	Layout Dimensions			
				B, Inches	R_c, Inches	R_t, Inches	R_b, Inches
15	18	11	1.1	4½	3½	10 1/16	19⅛
18	22	13	1.6	4¾	4	11⅞	37 1/16
21	25	16	2.2	5¼	4	12¾	33½
24	29	18	2.8	5½	4½	14¾	55
30	36	22	4.4	6¼	5	18¼	73¼
36	43	27	6.4	7	5½	21 9/16	91 9/16
42	50	31	8.7	8	6	25⅛	97¼
48	58	36	11.4	9¼	7	29⅛	115 11/16
54	65	40	14.3	10½	8	32¾	129 5/16
60	72	44	17.6	11¾	9	36 5/16	142 15/16
66	79	49	21.3	13¼	10	39¾	145½
72	85	54	25.3	14½	11	42⅝	154½

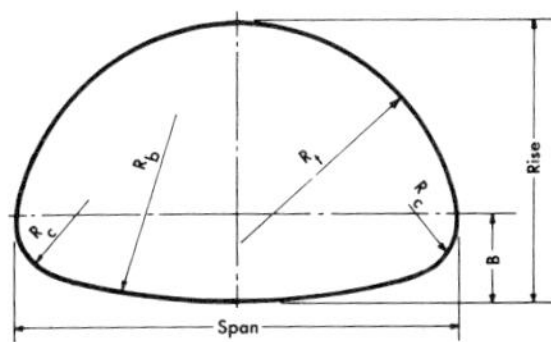

(B)

Diameter of Pipe of Equal Periphery, Inches	Span, Inches	Rise, Inches	Waterway Area, Sq Ft	Layout Dimensions			
				B, Inches	R_c, Inches	R_t, Inches	R_b, Inches
36	43	27	6.4	9¾	7¾	22½	54¾
42	50	31	8.7	11¼	9	26¼	67
48	58	36	11.4	13	10½	30½	82
54	65	40	14.3	14¾	12	34½	91¼
60	72	44	17.6	16¼	13¼	38½	98½
66	73	55	22	21	18	36¾	76¼
72	81	59	26	21½	18	40¾	92¾
78	87	63	31	22	18	43½	100½
84	95	67	35	22½	18	47¾	116
90	103	71	40	23	18	51¾	132½
96	112	75	46	23½	18	56¼	151¾
102	117	79	52	24	18	58¾	160½
108	128	83	58	24½	18	64½	185
114	137	87	64	25	18	69	201
120	142	91	71	25½	18	71¼	210

*Data in this table are subject to manufacturing tolerances.

SOURCE: *Handbook of Steel Drainage and Highway Construction Products*, American Iron and Steel Institute, Washington, D.C. 1971.

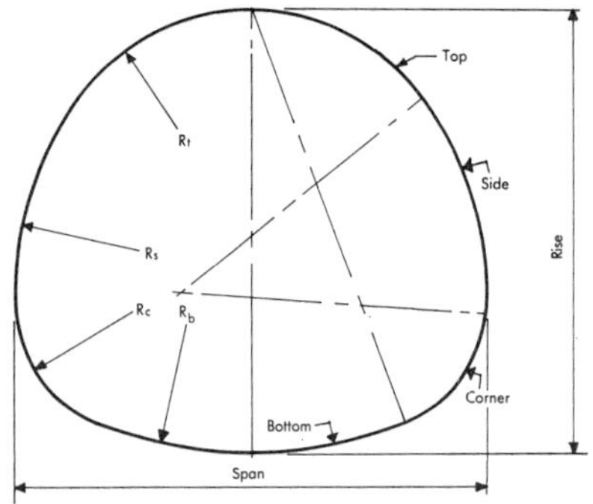

Span x Rise in Ft and In.		Periphery			Layout Dimensions in Inches			
Span	Rise	N	Pi	No. of Plates per Ring	R_t	R_s	R_c	R_b
5-8	5-9	24	72	6	27	53	18	Flat
5-8	6-6	26	78	6	29	75	18	Flat
5-9	7-4	28	84	6	28	95	18	Flat
5-10	7-8	29	87	7	30	112	18	Flat
5-10	8-2	30	90	6	28	116	18	Flat
12-2	11-0	47	141	8	68	93	38	136
12-11	11-2	49	147	9	74	92	38	148
13-2	11-10	51	153	11	73	102	38	161
13-10	12-2	53	159	11	77	106	38	168
14-1	12-10	55	165	11	77	115	38	183
14-6	13-5	57	171	11	78	131	38	174
14-10	14-0	59	177	11	79	136	38	193
15-6	14-4	61	183	12	83	139	38	201
15-8	15-0	63	189	12	82	151	38	212
16-4	15-5	65	195	12	86	156	38	217
16-5	16-0	67	201	12	88	159	38	271
16-9	16-3	68	204	12	89	168	38	246
17-3	17-0	70	210	12	90	174	47	214
18-4	16-11	72	216	12	99	157	47	248
19-1	17-2	74	222	13	105	156	47	262
19-6	17-7	76	228	13	107	158	47	295
20-4	17-9	78	234	13	114	155	47	316

All dimensions, to nearest whole number, are measured from inside crests.
Tolerances should be allowed for specification purposes. 6″ x 2″ Corrugations.

Fig. 13-16 Structural plate steel underpasses: sizes and layout details. *(American Iron and Steel Institute).*

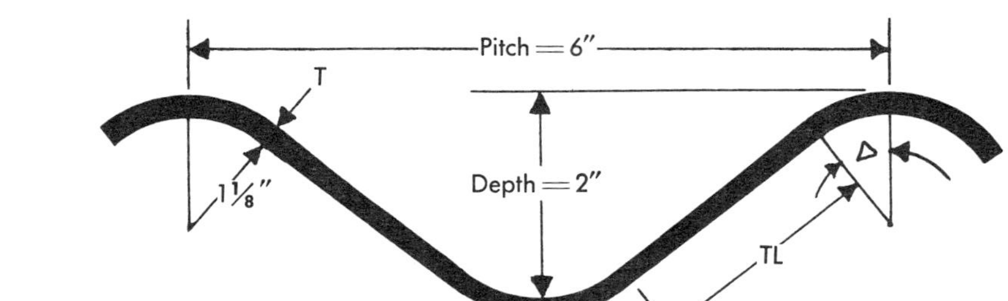

Per Foot of Section Width for Corrugation: **6 x 2 in.**

Radius of Curvature: 1⅛ in.

Specified Thickness In.	Uncoated Thickness *T* In.	Area of Section *A* Sq. In./Ft.	Tangent Length *TL* In.	Tangent Angle △ Degrees	Moment of Inertia(a) *I* In.⁴/Ft.	Section Modu us(a) *S* In.³/Ft.	Radius of Gyration *r* In.	Developed Width Factor(b)
0.109	0.1046	1.556	1.893	44.47	0.725	0.689	0.682	1.240
0.138	0.1345	2.003	1.861	44.73	0.938	0.879	0.684	1.241
0.168	0.1644	2.449	1.828	45.00	1.154	1.066	0.686	1.242
0.188	0.1838	2.739	1.807	45.18	1.296	1.187	0.688	1.242
0.218	0.2145	3.199	1.773	45.47	1.523	1.376	0.690	1.243
0.249	0.2451	3.658	1.738	45.77	1.754	1.562	0.692	1.244
0.280	0.2758	4.119	1.702	46.09	1.990	1.749	0.695	1.245

(a) Per foot of projection about the neutral axis.
To obtain *A*, *I*, or *S* per *inch* of width, divide by 12.
(b) Developed width factor measures the increase in profile length due to corrugating.
Dimensions are subject to manufacturing tolerances.

Fig. 13-17 Sectional properties of corrugated steel plates *(American Iron and Steel Institute).*

TABLE 13-5 Full-Flow Data for Corrugated Steel Pipe-Arches

Corrugations 6 x 2 in.

Corner Plates 15 pi **Radius (Rc) = 31 in.**

Span	Rise	Area	Hydraulic Radius
13′ 3″	9′ 4″	97Ft²	2.68′
13′ 6″	9′ 6″	102	2.74′
14′ 0″	9′ 8″	105	2.78′
14′ 2″	9′ 10″	109	2.83′
14′ 5″	10′ 0″	114	2.90′
14′ 11″	10′ 2″	118	2.94′
15′ 4″	10′ 4″	123	2.98′
15′ 7″	10′ 6″	127	3.04′
15′ 10″	10′ 8″	132	3.10′
16′ 3″	10′ 10″	137	3.14′
16′ 6″	11′ 0″	142	3.20′
17′ 0″	11′ 2″	146	3.24′
17′ 2″	11′ 4″	151	3.30′
17′ 5″	11′ 6″	157	3.36′
17′ 11″	11′ 8″	161	3.40′
18′ 1″	11′ 10″	167	3.45′
18′ 7″	12′ 0″	172	3.50′
18′ 9″	12′ 2″	177	3.56′
19′ 3″	12′ 4″	182	3.59′
19′ 6″	12′ 6″	188	3.65′
19′ 8″	12′ 8″	194	3.71′
19′ 11″	12′ 10″	200	3.77′
20′ 5″	13′ 0″	205	3.81′
20′ 7″	13′ 2″	211	3.87′

SOURCE: *Handbook of Steel Drainage and Highway Construction Products*, American Iron and Steel Institute, Washington, D.C. 1971.

TABLE 13-6 Part Circle Corrugated Steel Culverts: Minimum Thickness, Weights, and Dimensions

Specified Thickness in inches	Weight Per Foot, in Pounds																															
0.109	5.4	5.7	6.0	6.4	6.7	7.1	7.4	8.1	8.5	9.2	9.9	10.7	10.8	11.3	11.6	11.9	12.5	12.7	13.4	13.8	14.2	14.5	14.9	15.4	16.2	17.0	18.9	19.9	21.3	21.9	22.6	25.4
0.138	6.9	7.2	7.6	8.1	8.5	9.1	9.5	10.3	10.8	11.8	12.7	13.6	13.8	14.4	14.8	15.2	16.0	16.2	17.1	17.6	18.1	18.4	19.0	19.6	20.7	21.6	24.2	25.3	27.2	27.9	28.8	32.4
0.168	8.4	8.8	9.2	9.9	10.3	11.0	11.6	12.5	13.2	14.3	15.4	16.5	16.8	17.5	18.0	18.5	19.4	19.7	20.8	21.4	22.0	22.4	23.1	23.9	25.2	26.3	29.4	30.8	33.1	34.0	35.1	39.5
Base in In.	**Rise, in Inches**																															
12	2	2¾	3½	4⅛	4⅝	5⅛	5⅝																									
14	...	...	...	2¼	3¼	4⅛	4¾	5¾	6¼	7																						
16			...	...	...	...	3⅛	4¾	5⅜	6½	7⅜																					
18				...	...	...	...	2½	3¾	5½	6⅝	7½	7¾	8¼	8¾				Nominal Thickness 0.109 in. Min.													
20						...	...	...	...	2⅝	5½	6¾	7	7⅝	8	8⅜	9⅛	9⅜														
22							...	...	...	...	3¼	5⅜	5⅞	6⅝	7⅛	7⅝	8⅜	8¾	9½	10	10½	10¾										
24									...	...	...	...	3⅝	5	5¾	6⅜	7⅜	7¾	8¾	9¼	9¾	10⅛	10⅝	11¼								
26											...	...	...	...	3¼	4¼	6	6½	7⅝	8¼	9	9¼	9⅞	10½	11⅝	12⅜						
28												...	...	...	...	...	3½	4¼	6⅛	7	7⅞	8¼	9	9⅝	10⅞	11¾						
30														...	...	...	...	...	3⅝	5	6¼	6⅞	7¾	8½	10	11	13½	14⅝		Nominal Thickness 0.138 in. Min.		
32															...	...	...	...	...			4½	5⅞	7⅛	8⅞	10	12¾	13⅞				
34																	...	...	...	Nominal Thickness 0.168 in. Min.		...	...	5	7½	8¾	11⅞	13⅛	15	15¼		
36																			...			...	...	...	5	7⅛	10⅞	12¼	14⅜	15⅜	16⅝	
42																				...	...	...	...	...	...	...	5⅝	8½	11¼	12¼	13¼	17⅝
48																					...	...	...	...	...	...	...	...	...	6⅝	9⅝	14½

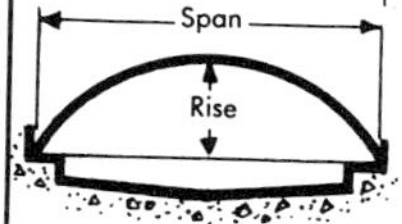

Note: Minimum gages shown on chart are for traffic conditions.
Thickness of 0.109 in. corrugated steel may be used under sidewalk areas for all sizes.
Part circles are made up in standard lengths of 25½ in., which, when lapped, make a length of 2 ft even.

SOURCE: *Handbook of Steel Drainage and Highway Construction Products*, American Iron and Steel Institute, Washington, D.C. 1971.

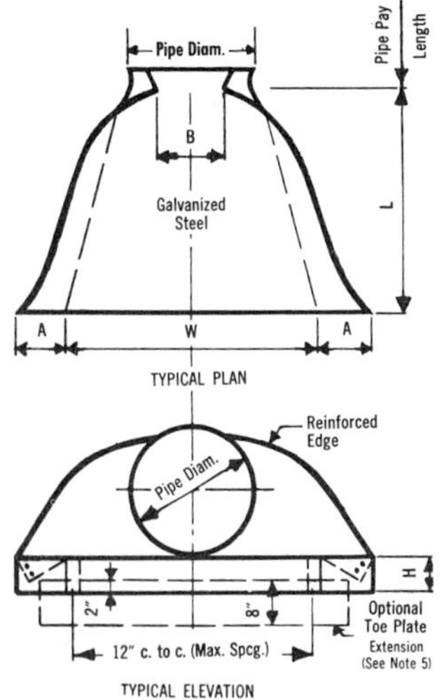

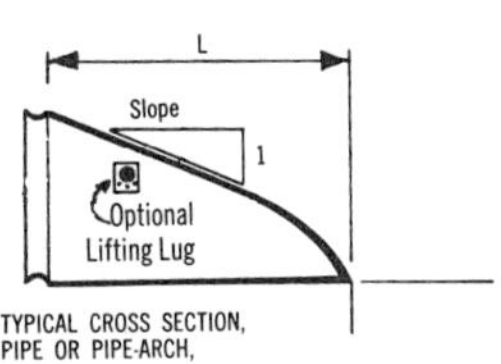

TYPICAL CROSS SECTION, PIPE OR PIPE-ARCH,

Fig. 13-18 Galvanized steel end sections for round pipe: dimensions and details *(American Iron and Steel Institute).*

Pipe Diam. in Inches	Metal Thickness in Inches	Dimensions in Inches					Approximate Slope	Body
		A ± 1″	B (max.)	H ± 1″	L ±1½″	W ± 2″		
12	.064	6	6	6	21	24	2½	1 Pc.
15	.064	7	8	6	26	30	2½	1 Pc.
18	.064	8	10	6	31	36	2½	1 Pc.
21	.064	9	12	6	36	42	2½	1 Pc.
24	.064	10	13	6	41	48	2½	1 Pc.
30	.079	12	16	8	51	60	2½	1 Pc.
36	.079	14	19	9	60	72	2½	2 Pc.
42	.109	16	22	11	69	84	2½	2 Pc.
48	.109	18	27	12	78	90	2¼	2 Pc.
54	.109	18	30	12	84	102	2	2 Pc.
60	.109	18	33	12	87	114	1¾	3 Pc.
66	.109	18	36	12	87	120	1½	3 Pc.
72	.109	18	39	12	87	126	1⅓	3 Pc.
78	.109	18	42	12	87	132	1¼	3 Pc.
84	.109	18	45	12	87	138	1⅙	3 Pc.

- All 3-piece bodies to have sides and center panels. Multiple panel bodies to have lap seams which are to be tightly joined by galvanized rivets or bolts.
- For 60 in. thru 84 in. sizes, reinforced edges to be supplemented with galvanized stiffener angles. The angles to be attached by galvanized nuts and bolts.

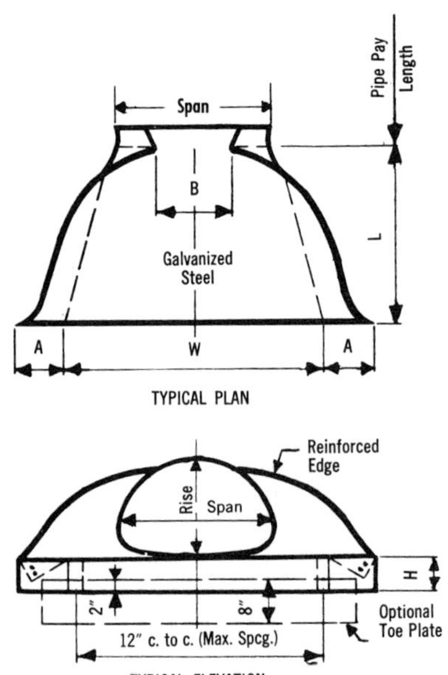

Pipe Pay
Length
Span
B
L
Galvanized
Steel
A
W
A
TYPICAL PLAN
Reinforced
Edge
Rise
Span
H
2″
8″
Optional
Toe Plate
12″ c. to c. (Max. Spcg.)
TYPICAL ELEVATION

Fig. 13-19 Galvanized steel end sections for pipe-arch: dimensions and details *(American Iron and Steel Institute)*.

Pipe-Arch in Inches		Metal Thickness in Inches	Dimensions in Inches					Approximate Slope	Body
Span	Rise		A ± 1″	B (max.)	H ± 1″	L ±1½″	W ± 2″		
18	11	.064	7	9	6	19	30	2½	1 Pc.
22	13	.064	7	10	6	23	36	2½	1 Pc.
25	16	.064	8	12	6	28	42	2½	1 Pc.
29	18	.064	9	14	6	32	48	2½	1 Pc.
36	22	.079	10	16	6	39	60	2½	1 Pc.
43	27	.079	12	18	8	46	75	2½	1 Pc.
50	31	.109	13	21	9	53	85	2½	2 Pc.
58	36	.109	18	26	12	63	90	2½	2 Pc.
65	40	.109	18	30	12	70	102	2¼	2 Pc.
72	44	.109	18	33	12	77	114	2¼	3 Pc.
79	49	.109	18	36	12	77	126	2	3 Pc.
85	54	.109	18	39	12	77	138	2	3 Pc.

- For the 79 in. by 49 in. and 85 in. by 54 in. sizes, reinforced edge to be supplemented by galvanized angles. The angles to be attached by galvanized bolts and nuts.
- Angle reinforcement will be placed under the center panel seams on the 79 in. by 49 in. and 85 in. by 54 in. sizes.
- Galvanized toe plate to be available as an accessory, when specified on the order and will be the same thickness as the End Section. Same applies to toe plate extension.
- Galvanized lifting lug available as an accessory when specified on the order.

TABLE 13-7 Height-of-Cover Limits for Structural Plate Pipe

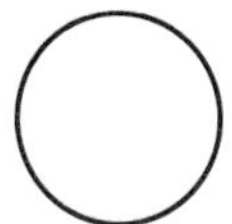

Height-of-Cover Limits for Structural Plate Pipe

H 20 LIVE LOAD **6″ x 2″** Corrugations

Diameter or Span		Min. Cover	Maximum Cover in Feet						
			Specified Thickness in Inches						
in Ft	in In.		0.109	.138	.168	.188	.218	.249	.280
5	60	12 inches	81	120	157	176	205	234	264
5.5	66		74	110	143	159	186	213	240
6	72		68	101	131	146	171	195	220
6.5	78		62	92	121	135	157	180	203
7	84		58	86	112	125	146	168	188
7.5	90		54	80	105	117	137	156	176
8	96		51	75	98	111	128	146	165
8.5	102	18 inches	48	71	92	103	120	137	155
9	108		45	67	87	97	114	130	146
9.5	114		43	63	82	92	108	123	139
10	120		40	60	78	87	102	117	132
10.5	126		39	57	74	83	97	112	126
11	132		37	54	71	79	93	106	120
11.5	138		35	52	68	76	89	102	114
12	144		34	50	65	73	85	97	110
12.5	150	24 inches	32	48	63	70	82	93	106
13	156		31	46	60	67	79	90	101
13.5	162		30	44	58	65	76	87	98
14	168		29	43	56	62	73	83	94
14.5	174		28	41	54	60	70	80	91
15	180		27	40	52	58	68	78	88
15.5	186		26	39	50	56	66	75	85
16	192		25	37	49	54	64	73	82
16.5	198	30 inches		36	47	53	62	71	80
17	204			35	45	51	60	68	77
17.5	210			34	43	49	57	65	74
18	216			33	42	47	55	63	71
18.5	222				40	45	52	60	68
19	228				38	43	50	58	65
19.5	234				37	41	48	55	62
20	240				35	40	47	53	60
20.5	246	36 inches				38	45	51	57
21	252					36	43	49	56
21.5	258					35	41	47	53
22	264						39	45	51
22.5	270						38	43	49
23	276						36	41	46
23.5	282							40	45
24	288	42 inches						38	43
24.5	294							36	41
25	300							35	39
25.5	306								37
26	312								35

SOURCE: *Handbook of Steel Drainage and Highway Construction Products,* American Iron and Steel Institute, Washington, D.C. 1971.

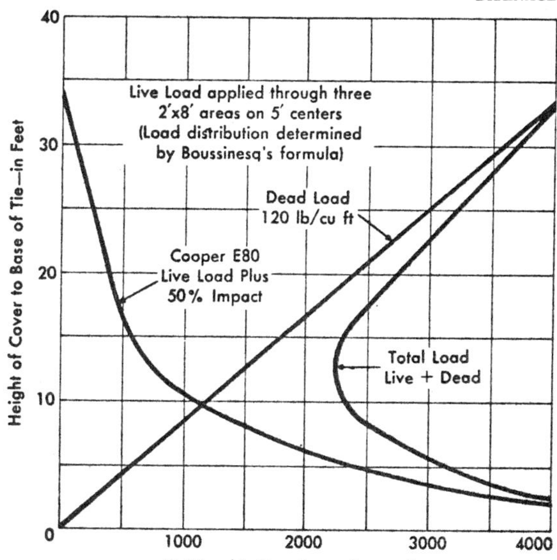

Railroad live load, Cooper E 80, combined with dead load is a minimum at about 12 ft. Load is applied through three 2 ft x 8 ft areas on 5 ft centers.

Fig. 13-20 Highway and railway live loads *(American Iron and Steel Institute).*

Highway H 20 Loading

Height of Cover (Ft)	Load, psf
1	1800
2	800
3	600
4	400
5	250
6	200
7	175
8	100

Railway E 80 Loading*[12]

Height of Cover (Ft)	Load, psf
2	3800
5	2400
8	1600
10	1100
12	800
15	600
20	300
30	100

*Neglect live load when less than 100 psf; use dead load only.

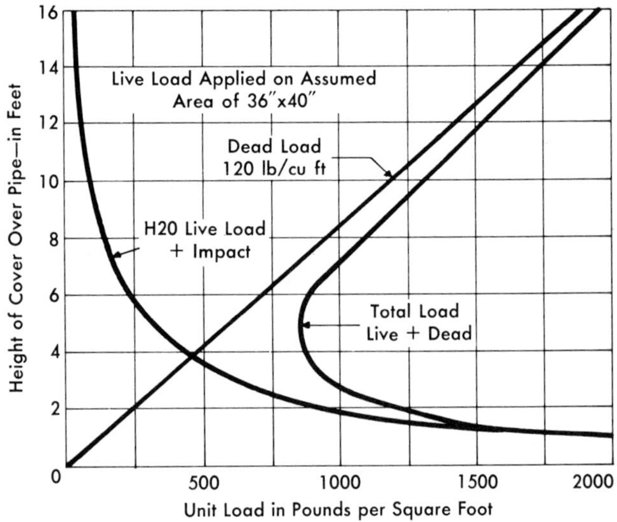

Fig. 13-21 Combined H-20 highway live load and dead load is a minimum at about 5 ft of cover, applied through a pavement 1 ft thick. *(American Iron and Steel Institute).*

TABLE 13-8 Cross-Sectional Areas for Drop-Inlet Culverts

Watershed Area in Acres	Cross-Sectional Area of Culvert in Square Feet					
	Rolling Land			Hilly Land		
	Cultivated C=1.0	Pasture 0.6	Woods 0.3	Cultivated 1.4	Pasture 0.8	Woods 0.4
1	1.9	1.1	0.6	2.7	1.5	0.8
2	2.1	1.3	0.6	2.9	1.7	0.8
4	2.5	1.5	0.8	3.5	2.0	1.0
6	2.9	1.7	0.9	4.1	2.3	1.2
8	3.4	2.0	1.0	4.8	2.7	1.4
10	3.8	2.3	1.1	5.3	3.0	1.5
15	4.8	2.9	1.4	6.7	3.8	1.9
20	5.8	3.5	1.7	8.1	4.6	2.3
30	7.8	4.7	2.3	10.9	6.2	3.1
40	9.7	5.8	2.9	13.6	7.8	3.9
50	11.5	6.9	3.5	16.1	9.2	4.6
75	15.9	9.5	4.8	22.3	12.7	6.4
100	20.0	12.0	6.0	28.0	16.0	8.0
125	23.8	14.3	7.1	33.3	19.0	9.5
150	27.3	16.4	8.2	38.2	21.9	10.9
200	33.7	20.2	10.1	47.2	27.0	13.5
250	39.4	23.6	11.8	55.2	31.5	15.8
300	44.4	26.6	13.3	62.2	35.5	17.8
350	48.9	29.3	14.7	68.5	39.1	19.6
400	53.0	31.8	15.9	74.2	42.4	21.2
500	60.0	36.0	18.0	84.0	48.0	24.0
600	65.8	39.5	19.7	92.1	52.6	26.3
700	70.8	42.5	21.2	99.1	56.6	28.3
800	75.0	45.0	22.5	105.0	60.0	30.0

Values computed by Ramser Formula, $a=c\left(130-\frac{77{,}000}{A+600}\right)$ .... (1)

Where a = cross-sectional area of culvert in sq ft.
A = watershed area in acres, c = coefficient depending on nature and type of watershed.
Formula not recommended for areas larger than given in table.
Use above values for vertical drop through culvert up to 5 ft.
Multiply above values by 0.71 for drop through culvert = 10 ft. and 0.58 for drop = 15 ft.
For fan or square shaped watersheds multiply above values by 1.25.
If side spillway of appreciable capacity is provided, reduction of culvert area may be made accordingly.

Example: A dam with a culvert having an inlet drop of 12 ft of corrugated steel pipe is to serve an ordinary watershed of 25 acres of rolling cultivated land. Interpolating in the second column of Table 6-2, between 20 and 30 acres, the cross-sectional area is 6.8 sq ft. From the footnotes of the same table, a value of 0.66 is obtained for a 12-ft drop through interpolation between values 0.71 and 0.58 for 10 and 15-ft drops. Multiply: 6.8 by 0.66 = 4.49 sq ft as the proper cross-sectional area. The nearest commercial size is a 30-in. diameter pipe with an area of 4.91 sq ft.

Size of the drop-inlet is generally larger than the culvert portion.
The top should be at least 1 ft below the emergency spillway level and 3 or 4 ft below the top of the dam.

SOURCE: *Handbook of Steel Drainage and Highway Construction Products*, American Iron and Steel Institute, Washington, D.C. 1971.

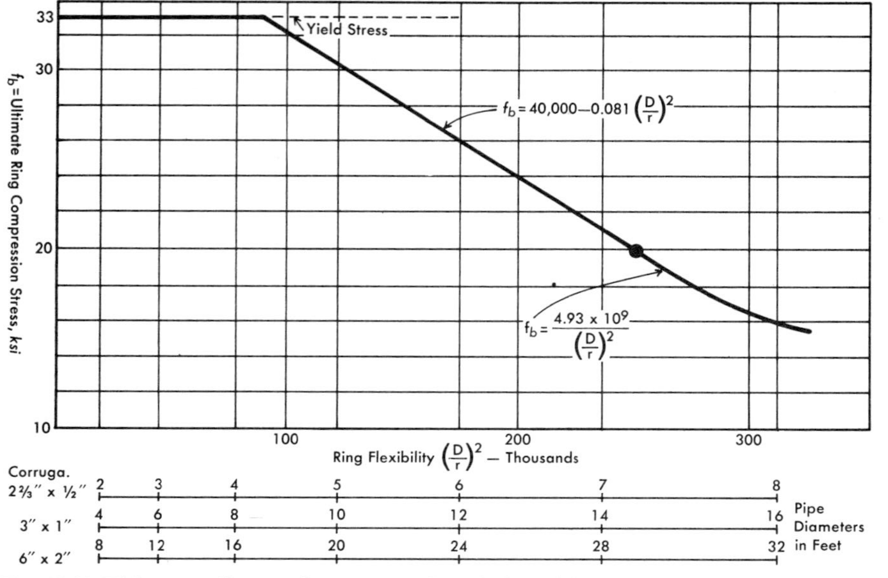

Fig. 13-22 Ultimate wall stress for corrugated steel pipe with compacted backfill (*American Iron and Steel Institute*).

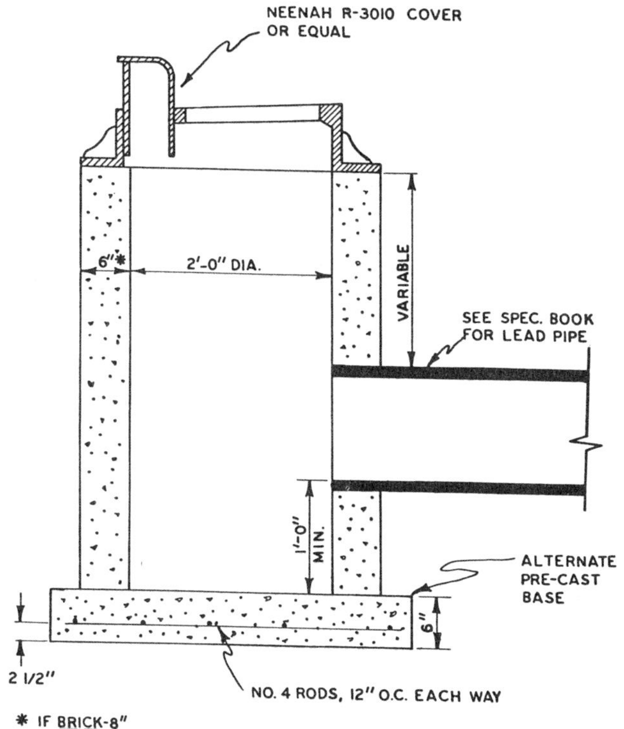

Fig. 13-23 Standard catch-basin detail. *(Morgan & Parmley, Ltd.)*

NOTES

Section Fourteen

▪

HYDRAULICS

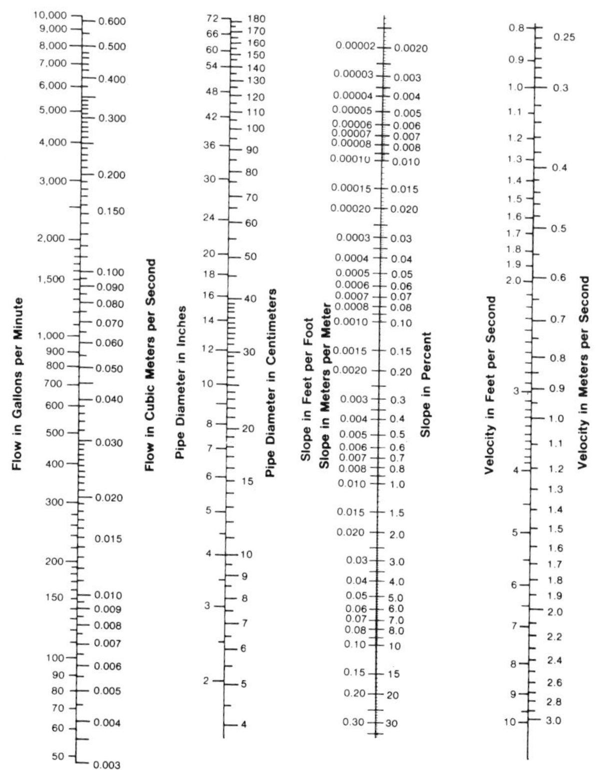

Chart based on the formula $Q = \frac{1.486}{n} \times AR^{\frac{2}{3}} \times S^{\frac{1}{2}}$ for pipe flowing full.

Fig. 14-1 Manning formula pipe flow chart (English/metric units) $n = 0.009$. *(Reprinted with permission,* Water & Sewage Works, *September 1977, 434 S. Wabash, Chicago, IL 60605.)*

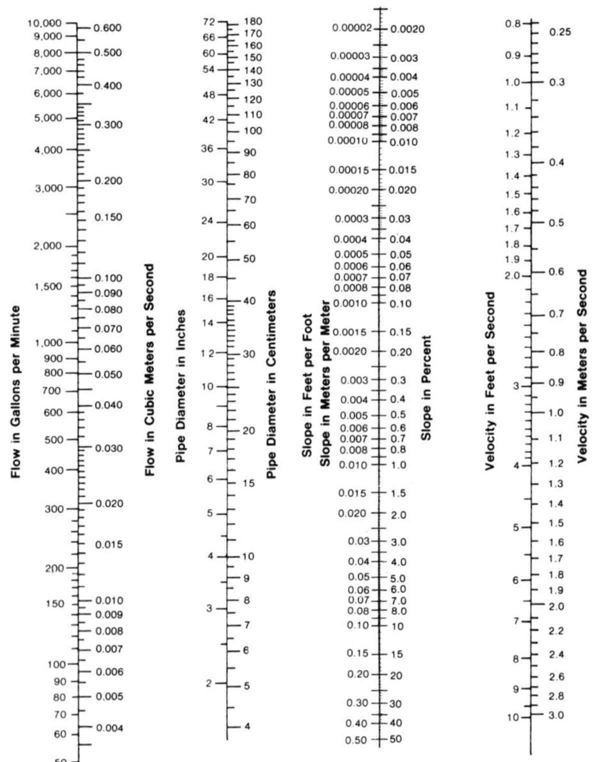

Chart based on the formula $Q=\frac{1.486}{n} \times AR^{\frac{2}{3}} \times S^{\frac{1}{2}}$ for pipe flowing full.

Fig. 14-2 Manning formula pipe flow chart (English/metric units) $n = 0.010$. *(Reprinted with permission,* Water & Sewage Works, *September 1977, 434 S. Wabash, Chicago, IL 60605.)*

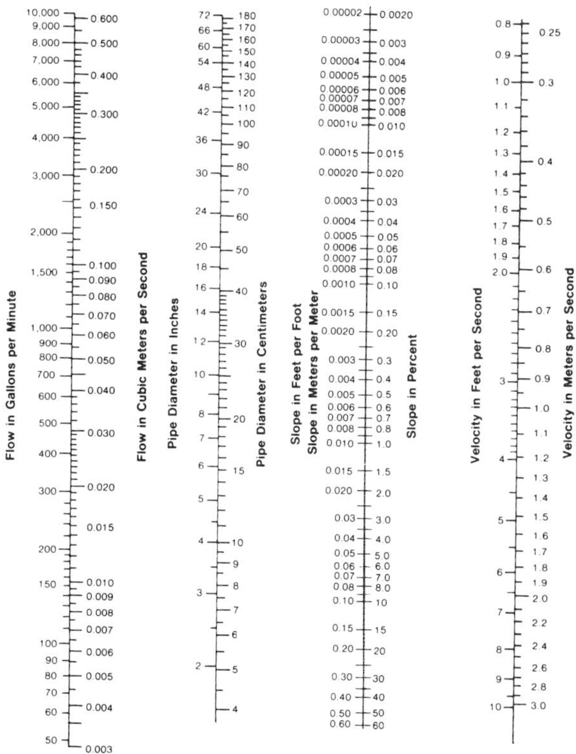

Chart based on the formula $Q = \frac{1.486}{n} \times AR^{\frac{2}{3}} \times S^{\frac{1}{2}}$ for pipe flowing full.

Fig. 14-3 Manning formula pipe flow chart (English/metric units) $n = 0.011$. *(Reprinted with permission,* Water & Sewage Works, *September 1977, 434 S. Wabash, Chicago, IL 60605.)*

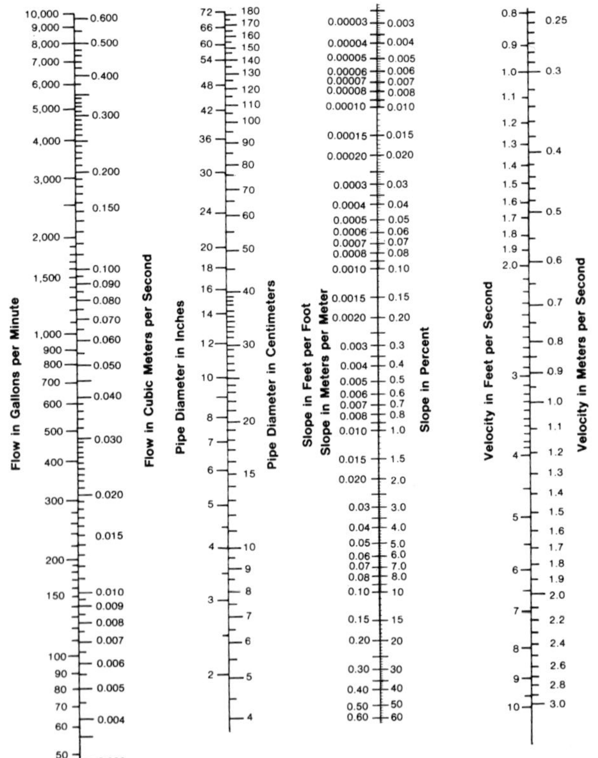

Chart based on the formula $Q = \frac{1.486}{n} \times AR^{\frac{2}{3}} \times S^{\frac{1}{2}}$ for pipe flowing full.

Fig. 14-4 Manning formula pipe flow chart (English/metric units) $n = 0.012$. *(Reprinted with permission,* Water & Sewage Works, *September 1977, 434 S. Wabash, Chicago, IL 60605.)*

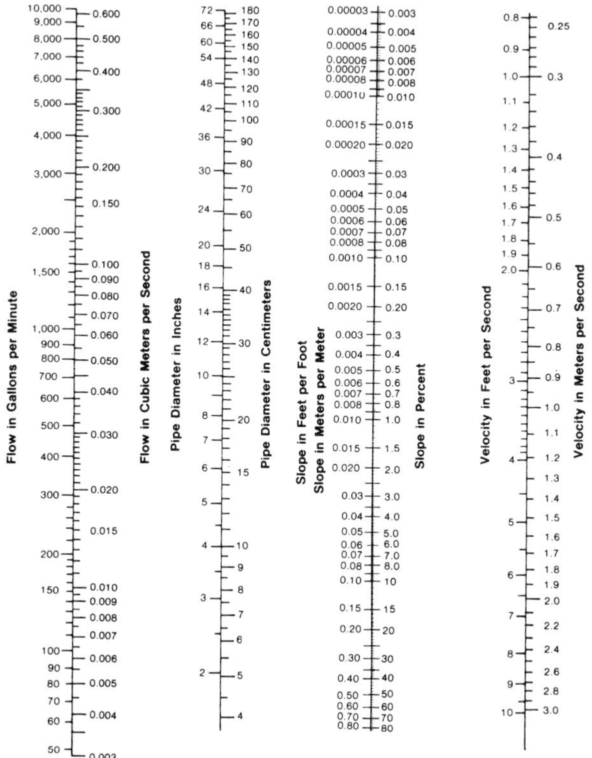

Chart based on the formula $Q = \frac{1.486}{n} \times AR^{\frac{2}{3}} \times S^{\frac{1}{2}}$ for pipe flowing full.

Fig. 14-5 Manning formula pipe flow chart (English/metric units) $n = 0.013$. *(Reprinted with permission,* Water & Sewage Works, *September 1977, 434 S. Wabash, Chicago, IL 60605.)*

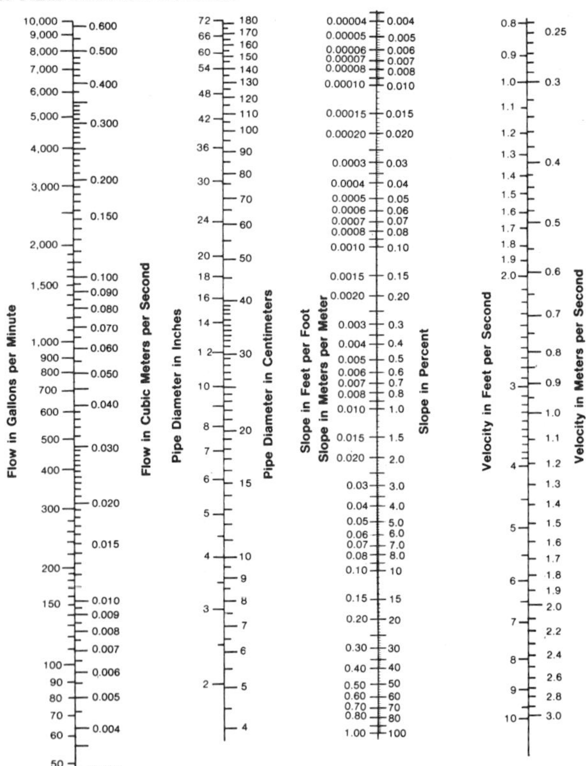

Chart based on the formula $Q = \frac{1.486}{n} \times AR^{\frac{2}{3}} \times S^{\frac{1}{2}}$ for pipe flowing full.

Fig. 14-6 Manning formula pipe flow chart (English/metric units) $n = 0.015$. *(Reprinted with permission,* Water & Sewage Works, *September 1977, 434 S. Wabash, Chicago, IL 60605.)*

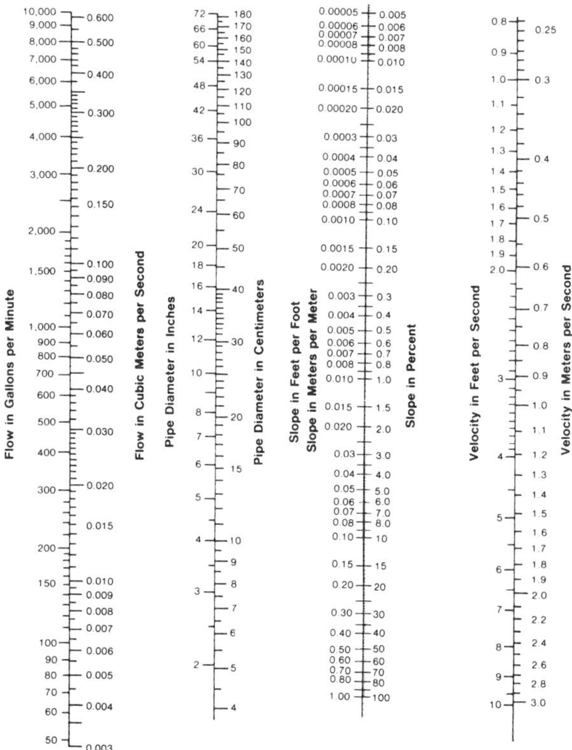

Chart based on the formula $Q = \frac{1.486}{n} \times AR^{\frac{2}{3}} \times S^{\frac{1}{2}}$ for pipe flowing full.

Fig. 14-7 Manning formula pipe flow chart (English/metric units) $n = 0.017$. *(Reprinted with permission,* Water & Sewage Works, *September 1977, 434 S. Wabash, Chicago, IL 60605.)*

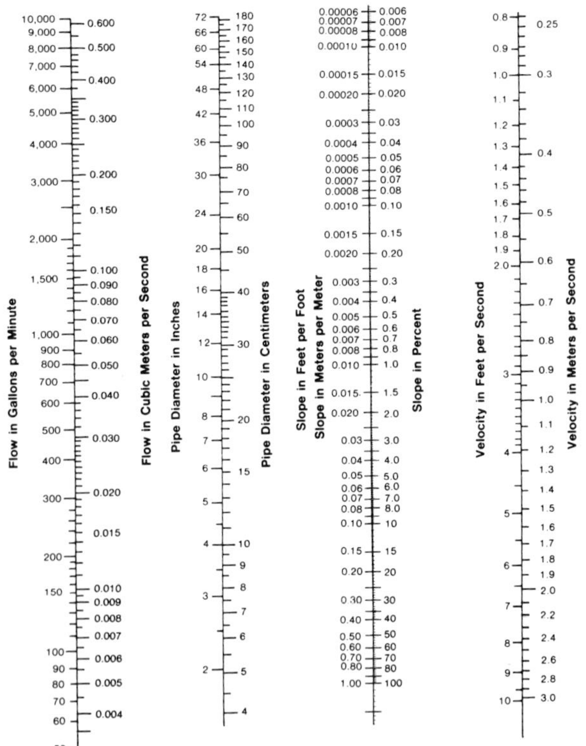

Chart based on the formula $Q = \frac{1.486}{n} \times AR^{\frac{2}{3}} \times S^{\frac{1}{2}}$ for pipe flowing full.

Fig. 14-8 Manning formula pipe flow chart (English/metric units) $n = 0.019$. *(Reprinted with permission,* Water & Sewage Works, *September 1977, 434 S. Wabash, Chicago, IL 60605.)*

TABLE 14-1 Velocity and Discharge of Circular Pipes
(Manning's Formula—Flow Full)

		SLOPE IN FEET PER THOUSAND FEET			Q (cfs)	
Diameter	v f/s	Slope			(Manning's n = 0.013)	
		(n = .009)	(n = .011)	(n = .013)	½ Full	Full
6″	1.5	1.32	1.97	2.76	0.147	0.294
	2.	2.35	3.50	4.90	0.196	0.392
	2.5	3.67	5.48	7.66	0.245	0.490
	3.	5.29	7.89	11.03	0.294	0.588
	3.5	7.14	10.61	14.88	0.343	0.686
8″	1.5	0.900	1.34	1.87	0.262	0.524
	2.	1.59	2.38	3.33	0.349	0.698
	2.5	2.50	3.73	5.21	0.437	0.873
	3.	3.60	5.37	7.50	0.525	1.05
	3.5	4.90	7.31	10.20	0.610	1.22
10″	1.5	0.676	1.00	1.41	0.409	0.818
	2.	1.20	1.79	2.50	0.545	1.09
	2.5	1.87	2.80	3.91	0.680	1.36
	3.	2.69	4.03	5.63	0.820	1.64
	3.5	3.67	5.48	7.66	0.955	1.91
12″	1.5	0.524	0.784	1.10	0.590	1.18
	2.	0.932	1.39	1.94	0.785	1.57
	2.5	1.45	2.17	3.04	0.980	1.96
	3.	2.10	3.12	4.37	1.18	2.36
	3.5	2.85	4.25	5.94	1.38	2.75
15″	1.5	0.388	0.581	0.812	0.920	1.84
	2.	0.686	1.03	1.44	1.23	2.45
	2.5	1.08	1.61	2.25	1.53	3.07
	3.	1.55	2.31	3.24	1.84	3.68
	3.5	2.12	3.16	4.41	2.15	4.29
18″	1.5	0.303	0.454	0.635	1.33	2.65
	2.	0.543	0.807	1.13	1.77	3.53
	2.5	0.847	1.27	1.77	2.21	4.42
	3.	1.23	1.82	2.55	2.65	5.30
	3.5	1.66	2.48	3.47	3.09	6.18
21″	1.5	0.250	0.369	0.520	1.81	3.61
	2.	0.441	0.655	0.918	2.41	4.81
	2.5	0.692	1.04	1.44	3.01	6.01
	3.	0.999	1.49	2.08	3.61	7.22
	3.5	1.35	2.03	2.83	4.21	8.42
24″	1.5	0.207	0.310	0.433	2.36	4.71
	2.	0.369	0.552	0.773	3.14	6.28
	2.5	0.576	0.864	1.20	3.93	7.86
	3.	0.835	1.25	1.74	4.71	9.42
	3.5	1.13	1.69	2.36	5.50	11.00
27″	1.5	0.180	0.266	0.372	2.98	5.96
	2.	0.317	0.471	0.660	3.98	7.95
	2.5	0.493	0.740	1.03	4.97	9.94
	3.	0.713	1.06	1.48	5.97	11.93
	3.5	0.973	1.45	2.03	6.96	13.92
30″	1.5	0.156	0.231	0.324	3.68	7.36
	2.	0.272	0.408	0.571	4.91	9.82
	2.5	0.428	0.640	0.894	6.14	12.27
	3.	0.620	0.924	1.29	7.37	14.73
	3.5	0.841	1.26	1.76	8.59	17.18
33″	1.5	0.135	0.202	0.282	4.46	8.91
	2.	0.243	0.361	0.506	5.94	11.88
	2.5	0.380	0.566	0.790	7.43	14.85
	3.	0.543	0.812	1.14	8.91	17.82
	3.5	0.740	1.11	1.54	10.40	20.79
36″	1.5	0.121	0.182	0.253	5.30	10.60
	2.	0.216	0.320	0.449	7.07	14.14
	2.5	0.335	0.502	0.702	8.84	17.67
	3.	0.484	0.724	1.01	10.60	21.20
	3.5	0.660	0.986	1.38	12.37	24.74
39″	1.5	0.108	0.161	0.225	6.23	12.45
	2.	0.193	0.289	0.404	8.30	16.60
	2.5	0.303	0.449	0.630	10.38	20.75
	3.	0.433	0.650	0.906	12.45	24.90
	3.5	0.590	0.882	1.23	14.53	29.05
42″	1.5	0.098	0.146	0.204	7.22	14.43
	2.	0.174	0.262	0.365	9.62	19.24
	2.5	0.272	0.408	0.571	12.03	24.05
	3.	0.396	0.590	0.824	14.43	28.86
	3.5	0.538	0.801	1.12	16.84	33.67

SOURCE: *Clay Pipe Engineering Manual,* © National Clay Pipe Institute, 1974, 1972, 1968, Crystal Lake, Ill.

TABLE 14-2 Commonly Used Velocity/Heads

$(v^2/2g)$

Velocity (feet per second)	Velocity head (feet)	Velocity (feet per second)	Velocity head (feet)
2.0	0.062	6.0	0.559
2.2	0.075	6.2	0.597
2.4	0.089	6.4	0.636
2.6	0.105	6.6	0.675
2.8	0.122	6.8	0.718
3.0	0.140	7.0	0.761
3.2	0.159	7.2	0.805
3.4	0.180	7.4	0.850
3.6	0.201	7.6	0.897
3.8	0.224	7.8	0.945
4.0	0.248	8.0	0.994
4.2	0.274	8.2	1.044
4.4	0.301	8.4	1.096
4.6	0.329	8.6	1.148
4.8	0.351	8.8	1.202
5.0	0.388	9.0	1.258
5.2	0.420	9.2	1.314
5.4	0.453	9.4	1.372
5.6	0.487	9.6	1.431
5.8	0.522	9.8	1.491

SOURCE: *Clay Pipe Engineering Manual,* © National Clay Pipe Institute, 1974, 1972, 1968, Crystal Lake, Ill.

TABLE 14-3 Frequently Used Constants

1 cfs =	0.646 Mgd	⅛ in =	0.01 ft (approx.)
1.55 cfs =	1 Mgd	1 ft^3 =	7.48 gal
1 Mgd =	694.44 gpm	1 gal =	0.134 ft^3
1 ft head =	0.433 psi	1 $mile^2$ =	640 acres
1 psi =	2.31 ft head	1 acre =	43,560 ft^2

SOURCE: *Clay Pipe Engineering Manual,* © National Clay Pipe Institute, 1974/1972/1968, Crystal Lake, Ill.

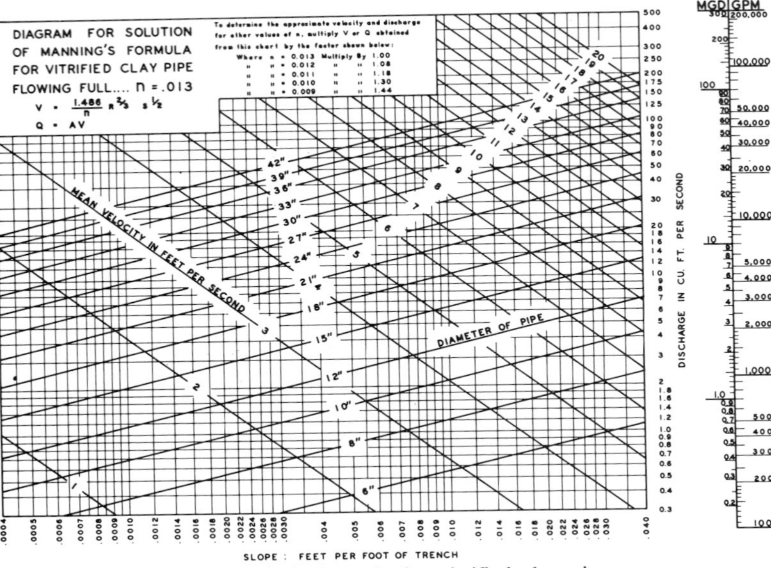

Fig. 14-9 Solution of Manning's formula for vitrified clay pipe—flowing full *(National Clay Pipe Institute)*.

TABLE 14-4 Hydraulic Properties of Clay Pipe

Diameter (inches)	Diameter (feet)	Area (square feet)	Hydraulic Radius (flowing full or half full) (feet)
4	0.333	0.087	0.083
6	0.500	0.196	0.125
8	0.667	0.349	0.167
10	0.833	0.545	0.208
12	1.000	0.785	0.250
15	1.250	1.227	0.313
18	1.500	1.767	0.375
21	1.750	2.405	0.437
24	2.000	3.142	0.500
27	2.250	3.976	0.563
30	2.500	4.909	0.625
33	2.750	5.940	0.688
36	3.000	7.068	0.750
39	3.250	8.300	0.813
42	3.500	9.621	0.875

SOURCE: *Clay Pipe Engineering Manual*, © National Clay Pipe Institute, 1974, 1972, 1968, Crystal Lake, Ill.

TABLE 14-5 Relative Carrying Capacities of Clay Pipe at Any Given Slope
(12-in pipe used as 100%)

Pipe Size (inches)	Relative Carrying Capacity in Per Cent	Pipe Size (inches)	Relative Carrying Capacity in Per Cent
4	5.3	21	445
6	15.7	24	635
8	34.0	27	870
10	61.4	30	1152
12	100	33	1484
15	181	36	1872
18	295	39	2319
		42	2820

SOURCE: *Clay Pipe Engineering Manual,* © National Clay Pipe Institute, 1974, 1972, 1968, Crystal Lake, Ill.

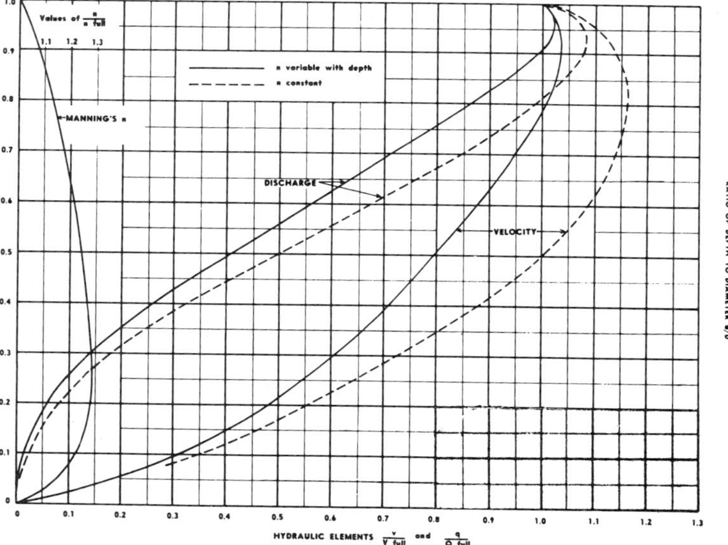

Fig. 14-10 Hydraulic properties of circular sewers *(National Clay Pipe Institute)*

TABLE 14-6 Discharge from Triangular Notch Weirs with End Contractions

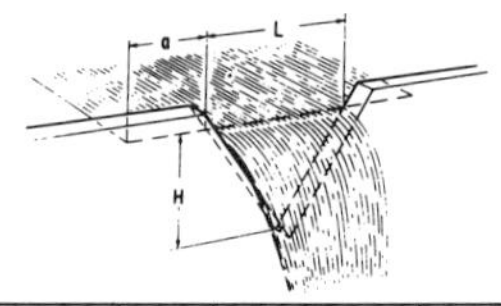

Head (H) in Inches	Flow in Gallons Per Min. 90° Notch	Flow in Gallons Per Min. 60° Notch	Head (H) in Inches	Flow in Gallons Per Min. 90° Notch	Flow in Gallons Per Min. 60° Notch	Head (H) in Inches	Flow in Gallons Per Min. 90° Notch	Flow in Gallons Per Min. 60° Notch
¼	0.07	0.04	6¼	214	124	14½	1,756	1,014
½	0.42	0.24	6½	236	136	15	1,912	1,104
¾	1.10	0.62	6¾	260	150	15½	2,073	1,197
1	2.19	1.27	7	284	164	16	2,246	1,297
1¼	3.83	2.21	7¼	310	179	16½	2,426	1,401
1½	6.05	3.49	7½	338	195	17	2,614	1,509
1¾	8.89	5.13	7¾	367	212	17½	2,810	1,623
2	12.4	7.16	8	397	229	18	3,016	1,741
2¼	16.7	9.62	8¼	429	248	18½	3,229	1,864
2½	21.7	12.5	8½	462	267	19	3,452	1,993
2¾	27.5	15.9	8¾	498	287	19½	3,684	2,127
3	34.2	19.7	9	533	308	20	3,924	2,266
3¼	41.8	24.1	9¼	571	330	20½	4,174	2,410
3½	50.3	29.0	9½	610	352	21	4,433	2,560
3¾	59.7	34.5	9¾	651	376	21½	4,702	2,715
4	70.2	40.5	10	694	401	22	4,980	2,875
4¼	81.7	47.2	10½	784	452	22½	5,268	3,041
4½	94.2	54.4	11	880	508	23	4,565	3,213
4¾	108	62.3	11½	984	568	23½	5,873	3,391
5	123	70.8	12	1,094	632	24	6,190	3,574
5¼	139	80.0	12½	1,212	700	24½	6,518	3,763
5½	156	89.9	13	1,337	772	25	6,855	3,958
5¾	174	100	13½	1,469	848			
6	193	112	14	1,609	929			

Based on formula:

$$Q = (C)\ (4/15)\ (L)\ (H)\ \sqrt{2gH}$$

in which Q = flow of water in cu. ft. per sec.

L = width of notch in ft. at H distance above apex.

H = head of water above apex of notch in ft.

C = constant varying with conditions, .57 being used for this table.

a = should be not less than ¾L.

For 90° notch the formula becomes

$$Q = 2.536H^{5/2}$$

For 60° notch the formula becomes

$$Q = 1.408H^{5/2}$$

SOURCE: *Clay Pipe Engineering Manual,* © National Clay Pipe Institute, 1974, 1972, 1968, Crystal Lake, Ill.

TABLE 14-7 Discharge from Rectangular Weirs with End Contractions

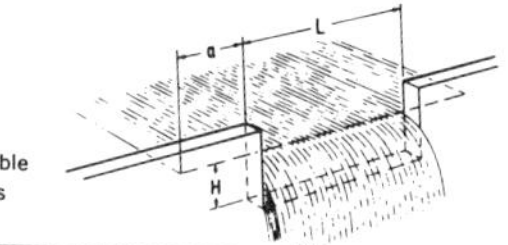

Figures in Table are in Gallons Per Minute

Head (H) in Inches	Length (L) of Weir in Feet				Head (H) in Inches	Length (L) of Weir in Feet		
	1	3	5	Additional g.p.m. for each ft. over 5 ft.		3	5	Additional g.p.m. for each ft. over 5 ft.
1/4	4.5	13.4	22.4	4.5	7 3/4	2,238	3,785	774
1/2	12.8	38.2	63.8	12.8	8	2,338	3,956	814
3/4	23.4	70.2	117.0	23.4	8 1/4	2,442	4,140	850
1	35.4	108	180	36.1	8 1/2	2,540	4,312	890
1 1/4	49.5	150	250	50.4	8 3/4	2,656	4,511	929
1 1/2	64.9	197	330	66.2	9	2,765	4,699	970
1 3/4	81.0	248	415	83.5	9 1/4	2,876	4,899	1,011
2	98.5	302	506	102	9 1/2	2,985	5,098	1,051
2 1/4	117	361	605	122	9 3/4	3,101	5,288	1,091
2 1/2	136	422	706	143	10	3,216	5,490	1,136
2 3/4	157	485	815	165	10 1/2	3,480	5,940	1,230
3	178	552	926	187	11	3,716	6,355	1,320
3 1/4	200	624	1,047	211	11 1/2	3,960	6,780	1,410
3 1/2	222	695	1,167	236	12	4,185	7,165	1,495
3 3/4	245	769	1,292	261	12 1/2	4,430	7,595	1,575
4	269	846	1,424	288	13	4,660	8,010	1,660
4 1/4	294	925	1,559	316	13 1/2	4,950	8,510	1,780
4 1/2	318	1,006	1,696	345	14	5,215	8,980	1,885
4 3/4	344	1,091	1,835	374	14 1/2	5,475	9,440	1,985
5	370	1,175	1,985	405	15	5,740	9,920	2,090
5 1/4	396	1,262	2,130	434	15 1/2	6,015	10,400	2,165
5 1/2	422	1,352	2,282	465	16	6,290	10,900	2,300
5 3/4	449	1,442	2,440	495	16 1/2	6,565	11,380	2,410
6	477	1,535	2,600	538	17	6,925	11,970	2,520
6 1/4		1,632	2,760	560	17 1/2	7,140	12,410	2,640
6 1/2		1,742	2,920	596	18	7,410	12,900	2,745
6 3/4		1,826	3,094	630	18 1/2	7,695	13,410	2,855
7		1,928	3,260	668	19	7,980	13,940	2,970
7 1/4		2,029	3,436	702	19 1/2	8,280	14,460	3,090
7 1/2		2,130	3,609	736				

$Q = 3.33\ (L - 0.2H)\ H^{1.5}$

in which

Q = cu. ft. of water flowing per second.

L = length of weir opening in feet (should be 4 to 8 times H).

H = head on weir in feet (to be measured at least 6 ft. back of weir opening).

a = should be at least 3H.

SOURCE: *Clay Pipe Engineering Manual,* © National Clay Pipe Institute, 1974, 1972, 1968, Crystal Lake, Ill.

TABLE 14-8 Energy-Loss Solutions by Manning's Formula for Pipe Flowing Full

Diam. in Inches	Area in Square Feet A	Hydraulic Radius R	$R^{2/3}$	$AR^{2/3}$	$\left(\frac{n}{1.486AR^{2/3}}\right)^2 \times 10^{-7}$				
					$n=0.012$	$n=0.015$	$n=0.019$	$n=0.021$	$n=0.024$
6	.196	.125	.250	.049	271,600.	424,420.	681,000.	831,940.	1,086,350.
8	.349	.167	.303	.106	58,000.	90,703.	145,509.	177,730.	232,164.
10	.545	.208	.351	.191	17,879.	27,936.	44,802.	54,707.	71,455.
12	.785	.250	.397	.312	6,698.	10,466.	17,797.	20,605.	26,791.
15	1.227	.3125	.461	.566	2,035.6	3,180.8	5,102.5	6,234.4	8,144.6
18	1.767	.375	.520	.919	772.2	1,206.5	1,935.5	2,364.7	3,088.7
21	2.405	.437	.576	1.385	340.00	531.24	852.60	1,041.0	1,359.98
24	3.142	.50	.630	1.979	166.5	260.04	417.31	510.20	666.39
30	4.909	.625	.731	3.588	50.7	79.126	127.01	155.12	202.54
36	7.069	.75	.825	5.832	19.20	29.953	48.071	58.713	76.691
42	9.621	.875	.915	8.803	8.40	13.148	21.096	25.773	33.667
48	12.566	1.00	1.00	12.566	4.130	6.452	10.353	12.647	16.521
54	15.904	1.125	1.082	17.208	2.202	3.440	5.520	6.741	8.817
60	19.635	1.25	1.16	22.777	1.257	1.965	3.337	3.848	5.030
66	23.758	1.375	1.236	29.365	0.756	1.182	1.895	2.316	3.026
72	28.274	1.50	1.310	37.039	0.475	0.743	1.192	1.456	1.902
78	33.183	1.625	1.382	45.859	0.310	0.485	0.777	0.950	1.241
84	38.485	1.75	1.452	55.880	0.209	0.326	0.524	0.640	0.835
90	44.179	1.875	1.521	67.196	0.144	0.226	0.362	0.442	0.578
96	50.266	2.00	1.587	79.772	0.102	0.160	0.257	0.314	0.410
108	63.617	2.25	1.717	109.230	0.055	0.085	0.137	0.167	0.219
114	70.882	2.375	1.780	126.170	0.041	0.064	0.103	0.125	0.164
120	78.54	2.5	1.842	144.671	0.031	0.049	0.078	0.098	0.125

Manning Flow Equation: $Q = \left(A \times \frac{1.486}{n} \times R^{2/3}\right) \times S^{1/2}$

Energy Loss $= S = Q^2\left(\frac{n}{1.486\,AR^{2/3}}\right)^2$

To find energy loss in pipe friction for a given Q, multiply Q^2 by figure shown under proper value of n.

SOURCE: *Handbook of Steel Drainage and Highway Construction Products*, American Iron and Steel Institute, Washington, D.C., 1971.

TABLE 14-9 Hydraulic Elements of Pipes

$\frac{d}{D}$	$\frac{A}{D^2}$	$\frac{Qn}{D^{8/3}S^{1/2}}$	$\frac{Qc}{D^{5/2}}$	$\frac{d}{D}$	$\frac{A}{D^2}$	$\frac{Qn}{D^{8/3}S^{1/2}}$	$\frac{Qc}{D^{5/2}}$
0.01	0.0013	0.00007	0.0006	0.51	0.4027	0.239	1.4494
0.02	0.0037	0.00031	0.0025	0.52	0.4127	0.247	1.5041
0.03	0.0069	0.00074	0.0055	0.53	0.4227	0.255	1.5598
0.04	0.0105	0.00138	0.0098	0.54	0.4327	0.263	1.6166
0.05	0.0147	0.00222	0.0153	0.55	0.4426	0.271	1.6741
0.06	0.0192	0.00328	0.0220	0.56	0.4526	0.279	1.7328
0.07	0.0242	0.00455	0.0298	0.57	0.4625	0.287	1.7924
0.08	0.0294	0.00604	0.0389	0.58	0.4724	0.295	1.8531
0.09	0.0350	0.00775	0.0491	0.59	0.4822	0.303	1.9147
0.10	0.0409	0.00967	0.0605	0.60	0.4920	0.311	1.9773
0.11	0.0470	0.01181	0.0731	0.61	0.5018	0.319	2.0410
0.12	0.0534	0.01417	0.0868	0.62	0.5115	0.327	2.1058
0.13	0.0600	0.01674	0.1016	0.63	0.5212	0.335	2.1717
0.14	0.0668	0.01952	0.1176	0.64	0 5308	0.343	2.2886
0.15	0.0739	0.0225	0.1347	0.65	0.5404	0.350	2.3068
0.16	0.0811	0.0257	0.1530	0.66	0.5499	0.358	2.3760
0.17	0.0885	0.0291	0.1724	0.67	0.5594	0.366	2.4465
0.18	0.0961	0.0327	0.1928	0.68	0.5687	0.373	2.5182
0.19	0.1039	0.0365	0.2144	0.69	0.5780	0.380	2.5912
0.20	0.1118	0.0406	0.2371	0.70	0.5872	0.388	2.6656
0.21	0.1199	0.0448	0.2609	0.71	0.5964	0.395	2.7416
0.22	0.1281	0.0492	0.2857	0.72	0.6054	0.402	2.8188
0.23	0.1365	0.0537	0.3116	0.73	0.6143	0.409	2.8977
0.24	0.1449	0.0585	0.3386	0.74	0.6231	0.416	2.9783
0.25	0.1535	0.0634	0.3667	0.75	0.6319	0.422	3.0606
0.26	0.1623	0.0686	0.3957	0.76	0.6405	0.429	3.1450
0.27	0.1711	0.0739	0.4259	0.77	0.6489	0.435	3.2314
0.28	0.1800	0.0793	0.4571	0.78	0.6573	0.441	3.3200
0.29	0.1890	0.0849	0.4893	0.79	0.6655	0.447	3.4111
0.30	0.1982	0.0907	0.5226	0.80	0.6736	0.453	3.5051
0.31	0.2074	0.0966	0.5569	0.81	0.6815	0.458	3.6020
0.32	0.2167	0.1027	0.5921	0.82	0.6893	0.463	3.7021
0.33	0.2260	0.1089	0.6284	0.83	0.6969	0.468	3.8062
0.34	0.2355	0.1153	0.6657	0.84	0.7043	0.473	3.9144
0.35	0.2450	0.1218	0.7040	0.85	0.7115	0.477	4.0276
0.36	0.2546	0.1284	0.7433	0.86	0.7186	0 481	4.1466
0.37	0.2642	0.1351	0.7836	0.87	0.7254	0.485	4.2722
0.38	0.2739	0.1420	0.8249	0.88	0.7320	0.488	4.4057
0.39	0.2836	0.1490	0.8672	0.89	0.7384	0.491	4.5486
0.40	0.2934	0.1561	0.9104	0.90	0.7445	0.494	4.7033
0.41	0.3032	0.1633	0.9546	0.91	0.7504	0.496	4.8724
0.42	0.3130	0.1705	0.9997	0.92	0.7560	0.497	5.0602
0.43	0.3229	0.1779	1.0459	0.93	0.7612	0.498	5.2727
0.44	0.3328	0.1854	1.0929	0.94	0.7662	0.498	5.5182
0.45	0.3428	0.1929	1.1410	0.95	0.7707	0.498	5.8119
0.46	0.3527	0.201	1.1900	0.96	0.7749	0.496	6.1785
0.47	0.3627	0.208	1.2400	0.97	0.7785	0.494	6.6695
0.48	0.3727	0.216	1.2908	0.98	0.7817	0.489	7.4063
0.49	0.3827	0.224	1.3427	0.99	0.7841	0.483	8.8261
0.50	0.3927	0.232	1.3956	1.00	0.7854	0.463	—

SOURCE: *Handbook of Steel Drainage and Highway Construction Products*, American Iron and Steel Institute, Washington, D.C., 1971.

NOTES

Section Fifteen

■

SEWAGE COLLECTION

DETERMINATION OF QUANTITY OF FLOW

Sanitary Sewage

Area to be served	×	*Density*	=	*Population*
200 Acres	×	20 population/acre	=	4000 persons
40 Acres	×	5 population/acre	=	200 persons
		Total tributary population	=	4200 persons

Average per capita flow = 100 gpd

4200 × 100	=	420,000 gpd
Industrial wastes	=	2,000,000 gpd
Commercial flow	=	10,000 gpd
Total avg. daily flow	=	2,430,000 gpd
Max. (peak) daily flow @ 200%	=	4,860,000 gpd
	=	4.86 mgd
(4.86 × 1.55)	=	7.53 cfs

1 mgd = 1.55 cfs; 0.646 mgd = 1 cfs

TYPICAL COMPUTATION SHEET

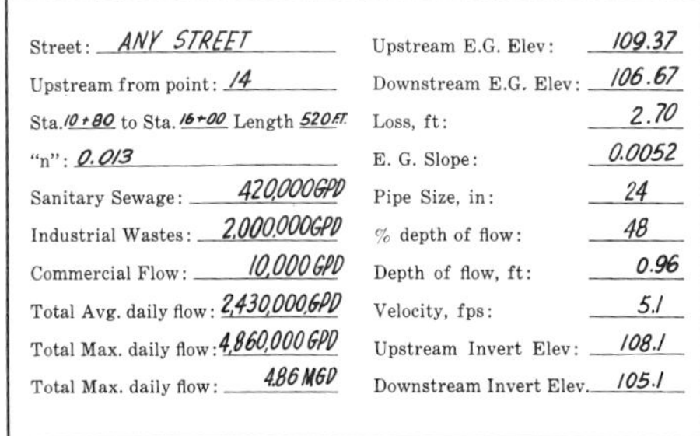

Street:	ANY STREET	Upstream E.G. Elev:	109.37
Upstream from point:	14	Downstream E.G. Elev:	106.67
Sta. 10+80 to Sta. 16+00 Length	520 FT.	Loss, ft:	2.70
"n":	0.013	E. G. Slope:	0.0052
Sanitary Sewage:	420,000 GPD	Pipe Size, in:	24
Industrial Wastes:	2,000,000 GPD	% depth of flow:	48
Commercial Flow:	10,000 GPD	Depth of flow, ft:	0.96
Total Avg. daily flow:	2,430,000 GPD	Velocity, fps:	5.1
Total Max. daily flow:	4,860,000 GPD	Upstream Invert Elev:	108.1
Total Max. daily flow:	4.86 MGD	Downstream Invert Elev.	105.1

Fig. 15-1 Determination of quantity of flow and typical computation sheet. (EPA guidelines require average per capita daily flow to be accurately determined per community.) *(National Clay Pipe Institute)*

TABLE 15-1 Velocity and Discharge in Sewers and Drainage Pipes

Based on Manning's Formula, pipes flowing full.

Diam. of Pipe in In.	N = 0.009 Velocity in feet per second	N = 0.009 Discharge in MGD	N = 0.013 Velocity in feet per second	N = 0.013 Discharge in MGD
SLOPE — **1.0** FEET PER HUNDRED FEET				
6	4.13	0.523	2.86	0.362
8	5.00	1.13	3.46	0.780
10	5.78	2.03	4.00	1.41
12	6.55	3.32	4.54	2.30
15	7.61	6.05	5.27	4.19
18	8.59	9.82	5.94	6.79
21	9.51	14.81	6.58	10.24
24	10.40	21.10	7.20	14.60
27	11.24	28.90	7.78	20.00
30	12.07	38.28	8.36	26.52
33	12.86	49.35	8.90	34.15
36	13.62	62.21	9.43	43.07
39	14.40	77.21	9.97	53.46
42	15.11	93.90	10.46	65.00

Diam. of Pipe in In.	N = 0.009 Velocity in feet per second	N = 0.009 Discharge in MGD	N = 0.013 Velocity in feet per second	N = 0.013 Discharge in MGD
SLOPE — **2.0** FEET PER HUNDRED FEET				
6	5.84	0.739	4.04	0.512
8	7.07	1.59	4.90	1.10
10	8.17	2.88	5.66	1.99
12	9.27	4.70	6.42	3.26
15	10.76	8.55	7.45	5.92
18	12.14	13.88	8.40	9.60
21	13.45	20.94	9.31	14.49
24	14.71	29.84	10.18	20.65
27	15.90	40.88	11.01	28.31
30	17.07	54.14	11.82	37.49
33	18.19	69.80	12.59	48.31
36	19.26	87.96	13.33	60.88
39	20.36	109.2	14.09	75.55
42	21.36	132.7	14.79	91.91

Diam. of Pipe in In.	N = 0.009 Velocity in feet per second	N = 0.009 Discharge in MGD	N = 0.013 Velocity in feet per second	N = 0.013 Discharge in MGD
SLOPE — **3.0** FEET PER HUNDRED FEET				
6	7.15	0.905	4.95	0.627
8	8.66	1.95	6.00	1.35
10	10.01	3.52	6.93	2.44
12	11.35	5.76	7.86	3.99
15	13.18	10.47	9.12	7.25
18	14.87	17.00	10.29	11.77
21	16.47	25.64	11.40	17.75
24	18.02	36.55	12.47	25.29
27	19.47	50.06	13.48	34.66
30	20.90	66.29	14.47	45.90
33	22.28	85.49	15.42	59.17
36	23.59	107.7	16.33	74.58
39	24.94	133.7	17.26	92.54
42	26.17	162.6	18.12	112.6

Diam. of Pipe in In.	N = 0.009 Velocity in feet per second	N = 0.009 Discharge in MGD	N = 0.013 Velocity in feet per second	N = 0.013 Discharge in MGD
SLOPE — **4.0** FEET PER HUNDRED FEET				
6	8.26	1.05	5.72	0.724
8	10.01	2.26	6.93	1.56
10	11.56	4.07	8.00	2.82
12	13.11	6.65	9.08	4.60
15	15.22	12.09	10.54	8.37
18	17.17	19.63	11.89	13.60
21	19.02	29.61	13.17	20.50
24	20.80	42.19	14.40	29.21
27	22.49	57.82	15.57	40.03
30	24.14	76.56	16.71	53.00
33	25.72	98.69	17.81	68.34
36	27.24	124.4	18.86	86.14
39	28.80	154.4	19.94	106.9
42	30.22	187.8	20.92	130.0

SOURCE: *Clay Pipe Engineering Manual,* © National Clay Pipe Institute 1974, 1972, 1968, Crystal Lake, Ill.

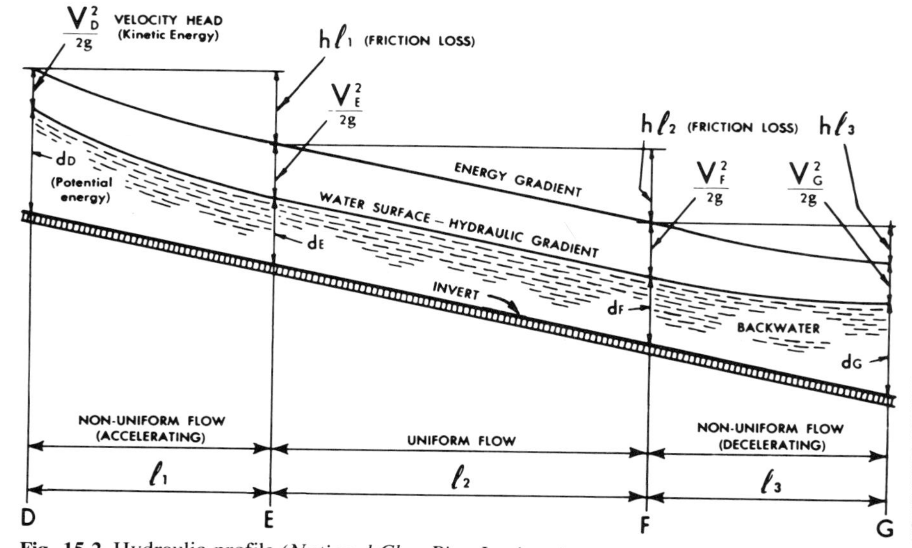

Fig. 15-2 Hydraulic profile *(National Clay Pipe Institute)*.

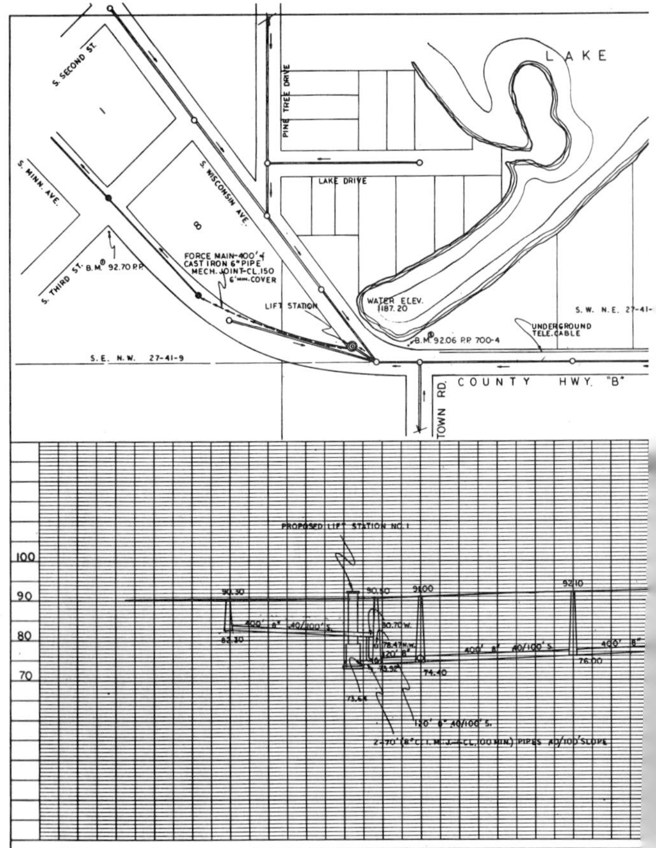

Fig. 15-3 Typical engineering layout of sanitary sewer: plan and

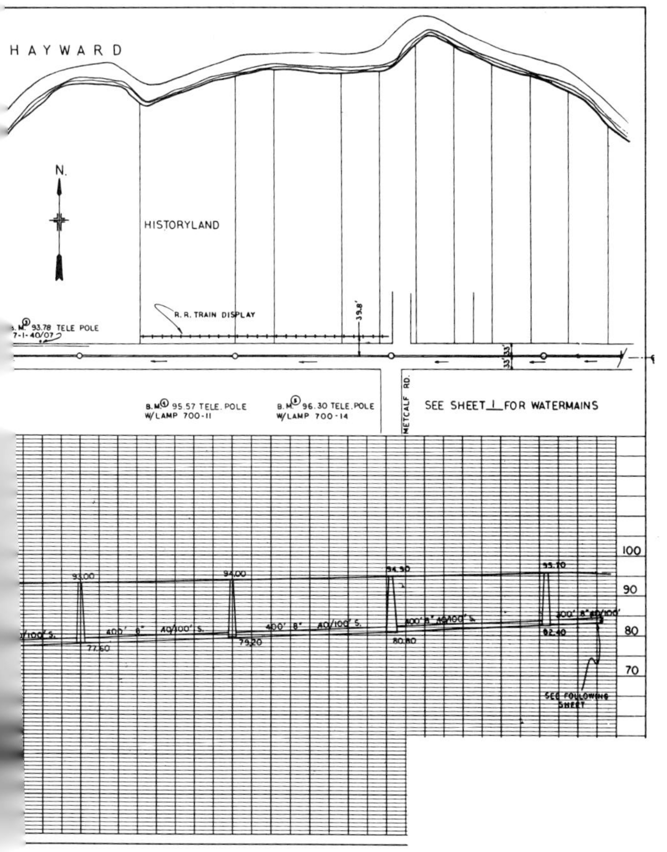

profile *(Morgan & Parmley, Ltd.).*

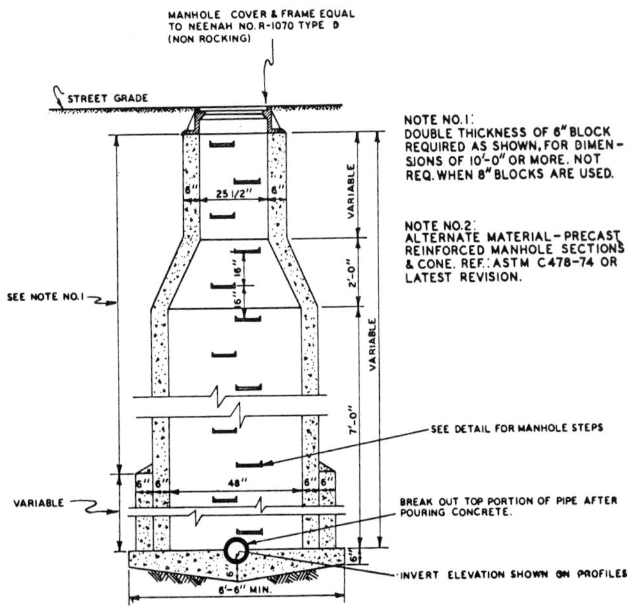

Fig. 15-4 Typical standard manhole design *(Morgan & Parmley, Ltd.)*.

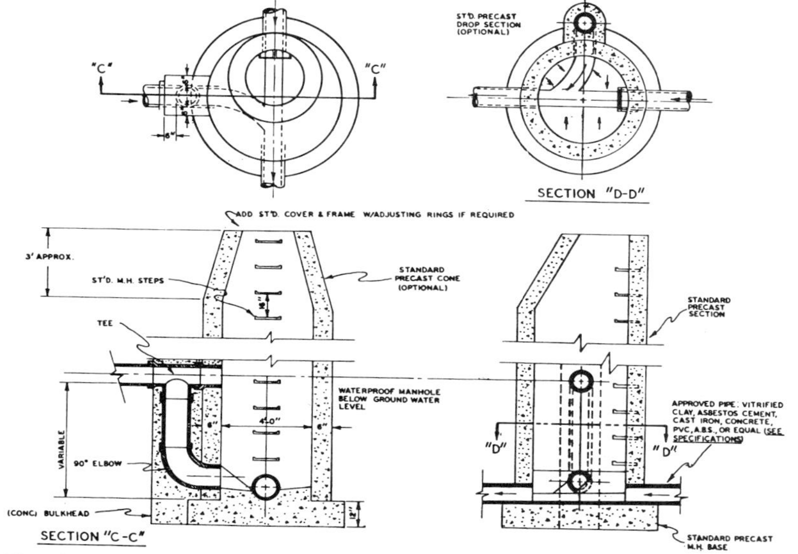

Fig. 15-5 Typical drop manhole design *(Morgan & Parmley, Ltd.).*

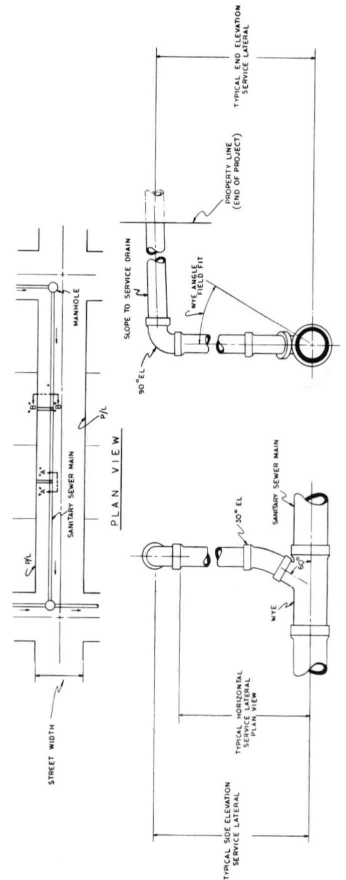

Fig. 15-6 Typical Y and service lateral details (*Morgan & Parmley, Ltd.*).

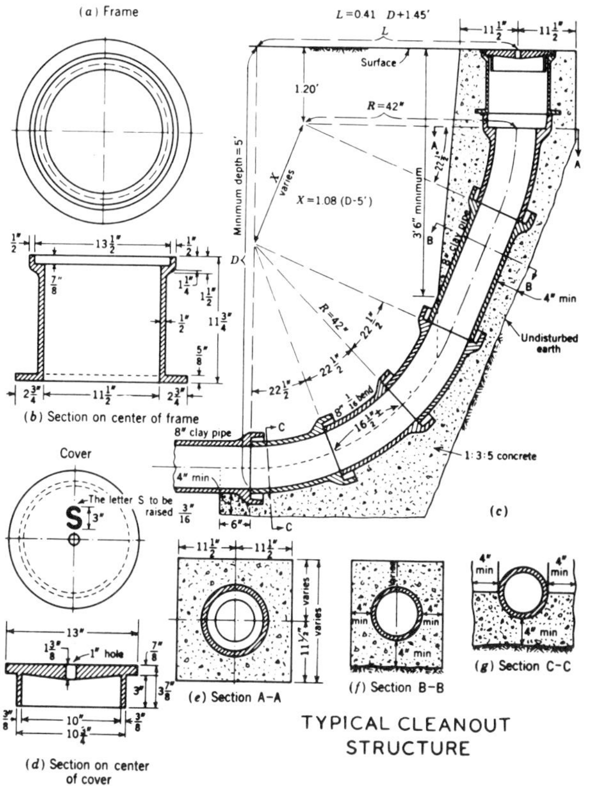

Fig. 15-7 Typical cleanout structure *(National Clay Pipe Institute)*.

DATE: AUG. 8, 1978 | GRADE SHEET | PAGE 1 OF 1

PROJECT: WITWER SANITARY SEWER EXTENSION

LOCATION: EASEMENT - BTWN. N. 2ND AVE. & N. 3RD. AVE.

FROM: IOWA STREET TO: WISCONSIN STREET

SLOPE: 0.40/100' OFFSET: 10 FT.

BENCH MARK ELEVATION: 1021.49 LOCATION: HYD. NUT @ IOWA & N. 2ND

	STATION	STAKE ELEVATION	GRADE ELEVATION	FILL	CUT	GRADE POLE@12	GRADE POLE	GRADE POLE
(EXIST'G.) M.H.	0 +00	17.59	7.83		9.76	2,24		
	+25	18.26	7.93		10.33	1,67		
	+50	18.27	8.03		10.24	1.76		
	+75	18.22	8.13		10.09	1.91		
	1 +00	18.07	8.23		9.84	2.16		
	+25	18.34	8.33		10.01	1.99		
	+50	19.04	8.43		10.61	1.39		
	+75	18.84	8.53		10.31	1.69		
	2 +00	18.77	8.63		10.14	1.86		
	+25	18.66	8.73		9.93	2.07		
	+50	18.80	8.83		9.97	2.03		
	+75	18.65	8.93		9.72	2.28		
	3 +00	19.16	9.03		10.13	1.87		
	+25	18.82	9.13		9.69	2.31		
	+50	19.01	9.23		9.78	2.22		
M.H.	+75	19.19	9.33		9.86	2.14		

NOTES:

Fig. 15-8 Sample grade sheet for sanitary sewer installation *(Morgan & Parmley, Ltd.)*.

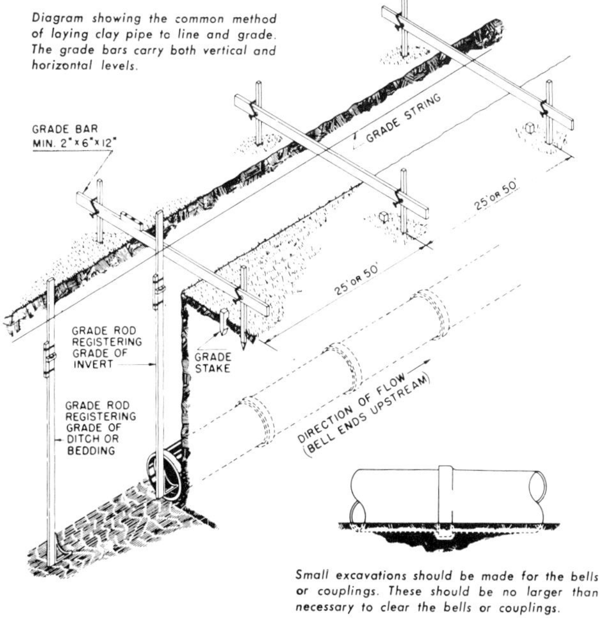

Fig. 15-9 Diagram showing common method of laying sanitary sewer pipe *(National Clay Pipe Institute).*

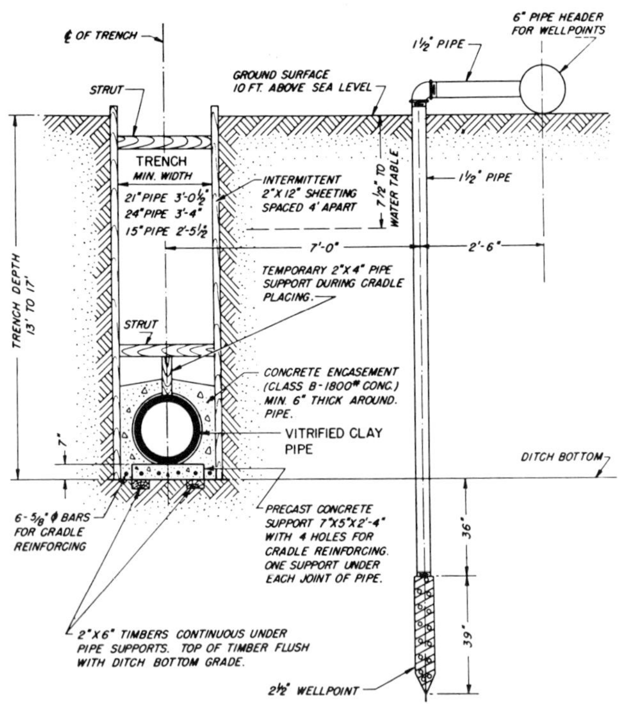

Fig. 15-10 Cross-sectional illustration of wellpoint system *(National Clay Pipe Institute).*

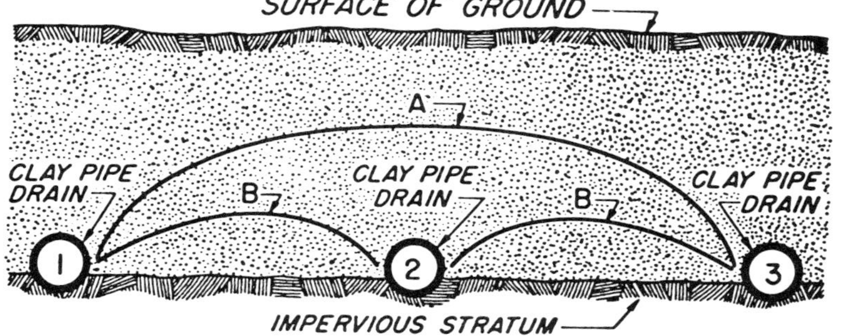

As previously shown, the amount by which ground water tables are lowered is influenced to a high degree by the spacing of the subdrains. Ground-water profiles follow the laws of hydraulic flow through porous mediums and resemble the shapes indicated in the above diagram. The ground water adjacent to an open conduit is always at the same elevation as the water in the conduit. The height to which the water rises between drains is a function of the fineness of the soil. Closer spacing between subdrains not only reduces the level of the ground water materially but also reduces, to a great degree, the amount of ground water remaining in the soil. This leaves the soil with a much greater storage capacity for subsequent percolation.

In the above diagram curve A shows the water-table for pipe lines 1 and 3 (omitting pipe 2.) Curve B shows the effect of the intermediate line. The area between curve A and B shows the increase in soil volume drained.

Fig. 15-11 Ground-water drainage profile *(National Clay Pipe Institute)*.

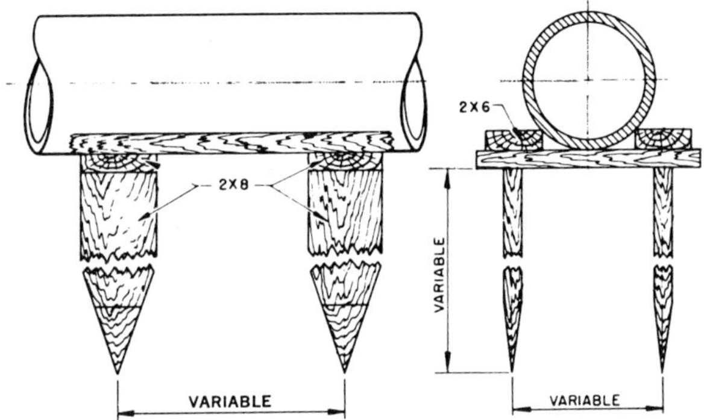

DWARF, OR SADDLE PILING FOR STABILIZING PIPE LINES

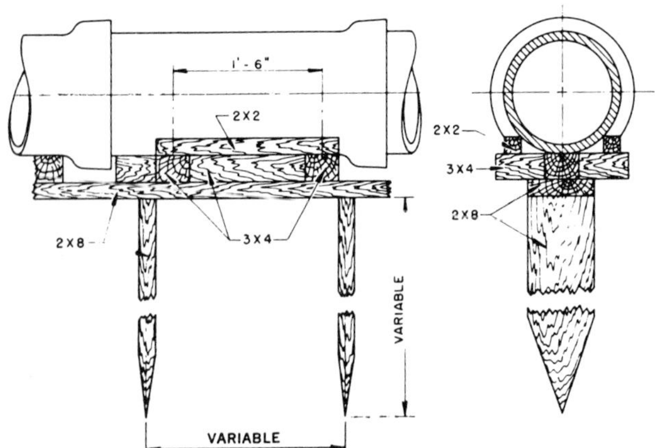

Fig. 15-12 Typical saddle piling detail *(National Clay Pipe Institute).*

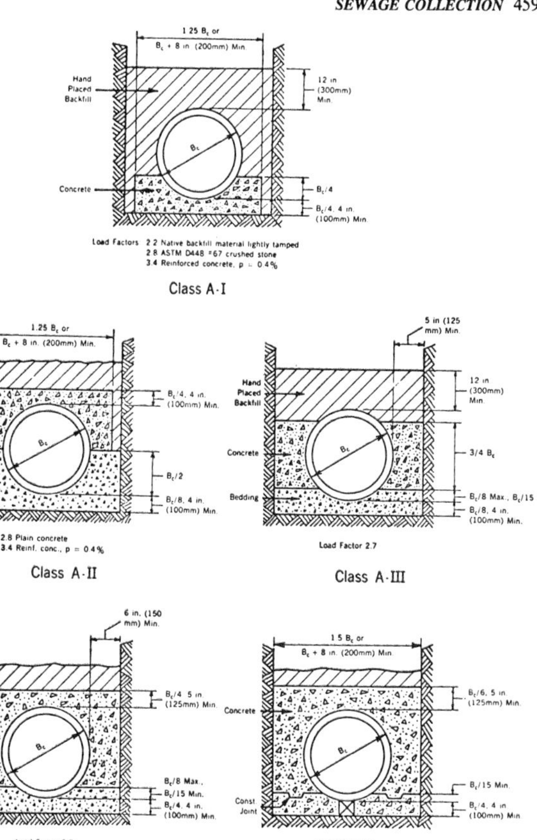

Fig. 15-13 Load factors—class A *(National Clay Pipe Institute).*

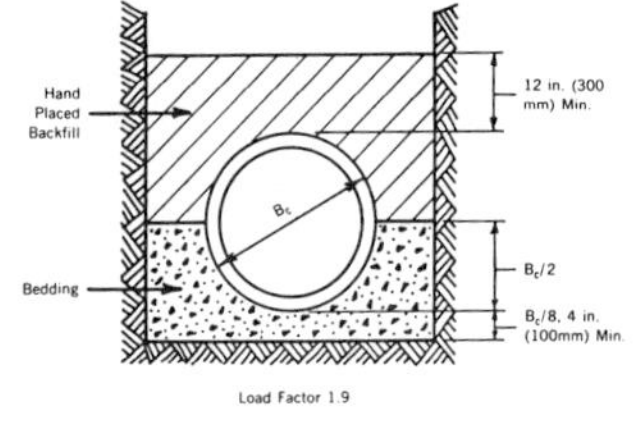

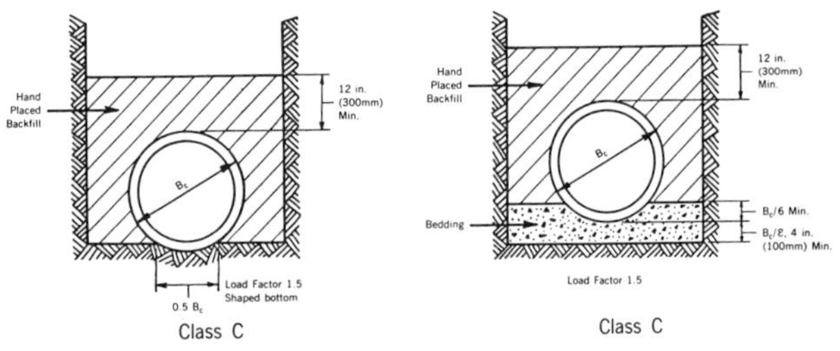

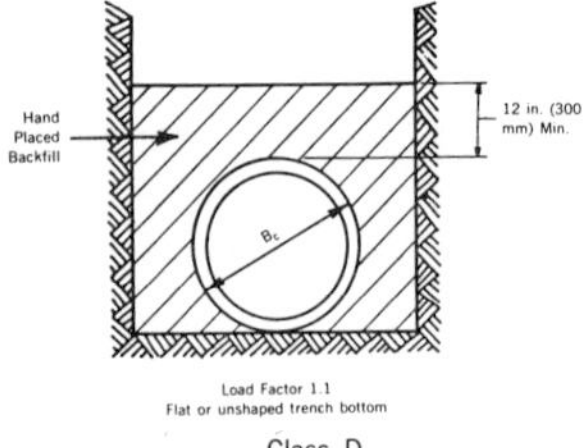

Fig. 15-14 Load factors—classes B, C, and D *(National Clay Pipe Institute).*

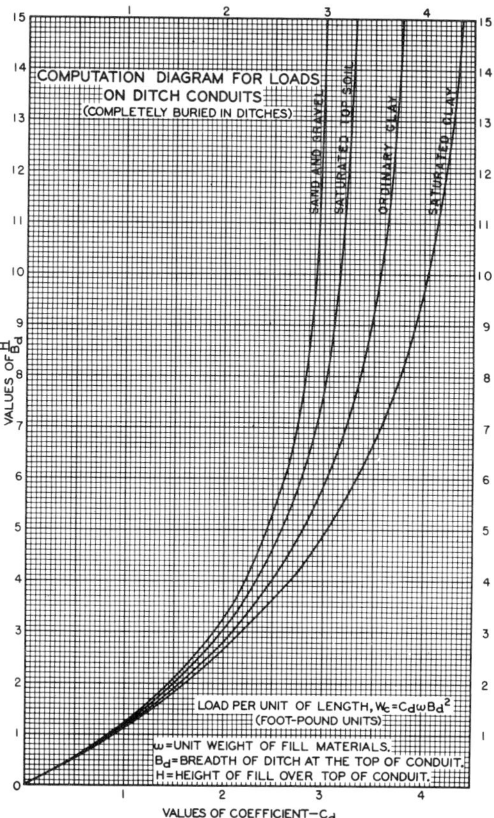

Fig. 15-15 Computation diagram for loads on ditch conduits *(National Clay Pipe Institute).*

TABLE 15-2 Percentage of Wheel Loads Transmitted to Underground Pipes*

Tabulated figures show percentage of wheel load applied to one lineal foot of pipe.

Depth of Backfill Over Top of Pipe in feet	Pipe Size Inches	6″	8″	10″	12″	15″	18″	21″	24″	27″	30″	33″	36″	39″	42″
	Outside Diam. of Pipe in Feet (Approx.)	.64	.81	1.0	1.2	1.5	1.8	2.1	2.4	2.7	3.0	3.3	3.5	3.9	4.2
1		12.8	15.0	17.3	20.0	22.6	24.8	26.4	27.2	28.0	28.6	29.0	29.4	29.8	29.9
2		5.7	7.0	8.3	9.6	11.5	13.2	15.0	15.6	16.8	17.8	18.7	19.5	20.0	20.5
3		2.9	3.6	4.3	5.2	6.4	7.5	8.6	9.3	10.2	11.1	11.8	12.5	12.9	13.5
4		1.7	2.1	2.5	3.1	3.9	4.6	5.3	5.8	6.5	7.2	7.9	8.5	8.8	9.2
5		1.2	1.4	1.7	2.1	2.6	3.1	3.6	3.9	4.4	4.9	5.3	5.8	6.1	6.4
6		0.8	1.0	1.1	1.4	1.8	2.1	2.5	2.8	3.1	3.5	3.8	4.2	4.3	4.4
7		0.5	0.7	0.8	1.0	1.3	1.6	1.9	2.1	2.3	2.6	2.9	3.2	3.3	3.5
8		0.4	0.5	0.6	0.8	1.0	1.2	1.4	1.6	1.8	2.0	2.2	2.3	2.5	2.6

*These figures make no allowance for impact.

SOURCE: *Clay Pipe Engineering Manual,* © National Clay Pipe Institute 1974, 1972, 1968, Crystal Lake, Ill.

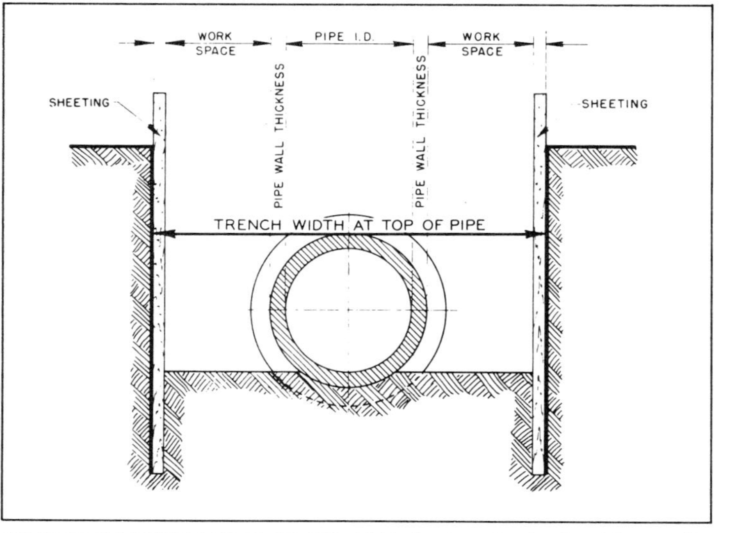

The "design trench width" equals the sum of the outside diameter of the pipe, the minimum working space on each side of the pipe, the thickness of sheeting (if used) on each side of the trench, and the allowable trench tolerance.

Fig. 15-16 Design trench width *(National Clay Pipe Institute)*.

TABLE 15-3 Loads on 8-in Vitrified Clay Pipe (pounds per linear foot) Caused by Backfilling

Bold print figures represent trench loads on pipe for each width of trench. Italicized figures represent maximum loads on pipe at and beyond transition width.

SAND AND GRAVEL

AVERAGE WT. = 110 LBS./CU. FT.

DEPTH OF BACKFILL OVER TOP OF PIPE (feet)	TRENCH WIDTH AT TOP OF PIPE (feet)									Transition Width
	1′6″	1′9″	2′0″	2′3″	2′6″	2′9″	3′0″	3′6″	4′0″	
5	**505**	**605**	*765*							*2′ 0″*
6	**540**	**680**	**840**	*930*						*2′ 2″*
7	**580**	**735**	**920**	*1085*						*2′ 3″*
8	**605**	**790**	**975**	**1170**	*1260*					*2′ 4″*
9	**630**	**825**	**1030**	**1245**	*1415*					*2′ 5″*
10	**655**	**860**	**1080**	**1300**	**1535**	*1580*				*2′ 7″*
11	**680**	**890**	**1115**	**1350**	**1595**	*1745*				*2′ 8″*
12	**695**	**920**	**1155**	**1395**	**1655**	*1910*				*2′ 9″*
13	**705**	**935**	**1175**	**1435**	**1705**	**2020**	*2075*			*2′10″*
14	**715**	**950**	**1195**	**1465**	**1750**	**2080**	*2235*			*2′11″*
15	**720**	**965**	**1215**	**1490**	**1790**	**2130**	*2400*			*3′ 0″*
16	**725**	**975**	**1240**	**1515**	**1825**	**2175**	**2475**	*2565*		*3′ 1″*
17	**725**	**980**	**1250**	**1530**	**1865**	**2215**	**2530**	*2730*		*3′ 2″*
18	**730**	**990**	**1260**	**1550**	**1890**	**2250**	**2580**	*2895*		*3′ 3″*
19	**730**	**995**	**1270**	**1570**	**1915**	**2275**	**2625**	*3060*		*3′ 4″*
20	**735**	**1000**	**1280**	**1585**	**1940**	**2305**	**2660**	*3225*		*3′ 4″*
21	**735**	**1000**	**1290**	**1595**	**1960**	**2330**	**2695**	*3385*		*3′ 5″*
22	**745**	**1005**	**1295**	**1610**	**1975**	**2350**	**2725**	*3550*		*3′ 6″*
23	**745**	**1010**	**1300**	**1625**	**1990**	**2365**	**2750**	**3620**	*3720*	*3′ 7″*
24	**750**	**1010**	**1305**	**1635**	**2010**	**2380**	**2780**	**3660**	*3885*	*3′ 8″*
25	**750**	**1020**	**1310**	**1645**	**2015**	**2395**	**2805**	**3695**	*4050*	*3′ 9″*
26	**760**	**1020**	**1320**	**1650**	**2025**	**2410**	**2825**	**3735**	*4210*	*3′10″*
27	**760**	**1020**	**1325**	**1650**	**2030**	**2420**	**2845**	**3760**	*4375*	*3′10″*
28	**770**	**1020**	**1325**	**1655**	**2035**	**2430**	**2860**	**3785**	*4540*	*3′11″*
29	**775**	**1025**	**1330**	**1655**	**2035**	**2435**	**2875**	**3805**	*4705*	*3′11″*
30	**780**	**1025**	**1330**	**1660**	**2040**	**2440**	**2895**	**3835**	*4870*	*4′ 0″*

Transition Width Column represents the critical width where trench loads reach their maximum and are equal to the embankment load.

SOURCE: *Clay Pipe Engineering Manual,* © National Clay Pipe Institute 1974, 1972, 1968, Crystal Lake, Ill.

TABLE 15-3 Loads on 8-in Vitrified Clay Pipe (pounds per linear foot) Caused by Backfilling *(cont.)*

SATURATED TOPSOIL

AVERAGE WT. = 115 LBS./CU. FT.

TRENCH WIDTH AT TOP OF PIPE (feet)									Transition Width	DEPTH OF BACKFILL OVER TOP OF PIPE (feet)
1′6″	1′9″	2′0″	2′3″	2′6″	2′9″	3′0″	3′6″	4′0″		
540	680	*780*							*2′ 0″*	5
605	740	920	*950*						*2′ 1″*	6
655	805	1000	*1125*						*2′ 2″*	7
690	855	1080	*1275*						*2′ 3″*	8
725	910	1145	1355	*1440*					*2′ 4″*	9
755	955	1190	1425	*1620*					*2′ 5″*	10
770	995	1235	1490	*1795*					*2′ 6″*	11
790	1025	1275	1540	1840	*1955*				*2′ 7″*	12
805	1045	1310	1585	1910	*2130*				*2′ 8″*	13
810	1070	1345	1625	1965	*2300*				*2′ 9″*	14
820	1080	1370	1660	2015	2345	*2460*			*2′10″*	15
830	1100	1390	1695	2060	2405	*2635*			*2′11″*	16
835	1110	1415	1725	2095	2445	*2805*			*3′ 0″*	17
840	1120	1430	1750	2130	2490	2885	*2980*		*3′ 1″*	18
845	1125	1450	1770	2155	2530	2945	*3140*		*3′ 2″*	19
850	1140	1460	1795	2185	2565	2990	*3300*		*3′ 2″*	20
850	1150	1470	1815	2210	2600	3035	*3475*		*3′ 3″*	21
850	1150	1485	1835	2230	2620	3070	*3645*		*3′ 4″*	22
855	1155	1495	1845	2245	2655	3105	*3805*		*3′ 5″*	23
865	1155	1500	1855	2255	2680	3140	*3980*		*3′ 5″*	24
865	1160	1505	1870	2265	2705	3170	*4140*		*3′ 6″*	25
865	1160	1520	1880	2275	2720	3200	4185	*4300*	*3′ 7″*	26
870	1165	1520	1885	2290	2735	3220	4225	*4475*	*3′ 8″*	27
875	1165	1520	1890	2300	2765	3235	4255	*4635*	*3′ 8″*	28
875	1165	1520	1900	2305	2775	3255	4280	*4805*	*3′ 9″*	29
875	1165	1520	1905	2310	2785	3265	4300	*4970*	*3′10″*	30

TABLE 15-3 Loads on 8-in Vitrified Clay Pipe (pounds per linear foot) Caused by Backfilling *(cont.)*

DRY CLAY

AVERAGE WT. = 120 LBS./CU. FT.

DEPTH OF BACKFILL OVER TOP OF PIPE (feet)	TRENCH WIDTH AT TOP OF PIPE (feet)								Transition Width
	1′6″	1′9″	2′0″	2′3″	2′6″	2′9″	3′0″	3′6″	
5	600	730	*770*						*1′10″*
6	670	835	*935*						*1′11″*
7	730	920	*1105*						*2′ 0″*
8	780	985	1190	*1270*					*2′ 1″*
9	820	1045	1270	*1440*					*2′ 2″*
10	860	1100	1350	*1610*					*2′ 3″*
11	890	1140	1415	1675	*1775*				*2′ 4″*
12	910	1175	1470	1745	*1945*				*2′ 5″*
13	935	1205	1520	1810	*2110*				*2′ 6″*
14	950	1230	1555	1870	2210	*2280*			*2′ 7″*
15	960	1255	1590	1920	2280	*2450*			*2′ 8″*
16	970	1270	1620	1970	2330	*2615*			*2′ 8″*
17	985	1290	1650	2010	2375	*2665*			*2′ 9″*
18	995	1310	1675	2040	2430	2850	*2950*		*2′10″*
19	1000	1320	1690	2075	2470	2905	*3120*		*2′10″*
20	1010	1330	1710	2105	2510	2950	*3290*		*2′11″*
21	1015	1345	1720	2135	2550	3000	*3455*		*3′ 0″*
22	1020	1355	1740	2155	2585	3040	3540	*3625*	*3′ 1″*
23	1025	1360	1750	2170	2615	3080	3590	*3790*	*3′ 1″*
24	1030	1370	1765	2190	2640	3115	3625	*3960*	*3′ 2″*
25	1030	1375	1770	2210	2665	3150	3660	*4130*	*3′ 3″*
26	1030	1380	1775	2220	2690	3180	3685	*4295*	*3′ 3″*
27	1040	1385	1780	2230	2705	3210	3710	*4465*	*3′ 4″*
28	1040	1390	1790	2245	2725	3235	3730	*4630*	*3′ 5″*
29	1040	1390	1795	2250	2735	3260	3755	*4800*	*3′ 5″*
30	1040	1400	1800	2255	2750	3275	3780	*4970*	*3′ 6″*

TABLE 15-3 Loads on 8-in Vitrified Clay Pipe (pounds per linear foot) Caused by Backfilling *(cont.)*

WET CLAY

AVERAGE WT. = 130 LBS./CU. FT.

TRENCH WIDTH AT TOP OF PIPE (feet)								Transition Width	DEPTH OF BACKFILL OVER TOP OF PIPE (feet)
1'6"	1'9"	2'0"	2'3"	2'6"	2'9"	3'0"	3'6"		
700	*805*							*1' 8"*	5
780	*980*							*1' 9"*	6
860	1065	*1155*						*1'10"*	7
925	1155	*1335*						*1'11"*	8
990	1235	1455	*1510*					*2' 1"*	9
1025	1300	1560	*1685*					*2' 2"*	10
1065	1350	1650	*1860*					*2' 2"*	11
1105	1405	1730	*2035*					*2' 3"*	12
1130	1455	1805	2160	*2210*				*2' 4"*	13
1155	1495	1870	2250	*2385*				*2' 4"*	14
1185	1535	1925	2315	*2560*				*2' 5"*	15
1195	1560	1975	2365	*2735*				*2' 6"*	16
1220	1585	2015	2430	2860	*2910*			*2' 7"*	17
1235	1610	2055	2485	2940	*3090*			*2' 7"*	18
1250	1640	2080	2520	3015	*3265*			*2' 8"*	19
1260	1665	2120	2560	3070	*3440*			*2' 8"*	20
1275	1675	2140	2600	3120	*3615*			*2' 9"*	21
1280	1690	2160	2625	3170	3690	*3790*		*2'10"*	22
1285	1710	2185	2650	3210	3755	*3965*		*2'10"*	23
1300	1725	2210	2680	3250	3810	*4140*		*2'11"*	24
1315	1735	2225	2705	3290	3850	*4315*		*2'11"*	25
1320	1740	2235	2730	3330	3895	*4490*		*3' 0"*	26
1325	1755	2250	2755	3355	3925	4570	*4665*	*3' 1"*	27
1335	1770	2260	2770	3380	3965	4615	*4835*	*3' 1"*	28
1335	1770	2270	2780	3405	4000	4665	*5020*	*3' 2"*	29
1340	1775	2275	2790	3430	4030	4705	*5185*	*3' 2"*	30

TABLE 15-4 Loads on 12-in Vitrified Clay Pipe (pounds per linear foot) Caused by Backfilling

Bold print figures represent trench loads on pipe for each width of trench. Italicized figures represent maximum loads on pipe at and beyond transition width.

SAND AND GRAVEL

AVERAGE WT. = 110 LBS./CU. FT.

DEPTH OF BACKFILL OVER TOP OF PIPE (feet)	TRENCH WIDTH AT TOP OF PIPE (feet)										Transition Width
	1'9"	2'0"	2'3"	2'6"	2'9"	3'0"	3'6"	4'0"	4'6"	5'0"	
5	605	745	865	1010	*1065*						2' 7"
6	680	840	995	1160	1280	*1320*					2'10"
7	735	920	1090	1280	1415	*1560*					2'10"
8	790	975	1170	1380	1545	1735	*1800*				3' 1"
9	825	1030	1245	1465	1655	1870	*2045*				3' 2"
10	860	1080	1300	1535	1760	1990	*2265*				3' 4"
11	890	1115	1350	1595	1850	2100	*2510*				3' 5"
12	920	1155	1395	1655	1940	2200	*2750*				3' 6"
13	935	1175	1435	1705	2020	2275	2890	*2980*			3' 7"
14	950	1195	1465	1750	2080	2350	3000	*3220*			3' 8"
15	965	1215	1490	1790	2130	2415	3115	*3445*			3' 9"
16	975	1240	1515	1825	2175	2475	3205	*3695*			3'10"
17	980	1250	1530	1865	2215	2530	3285	*3925*			3'11"
18	990	1260	1550	1890	2250	2580	3355	4120	*4180*		4' 1"
19	995	1270	1570	1915	2275	2625	3420	4210	*4415*		4' 3"
20	1000	1280	1585	1940	2305	2660	3475	4280	*4660*		4' 3"
21	1000	1290	1595	1960	2330	2695	3525	4365	*4895*		4' 4"
22	1005	1295	1610	1975	2350	2725	3575	4440	*5145*		4' 5"
23	1010	1300	1625	1990	2365	2750	3620	4510	*5370*		4' 5"
24	1010	1305	1635	2010	2380	2780	3660	4570	*5610*		4' 6"
25	1020	1310	1645	2015	2395	2805	3695	4630	5665	*5845*	4' 7"
26	1020	1320	1650	2025	2410	2825	3735	4685	5740	*6085*	4' 8"
27	1020	1325	1650	2030	2420	2845	3760	4740	5810	*6325*	4' 9"
28	1020	1325	1655	2035	2430	2860	3785	4795	5880	*6575*	4'10"
29	1025	1330	1655	2035	2435	2875	3805	4850	5940	*6805*	4'11"
30	1025	1330	1660	2040	2440	2895	3835	4905	5995	*7040*	4'11"

Transition Width Column represents the critical width where trench loads reach their maximum and are equal to the embankment load.

SOURCE: *Clay Pipe Engineering Manual,* © National Clay Pipe Institute 1974, 1972, 1968, Crystal Lake, Ill.

TABLE 15-4 Loads on 12-in Vitrified Clay Pipe (pounds per linear foot) Caused by Backfilling ***(cont.)***

SATURATED TOPSOIL

AVERAGE WT. = 115 LBS./CU. FT.

TRENCH WIDTH AT TOP OF PIPE (feet)											
1'9"	2'0"	2'3"	2'6"	2'9"	3'0"	3'6"	4'0"	4'6"	5'0"	Transition Width	DEPTH OF BACKFILL OVER TOP OF PIPE (feet)
680	805	945	*1060*							*2' 6"*	5
740	920	1070	1220	*1300*						*2' 8"*	6
805	1000	1175	1355	*1540*						*2' 9"*	7
855	1080	1265	1470	1690	*1795*					*2'11"*	8
910	1145	1355	1575	1830	*2035*					*3' 0"*	9
955	1190	1425	1680	1930	2180	*2295*				*3' 2"*	10
995	1235	1490	1760	2025	2300	*2530*				*3' 3"*	11
1025	1275	1540	1840	2130	2410	*2770*				*3' 4"*	12
1045	1310	1585	1910	2215	2490	*3015*				*3' 5"*	13
1070	1345	1625	1965	2275	2590	*3265*				*3' 6"*	14
1080	1370	1660	2015	2345	2670	3405	*3520*			*3' 7"*	15
1100	1390	1695	2060	2405	2755	3495	*3770*			*3' 8"*	16
1110	1415	1725	2095	2445	2820	3590	*4015*			*3' 9"*	17
1120	1430	1750	2130	2490	2885	3680	*4255*			*3'10"*	18
1125	1450	1770	2155	2530	2945	3760	*4495*			*3'11"*	19
1140	1460	1795	2185	2565	2990	3840	*4750*			*4' 0"*	20
1150	1470	1815	2210	2600	3035	3935	4840	*5005*		*4' 1"*	21
1150	1485	1835	2230	2620	3070	3990	4945	*5245*		*4' 2"*	22
1155	1495	1845	2245	2655	3105	4050	5025	*5495*		*4' 3"*	23
1155	1500	1855	2255	2680	3140	4100	5105	*5750*		*4' 4"*	24
1160	1505	1870	2265	2705	3170	4140	5175	*5990*		*4' 5"*	25
1160	1520	1880	2275	2720	3200	4185	5245	*6235*		*4' 5"*	26
1165	1520	1885	2290	2735	3220	4225	5315	*6475*		*4' 6"*	27
1165	1520	1890	2300	2765	3235	4255	5380	6545	*6720*	*4' 7"*	28
1165	1520	1900	2305	2775	3255	4280	5435	6625	*6960*	*4' 8"*	29
1165	1520	1905	2310	2785	3265	4300	5475	6695	*7195*	*4' 8"*	30

TABLE 15-4 Loads on 12-in Vitrified Clay Pipe (pounds per linear foot) Caused by Backfilling *(cont.)*

DRY CLAY

AVERAGE WT. = 120 LBS./CU. FT.

DEPTH OF BACKFILL OVER TOP OF PIPE (feet)	TRENCH WIDTH AT TOP OF PIPE (feet)									Transition Width
	1'9"	2'0"	2'3"	2'6"	2'9"	3'0"	3'6"	4'0"	4'6"	
5	730	875	1025	*1080*						2' 5"
6	835	985	1150	1295	*1320*					2' 7"
7	920	1085	1270	1450	*1560*					2' 8"
8	985	1190	1390	1595	*1800*					2' 9"
9	1045	1270	1490	1730	1970	*2040*				2'10"
10	1100	1350	1585	1850	2110	*2280*				2'11"
11	1140	1415	1675	1955	2240	*2520*				3' 0"
12	1175	1470	1745	2050	2350	2680	*2765*			3' 1"
13	1205	1520	1810	2135	2460	2810	*3010*			3' 2"
14	1230	1555	1870	2210	2555	2915	*3260*			3' 3"
15	1255	1590	1920	2280	2650	3025	*3505*			3' 4"
16	1270	1620	1970	2330	2725	3110	*3750*			3' 5"
17	1290	1650	2010	2375	2785	3190	*3990*			3' 6"
18	1310	1675	2040	2430	2850	3275	4170	*4235*		3' 7"
19	1320	1690	2075	2470	2905	3355	4285	*4480*		3' 8"
20	1330	1710	2105	2510	2950	3420	4390	*4730*		3' 8"
21	1345	1720	2135	2550	3000	3480	4490	*4970*		3' 9"
22	1355	1740	2155	2585	3040	3540	4570	*5215*		3'10"
23	1360	1750	2170	2615	3080	3590	4650	*5455*		3'10"
24	1370	1765	2190	2640	3115	3625	4730	*5695*		3'11"
25	1375	1770	2210	2665	3150	3660	4800	*5935*		4' 0"
26	1380	1775	2220	2690	3180	3685	4850	6020	*6175*	4' 2"
27	1385	1780	2230	2705	3210	3710	4895	6110	*6420*	4' 4"
28	1390	1790	2245	2725	3235	3730	4945	6185	*6665*	4' 5"
29	1390	1795	2250	2735	3260	3755	4980	6265	*6910*	4' 6"
30	1400	1800	2255	2750	3275	3780	5005	6330	*7150*	4' 6"

TABLE 15-4 Loads on 12-in Vitrified Clay Pipe (pounds per linear foot) Caused by Backfilling *(cont.)*

WET CLAY

AVERAGE WT. = 130 LBS./CU. FT.

TRENCH WIDTH AT TOP OF PIPE (feet)								Transition Width	DEPTH OF BACKFILL OVER TOP OF PIPE (feet)
1′9″	2′0″	2′3″	2′6″	2′9″	3′0″	3′6″	4′0″		
830	1010	*1125*						2′ 2″	5
960	1130	1320	*1390*					2′ 4″	6
1065	1250	1470	*1650*					2′ 5″	7
1155	1360	1600	*1885*					2′ 6″	8
1235	1455	1730	2030	*2145*				2′ 8″	9
1300	1560	1845	2165	*2390*				2′ 9″	10
1350	1650	1965	2290	2580	*2650*			2′10″	11
1405	1730	2080	2390	2705	*2910*			2′11″	12
1455	1805	2160	2495	2845	*3170*			2′11″	13
1495	1870	2250	2600	2975	*3420*			3′ 0″	14
1535	1925	2315	2705	3110	3550	*3665*		3′ 1″	15
1560	1975	2365	2780	3210	3675	*3925*		3′ 1″	16
1585	2015	2430	2860	3315	3785	*4185*		3′ 2″	17
1610	2055	2485	2940	3405	3900	*4445*		3′ 3″	18
1640	2080	2520	3015	3495	4005	*4705*		3′ 4″	19
1665	2120	2560	3070	3575	4095	*4965*		3′ 5″	20
1675	2140	2600	3120	3640	4180	*5225*		3′ 6″	21
1690	2160	2625	3170	3690	4250	5410	*5475*	3′ 7″	22
1710	2185	2650	3210	3755	4330	5525	*5735*	3′ 7″	23
1725	2210	2680	3250	3810	4395	5605	*5980*	3′ 8″	24
1735	2225	2705	3290	3850	4445	5705	*6240*	3′ 8″	25
1740	2235	2730	3330	3895	4510	5800	*6500*	3′ 9″	26
1755	2250	2755	3355	3925	4570	5875	*6760*	3′10″	27
1770	2260	2770	3380	3965	4615	5955	*7020*	3′11″	28
1770	2270	2780	3405	4000	4665	6030	*7280*	3′11″	29
1775	2275	2790	3430	4030	4705	6085	*7540*	4′ 0″	30

TABLE 15-5 Loads on 18-in Vitrified Clay Pipe (pounds per linear foot) Caused by Backfilling

Bold print figures represent trench loads on pipe for each width of trench. Italicized figures represent maximum loads on pipe at and beyond transition width.

SAND AND GRAVEL

AVERAGE WT. = 110 LBS./CU. FT.

DEPTH OF BACKFILL OVER TOP OF PIPE (feet)	TRENCH WIDTH AT TOP OF PIPE (feet)										Transition Width
	2'3"	2'6"	2'9"	3'0"	3'6"	4'0"	4'6"	5'0"	6'0"	7'0"	
5	865	1010	1140	1250	*1530*						*3' 5"*
6	995	1160	1280	1430	1760	*1870*					*3' 8"*
7	1090	1280	1415	1595	1960	*2220*					*3'10"*
8	1170	1380	1545	1735	2145	2530	*2585*				*4' 1"*
9	1245	1465	1655	1870	2340	2760	*2925*				*4' 2"*
10	1300	1535	1760	1990	2505	2970	*3280*				*4' 4"*
11	1350	1595	1850	2100	2650	3170	*3630*				*4' 5"*
12	1395	1655	1940	2200	2780	3345	*3970*				*4' 6"*
13	1435	1705	2020	2275	2890	3525	4180	*4325*			*4' 7"*
14	1465	1750	2080	2350	3000	3660	4355	*4675*			*4' 8"*
15	1490	1790	2130	2415	3115	3805	4520	*5015*			*4'10"*
16	1515	1825	2175	2475	3205	3915	4675	*5390*			*4'11"*
17	1530	1865	2215	2530	3285	4050	4815	5645	*5730*		*5' 1"*
18	1550	1890	2250	2580	3355	4120	4940	5810	*6070*		*5' 3"*
19	1570	1915	2275	2625	3420	4210	5070	5950	*6435*		*5' 4"*
20	1585	1940	2305	2660	3475	4280	5180	6085	*6785*		*5' 5"*
21	1595	1960	2330	2695	3525	4365	5290	6215	*7150*		*5' 6"*
22	1610	1975	2350	2725	3575	4440	5395	6370	*7490*		*5' 7"*
23	1625	1990	2365	2750	3620	4510	5500	6490	*7855*		*5' 8"*
24	1635	2010	2380	2780	3660	4570	5585	6590	*8195*		*5' 9"*
25	1645	2015	2395	2805	3695	4630	5665	6690	*8545*		*5'10"*
26	1650	2025	2410	2825	3735	4685	5740	6795	*8910*		*5'11"*
27	1650	2030	2420	2845	3760	4740	5810	6905	*9270*		*6' 0"*
28	1655	2035	2430	2860	3785	4795	5880	6995	9440	*9625*	*6' 1"*
29	1655	2035	2435	2875	3805	4850	5940	7095	8570	*9965*	*6' 2"*
30	1660	2040	2440	2895	3835	4905	5995	7170	9690	*10340*	*6' 3"*

Transition Width Column represents the critical width where trench loads reach their maximum and are equal to the embankment load.

SOURCE: *Clay Pipe Engineering Manual,* © National Clay Pipe Institute 1974, 1972, 1968, Crystal Lake, Ill.

TABLE 15-5 Loads on 18-in Vitrified Clay Pipe (pounds per linear foot) Caused by Backfilling *(cont.)*

SATURATED TOPSOIL

AVERAGE WT. = 115 LBS./CU. FT.

TRENCH WIDTH AT TOP OF PIPE (feet)									Transition Width	DEPTH OF BACKFILL OVER TOP OF PIPE (feet)
2′3″	2′6″	2′9″	3′0″	3′6″	4′0″	4′6″	5′0″	6′0″		
945	1080	1240	1380	*1530*					*3′ 3″*	5
1070	1220	1405	1560	*1875*					*3′ 6″*	6
1175	1355	1540	1735	2130	*2255*				*3′ 8″*	7
1265	1470	1690	1900	2325	*2610*				*3′10″*	8
1355	1575	1830	2035	2530	*2990*				*4′ 0″*	9
1425	1680	1930	2180	2705	3220	*3325*			*4′ 2″*	10
1490	1760	2025	2300	2875	3440	*3680*			*4′ 3″*	11
1540	1840	2130	2410	3020	3635	*4050*			*4′ 4″*	12
1585	1910	2215	2490	3165	3820	*4405*			*4′ 5″*	13
1625	1965	2275	2590	3280	3970	*4775*			*4′ 6″*	14
1660	2015	2345	2670	3405	4130	4945	*5140*		*4′ 8″*	15
1695	2060	2405	2755	3495	4270	5105	*5495*		*4′ 9″*	16
1725	2095	2445	2820	3590	4405	5280	*5855*		*4′10″*	17
1750	2130	2490	2885	3680	4530	5430	*6235*		*4′11″*	18
1770	2155	2530	2945	3760	4645	5570	*6600*		*5′ 0″*	19
1795	2185	2565	2990	3840	4750	5705	6695	*6960*	*5′ 1″*	20
1815	2210	2600	3035	3935	4840	5835	6865	*7325*	*5′ 2″*	21
1835	2230	2620	3070	3990	4945	5955	7015	*7705*	*5′ 3″*	22
1845	2245	2655	3105	4050	5025	6070	7190	*8060*	*5′ 4″*	23
1855	2255	2680	3140	4100	5105	6175	7335	*8420*	*5′ 5″*	24
1870	2265	2705	3170	4140	5175	6280	7465	*8780*	*5′ 6″*	25
1880	2275	2720	3200	4185	5245	6370	7590	*9145*	*5′ 7″*	26
1885	2290	2735	3220	4225	5315	6465	7705	*9510*	*5′ 8″*	27
1890	2300	2765	3235	4255	5380	6545	7820	*9865*	*5′ 9″*	28
1900	2305	2775	3255	4280	5435	6625	7935	*10235*	*5′10″*	29
1905	2310	2785	3265	4300	5475	6695	8050	*10610*	*5′11″*	30

TABLE 15-5 Loads on 18-in Vitrified Clay Pipe (pounds per linear foot) Caused by Backfilling *(cont.)*

DRY CLAY

AVERAGE WT. = 120 LBS./CU. FT.

DEPTH OF BACKFILL OVER TOP OF PIPE (feet)	TRENCH WIDTH AT TOP OF PIPE (feet)									Transition Width
	2'3"	2'6"	2'9"	3'0"	3'6"	4'0"	4'6"	5'0"	6'0"	
5	1025	1130	1320	1475	*1500*					*3' 1"*
6	1150	1295	1495	1680	*1870*					*3' 4"*
7	1270	1450	1670	1885	*2245*					*3' 6"*
8	1390	1595	1825	2065	2515	*2605*				*3' 7"*
9	1490	1730	1970	2245	2735	*2965*				*3' 9"*
10	1585	1850	2110	2400	2940	*3335*				*3'10"*
11	1675	1955	2240	2550	3130	*3695*				*3'11"*
12	1745	2050	2350	2680	3320	*4055*				*4' 0"*
13	1810	2135	2460	2810	3485	4225	*4415*			*4' 2"*
14	1870	2210	2555	2915	3650	4415	*4775*			*4' 3"*
15	1920	2280	2650	3025	3785	4595	*5135*			*4' 4"*
16	1970	2330	2725	3110	3925	4765	*5510*			*4' 5"*
17	2010	2375	2785	3190	4055	4930	*5870*			*4' 6"*
18	2040	2430	2850	3275	4170	5090	6010	*6230*		*4' 7"*
19	2075	2470	2905	3355	4285	5230	6205	*6600*		*4' 8"*
20	2105	2510	2950	3420	4390	5375	6385	*6960*		*4' 9"*
21	2135	2550	3000	3480	4490	5495	6550	*7320*		*4'10"*
22	2155	2585	3040	3540	4570	5610	6710	*7680*		*4'11"*
23	2170	2615	3080	3590	4650	5720	6850	*8040*		*5' 0"*
24	2190	2640	3115	3625	4730	5825	6995	8220	*8400*	*5' 1"*
25	2210	2665	3150	3660	4800	5930	7140	8390	*8760*	*5' 2"*
26	2220	2690	3180	3685	4850	6020	7250	8555	*9120*	*5' 3"*
27	2230	2705	3210	3710	4895	6110	7355	8710	*9480*	*5' 3"*
28	2245	2725	3235	3730	4945	6185	7465	8855	*9840*	*5' 4"*
29	2250	2735	3260	3755	4980	6265	7570	9010	*10200*	*5' 5"*
30	2255	2750	3275	3780	5005	6330	7680	9155	*10560*	*5' 6"*

TABLE 15-5 Loads on 18-in Vitrified Clay Pipe (pounds per linear foot) Caused by Backfilling ***(cont.)***

WET CLAY

AVERAGE WT. = 130 LBS./CU. FT.

TRENCH WIDTH AT TOP OF PIPE (feet)									Transition Width	DEPTH OF BACKFILL OVER TOP OF PIPE (feet)
2′3″	2′6″	2′9″	3′0″	3′6″	4′0″	4′6″	5′0″	6′0″		
1155	1315	1495	*1560*						*2′11″*	5
1320	1545	1690	1885	*1935*					*3′ 1″*	6
1470	1730	1870	2120	*2315*					*3′ 3″*	7
1600	1885	2055	2340	*2690*					*3′ 5″*	8
1730	2030	2230	2555	*3070*					*3′ 6″*	9
1845	2165	2390	2750	3380	*3445*				*3′ 7″*	10
1965	2290	2580	2940	3600	*3820*				*3′ 9″*	11
2080	2390	2705	3120	3820	*4200*				*3′10″*	12
2160	2495	2845	3275	4030	*4575*				*3′11″*	13
2250	2600	2975	3420	4225	*4955*				*3′11″*	14
2315	2705	3110	3550	4420	*5330*				*4′ 0″*	15
2365	2780	3210	3675	4590	5565	*5705*			*4′ 1″*	16
2430	2860	3315	3785	4745	5760	*6085*			*4′ 2″*	17
2485	2940	3405	3900	4890	5980	*6460*			*4′ 3″*	18
2520	3015	3495	4005	5030	6150	*6840*			*4′ 4″*	19
2560	3070	3575	4095	5175	6320	*7215*			*4′ 5″*	20
2600	3120	3640	4180	5290	6485	*7590*			*4′ 6″*	21
2625	3170	3690	4250	5410	6630	7850	*7970*		*4′ 7″*	22
2650	3210	3755	4330	5525	6760	8045	*8345*		*4′ 7″*	23
2680	3250	3810	4395	5605	6890	8230	*8725*		*4′ 8″*	24
2705	3290	3850	4445	5705	7020	8400	*9100*		*4′ 9″*	25
2730	3330	3895	4510	5800	7125	8580	*9475*		*4′10″*	26
2755	3355	3925	4570	5875	7230	8725	*9855*		*4′11″*	27
2770	3380	3965	4615	5955	7330	8890	*10230*		*4′11″*	28
2780	3405	4000	4665	6030	7425	9060	*10610*		*5′ 0″*	29
2790	3430	4030	4705	6085	7500	9190	10830	*10985*	*5′ 1″*	30

TABLE 15-6 Loads on 24-in Vitrified Clay Pipe (pounds per linear foot) Caused by Backfilling

Bold print figures represent trench loads on pipe for each width of trench. Italicized figures represent maximum loads on pipe at and beyond transition width.

SAND AND GRAVEL

AVERAGE WT. = 110 LBS./CU. FT.

DEPTH OF BACKFILL OVER TOP OF PIPE (feet)	TRENCH WIDTH AT TOP OF PIPE (feet)								Transition Width
	3'0"	3'6"	4'0"	4'6"	5'0"	6'0"	7'0"	8'0"	
5	**1250**	**1540**	**1805**	*1980*					*4' 3"*
6	**1430**	**1760**	**2055**	*2420*					*4' 6"*
7	**1595**	**1960**	**2310**	**2715**	*2895*				*4' 9"*
8	**1735**	**2145**	**2530**	**3005**	*3335*				*5' 0"*
9	**1870**	**2340**	**2760**	**3265**	**3675**	*3815*			*5' 2"*
10	**1990**	**2505**	**2970**	**3520**	**3980**	*4290*			*5' 3"*
11	**2100**	**2650**	**3170**	**3750**	**4300**	*4750*			*5' 4"*
12	**2200**	**2780**	**3345**	**3995**	**4620**	*5205*			*5' 6"*
13	**2275**	**2890**	**3525**	**4180**	**4860**	*5685*			*5' 8"*
14	**2350**	**3000**	**3660**	**4355**	**5095**	*6150*			*5'10"*
15	**2415**	**3115**	**3805**	**4520**	**5300**	*6600*			*5'11"*
16	**2475**	**3205**	**3915**	**4675**	**5480**	**6985**	*7085*		*6' 1"*
17	**2530**	**3285**	**4050**	**4815**	**5645**	**7240**	*7535*		*6' 2"*
18	**2580**	**3355**	**4120**	**4940**	**5810**	**7480**	*8030*		*6' 3"*
19	**2625**	**3420**	**4210**	**5070**	**5950**	**7735**	*8470*		*6' 4"*
20	**2660**	**3475**	**4280**	**5180**	**6085**	**7940**	*8945*		*6' 5"*
21	**2695**	**3525**	**4365**	**5290**	**6215**	**8150**	*9405*		*6' 7"*
22	**2725**	**3575**	**4440**	**5395**	**6370**	**8395**	*9900*		*6' 8"*
23	**2750**	**3620**	**4510**	**5500**	**6490**	**8590**	*10360*		*6' 9"*
24	**2780**	**3660**	**4570**	**5585**	**6590**	**8780**	*10800*		*6'10"*
25	**2805**	**3695**	**4630**	**5665**	**6690**	**8965**	*11310*		*7' 0"*
26	**2825**	**3735**	**4685**	**5740**	**6795**	**9140**	**11550**	*11750*	*7' 1"*
27	**2845**	**3760**	**4740**	**5810**	**6905**	**9295**	**11770**	*12210*	*7' 2"*
28	**2860**	**3785**	**4795**	**5880**	**6995**	**9440**	**11970**	*12705*	*7' 3"*
29	**2875**	**3805**	**4850**	**5940**	**7095**	**9570**	**12175**	*13145*	*7' 4"*
30	**2895**	**3835**	**4905**	**5995**	**7170**	**9690**	**12375**	*13640*	*7' 5"*

Transition Width Column represents the critical width where trench loads reach their maximum and are equal to the embankment load.

SOURCE: *Clay Pipe Engineering Manual,* © National Clay Pipe Institute 1974, 1972, 1968, Crystal Lake, Ill.

TABLE 15-6 Loads on 24-in Vitrified Clay Pipe (pounds per linear foot) Caused by Backfilling *(cont.)*

SATURATED TOPSOIL

AVERAGE WT. = 115 LBS./CU. FT.

TRENCH WIDTH AT TOP OF PIPE (feet)								Transition Width	DEPTH OF BACKFILL OVER TOP OF PIPE (feet)
3′0″	3′6″	4′0″	4′6″	5′0″	6′0″	7′0″	8′0″		
1380	1680	*1900*						*3′11″*	5
1560	1900	2220	*2360*					*4′ 3″*	6
1735	2130	2495	*2840*					*4′ 6″*	7
1900	2325	2760	3210	*3325*				*4′ 8″*	8
2035	2530	3000	3495	*3855*				*4′10″*	9
2180	2705	3220	3795	*4325*				*5′ 0″*	10
2300	2875	3440	4065	4715	*4785*			*5′ 2″*	11
2410	3020	3635	4290	5005	*5280*			*5′ 4″*	12
2490	3165	3820	4525	5245	*5750*			*5′ 6″*	13
2590	3280	3970	4740	5465	*6210*			*5′ 8″*	14
2670	3405	4130	4945	5695	*6715*			*5′10″*	15
2755	3495	4270	5105	5900	*7215*			*5′10″*	16
2820	3590	4405	5280	6095	*7735*			*5′11″*	17
2885	3680	4530	5430	6300	*8225*			*6′ 0″*	18
2945	3760	4645	5570	6485	8510	*8685*		*6′ 2″*	19
2990	3840	4750	5705	6695	8740	*9145*		*6′ 3″*	20
3035	3935	4840	5835	6865	9000	*9660*		*6′ 4″*	21
3070	3990	4945	5955	7015	9230	*10120*		*6′ 5″*	22
3105	4050	5025	6070	7190	9460	*10640*		*6′ 6″*	23
3140	4100	5105	6175	7335	9660	*11100*		*6′ 7″*	24
3170	4140	5175	6280	7465	9860	*11560*		*6′ 8″*	25
3200	4185	5245	6370	7590	10065	*12075*		*6′ 9″*	26
3220	4225	5315	6465	7705	10235	*12535*		*6′10″*	27
3235	4255	5380	6545	7820	10410	*13055*		*7′ 0″*	28
3255	4280	5435	6625	7935	10550	13340	*13515*	*7′ 1″*	29
3265	4300	5475	6695	8050	10695	13570	*13975*	*7′ 2″*	30

TABLE 15-6 Loads on 24-in Vitrified Clay Pipe (pounds per linear foot) Caused by Backfilling ***(cont.)***

DRY CLAY

AVERAGE WT. = 120 LBS./CU. FT.

DEPTH OF BACKFILL OVER TOP OF PIPE (feet)	TRENCH WIDTH AT TOP OF PIPE (feet)							Transition Width
	3'0"	3'6"	4'0"	4'6"	5'0"	6'0"	7'0"	
5	1475	1765	*1895*					*3' 8"*
6	1680	2030	*2375*					*4' 0"*
7	1885	2280	2700	*2855*				*4' 2"*
8	2065	2515	3000	*3335*				*4' 4"*
9	2245	2735	3275	*3815*				*4' 6"*
10	2400	2940	3540	4130	*4295*			*4' 8"*
11	2550	3130	3780	4440	*4775*			*4'10"*
12	2680	3320	4010	4705	*5255*			*4'11"*
13	2810	3485	4225	4955	*5735*			*5' 0"*
14	2915	3650	4415	5185	6000	*6215*		*5' 2"*
15	3025	3785	4595	5400	6265	*6695*		*5' 3"*
16	3110	3925	4765	5615	6515	*7175*		*5' 4"*
17	3190	4055	4930	5810	6745	*7655*		*5' 6"*
18	3275	4170	5090	6010	6950	*8135*		*5' 8"*
19	3355	4285	5230	6205	7130	*8615*		*5' 9"*
20	3420	4390	5375	6385	7450	*9095*		*5'10"*
21	3480	4490	5495	6550	7655	*9575*		*5'11"*
22	3540	4570	5610	6710	7850	*10055*		*5'11"*
23	3590	4650	5720	6850	8030	10500	*10535*	*6' 1"*
24	3625	4730	5825	6995	8220	10775	*11015*	*6' 2"*
25	3660	4800	5930	7140	8390	11015	*11495*	*6' 3"*
26	3685	4850	6020	7250	8555	11245	*11975*	*6' 3"*
27	3710	4895	6110	7355	8710	11470	*12455*	*6' 4"*
28	3730	4945	6185	7465	8855	11675	*12935*	*6' 5"*
29	3755	4980	6265	7570	9010	11880	*13415*	*6' 6"*
30	3780	5005	6330	7680	9155	12085	*13895*	*6' 7"*

TABLE 15-6 Loads on 24-in Vitrified Clay Pipe (pounds per linear foot) Caused by Backfilling *(cont.)*

WET CLAY

AVERAGE WT. = 130 LBS./CU. FT.

TRENCH WIDTH AT TOP OF PIPE (feet)							Transition Width	DEPTH OF BACKFILL OVER TOP OF PIPE (feet)
3′0″	3′6″	4′0″	4′6″	5′0″	6′0″	7′0″		
1640	*1935*						*3′ 6″*	5
1885	2290	*2455*					*3′ 9″*	6
2120	2585	*2950*					*4′ 0″*	7
2340	2860	3340	*3445*				*4′ 2″*	8
2555	3120	3680	*3965*				*4′ 4″*	9
2750	3380	3980	*4470*				*4′ 5″*	10
2940	3600	4275	*4965*				*4′ 6″*	11
3120	3820	4550	5330	*5460*			*4′ 8″*	12
3275	4030	4835	5630	*5955*			*4′ 9″*	13
3420	4225	5095	5940	*6450*			*4′10″*	14
3550	4420	5330	6215	*6955*			*4′11″*	15
3675	4590	5565	6460	*7435*			*5′ 0″*	16
3785	4745	5760	6720	7735	*7945*		*5′ 1″*	17
3900	4890	5980	6970	8035	*8450*		*5′ 2″*	18
4005	5030	6150	7230	8320	*8945*		*5′ 3″*	19
4095	5175	6320	7475	8620	*9425*		*5′ 4″*	20
4180	5290	6485	7720	8905	*9945*		*5′ 5″*	21
4250	5410	6630	7850	9165	*10465*		*5′ 6″*	22
4330	5525	6760	8045	9410	*10960*		*5′ 7″*	23
4395	5605	6890	8230	9645	*11455*		*5′ 8″*	24
4445	5705	7020	8400	9865	*11960*		*5′ 9″*	25
4510	5800	7125	8580	10075	*12455*		*5′10″*	26
4570	5875	7230	8725	10285	*12950*		*5′11″*	27
4615	5955	7330	8890	10465	*13455*		*5′11″*	28
4665	6030	7425	9060	10645	*13935*		*6′ 0″*	29
4705	6085	7500	9190	10830	14120	*14455*	*6′ 2″*	30

TABLE 15-7 Loads on 36-in Vitrified Clay Pipe (pounds per linear foot) Caused by Backfilling

Bold print figures represent trench loads on pipe for each width of trench. Italicized figures represent maximum loads on pipe at and beyond transition width.

SAND AND GRAVEL

AVERAGE WT. = 110 LBS./CU. FT.

DEPTH OF BACKFILL OVER TOP OF PIPE (feet)	TRENCH WIDTH AT TOP OF PIPE (feet)							Transition Width
	4'6"	5'0"	6'0"	7'0"	8'0"	9'0"	10'0"	
5	2100	2345	*2410*					*5' 2"*
6	2420	2695	*3100*					*5' 9"*
7	2715	3025	3785	*3815*				*6' 1"*
8	3005	3355	4215	*4490*				*6' 4"*
9	3265	3675	4675	*5170*				*6' 8"*
10	3520	3980	5095	*5885*				*6'10"*
11	3750	4300	5480	*6545*				*7' 0"*
12	3995	4620	5830	7060	*7240*			*7' 2"*
13	4180	4860	6160	7525	*7920*			*7' 4"*
14	4355	5095	6445	7940	*8635*			*7' 6"*
15	4520	5300	6720	8360	*9350*			*7' 8"*
16	4675	5480	6985	8745	*10065*			*7'10"*
17	4815	5645	7240	9095	*10835*			*8' 0"*
18	4940	5810	7480	9425	11220	*11495*		*8' 2"*
19	5070	5950	7735	9745	11605	*12210*		*8' 4"*
20	5180	6085	7940	10045	11970	*12870*		*8' 5"*
21	5290	6215	8150	10305	12320	*13640*		*8' 7"*
22	5395	6370	8395	10580	12670	*14300*		*8'10"*
23	5500	6490	8590	10835	13035	*15015*		*8'11"*
24	5585	6590	8780	11075	13345	*15695*		*9' 0"*
25	5665	6690	8965	11330	13660	16205	*16390*	*9' 1"*
26	5740	6795	9140	11550	13980	16610	*17105*	*9' 2"*
27	5810	6905	9295	11770	14300	16975	*17765*	*9' 4"*
28	5880	6995	9440	11970	14620	17325	*18535*	*9' 6"*
29	5940	7095	9570	12175	14915	17655	*19250*	*9' 7"*
30	5995	7170	9690	12375	15235	17930	*19910*	*9' 8"*

Transition Width Column represents the critical width where trench loads reach their maximum and are equal to the embankment load.

SOURCE: *Clay Pipe Engineering Manual,* © National Clay Pipe Institute 1974, 1972, 1968, Crystal Lake, Ill.

TABLE 15-7 Loads on 36-in Vitrified Clay Pipe (pounds per linear foot) Caused by Backfilling *(cont.)*

SATURATED TOPSOIL

AVERAGE WT. = 115 LBS./CU. FT.

TRENCH WIDTH AT TOP OF PIPE (feet)							Transition Width	DEPTH OF BACKFILL OVER TOP OF PIPE (feet)
4'6"	5'0"	6'0"	7'0"	8'0"	9'0"	10'0"		
2245	*2475*						*5' 0"*	5
2565	2935	*3165*					*5' 5"*	6
2900	3335	*3875*					*5' 9"*	7
3210	3680	*4600*					*6' 0"*	8
3495	4060	5005	*5290*				*6' 3"*	9
3795	4405	5405	*5980*				*6' 6"*	10
4065	4715	5835	*6730*				*6' 9"*	11
4290	5005	6210	*7420*				*7' 0"*	12
4525	5245	6615	7995	*8140*			*7' 1"*	13
4740	5465	6960	8455	*8855*			*7' 3"*	14
4945	5695	7305	8855	*9605*			*7' 5"*	15
5105	5900	7620	9285	*10350*			*7' 7"*	16
5280	6095	7935	9690	*11040*			*7' 9"*	17
5430	6300	8220	10065	*11730*			*7'11"*	18
5570	6485	8510	10465	*12480*			*8' 0"*	19
5705	6695	8740	10810	12940	*13170*		*8' 1"*	20
5835	6865	9000	11155	13340	*13915*		*8' 3"*	21
5955	7015	9230	11500	13800	*14665*		*8' 5"*	22
6070	7190	9460	11790	14145	*15355*		*8' 6"*	23
6165	7335	9660	12075	14550	*16100*		*8' 8"*	24
6280	7465	9860	12360	14950	*16790*		*8' 9"*	25
6370	7590	10065	12650	15240	*17540*		*8'10"*	26
6465	7705	10235	12880	15585	*18230*		*8'11"*	27
6545	7820	10410	13110	15930	*18860*		*9' 0"*	28
6625	7935	10550	13340	16275	19365	*19585*	*9' 1"*	29
6695	8050	10695	13570	16505	19755	*20265*	*9' 3"*	30

TABLE 15-7 Loads on 36-in Vitrified Clay Pipe (pounds per linear foot) Caused by Backfilling *(cont.)*

DRY CLAY

AVERAGE WT. = 120 LBS./CU. FT.

DEPTH OF BACKFILL OVER TOP OF PIPE (feet)	TRENCH WIDTH AT TOP OF PIPE (feet)						Transition Width
	4'6"	5'0"	6'0"	7'0"	8'0"	9'0"	
5	2340	*2375*					*4' 6"*
6	2735	*3095*					*5' 0"*
7	3095	3505	*3815*				*5' 5"*
8	3455	3925	*4535*				*5' 9"*
9	3815	4320	*5255*				*6' 0"*
10	4130	4705	5760	*5975*			*6' 2"*
11	4440	5050	6230	*6695*			*6' 4"*
12	4705	5390	6660	*7415*			*6' 6"*
13	4955	5710	7090	*8135*			*6' 8"*
14	5185	6000	7500	*8855*			*6'10"*
15	5400	6265	7895	*9575*			*6'11"*
16	5615	6515	8280	10140	*10295*		*7' 1"*
17	5810	6745	8675	10595	*11015*		*7' 2"*
18	6010	6950	9010	11040	*11735*		*7' 3"*
19	6205	7130	9350	11460	*12455*		*7' 5"*
20	6385	7450	9650	11880	*13175*		*7' 6"*
21	6550	7655	9960	12275	*13885*		*7' 8"*
22	6710	7850	10235	12650	*14605*		*7'10"*
23	6850	8030	10500	13020	*15325*		*7'11"*
24	6995	8220	10775	13380	*16045*		*8' 0"*
25	7140	8390	11015	13715	16440	*16750*	*8' 2"*
26	7250	8555	11245	14040	16860	*17470*	*8' 3"*
27	7355	8710	11470	14365	17255	*18190*	*8' 4"*
28	7465	8855	11675	14650	17640	*18900*	*8' 5"*
29	7570	9010	11880	14940	18000	*19610*	*8' 7"*
30	7680	9155	12085	15180	18360	*20315*	*8' 8"*

TABLE 15-7 Loads on 36-in Vitrified Clay Pipe (pounds per linear foot) Caused by Backfilling *(cont.)*

WET CLAY

AVERAGE WT. = 130 LBS./CU. FT.

TRENCH WIDTH AT TOP OF PIPE (feet)					Transition Width	DEPTH OF BACKFILL OVER TOP OF PIPE (feet)
4'6"	5'0"	6'0"	7'0"	8'0"		
2470					4' 4"	5
3070	*3250*				4' 9"	6
3485	*3965*				5' 0"	7
3875	4395	*4745*			5' 4"	8
4290	4860	*5485*			5' 8"	9
4640	5280	*6240*			5'11"	10
5005	5720	*6970*			6' 0"	11
5330	6110	7565	*7735*		6' 2"	12
5630	6435	8060	*8465*		6' 4"	13
5940	6760	8515	*9205*		6' 6"	14
6215	7110	8995	*9945*		6' 8"	15
6460	7425	9440	*10710*		6'10"	16
6720	7735	9840	*11440*		6'11"	17
6970	8035	10245	*12195*		6'11"	18
7230	8320	10635	*12935*		7' 0"	19
7475	8620	11025	13480	*13665*	7' 2"	20
7720	8905	11390	13935	*14430*	7' 4"	21
7850	9165	11765	14380	*15160*	7' 5"	22
8045	9410	12115	14860	*15910*	7' 5"	23
8230	9645	12430	15275	*16640*	7' 6"	24
8400	9865	12755	15705	*17410*	7' 8"	25
8580	10075	13065	16120	*18120*	7' 9"	26
8725	10285	13350	16575	*18850*	7'10"	27
8890	10465	13635	16980	*19605*	7'11"	28
9060	10645	13895	17405	*20345*	7'11"	29
9190	10830	14120	17810	*21060*	8' 0"	30

NOTES

Section Sixteen

▪

WATER SUPPLY, WATER STORAGE, AND FIRE PROTECTION

TABLE 16-1 Planning Guide for Water Use

Types of establishments	Gallons per day
Airports (per passenger)	3-5
Apartments, multiple family (per resident)	60
Bath houses (per bather)	10
Camps:	
Construction, semipermanent (per worker)	50
Day with no meals served (per camper)	15
Luxury (per camper)	100-150
Resorts, day and night, with limited plumbing (per camper)	50
Tourist with central bath and toilet facilities (per person)	35
Cottages with seasonal occupancy (per resident)	50
Courts, tourist with individual bath units (per person)	50
Clubs:	
Country (per resident member)	100
Country (per nonresident member present)	25
Dwellings:	
Boardinghouses (per boarder)	50
Additional kitchen requirements for nonresident boarders	10
Luxury (per person)	100-150
Multiple-family apartments (per resident)	40
Rooming houses (per resident)	60
Single family (per resident)	50-75
Estates (per resident)	100-150
Factories (gallons per person per shift)	15-35
Highway rest area (per person)	5
Hotels with private baths (2 persons per room)	60
Hotels without private baths (per person)	50
Institutions other than hospitals (per person)	75-125
Hospitals (per bed)	250-400
Laundries, self-serviced (gallons per washing, i.e., per customer)	50
Livestock (per animal):	
Cattle (drinking)	12
Dairy (drinking and servicing)	35
Goat (drinking)	2
Hog (drinking)	4
Horse (drinking)	12
Mule (drinking)	12
Sheep (drinking)	2
Steer (drinking)	12
Motels with bath, toilet, and kitchen facilities (per bed space)	50
With bed and toilet (per bed space)	40
Parks:	
Overnight with flush toilets (per camper)	25
Trailers with individual bath units, no sewer connection (per trailer)	25
Trailers with individual baths, connected to sewer (per person)	50
Picnic:	
With bathhouses, showers, and flush toilets (per picnicker)	20
With toilet facilities only (gallons per picnicker)	10
Poultry:	
Chickens (per 100)	5-10
Turkeys (per 100)	10-18

SOURCE: US/EPA

TABLE 16-1 Planning Guide for Water Use *(cont.)*

Types of establishments	Gallons per day
Restaurants with toilet facilities (per patron)	7-10
Without toilet facilities (per patron)	2½-3
With bars and cocktail lounge (additional quantity per patron)	2
Schools:	
Boarding (per pupil)	75-100
Day with cafeteria, gymnasiums, and showers (per pupil)	25
Day with cafeteria but no gymnasiums or showers (per pupil)	20
Day without cafeteria, gymnasiums, or showers (per pupil)	15
Service stations (per vehicle)	10
Stores (per toilet room)	400
Swimming pools (per swimmer)	10
Theaters:	
Drive-in (per car space)	5
Movie (per auditorium seat)	5
Workers:	
Construction (per person per shift)	50
Day (school or offices per person per shift)	15

TABLE 16-2 Rates of Flow for Certain Plumbing, Household, and Farm Fixtures

Location	Flow pressure[1] —pounds per square inch (psi)	Flow rate—gallons per minute (gpm)
Ordinary basin faucet	8	2.0
Self-closing basin faucet	8	2.5
Sink faucet, 3/8 inch	8	4.5
Sink faucet, 1/2 inch	8	4.5
Bathtub faucet	8	6.0
Laundry tub faucet, 1/2 inch	8	5.0
Shower	8	5.0
Ball-cock for closet	8	3.0
Flush valve for closet	15	[2] 15-40
Flushometer valve for urinal	15	15.0
Garden hose (50 ft., 3/4-inch sill cock)	30	5.0
Garden hose (50 ft., 5/8-inch outlet)	15	3.33
Drinking fountains	15	.75
Fire hose 1-1/2 inches, 1/2-inch nozzle	30	40.0

[1] Flow pressure is the pressure in the supply near the faucet or water outlet while the faucet or water outlet is wide open and flowing.

[2] Wide range due to variation in design and type of closet flush valves.

SOURCE: US/EPA

TABLE 16-3 Suitability of Well Construction Methods

Characteristics	Dug	Bored	Driven	Drilled			Jetted
				Percussion	Rotary		
					Hydraulic	Air	
Range of practical depths (general order of magnitude)	0-50 feet	0-100 feet	0-50 feet	0-1,000 feet	0-1,000 feet	0-750 feet	0-100 feet
Diameter	3-20 feet	2-30 inches	1¼-2 inches	4-18 inches	4-24 inches	4-10 inches	2-12 inches
Type of geologic formation:							
Clay	Yes	Yes	Yes	Yes	Yes	No	Yes
Silt	Yes	Yes	Yes	Yes	Yes	No	Yes
Sand	Yes	Yes	Yes	Yes	Yes	No	Yes
Gravel	Yes	Yes	Fine	Yes	Yes	No	¼-inch pea gravel
Cemented gravel	Yes	No	No	Yes	Yes	No	No
Boulders	Yes	Yes, if less than well diameter	No	Yes, when in firm bedding	(Difficult)	No	No
Sandstone	Yes, if soft and/or fractured	Yes, if soft and/or fractured	Thin layers only	Yes	Yes	Yes	No
Limestone			No	Yes	Yes	Yes	No
Dense igneous rock	No	No	No	Yes	Yes	Yes	No

[1]The ranges of values in this table are based upon general conditions. They may be exceeded for specific areas or conditions.

SOURCE: US/EPA

Quantities* of calcium hypochlorite, 70 percent (rows A) and liquid household bleach, 5.25 percent (rows B) required for water well disinfection

Depth of water in well (ft.)		Well diameter, (in)															
		2	3	4	5	6	8	10	12	16	20	24	28	32	36	42	48
5	A	1T	1T	1T	1T	1T	1T	2T	3T	5T	6T	3 oz.	4 oz.	5 oz.	7 oz.	9 oz.	12 oz.
	B	1C	1C	1C	1C	1C	1C	1C	1C	2C	4C	1Q	2Q	3Q	3Q	4Q	5Q
10	A	1T	1T	1T	1T	1T	2T	3T	5T	8T	4 oz.	6 oz.	8 oz.	10 oz.	13 oz.	1½ lb.	1½ lb.
	B	1C	1C	1C	1C	1C	1C	2C	2C	1Q	2Q	3Q	4Q	4Q	6Q	8Q	2½G
15	A	1T	1T	1T	1T	2T	3T	5T	8T	4 oz.	6 oz.	9 oz.	12 oz.	1 lb.	1½ lb.	1½ lb.	2 lb.
	B	1C	1C	1C	1C	1C	2C	3C	4C	2Q	2½Q	4Q	5Q	6Q	2G	3G	4G
20	A	1T	1T	1T	2T	3T	4T	6T	3 oz.	5 oz.	8 oz.						
	B	1C	1C	1C	1C	1C	2C	4C	1Q	2½Q	3½Q						
30	A	1T	1T	2T	3T	4T	6T	3 oz.	4 oz.	8 oz.	12 oz.						
	B	1C	1C	1C	1C	2C	4C	1½Q	2Q	4Q	5Q						
40	A	1T	1T	2T	4T	6T	8T	4 oz.	6 oz.	10 oz.	1 lb.						
	B	1C	1C	1C	2C	2C	1Q	2Q	2½Q	4½Q	7Q						
60	A	1T	2T	3T	5T	8T	4 oz.	6 oz.	9 oz.								
	B	1C	1C	2C	3C	4C	2Q	3Q	4Q								
80	A	1T	3T	4T	7T	9T	5 oz.	8 oz.	12 oz.								
	B	1C	1C	2C	4C	1Q	2Q	3½Q	5Q								
100	A	2T	3T	5T	8T	4 oz.	7 oz.	10 oz.	1 lb.								
	B	1C	2C	3C	1Q	1½Q	2½Q	4Q	6Q								
150	A	3T	5T	8T	4 oz.	6 oz.	10 oz.	1 lb.	1½ lb.								
	B	2C	2C	4C	2Q	2½Q	4Q	6Q	2½G								

*Quantities are indicated as: T = tablespoons; oz = ounces (by weight); C = cups; lb = pounds; Q = quarts; G = gallons.

NOTE: Figures corresponding to Rows A are amounts of solid calcium hypochlorite required; those corresponding to rows B are amounts of liquid household bleach. For cases lying in shaded area, add 5 gallons of chlorinated water, as final step, to force solution into formation. For those in non-shaded area, add 10 gallons of chlorinated water.

Fig. 16-1 Water-well disinfection *(US/EPA)*

TABLE 16-4 Limits of Basic Impurities in Domestic Water Supplies*

Impurity	*Limit ppm (mg/L)*
Turbidity	10
Color	20
Lead	0.1
Fluoride	1.0
Copper	3.0
Iron + Manganese	0.3
Magnesium	125
Total solids	500
Total hardness (Calcium + Magnesium salts)	100

*The major requirement for a domestic water supply is freedom from pathogenic bacteria. (For a complete standard, refer to the current issue of the U.S. Public Health Service publication, "Drinking Water Standards.")

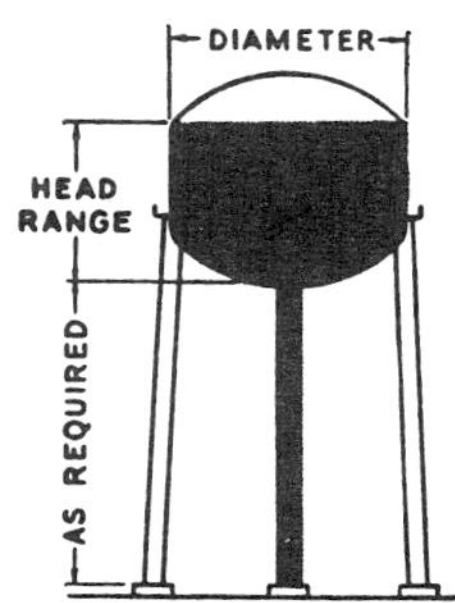

Capacity	Diameter	Range	No. Cols.	Rad. Bev.
50,000	22.0′	19.21′	4	1:12
60,000	24.0	19.65	4	1:12
75,000	26.0	21.18	4	1:12
100,000	30.0	21.08	4	1:12
125,000	32.0	23.18	4	1:12
150,000	34.0	24.81	4	1:12
200,000	36.0	29.38	4	1:12
250,000	40.0	30.87	5 & 6	1:20
300,000	44.0	30.75	5 & 6	1:20
400,000	48.0	34.28	6	1:20
500,000	50.0	39.07	6 & 8	1:20
500,000	55.0	29.92	6 & 8	Vert
750,000	60.0	41.64	8	Vert
1,000,000	65.0	47.10	10	Vert

Fig. 16-2 Double ellipsoidal tanks *(Pittsburgh-DesMoines Steel Company)*.

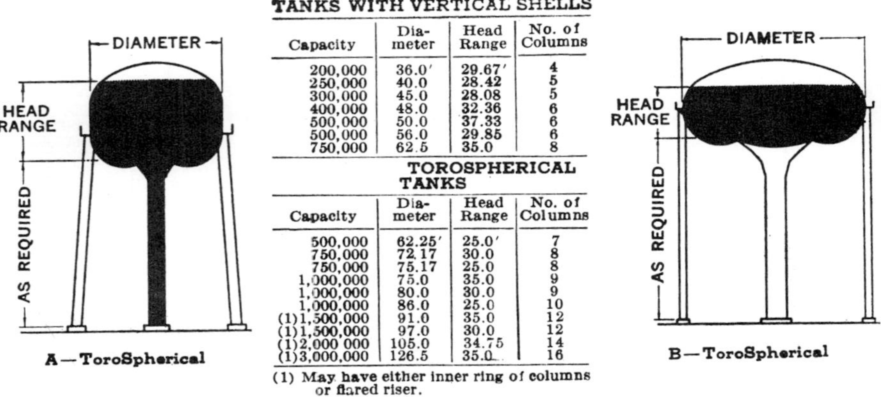

TOROSPHERICAL TANKS WITH VERTICAL SHELLS

Capacity	Diameter	Head Range	No. of Columns
200,000	36.0′	29.67′	4
250,000	40.0	28.42	5
300,000	45.0	28.08	5
400,000	48.0	32.36	6
500,000	50.0	37.33	6
500,000	56.0	29.85	6
750,000	62.5	35.0	8

TOROSPHERICAL TANKS

Capacity	Diameter	Head Range	No. of Columns
500,000	62.25′	25.0′	7
750,000	72.17	30.0	8
750,000	75.17	25.0	8
1,000,000	75.0	35.0	9
1,000,000	80.0	30.0	9
1,000,000	86.0	25.0	10
(1) 1,500,000	91.0	35.0	12
(1) 1,500,000	97.0	30.0	12
(1) 2,000,000	105.0	34.75	14
(1) 3,000,000	126.5	35.0	16

(1) May have either inner ring of columns or flared riser.

Fig. 16-3 Torospherical tanks *(Pittsburgh-DesMoines Steel Company).*

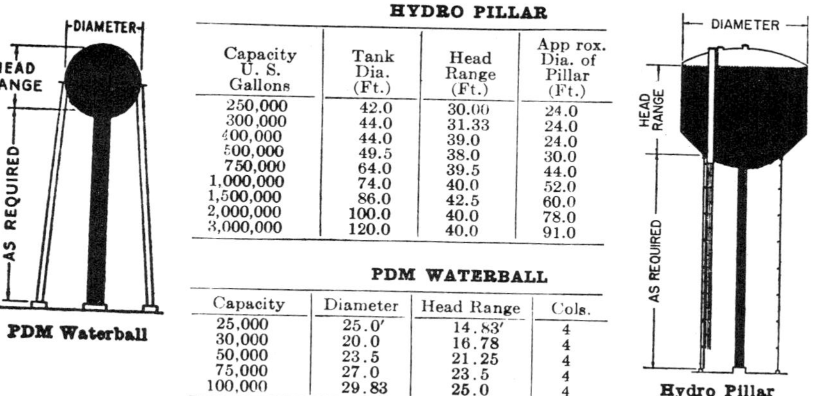

HYDRO PILLAR

Capacity U. S. Gallons	Tank Dia. (Ft.)	Head Range (Ft.)	Approx. Dia. of Pillar (Ft.)
250,000	42.0	30.00	24.0
300,000	44.0	31.33	24.0
400,000	44.0	39.0	24.0
500,000	49.5	38.0	30.0
750,000	64.0	39.5	44.0
1,000,000	74.0	40.0	52.0
1,500,000	86.0	42.5	60.0
2,000,000	100.0	40.0	78.0
3,000,000	120.0	40.0	91.0

PDM WATERBALL

Capacity	Diameter	Head Range	Cols.
25,000	25.0′	14.83′	4
30,000	20.0	16.78	4
50,000	23.5	21.25	4
75,000	27.0	23.5	4
100,000	29.83	25.0	4

Fig. 16-4 Hydro pillar and PDM Waterball *(Pittsburgh-DesMoines Steel Company)*.

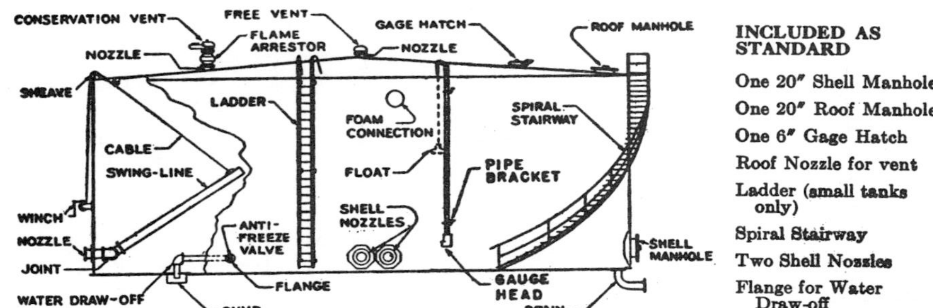

Fig. 16-5 Usual accessories and fittings on standard cone-roof tanks *(Pittsburgh-DesMoines Steel Company)*.

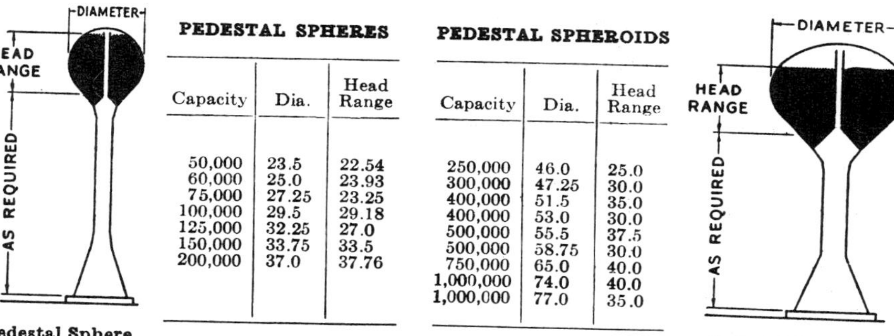

PEDESTAL SPHERES

Capacity	Dia.	Head Range
50,000	23.5	22.54
60,000	25.0	23.93
75,000	27.25	23.25
100,000	29.5	29.18
125,000	32.25	27.0
150,000	33.75	33.5
200,000	37.0	37.76

PEDESTAL SPHEROIDS

Capacity	Dia.	Head Range
250,000	46.0	25.0
300,000	47.25	30.0
400,000	51.5	35.0
400,000	53.0	30.0
500,000	55.5	37.5
500,000	58.75	30.0
750,000	65.0	40.0
1,000,000	74.0	40.0
1,000,000	77.0	35.0

Fig. 16-6 Pedestal spheres and spheroids *(Pittsburgh-DesMoines Steel Company)*.

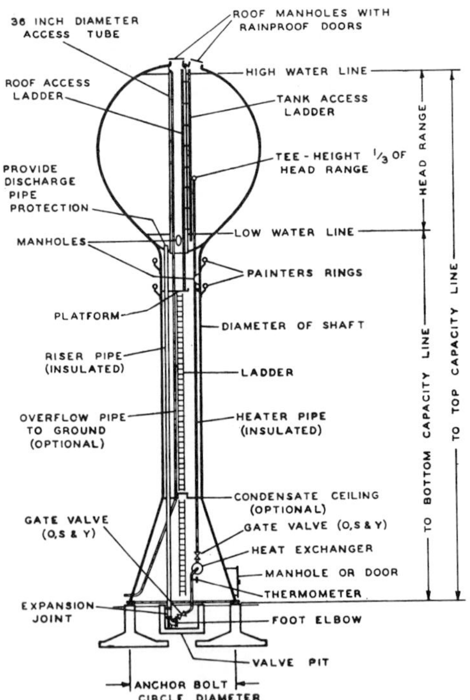

Fig. 16-7 Typical details of pedestal tank *(Reproduced with permission from NFPA 22, Water Tanks for Private Fire Protection, Copyright © 1976, National Fire Protection Association, Boston, MA 02210.)*

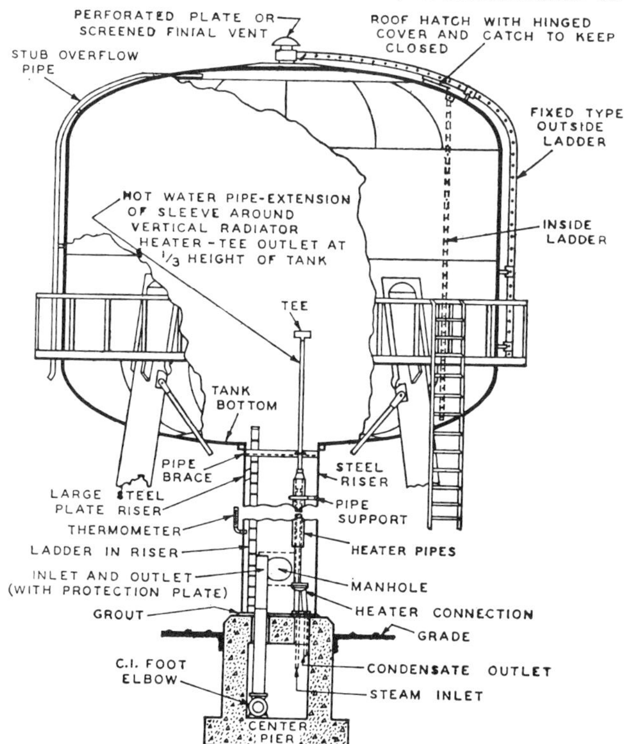

Fig. 16-8 Typical tower-supported double ellipsoidal tank *(Reproduced with permission from NFPA 22, Water Tanks for Private Fire Protection, Copyright © 1976, National Fire Protection Association, Boston, MA 02210.)*

TABLE 16-5 Average Efficiencies of Tank Heaters and Heat Content of Fuels

Type of Heater	Average Efficiency	Fuel	Average Heat Content Btu
Steam heaters, coils, etc.	95%	Anthracite Coal, *per lb.*	13300
Electric heaters	95%	Bituminous Coal, high grade, *per lb.*	13200
Boilers (Fuel Oil)	70%	Bituminous Coal, medium grade, *per lb.*	12000
Boilers (Coal)	70%	Bituminous, Coal, low grade, *per lb.*	10300
Gas Water Heaters	70%	Fuel Oil, *per lb.*	19000
Coal-burning Water Heaters	40% to 60%	Gas, Natural, *per cu. ft.*	1100
		Gas, Artificial, *per cu. ft.*	600
		Electricity, *per kw. hour*	3415

SOURCE: Reproduced with permission from NFPA 22, Water Tanks for Private Fire Protection, Copyright © 1976, National Fire Protection Association, Boston, MA 02210.

TABLE 16-6 Thousands of Btu Loss per Hour from Elevated Steel Tanks

Based on minimum water temperature of 42°F and a wind velocity of 12 mph.

To determine capacity of heater needed, find the minimum mean atmospheric temperature for one day from the Isothermal Map, Figure 10-1.4, and note the corresponding heat loss below.

Atmospheric Temperature Deg. F	Heat (Btu) Loss Per Sq. Ft. Tank Radiating Surface	TANK CAPACITIES—THOUSANDS U. S. GALLONS									Add Btu Per Lineal Ft. Uninsulated Steel Riser	
		25	30	40	50	75	100	150	200	250		
		Square feet of tank surface*										
		1210	1325	1550	1800	2370	2845	3705	4470	5240	3 ft. dia.	4 ft. dia.
		Btu Lost per hour, Thousands										
35	32.3	40	43	51	59	77	92	120	145	168	50	69
30	46.1	56	62	72	83	110	132	171	207	242	144	192
25	61.5	75	82	96	111	146	175	228	275	323	255	340
20	77.2	94	103	120	139	183	220	287	346	405	380	506
15	93.6	114	125	146	169	222	267	347	419	491	519	692
10	110.9	135	147	172	200	263	316	411	496	582	670	893
5	128.9	156	171	200	233	306	367	478	577	676	820	1092
0	148.5	180	197	231	268	352	423	551	664	779	982	1309
−5	168.7	205	224	262	304	400	480	626	755	884	1152	1536
−10	190.7	231	253	296	344	452	543	707	853	1000	1329	1771
−15	213.2	258	283	331	384	506	607	790	954	1118	1515	2020
−20	236.8	287	314	368	427	562	674	878	1059	1241	1718	2291
−25	262.3	318	348	407	473	622	747	972	1173	1375	1926	2568
−30	288.1	349	382	447	519	683	820	1068	1288	1510	2145	2860
−35	316.0	383	419	490	569	749	900	1171	1413	1656	2381	3174
−40	344.0	417	456	534	620	816	979	1275	1538	1803	2620	3494
−50	405.6	491	538	629	731	962	1154	1503	1814	2126	3139	4186
−60	470.8	570	624	730	848	1116	1340	1745	2105	2467	3702	4936

*These numbers are square feet of tank radiating surfaces used for each capacity to compute the tabulated heat loss values and are typical for tanks with D/4 ellipsoidal roofs and bottoms.

NOTE TO TABLE Heat loss for a given capacity with a different tank radiating surface than shown above shall be obtained by multiplying the radiating surface by the tabulated heat loss per square foot for the atmospheric temperature involved. The minimum radiaton surface area shall be the wetted tank steel surface area plus the top water surface area. For tanks with large steel plate risers the heat loss from the riser shall be added to that from the tank. The riser loss per linear foot shall be as tabulated above.

SOURCE: Reproduced with permission from NFPA 22, Water Tanks for Private Fire Protection, Copyright © 1976, National Fire Protection Association, Boston, MA 02210.

TABLE 16-7 Thousands of Btu Loss per Hour from Steel Suction Tanks and Standpipes

Based on minimum water temperature of 42°F and a wind velocity of 12 mph.

To determine capacity of heater needed, find the minimum mean atmospheric temperature for one day from the Isothermal Map, Figure 10-1.4, and note the corresponding heat loss below.

Atmospheric Temperature Deg. F	Heat (Btu) Loss Per Sq. Ft. Tank Radiating Surface	TANK CAPACITIES—THOUSANDS U. S. GALLONS									
		100	125	150	200	250	300	400	500	750	1000
		Square feet of tank surface*									
		2610	3030	3505	4175	4795	5360	6375	7355	9650	11740
		Btu Lost per hour, Thousands									
35	32.3	85	98	114	135	155	175	206	238	312	380
30	46.1	121	140	162	193	222	248	294	340	445	542
25	61.5	161	187	216	257	295	330	393	453	594	722
20	77.2	202	234	271	323	371	414	493	568	745	907
15	93.6	245	284	329	391	449	502	597	689	904	1099
10	110.9	290	337	389	463	532	595	707	816	1071	1302
5	128.9	337	391	452	539	619	691	822	949	1244	1514
0	148.5	388	450	521	620	713	796	947	1093	1434	1744
−5	168.7	441	512	592	705	809	905	1076	1241	1628	1981
−10	190.7	498	578	669	797	915	1023	1216	1403	1841	2239
−15	213.2	557	646	748	891	1023	1143	1360	1569	2058	2503
−20	236.8	619	718	830	989	1136	1270	1510	1742	2286	2781
−25	262.3	685	795	920	1096	1258	1406	1673	1930	2532	3080
−30	288.1	752	873	1010	1203	1382	1545	1837	2119	2781	3383
−35	316.0	825	958	1108	1320	1516	1694	2015	2325	3050	3710
−40	344.0	898	1043	1206	1437	1650	1844	2193	2531	3320	4039
−50	405.6	1059	1229	1422	1694	1945	2175	2586	2984	3915	4762
−60	470.8	1229	1427	1651	1966	2258	2524	3002	3463	4544	5528

*These numbers are square feet of radiating surface used for each capacity to compute the tabulated heat loss values and are typical for cone roof reservoirs on grade.

NOTE TO TABLE Heat loss for a given capacity with a different radiating surface than shown above shall be obtained by multiplying the radiating surface by the tabulated heat loss per square foot for the atmospheric temperature involved. The minimum radiation surface area shall be the wetted surface exposed to atmosphere plus the top water surface area. No heat loss shall be figured for tank bottoms resting on grade.

SOURCE: Reproduced with permission from NFPA 22, Water Tanks for Private Fire Protection, Copyright © 1976, National Fire Protection Association, Boston, MA 02210.

TABLE 16-8 Minimum Size of Circulating Pipes for Elevated Steel Tanks

Minimum One-Day mean Temp. Degrees F.	15,000	20,000	25,000	30,000	40,000	50,000	60,000	75,000	100,000	150,000
+10	2	2	2	2	2	2	2	2	2	2½
+ 5	2	2	2	2	2	2	2	2	2	2½
0	2	2	2	2	2	2	2	2	2½	2½
− 5	2	2	2	2	2	2	2	2	2½	2½
−10	2	2	2	2	2	2	2	2½	2½	2½
−15	2	2	2	2	2	2	2½	2½	2½	3
−20	2	2	2	2	2	2½	2½	2½	2½	3
−25	2	2	2	2	2½	2½	2½	2½	3	3
−30	2	2	2	2	2½	2½	2½	2½	3	3
−35	2	2	2	2½	2½	2½	2½	3	3	3
−40	2	2	2	2½	2½	2½	2½	3	3	3

SOURCE: Reproduced with permission from NFPA 22, Water Tanks for Private Fire Protection, Copyright © 1976, National Fire Protection Association, Boston, MA 02210.

TABLE 16-9 Approximate Heat Transfer from Coils and Pipe Radiators When Coldest Water Is Just Safely Above Freezing

Steam Pressure Lbs. per Sq. In.	Heat Transfer (Steam to Water) Btu per hr. per sq. ft.
10	19500
15	22000
20	24500
30	29500
40	34500
50	39000

NOTE.—These values should not be used to determine the area of coil needed in a circulating heater.

SOURCE: Reproduced with permission from NFPA 22, Water Tanks for Private Fire Protection, Copyright © 1976, National Fire Protection Association, Boston, MA 02210.

TABLE 16-10 Square Feet of Heating Surface in Coils or Pipe Radiators

	Size of Pipe—Inches										
Length Feet	¾	1	1¼	1½	2	2½	3	3½	4	5	6
1	.275	.346	.434	.494	.622	.753	.916	1.048	1.175	1.455	1.739
10	2.7	3.5	4.3	4.9	6.2	7.5	9.2	10.5	11.8	14.6	17.4
15	4.1	5.2	6.5	7.4	9.3	11.3	13.7	15.7	17.6	21.8	26.1
20	5.5	6.9	8.7	9.9	12.5	15.0	18.3	21.0	23.5	29.1	34.8
25	6.9	8.6	10.9	12.3	15.6	18.8	22.9	26.2	29.3	36.3	43.5
30	8.3	10.4	13.0	14.8	18.7	22.5	27.5	31.4	35.3	43.6	52.1
35	9.6	12.1	15.2	17.3	21.8	26.3	32.0	36.7	41.1	50.9	60.8
40	11.0	13.8	17.4	19.8	24.9	30.1	36.6	41.9	47.0	58.2	69.5
45	12.4	15.6	19.5	22.2	28.0	33.8	41.2	47.2	52.9	65.5	78.2
50	13.8	17.3	21.7	24.7	31.1	37.6	45.8	52.4	58.7	72.7	87.0
55	15.1	19.1	23.8	27.2	34.2	41.4	50.4	57.6	64.6	80.0	95.7
60	16.5	20.8	26.0	29.6	37.3	45.2	55.0	62.8	70.5	87.3	104.3
65	17.9	22.5	28.2	32.1	40.4	49.0	59.5	68.1	76.4	94.6	
70	19.2	24.2	30.4	34.6	43.5	52.7	64.1	73.3	82.3	101.9	
75	20.6	26.0	32.6	37.1	46.6	56.5	68.7	78.5	88.2		
80	22.0	27.7	34.7	39.5	49.8	60.2	73.2	83.8	93.0		
85	23.4	29.4	36.8	42.0	52.9	63.0	77.8		99.9		
90	24.7	31.2	39.0	44.5	56.0	67.8	82.4		105.8		
95	26.1	32.9	41.2	46.9	59.1	71.5	87.0				
100	27.5	34.6	43.4	49.4	62.2	75.3	91.6				

SOURCE: Reproduced with permission from NFPA 22, Water Tanks for Private Fire Protection, Copyright © 1976, National Fire Protection Association, Boston, MA 02210.

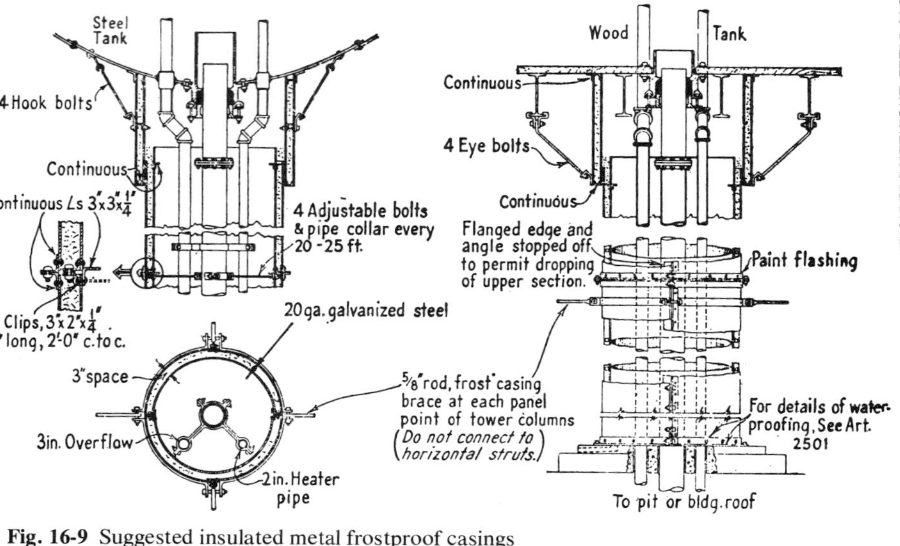

Fig. 16-9 Suggested insulated metal frostproof casings *(Reproduced with permission from NFPA 22, Water Tanks for Private Fire Protection, Copyright © 1976, National Fire Protection Association, Boston, MA 02210.)*

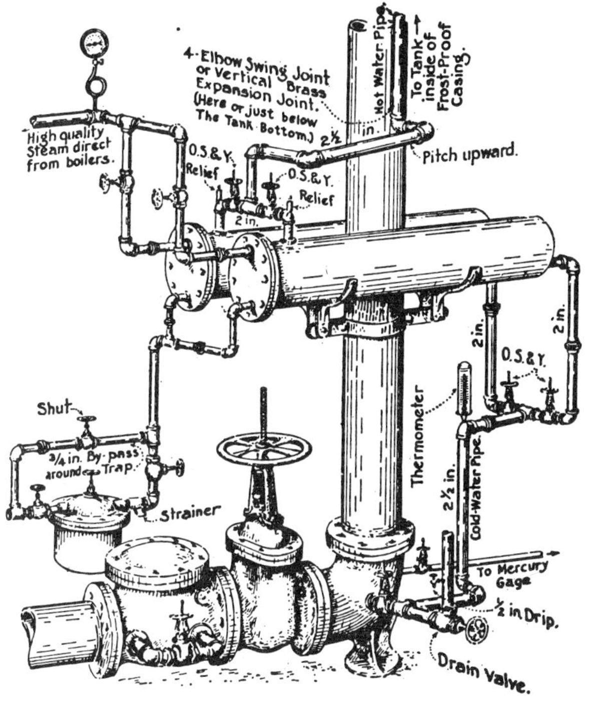

Fig. 16-10 Arrangements of multiple heaters
(Reproduced with permission from NFPA 22, Water Tanks for Private Fire Protection, Copyright © 1976, National Fire Protection Association, Boston, MA 02210.)

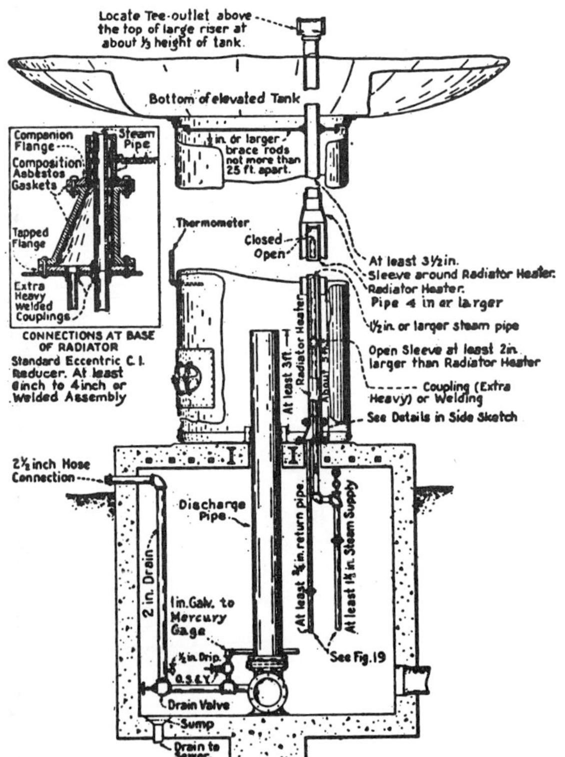

Fig. 16-11 Arrangement of a radiator-heater for a tank with a large riser
(Reproduced with permission from NFPA 22, Water Tanks for Private Fire Protection, Copyright © 1976, National Fire Protection Association, Boston, MA 02210.)

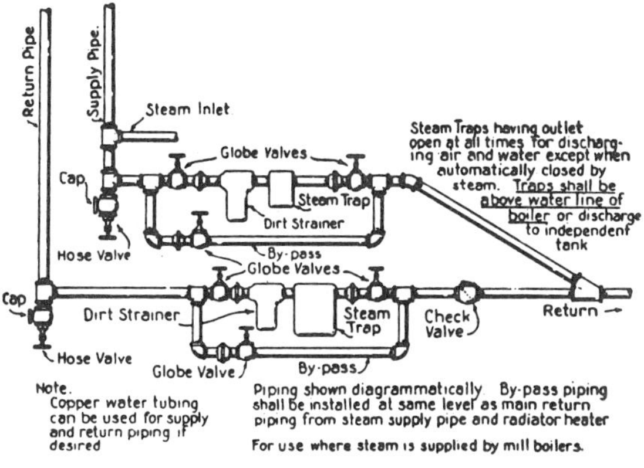

Fig. 16-12 Arrangement for removing condensate from tank radiator heaters
(Reproduced with permission from NFPA 22, Water Tanks for Private Fire Protection, Copyright © 1976, National Fire Protection Association, Boston, MA 02210.)

TABLE 16-11 Cylindrical Tanks

Dia., ft-in.	Vol. per lin ft, cu ft	U.S. gal per lin ft	Dia., ft-in.	Vol. per lin ft, cu ft	U.S. gal per lin ft
1-0	0.785	5.87	13-6	143.14	1,070.80
1-6	1.767	13.22	14-0	153.94	1,151.50
2-0	3.142	23.50	14-6	165.13	1,235.30
2-6	4.909	36.72	15-0	176.71	1,321.90
3-0	7.069	52.88	15-6	188.69	1,411.50
3-6	9.621	71.97	16-0	201.06	1,504.10
4-0	12.566	94	16-6	213.82	1,599.50
4-6	15.90	118.97	17-0	226.98	1,697.90
5-0	19.63	146.88	17-6	240.53	1,799.30
5-6	23.76	177.72	18-0	254.47	1,903.60
6-0	28.27	211.51	18-6	268.80	2,010.80
6-6	33.18	248.23	19-0	283.53	2,120.90
7-0	38.48	287.88	19-6	298.65	2,234
7-6	44.18	330.48	20-0	314.16	2,350.10
8-0	50.27	376.01	20-6	330.06	2,469.10
8-6	56.75	424.48	21-0	346.36	2,591
9-0	63.62	475.89	21-6	363.05	2,715.80
9-6	70.88	530.24	22-0	380.13	2,843.60
10-0	78.54	587.52	22-6	397.61	2,974.30
10-6	86.59	670.74	23-0	415.48	3,108
11-0	95.03	710.90			
11-6	103.87	776.99			
12-0	113.10	846.04			
12-6	122.72	918			
13-0	132.73	992.91			

SOURCE: N. Foster, *Practical Tables for Building Construction*. Copyright © 1963 by McGraw-Hill, Inc., New York. (Used with permission of McGraw-Hill Book Co.)

DESCRIPTION:

WHEN YOU WANT TO DETERMINE THE DISTANCE TO WATER IN A WELL, INSTALL A SMALL COPPER OR PLASTIC TUBE OF KNOWN LENGTH, APPROXIMATELY 15 TO 20 FEET BELOW THE PUMPING LEVEL. ALL JOINTS MUST BE AIR TIGHT. CONNECT AIR TUBE TO A PRESSURE GAUGE AND AN AIR PUMP. SEE ILLUSTRATION.

MANUALLY PUMP AIR INTO THE AIR TUBE UNTIL THE REGISTERED AIR PRESSURE REMAINS CONSTANT. THIS WILL INDICATE THAT ALL WATER IN THE AIR TUBE HAS BEEN EXPELLED. GAUGE READING WILL SHOW PRESSURE REQUIRED TO SUPPORT A WATER COLUMN OF A HEIGHT EQUAL TO DEPTH THAT THE AIR TUBE IS SUBMERGED.

A= DEPTH TO WATER IN FEET (REQUIRED)
B= AIR LINE WATER PRESSURE IN FEET OF WATER
(IF GAUGE READS IN POUNDS, MULTIPLY BY 2.31)
C= KNOWN LENGTH OF AIR LINE IN FEET

FORMULA:

A = C - B

EXAMPLE:

KNOWN AIR LINE IS 120 FEET LONG, MEASURED FROM CENTERLINE OF GAUGE. GAUGE READING IS 15 POUNDS.

SOLVE:

C= 120 FEET
B= 15 x 2.31 = 34.65 FEET
A= C - B = 120 - 34.65 = 85.35

THUS:

DEPTH TO WATER = 85.35 FEET

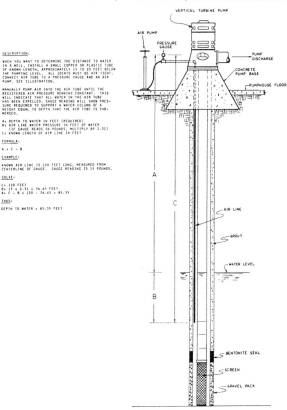

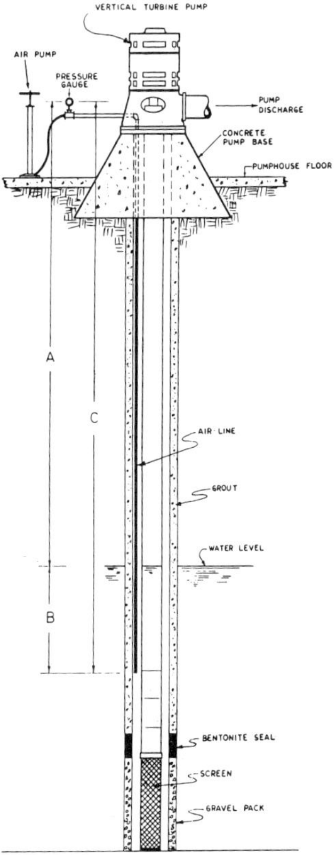

Fig. 16-13 Measurement of well-water depth by air-line method

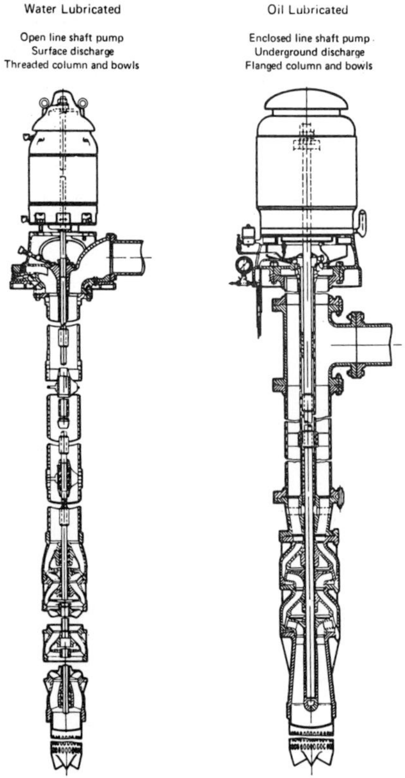

Water Lubricated

Open line shaft pump
Surface discharge
Threaded column and bowls

Oil Lubricated

Enclosed line shaft pump
Underground discharge
Flanged column and bowls

Fig. 16-14 Water-lubricated and oil-lubricated shaft pumps *(Reproduced with permission from NFPA 20, Installation of Centrifugal Fire Pumps, Copyright © 1976, National Fire Protection Association, Boston, MA 02210.)*

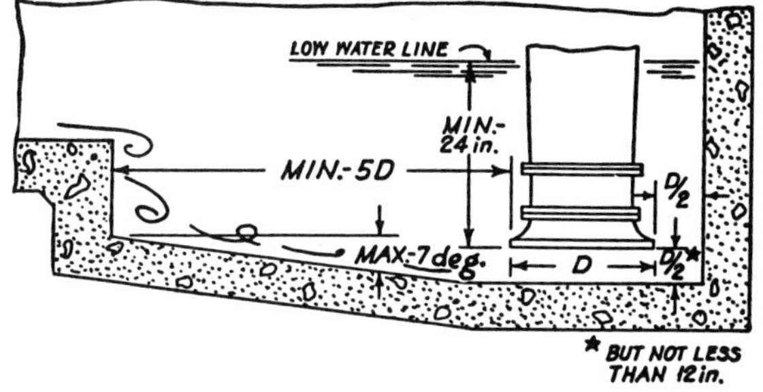

Fig. 16-15 Wet pit sump design

(Reproduced with permission from NFPA 20, Installation of Centrifugal Fire Pumps, Copyright © 1976, National Fire Protection Association, Boston, MA 02210.)

TABLE 16-12 Summary of Pump Data

Pump Rating GPM	(l/min)	Minimum Pipe Sizes (Nominal): Suction In.	(mm)	Discharge In.	(mm)	Relief Valve In.	(mm)	Relief Valve Discharge In.	(mm)	Meter Device In.	(mm)	Number and Size of Hose Valves In.	(mm)	Hose Header Supply In.	(mm)
25	(95)	1	(25)	1	(25)	¾	(19)	1	(25)	1¼	(32)	1 — 1½	(38)	1	(25)
50	(189)	1½	(38)	1¼	(32)	1¼	(32)	1½	(38)	2	(51)	1 — 1½	(38)	1¼	(32)
100	(378)	2	(51)	2	(51)	1½	(38)	2	(51)	2½	(63)	2 — 1½	(38)	2	(38)
150	(568)	2½	(63)	2½	(63)	2	(51)	2½	(63)	3	(76)	1 — 2½	(63)	2½	(63)
200	(757)	3	(76)	3	(76)	2	(51)	2½	(63)	3	(76)	1 — 2½	(63)	2½	(63)
250	(946)	3½	(89)	3	(76)	2	(51)	2½	(63)	3½	(89)	1 — 2½	(63)	3	(76)
300	(1135)	4	(102)	4	(102)	2½	(63)	3½	(89)	3½	(89)	1 — 2½	(63)	3	(76)
400	(1514)	4	(102)	4	(102)	3	(76)	5	(127)	4	(102)	2 — 2½	(63)	4	(102)
450	(1703)	5	(127)	5	(127)	3	(76)	5	(127)	4	(102)	2 — 2½	(63)	4	(102)
500	(1892)	5	(127)	5	(127)	3	(76)	5	(127)	5	(127)	2 — 2½	(63)	4	(102)
750	(2839)	6	(152)	6	(152)	4	(102)	6	(152)	5	(127)	3 — 2½	(63)	6	(152)
1000	(3785)	8	(203)	6	(152)	4	(102)	8	(203)	6	(152)	4 — 2½	(63)	6	(152)
1250	(4731)	8	(203)	8	(203)	6	(152)	8	(203)	6	(152)	6 — 2½	(63)	8	(203)
1500	(5677)	8	(203)	8	(203)	6	(152)	8	(203)	8	(203)	6 — 2½	(63)	8	(203)
2000	(7570)	10	(254)	10	(254)	6	(152)	10	(254)	8	(203)	6 — 2½	(63)	8	(203)
2500	(9462)	10	(254)	10	(254)	6	(152)	10	(254)	8	(203)	8 — 2½	(63)	10	(254)
3000	(11 355)	12	(305)	12	(305)	8	(203)	12	(305)	8	(203)	12 — 2½	(63)	10	(254)
3500	(13 247)	12	(305)	12	(305)	8	(203)	12	(305)	10	(254)	12 — 2½	(63)	12	(305)
4000	(15 140)	14	(356)	12	(305)	8	(203)	14	(356)	10	(254)	16 — 2½	(63)	12	(305)
4500	(17 032)	16	(406)	14	(356)	8	(203)	14	(356)	10	(254)	16 — 2½	(63)	12	(305)

SOURCE: Reproduced with permission from NFPA 20, Installation of Centrifugal Fire Pumps, Copyright © 1976, National Fire Protection Association, Boston, MA 02210.

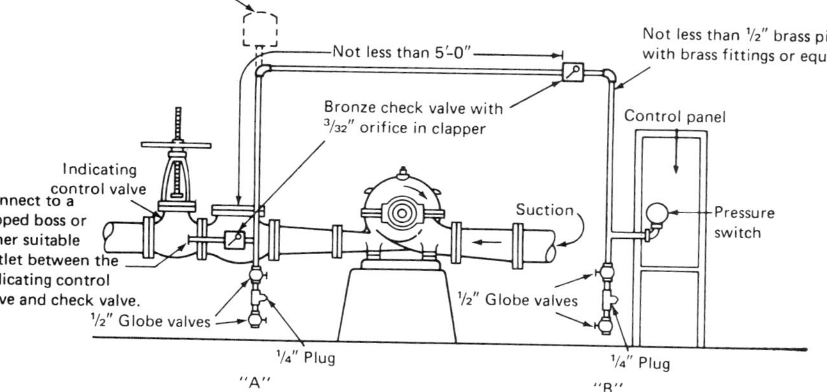

Fig. 16-16 Pipe connection for automatic pressure switch *(Reproduced with permission from NFPA 20, Installation of Centrifugal Fire Pumps, Copyright © 1976, National Fire Protection Association, Boston, MA 02210.)*

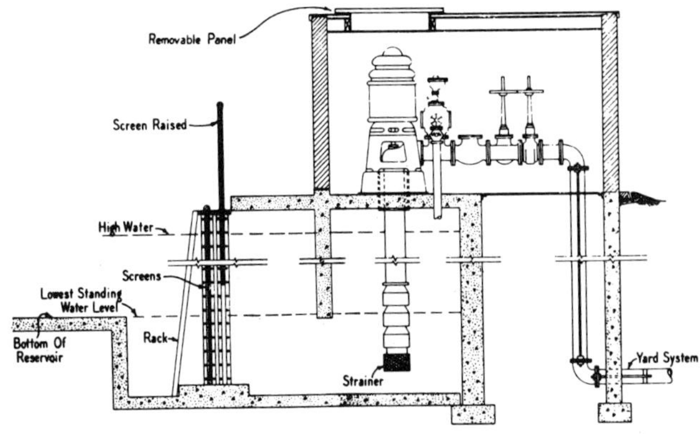

NOTE: The distance between the bottom of the strainer and the bottom of the wet pit should be one-half of the pump bowl diameter but not less than 12 inches (3 dm).

Fig. 16-17 Vertical shaft turbine-type pump installation in a wet pit *(Reproduced with permission from NFPA 20, Installation of Centrifugal Fire Pumps, Copyright © 1976, National Fire Protection Association, Boston, MA 02210.)*

TABLE 16-13 Horsepower, Locked Rotor Current, Motor Designation

Rated Horsepower†	Locked Rotor Current Three-Phase 230 Volts‡ (Amps)	Motor Designation (NEC Code Letter) 230 Volts A to and Including§
5	92	J
7½	127	H
10	162	H
15	232	G
20	290	G
25	365	G
30	435	G
40	580	G
50	725	G
60	870	G
75	1,085	G
100	1,450	G
125	1,815	G
150	2,170	G
200	2,900	G
250	3,650	G
300	4,400	G
350	5,100	G
400	5,800	G
450	6,500	G
500	7,250	G

†For motors of larger horsepower, refer to the manufacturer for locked rotor current.

‡The locked rotor currents for 230 volt motors are approximately six times the full load current. The corresponding values of locked rotor current for motors rated at other voltages shall be determined by multiplication of the values shown by the following factors:

Rated Voltage	**Factor**
200	1.15
208	1.1
460	0.5
575	0.4
Any other voltage	Ratio of 230 volts to the rated voltage

Example: A 15 horsepower, 460 volt motor would have a value of 116 amperes.

§ Code letters of motors rated for all other voltages shall conform with those shown for 230 volts.

SOURCE: Reproduced with permission from NFPA 20, Installation of Centrifugal Fire Pumps, Copyright © 1976, National Fire Protection Association, Boston, MA 02210.

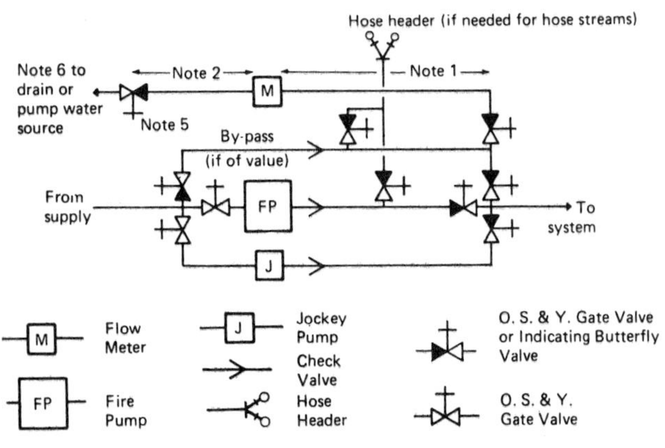

Fig. 16-18 Diagrams for measuring fire pump water flow with meter *(Reproduced with permission from NFPA 20, Installation of Centrifugal Fire Pumps, Copyright © 1976, National Fire Protection Association, Boston, MA 02210.)*

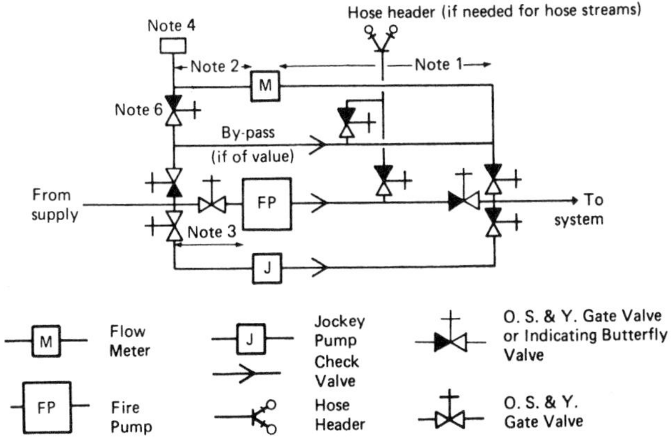

Fig. 16-18 Diagrams for measuring fire pump water flow with meter *(cont.)*

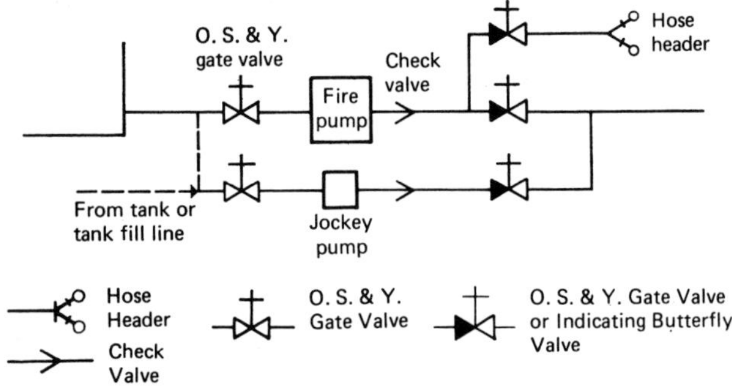

NOTE (a): A jockey pump is usually required with automatically controlled pumps.

NOTE (b): Jockey pump suction may come from the tank filling supply line. This would allow high pressure to be maintained on the fire protection system even when the supply tank may be empty for repairs.

Fig. 16-19 Jockey pump installation with fire pump—Diagram *(Reproduced with permission from NFPA 20, Installation of Centrifugal Fire Pumps, Copyright © 1976, National Fire Protection Association, Boston, MA 02210.)*

Abbreviations and Symbols. The following standard abbreviations and symbols shall be used on the calculation form.

TABLE 16-14 Abbreviations and Symbols for Fire Protection Calculations

Symbol or Abbreviation	*Item*
P	Pressure in psi
gpm	U.S. gallons per minute
q	Flow increment in gpm to be added at a specific location
Q	Summation of flow in gpm at a specific location
P_t	Total pressure in psi at a point in a pipe
P_f	Pressure loss due to friction between points indicated in location column
P_e	Pressure due to elevation difference between indicated points. This can be a plus value or a minus value. Where minus, the (−) shall be used; where plus, no sign need be indicated.
P_v	Velocity pressure in psi at a point in a pipe
P_n	Normal pressure in psi at a point in a pipe
E	90° Ell
EE	45° Ell
Lt.E	Long Turn Elbow
Cr	Cross
T	Tee—flow turned 90°
GV	Gate Valve
BV	Butterfly Valve
DelV	Deluge Valve
DPV	Dry-Pipe Valve
ALV	Alarm Valve
CV	Swing Check Valve
WCV	Butterfly (Wafer) Check Valve
St.	Strainer
psi	Pounds per square inch
v	Velocity of water in pipe in feet per second

SOURCE: Reproduced with permission from NFPA 13, Installation of Sprinkler Systems, Copyright © 1976 National Fire Protection Association, Boston, MA 02210.

TABLE 16-15 Possible Causes of Fire Pump Troubles

	SUCTION															PUMP
Fire Pump Troubles	Suction lift too high.	Foot valve too small, partially obstructed or of inferior design causing excessive suction head loss.	Air drawn into suction connection through foot valve not submerged enough.	Air drawn into suction connection through leak.	Suction connection obstructed.	Air pocket in suction pipe.	Hydraulic cavitation from too high suction lift.	Well collapsed or serious misalignment.	Stuffing box too tight or packing improperly installed, worn, defective, too tight, or incorrect type.	Water-seal or pipe to seal obstructed.	Air leak into pump through stuffing boxes.	Impeller obstructed.	Wearing rings worn.	Impeller damaged.	Wrong diameter impeller.	Actual net head lower than rated.
	1	2	3	4	5	6	7	8	9	10	11	12	13	14	15	16
Excessive Leakage at Stuffing Box									X							
Pump or Driver Overheats								X	X	X		X			X	
Pump Unit Will Not Start								X	X							
No Water Discharge	X			X	X	X						X				
Pump is Noisy or Vibrates	X	X	X				X	X	X			X		X		
Too Much Power Required								X	X			X	X		X	
Disch. Press. Not Constant For Same gpm.	X	X	X	X					X	X	X					
Pump Loses Suction After Starting	X	X	X	X	X	X				X	X					
Insuf-ficient Water Discharge	X	X	X	X	X	X	X			X	X	X	X	X	X	X
Disch. Press. Too Low For gpm. Discharge	X	X	X	X	X	X	X		X	X	X	X	X	X	X	X

SOURCE: Reproduced with permission from NFPA 20, Installation of Centrifugal Fire Pumps, Copyright © 1976, National Fire Protection Association, Boston, MA 02210.

PUMP								DRIVER PUMP and/or					DRIVER						
Casing gasket defective permitting internal leakage. (Multi-stage pumps only.)	Pressure gage on top of pump casing.	Incorrect impeller adjustment. (Vertical pumps only.)	Impellers locked.	Pump is frozen.	Pump shaft or shaft sleeve scored, bent, or worn.	Pump not primed.	Seal ring improperly located in stuffing box, preventing water from entering space to form seal.	Excess bearing friction due to wear, dirt, rusting, failure, or improper installation.	Rotating element binds against stationary element.	Pump and driver misaligned. Shaft running off center because of worn bearings or misalignment.	Foundation not rigid.	Engine cooling system obstructed. Heat exchanger or cooling water system too small. Cooling pump faulty.	Faulty driver.	Lack of lubrication.	Speed too low.	Wrong direction of rotation.	Speed too high.	Rated motor voltage different from line voltage, i.e. 220 or 440 volt motor on 208 or 416 volt line.	Faulty electric circuit, obstructed fuel system or obstructed steam pipe, or dead battery.
17	18	19	20	21	22	23	24	25	26	27	28	29	30	31	32	33	34	35	36
					X					X									
		X			X	X	X	X	X	X	X	X		X		X	X	X	
		X	X	X				X					X	X					X
						X													
					X			X	X	X	X			X					
X		X			X			X	X	X	X			X		X	X	X	
							X												
X		X													X	X		X	
X	X														X	X		X	

TABLE 16-16 Guide to Water Supply Requirements for Pipe Schedule Sprinkler Systems

Occupancy Classification	Residual Pressure Required (See Note 1)	Acceptable Flow at Base of Riser (See Note 2)	Duration in Minutes (See Note 4)
Light Hazard	15 psi	500–750 gpm (See Note 3)	30–60
Ordinary Hazard (Group 1)	15 psi or higher	700–1000 gpm	60–90
Ordinary Hazard (Group 2)	15 psi or higher	850–1500 gpm	60–90
Ordinary Hazard (Group 3)	Pressure and flow requirements for sprinklers and hose streams to be determined by authority having jurisdiction.		60–120
Warehouses	Pressure and flow requirements for sprinklers and hose streams to be determined by authority having jurisdiction. See Chapter 7 and NFPA 231–1974 and NFPA 231C–1975.		
High-Rise Buildings	Pressure and flow requirements for sprinklers and hose streams to be determined by authority having jurisdiction. See Chapter 8.		
Extra Hazard	Pressure and flow requirements for sprinklers and hose streams to be determined by authority having jurisdiction.		

Notes:

1. The pressure required at the base of the sprinkler riser(s) is defined as the residual pressure required at the elevation of the highest sprinkler plus the pressure required to reach this elevation.

2. The lower figure is the minimum flow including hose streams ordinarily acceptable for pipe schedule sprinkler systems. The higher flow should normally suffice for all cases under each group.

3. The requirement may be reduced to 250 gpm if building area is limited by size or compartmentation or if building (including roof) is noncombustible construction.

4. The lower duration figure is ordinarily acceptable where remote station water flow alarm service or equivalent is provided. The higher duration figure should normally suffice for all cases under each group.

SOURCE: Reproduced with permission from NFPA 13, Installation of Sprinkler Systems, Copyright © 1976 National Fire Protection Association, Boston, MA 02210.

TABLE 16-17 Equivalent Pipe Length Chart

Fittings and Valves	*Fittings and Valves Expressed in Equivalent Feet of Pipe.*													
	¾ in.	*1 in.*	*1¼ in.*	*1½ in.*	*2 in.*	*2½ in.*	*3 in.*	*3½ in.*	*4 in.*	*5 in.*	*6 in.*	*8 in.*	*10 in.*	*12 in.*
45° Elbow	1	1	1	2	2	3	3	3	4	5	7	9	11	13
90° Standard Elbow	2	2	3	4	5	6	7	8	10	12	14	18	22	27
90° Long Turn Elbow	1	2	2	2	3	4	5	5	6	8	9	13	16	18
Tee or Cross (Flow Turned 90°)	3	5	6	8	10	12	15	17	20	25	30	35	50	60
Butterfly Valve	–	–	–	–	6	7	10	–	12	9	10	12	19	21
Gate Valve	–	–	–	–	1	1	1	1	2	2	3	4	5	6
Swing Check*	–	5	7	9	11	14	16	19	22	27	32	45	55	65

Use with Hazen and Williams' C = 120 only. For other values of C, the figures in Table 7-4.2 should be multiplied by the factors indicated below:

Value of C	100	120	130	140	150
Multiplying factor	0.713	1.00	1.16	1.33	1.51

(This is based upon the friction loss through the fitting being independent of the C factor applicable to the piping.)

Specific friction loss values or equivalent pipe lengths for alarm valves, dry-pipe valves, deluge valves, strainers and other devices shall be made available to the authority having jurisdiction.

*Due to the variations in design of swing check valves, the pipe equivalents indicated in the above chart to be considered average.

SOURCE: Reproduced with permission from NFPA 13, Installation of Sprinkler Systems, Copyright © 1976 National Fire Protection Association, Boston, MA 02210.

TABLE 16-18 Maximum Number of Sprinklers Supplied on Line

Branch line sizes on pipe schedule systems shall be as follows:

Maximum Number of Sprinklers Supplied on Line

Size of Pipe Inches	Orifice Size — Inches						
	¼	5/16	⅜	7/16	½	⅝	¾
1	4	3	2	2	1	1	1
1¼	8	6	4	3	2	2	1
1½		9	6	4	3	3	2
2				5	4	4	3

Risers and feed main sizes on pipe schedule systems shall be as follows for central feed risers:

Pipe Size	Number of Sprinklers		
	⅜" or smaller orifice	½" orifice	¾" orfiice
1½	6	3	2
2	10	5	4
2½	18	9	7
3	32	16	12
3½	48	24	17
4	65	33	24
5	120	60	43
6		100	70

SOURCE: Reproduced with permission from NFPA 13, Installation of Sprinkler Systems, Copyright © 1976 National Fire Protection Association, Boston, MA 02210.

TABLE 16-19 Summary of Spacing Rules for Sprinkler Heads

Maximum Distance of Deflectors Below Ceiling (Inches)				
TYPE OF CONSTRUCTION	in bays*		under beams	
	comb.	noncomb.	comb.	noncomb.
SMOOTH CEILING	10	12	14	16
BEAM AND GIRDER	16	16	20	20
Panel up to 300 sq. ft.	18	18	22	22
BAR JOISTS	10	12	—	—
OPEN WOOD JOISTS—center 3′ or less (see para. 4-3.4 for centers over 3′)	6	—	—	—

Minimum below ceiling is 1″.
Minimum below beams 1″, maximum 4″. Do not exceed maximum below ceiling.
*Not more than 4″ below beams where lines run across beams.

Maximum Coverage Per Sprinkler:

LIGHT HAZARD — 200 sq. ft. smooth ceiling and beam and girder construction
130 sq. ft. open wood joist
168 sq. ft. all other types of construction

ORDINARY HAZARD — 130 sq. ft. all types of construction except
100 sq. ft. high piled storage (see para. 4-2.2.2)

EXTRA HAZARD — 90 sq. ft. all types of construction

Direction of Lines: Either direction to facilitate hanging except:
Across beams for beams on girders 3′ to 7½′ on centers
Across joists for wood joists (open or sheathed) and bar joists (through or under)

Maximum Spacing Between Lines and Sprinklers:

LIGHT AND ORDINARY HAZARD — 15 ft. except 12 ft. for high piled storage (see para. 4-2.1.2)

EXTRA HAZARD — 12 ft.

SOURCE: Reproduced with permission from NFPA 13, Installation of Sprinkler Systems, Copyright © 1976 National Fire Protection Association, Boston, MA 02210.

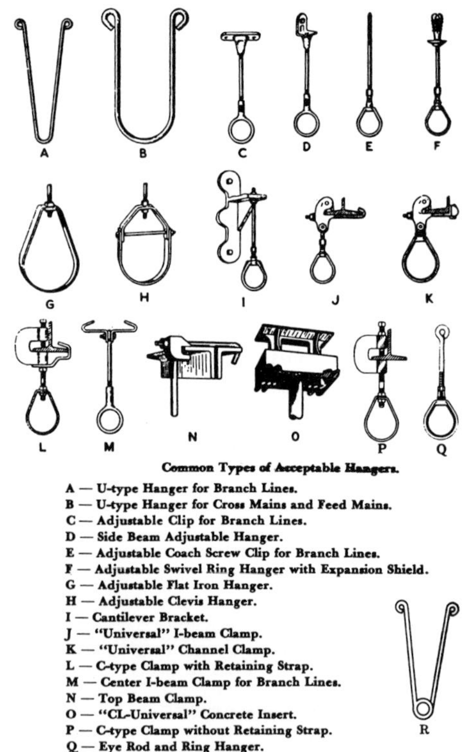

Fig. 16-20 Common types of acceptable hangers for pipe *(Reproduced with permission from NFPA 13, Installation of Sprinkler Systems, Copyright © 1976, National Fire Protection Association, Boston, MA 02210.)*

Item	*Max. Length* *l/r = 200*	*Item*	*Max. Length* *l/r = 200*
ANGLES		FLATS	
1½ x 1½ x ¼ in.	4 ft. 10 in.	1½ x ¼ in.	1 ft. 2 in.
2 x 2 x ¼ in.	6 ft. 6 in.	2 x ¼ in.	1 ft. 2 in.
2½ x 2 x ¼ in.	7 ft. 0 in.	2 x ⅜ in.	1 ft. 9 in.
2½ x 2½ x ¼ in.	8 ft. 2 in.	PIPE	
3 x 2½ x ¼ in.	8 ft. 10 in.	1 in.	7 ft. 0 in.
3 x 3 x ¼ in.	9 ft. 10 in.	1¼ in.	9 ft. 0 in.
RODS		1½ in.	10 ft. 4 in.
¾ in.	3 ft. 1 in.	2 in.	13 ft. 1 in.
⅞ in.	3 ft. 7 in.		

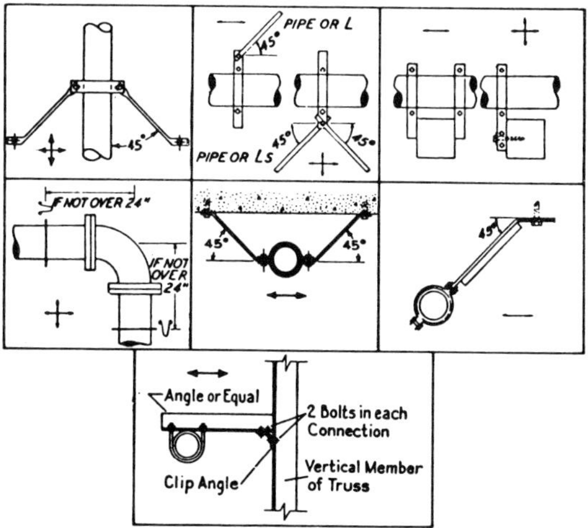

Fig. 16-21 Acceptable types of sway bracing *(Reproduced with permission from NFPA 13, Installation of Sprinkler Systems, Copyright © 1976, National Fire Protection Association, Boston, MA 02210.)*

The following practices shall be observed when installing high-temperature sprinklers, unless maximum expected temperatures are otherwise determined or unless high-temperature sprinklers are used throughout.

(a) Sprinklers near unit heaters. Where steam pressure is not more than 15 pounds per square inch, sprinklers in the heater zone shall be high and sprinklers in the danger zone [shall be] intermediate temperature classification.
(b) Sprinklers located within 12 inches to one side or 30 inches above an uncovered steam main, heating coil, or radiator shall be intermediate temperature classification.
(c) Sprinklers within 7 feet of a low-pressure blowoff valve which discharges free in a large room shall be high-temperature classification.
(d) Sprinklers under glass or plastic skylights exposed to the direct rays of the sun shall be intermediate temperature classification.

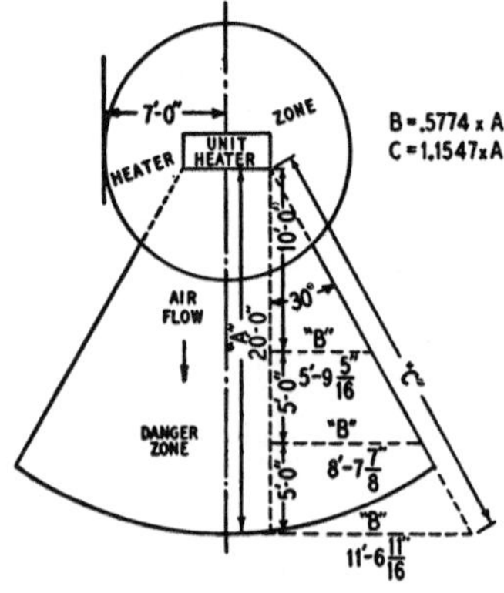

Fig. 16-22 Heat and danger zones at unit heaters *(Reproduced with permission from NFPA 13, Installation of Sprinkler Systems, Copyright © 1976, National Fire Protection Association, Boston, MA 02210.)*

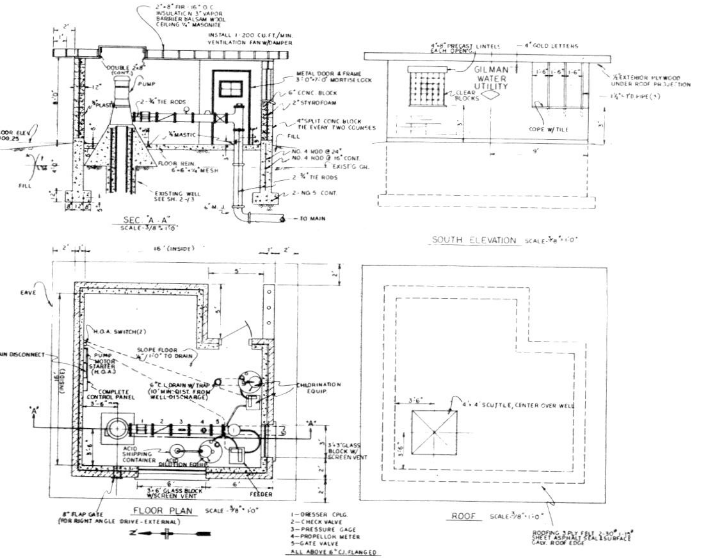

Fig. 16-23 Typical small municipal pump house supplying potable water *(Morgan & Parmley, Ltd.)*

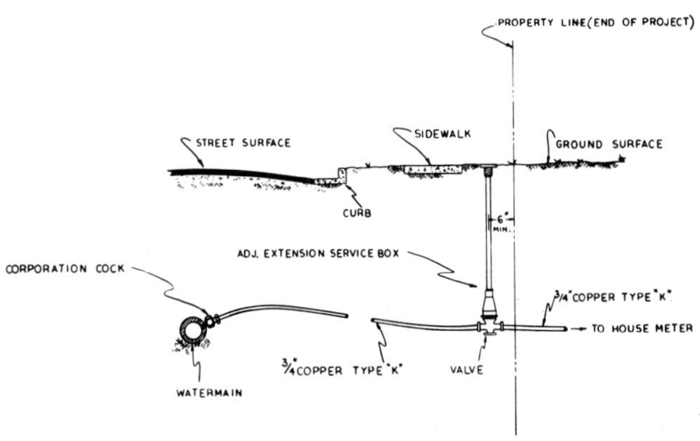

Fig. 16-24 Typical water service detail

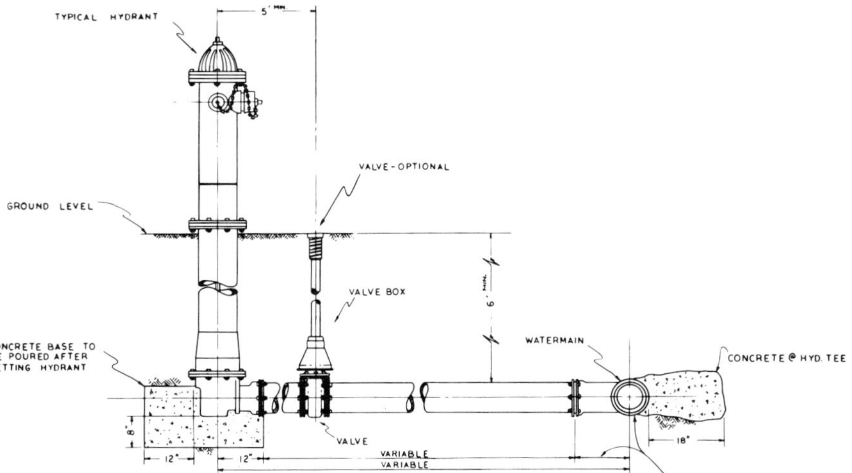

Fig. 16-25 Typical municipal fire hydrant

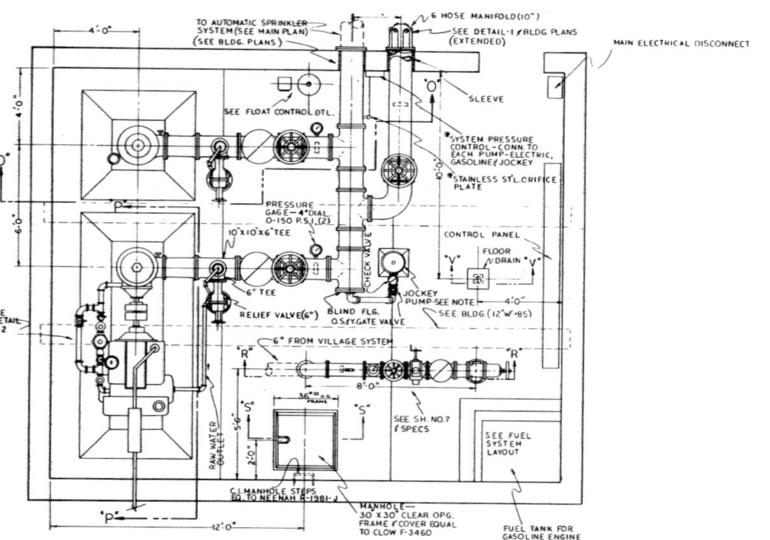

Fig. 16-26 Typical layout of pump house for fire pumps and related equipment *(Morgan & Parmley, Ltd.)*

NOTES

NOTES

Section Seventeen

▪

GEOLOGY

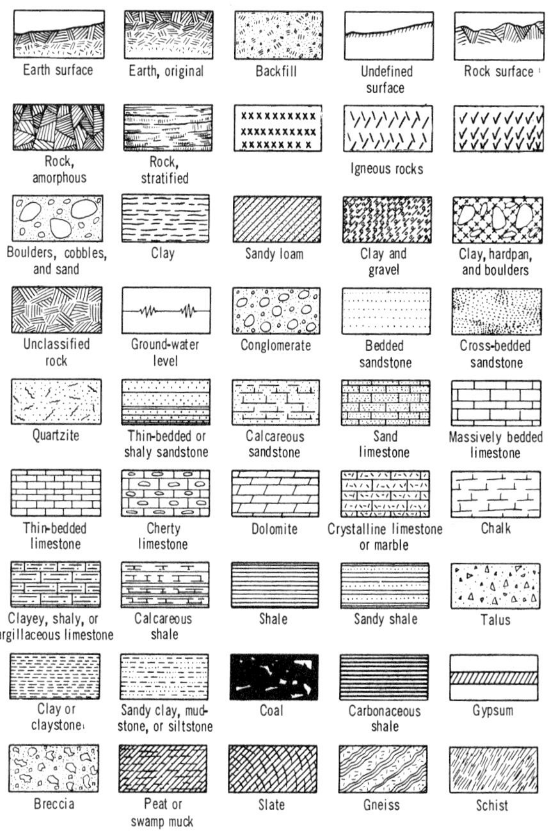

Fig. 17-1 Standard geologic symbols *(Bureau of Mines, Department of the Interior)*.

DESCRIPTION	PROPOSED	COMPLETED
Drill hole (up to and incl 6 inches)	DH-2	DH-2
Drill hole, large-diam (above 6 inches)	DH-2	DH-2
Angle drill hole	DH-2	DH-2
Auger hole (up to and incl 6 in)	AH-2	AH-2
Auger hole, large-diam (above 6 in)	AH-2	AH-2
Test pit	TP-2	TP-2
Test shaft	TS-2	TS-2
Test trench	TT-2	TT-2
Test drift	TD-2	TD-2

When advantageous, indicate special character of drill hole or sampling by these abbreviations in parentheses, following drill-hole number.

DESIGNATIONS	EXAMPLES
DN = Denison samples	DH-2 (DN)
DS = Drive-sample hole	DH-3 (DS)
CX = Calyx-drill hole	DH-4 (CX)
CD = Churn-drill hole	DH-5 (CD)
WB = Wash boring	DH-6 (WB)

Fig. 17-2 Standard symbols for geologic and materials investigations *(Bureau of Mines, Department of the Interior).*

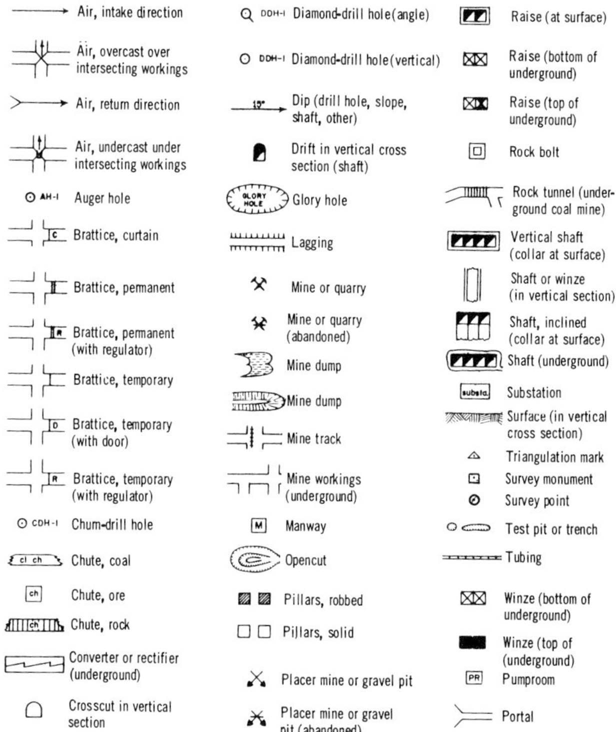

Fig. 17-3 Standard mining symbols *(Bureau of Mines, Department of the Interior).*

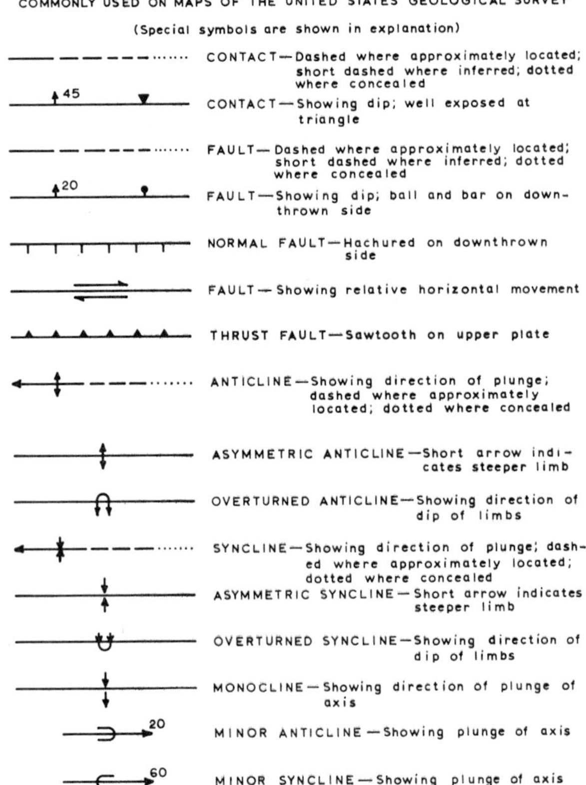
COMMONLY USED ON MAPS OF THE UNITED STATES GEOLOGICAL SURVEY
(Special symbols are shown in explanation)
CONTACT—Dashed where approximately located; short dashed where inferred; dotted where concealed
45
CONTACT—Showing dip; well exposed at triangle
FAULT—Dashed where approximately located; short dashed where inferred; dotted where concealed
20
FAULT—Showing dip; ball and bar on downthrown side
NORMAL FAULT—Hachured on downthrown side
FAULT—Showing relative horizontal movement
THRUST FAULT—Sawtooth on upper plate
ANTICLINE—Showing direction of plunge; dashed where approximately located; dotted where concealed
ASYMMETRIC ANTICLINE—Short arrow indicates steeper limb
OVERTURNED ANTICLINE—Showing direction of dip of limbs
SYNCLINE—Showing direction of plunge; dashed where approximately located; dotted where concealed
ASYMMETRIC SYNCLINE—Short arrow indicates steeper limb
OVERTURNED SYNCLINE—Showing direction of dip of limbs
MONOCLINE—Showing direction of plunge of axis
20
MINOR ANTICLINE—Showing plunge of axis
60
MINOR SYNCLINE—Showing plunge of axis

Fig. 17-4 Geologic map symbols *(USGS)*.

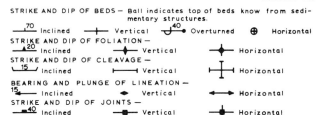

STRIKE AND DIP OF BEDS — Ball indicates top of beds know from sedimentary structures.

70 Inclined   Vertical   40 Overturned   Horizontal

STRIKE AND DIP OF FOLIATION —

20 Inclined   Vertical   Horizontal

STRIKE AND DIP OF CLEAVAGE —

15 Inclined   Vertical   Horizontal

BEARING AND PLUNGE OF LINEATION —

15 Inclined   Vertical   Horizontal

STRIKE AND DIP OF JOINTS —

40 Inclined   Vertical   Horizontal

NOTE: Planar symbols (strike and dip of beds, foliation or schistosity and cleavage) may be combined with linear symbols to record data observed at same locality by superimposed symbols at point of observation. Coexisting planar symbols are shown intersecting at point of observation.

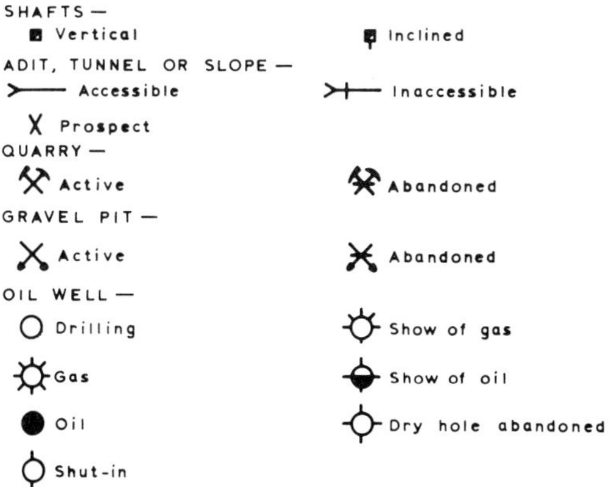

Fig. 17-4 Geologic map symbols *(cont.) (USGS).*

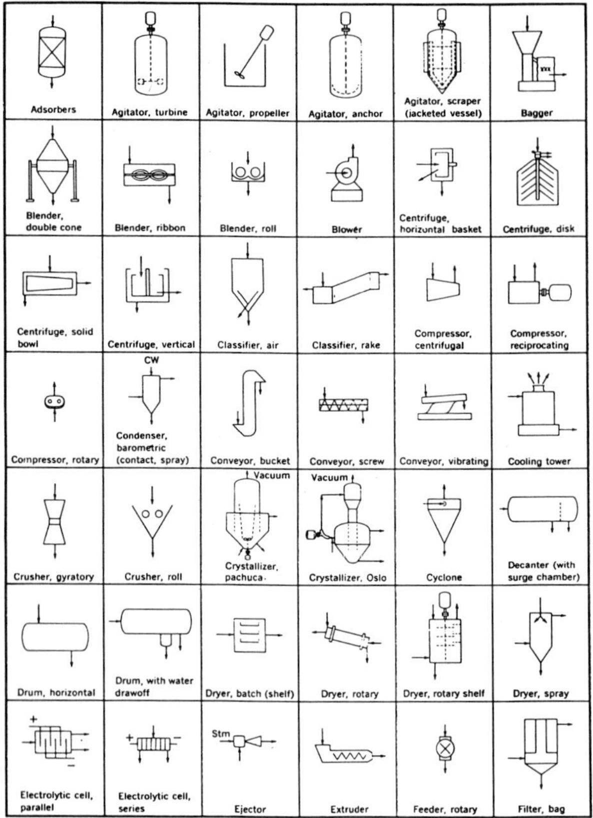

Fig. 17-5 Symbols for equipment commonly used in chemical-processing facilities *(Bureau of Mines, Department of the Interior)*.

Filter, leaf	Filter, oil bath	Filter press	Filter, rotary	Furnace, arc	Furnace, fired
Flaker	Gas holder	Grinder, ball mill	Heat exchanger, air cooled	Heat exchanger, bayonet	Heat exchanger, kettle
Heat exchanger, shell and tube	Heat exchanger, spray	Hopper	Kettle	Kiln	Kneader
Mist eliminator, brink	Mist eliminator, mesh	Mixer-settler	Montejus (blowing egg)	Prill tower	Pug mill
Pump, centrifugal	Pump, gear	Pump, in line	Pump, proportioning	Pump, reciprocating	Pump, submerged
Scale and truck	Scale and hopper	Screen	Sphere	Stack	Tank, cone-roof
Tank, floating roof	Thickener	Tower, disk and donut	Tower, packed	Tower, tray	Turbine

Fig. 17-5 *(cont.)*

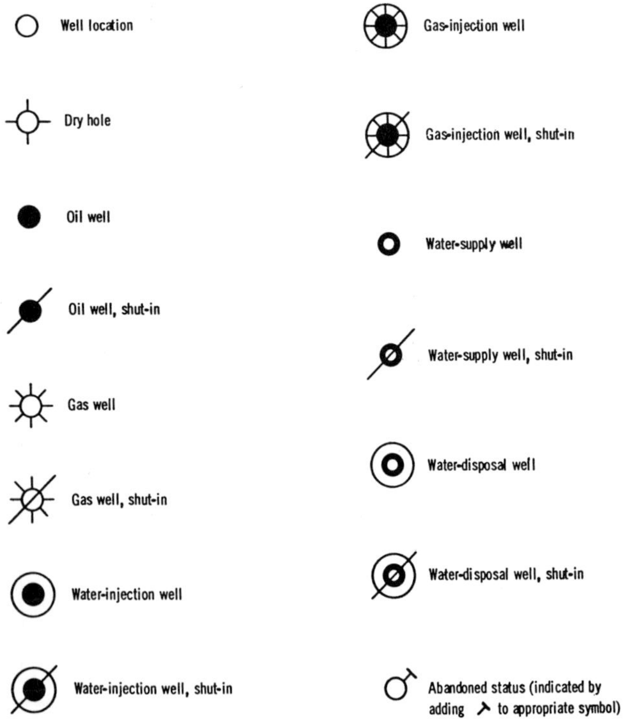

Well location

Dry hole

Oil well

Oil well, shut-in

Gas well

Gas well, shut-in

Water-injection well

Water-injection well, shut-in

Gas-injection well

Gas-injection well, shut-in

Water-supply well

Water-supply well, shut-in

Water-disposal well

Water-disposal well, shut-in

Abandoned status (indicated by adding ⅄ to appropriate symbol)

Note: Indicate abandoned status for shut-in wells by adding a crossbar to the long spoke that passes through the symbol. Thus, (gas well, shut-in) becomes (gas well, shut-in and abandoned).

Fig. 17-6 Standard petroleum symbols *(Bureau of Mines, Department of the Interior).*

NOTES

NOTES

Section Eighteen

■

ENERGY AND FUELS

TABLE 18-1 Fuel Heat Values and Weights of Wood

Type	*Average Wt/cord,* lb*	*Btu/cord†*	*Btu/lb*	*Remarks*
Hickory	3595	30,600,000	8510	Highest heat value
Hard maple	3075	29,000,000	9430	High heat value
Beech	3240	27,800,000	8580	
White oak	3750	27,700,000	7380	
Red oak	3240	26,300,000	8110	
Birch	3000	26,200,000	8730	
Elm	2750	24,500,000	8900	Hard to split
Tamarack	2500	24,010,000	9600	
Soft maple	2500	24,000,000	9600	Acceptable
Cherry	2550	23,500,000	9210	Hard to find
Ash	2950	22,600,000	7660	Marginal heat value
Spruce	2100	18,100,000	8610	
Hemlock	2100	17,910,000	8520	
White pine	1800	17,900,000	9940	
Aspen	1900	17,700,000	9310	
Basswood	1900	17,001,000	8940	

*Cord: Wood cut for fuel or pulpwood (128 ft³) as arranged in a stack 4′ x4′ x8′. "Fireplace" or "face" cords are only about one-third of a true cord.
†Btu: The British thermal unit is the quanity of heat required to raise the temperature of one pound of water one degree Fahrenheit.

TABLE 18-2 Fuel Heat Values for Petrofuels

Type	*Btu/lb*	*Btu/gal or ft³*
No. 2 fuel oil	20,571	144,000/gal
Natural gas	18,000	1030/ft³
Propane	21,564	2572/ft³
Butane	21,440	3200/ft³

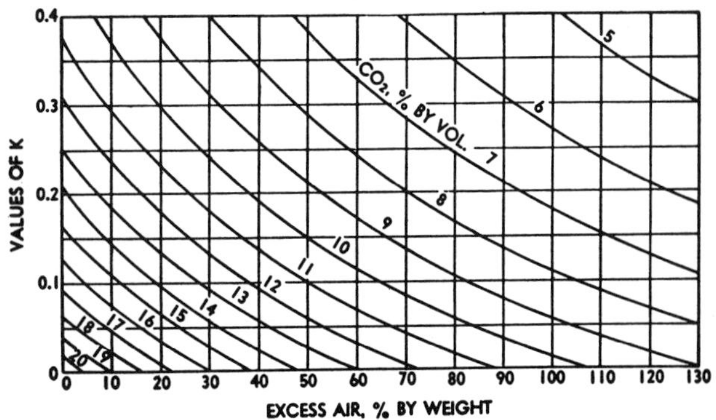

The method of converting excess air to CO_2 or converting CO_2 to excess air, is as follows:

1. When the fuel analysis is known, calculate K in accordance with Equation 1, otherwise obtain an average value from table below.
2. With K, obtain either the excess air or CO_2 from chart on the next page or, when N_2 is high, or K is negative, or when greater accuracy is required, make the conversion in accordance with Equation 2.

Eq. 1: $K = \dfrac{H_2 - 0.125\,O_2}{C + 0.375\,S}$

Eq. 2: $CO_2\,\% = \dfrac{20.86}{3K\left(\frac{x}{100}\right) + 2.374K + \left(1 + \frac{x}{100}\right) + \frac{0.0894\,N_2}{C + 0.375\,S}}$

Where: (Fuel analysis on as fired or dry basis)
- H_2 = Total hydrogen in fuel, % by wt.
- O_2 = Oxygen in fuel, % by wt.
- C = Total carbon in fuel, % by wt.
- S = Sulphur in fuel, % by wt.
- N_2 = Nitrogen in fuel, % by wt.
- CO_2 = Carbon dioxide in flue gas, % by vol.
- X = Excess air in flue gas, % by wt.

Typical Values of K

Fuel	K
Coal, Average	0.056
West Va. (13,500 Btu/lb)	0.055
Kincaid	0.051
New River	0.047
Anthracite	0.019
Colo. Lignite	0.033
Oil, Range:	0.102 to 0.149
Oil, For 18,000 Btu/lb	0.123
Acid Sludge	0.156
Coke-Oven Gas	0.487
Natural Gas	0.331
Blast-Furnace Gas	—0.207
Petroleum Coke	0.051
Coke, Range:	0.004 to 0.008
Wood	0.016
Bagasse	0.008

Fig. 18-1 Excess air chart—CO_2 conversion *(Babcock & Wilcox).*

18-1 THEORETICAL AIR REQUIRED FOR COMBUSTION OF COAL OR OIL WITH COKE RATES AND GAS THERMAL VALUES AT TYPICAL BLAST FURNACES*

$$\text{Theoretical Air, Lb/10,000 Btu as fired} = 144 \times \frac{8.01C + 23.86\left(H_2 - \frac{O_2}{8}\right) + 3S}{\text{Btu/lb}}$$

$$\text{Theoretical Air, Lb/Lb Fuel A.F.} = \frac{11.53C + 34.36\left(H_2 - \frac{O_2}{8}\right) + 4.32S}{100}$$

Where C, H_2, O_2, and S equal, respectively, the per cent by weight of Carbon, Hydrogen, Oxygen and Sulphur from the ultimate analysis; Btu/lb equals the heat value of the fuel as fired.

Coke Rates and Gas Thermal Values at Typical Blast Furnaces

Lb Coke Per Net Ton of Pig	Average Gross Btu/Cu Ft of Dry Gas
1600 to 1700	87 Lean
1701 to 2000	92 Medium
above 2000	98 Rich

*Reproduced by permission of The Babcock & Wilcox Company.

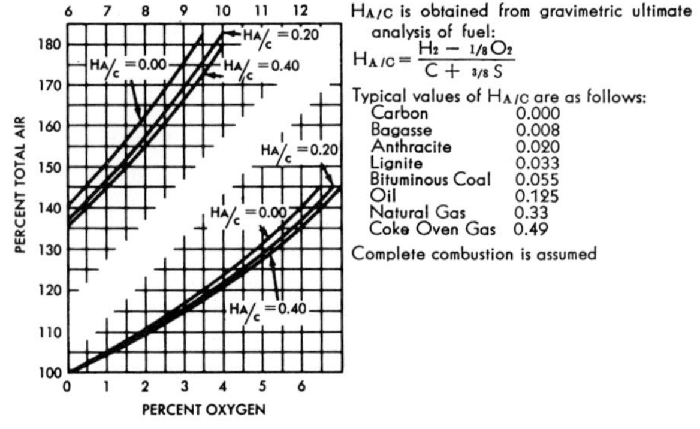

Fig. 18-2 Percent total air from oxygen in flue gas *(Babcock & Wilcox)*.

18-2 SO_2, SO_3 EMISSIONS FROM BOILERS FIRING COAL OR OIL*

SO_2:

$$\text{lb } SO_2 \text{ produced/lb fuel} = K\left(\frac{\text{lb sulfur}}{\text{lb fuel}}\right)\left(\frac{64 \text{ lb } SO_2\text{/mole}}{32 \text{ lb sulfur/mole}}\right)$$

$$\text{lb } SO_2 \text{ produced/hr} = K\left(\frac{\text{lb fuel}}{\text{hr}}\right)\left(\frac{\%\text{ sulfur in fuel} \times 2}{100}\right)$$

$$\%\ SO_2 \text{ by weight} = \frac{\text{lb } SO_2\text{/hr} \times 10^2}{\text{lb flue gas/hr}}$$

$$\text{ppm } SO_2 \text{ by weight} = (\%\ SO_2 \text{ by weight}) \times 10^4$$

$$\%\ SO_2 \text{ by volume} = \frac{100\ (\%\ SO_2 \text{ by weight}/64)}{\dfrac{(\%\ SO_2 \text{ by weight})}{64} + \dfrac{100 - (\%\ SO_2 \text{ by weight})}{M}}$$

$$\text{ppm } SO_2 \text{ by volume} = \%\ SO_2 \text{ by volume} \times 10^4$$

where K = 0.95 cyclone furnace firing
0.97 dry bottom PC firing
0.99 oil firing

M = molecular weight of flue gas
30.2 coal firing
29.0 oil firing

SO_3:

SO_3 quantity = 1% of SO_2 quantity (coal firing)
= 2% of SO_2 quantity (oil firing)

*Reproduced by permission of The Babcock & Wilcox Company.

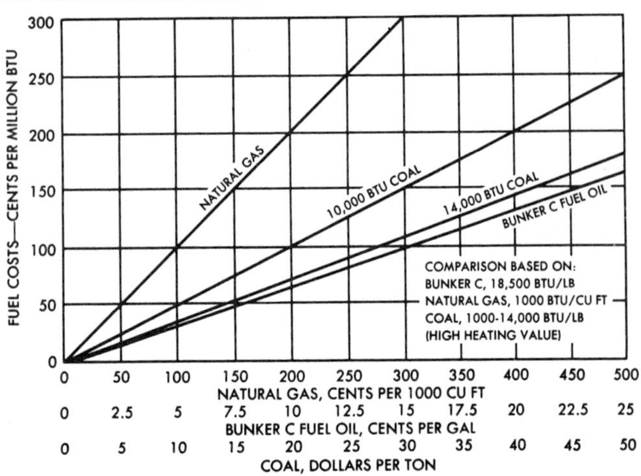

Fig 18-3 Comparative costs, in cents per million Btu, of different fuels *(Babcock & Wilcox)*.

TABLE 18-3 Typical as-Fired Analyses of Various Bituminous Coals

STATE		Alabama		Illinois	
COUNTIES		Jefferson, Tuscaloosa	Walker, Shelby, St. Clair, Jefferson, Bibb		
DISTRICTS OR SEAMS		Blue Creek, Nickel Plate, Jagger	Mary Lee, America, Jagger, Black Creek, Pratt, Clark, Helena, Harkness, Thompson, Gholson, Nunally	Southern Illinois	Northern Illinois
A.S.T.M. Rank		Class II — Group 2	Class II — Group 3	Class II — Group 4	Class II — Group 5
% By Wt Prox Anal	Moisture	3.00	3.50	9.00	12.0
	VM	24.03	33.54	33.86	37.6
	FC	64.97	54.96	49.14	41.4
	Ash	8.00	8.00	8.00	9.0
	Total	100.00	100.00	100.00	100.00
% By Wt Ult Anal	Ash	8.00	8.00	8.00	9.00
	S	0.62	1.06	1.91	3.79
	H_2	4.63	4.87	4.48	4.26
	C	78.50	74.79	67.40	61.78
	Moisture	3.00	3.50	9.00	12.00
	N_2	1.58	1.46	1.31	1.21
	O_2	3.67	6.32	7.90	7.96
	Total	100.00	100.00	100.00	100.00
Heat Value, Btu/lb		13,850	13,350	12,050	11,200
(Fuel-Ash), lb/10,000 Btu		0.664	0.689	0.764	0.812
Theor. Air, lb/10,000 Btu		7.580	7.570	7.560	7.560
Total H_2O, lb/10,000 Btu		0.323	0.355	0.409	0.449

SOURCE: *Useful Tables*, 13th ed., The Babcock & Wilcox Company, New York, 1978.

TABLE 18-3 Typical as-Fired Analyses of Various Bituminous Coals *(cont.)*

STATE		Indiana	Iowa	Kansas		Kentucky
COUNTIES						
DISTRICTS OR SEAMS				Cherokee, Southeastern Kansas	Leavenworth, Northeastern Kansas	Eastern Kentucky
A.S.T.M. Rank		Class II — Group 5	Class II — Group 5	Class II — Group 4	Class II — Group 5	Class II — Group 3
% By Wt Prox Anal	Moisture	10.80	14.69	7.00	12.00	3.50
	VM	37.30	34.52	32.76	35.72	36.14
	FC	41.20	39.81	51.24	40.28	55.36
	Ash	10.70	10.98	9.00	12.00	5.00
	Total	100.00	100.00	100.00	100.00	100.00
% By Wt Ult Anal	Ash	10.70	10.98	9.00	12.00	5.00
	S	4.20	4.03	3.27	4.10	0.92
	H_2	4.35	4.05	4.54	4.09	5.12
	C	61.52	57.48	69.89	60.65	77.13
	Moisture	10.80	14.69	7.00	12.00	3.50
	N_2	1.25	1.15	1.37	1.16	1.49
	O_2	7.18	7.62	4.93	6.00	6.84
	Total	100.00	100.00	100.00	100.00	100.00
Heat Value, Btu/lb		11,300	11,050	12,650	10,950	13,750
(Fuel-Ash), lb/10,000 Btu		0.790	0.806	0.719	0.804	0.691
Theor. Air, lb/10,000 Btu		7.560	7.560	7.570	7.560	7.570
Total H_2O, lb/10,000 Btu		0.440	0.463	0.377	0.447	0.361

TABLE 18-3 Typical as-Fired Analyses of Various Bituminous Coals *(cont.)*

STATE		Kentucky	Maryland			Missouri
COUNTIES			Allegany	Garrett	Garrett	
DISTRICTS OR SEAMS		Western Kentucky	George's Creek	Upper Potomac	Castleman, Youghio-gheny	Barton, Bates
A.S.T.M. Rank		Class II — Group 4	Class II — Group 1	Class II — Group 1	Class II — Group 2	Class II — Group 4
% By Wt Prox Anal	Moisture	7.50	2.5	2.50	3.50	8.0
	VM	36.74	18.8	17.16	22.19	32.0
	FC	46.76	70.7	70.84	64.81	48.0
	Ash	9.00	8.0	9.50	9.50	12.0
	Total	100.00	100.0	100.00	100.00	100.0
% By Wt Ult Anal	Ash	9.00	8.00	9.50	9.50	12.00
	S	2.92	0.98	1.67	2.44	3.76
	H_2	4.50	4.30	4.05	4.35	4.40
	C	67.24	79.57	78.41	76.04	65.60
	Moisture	7.50	2.50	2.50	3.50	8.00
	N_2	1.32	1.65	1.64	1.56	1.30
	O_2	7.52	3.00	2.23	2.61	4.94
	Total	100.00	100.00	100.00	100.00	100.00
Heat Value, Btu/lb		12,100	14,000	13,750	13,500	12,000
(Fuel-Ash), lb/10,000 Btu		0.752	0.657	0.658	0.670	0.733
Theor. Air, lb/10,000 Btu		7.560	7.600	7.610	7.590	7.570
Total H_2O, lb/10,000 Btu		0.397	0.294	0.283	0.316	0.397

TABLE 18-3 Typical as-Fired Analyses of Various Bituminous Coals *(cont.)*

STATE		Missouri	Ohio	Pennsylvania		
COUNTIES				Huntington, Bedford	Indiana, Cambria, Clearfield	Western Penna
DISTRICTS OR SEAMS		Bevier, Lexington		Broadtop		
A.S.T.M. Rank		Class II — Group 5	Class II — Group 4	Class II — Group 1	Class II — Group 2	Class II — Group 3
% By Wt Prox Anal	Moisture	10.00	5.00	2.00	3.00	3.00
	VM	35.65	37.07	16.28	24.92	34.26
	FC	42.35	48.93	71.72	64.08	54.74
	Ash	12.00	9.00	10.00	8.00	8.00
	Total	100.00	100.00	100.00	100.00	100.00
% By Wt Ult Anal	Ash	12.00	9.00	10.00	8.00	8.00
	S	3.90	3.18	1.58	1.96	1.78
	H_2	4.29	4.74	4.13	4.63	4.98
	C	61.23	69.41	79.03	77.87	75.30
	Moisture	10.00	5.00	2.00	3.00	3.00
	N_2	1.21	1.36	1.65	1.57	1.46
	O_2	7.37	7.31	1.61	2.97	5.48
	Total	100.00	100.00	100.00	100.00	100.00
Heat Value, Btu/lb		11,150	12,550	13,800	13,850	13,500
(Fuel-Ash), lb/10,000 Btu		0.789	0.725	0.652	0.664	0.681
Theor. Air, lb/10,000 Btu		7.560	7.560	7.610	7.580	7.570
Total H_2O, lb/10,000 Btu		0.436	0.380	0.284	0.323	0.354

TABLE 18-3 Typical as-Fired Analyses of Various Bituminous Coals (cont.)

STATE		Tennessee		Virginia		
COUNTIES		Campbell, Claiborne, Scott, Anderson, Morgan, Fentress, Overton, White	Marion, Grundy, Sequatchie, Roane, Bledsoe, Hamilton, Cumberland, Rhea	Tazewell, Buchanan	Tazewell, Dickenson, Buchanan, Henrico	Wise, Lee, Scott, Russell
DISTRICTS OR SEAMS				Pocahontas No. 3		
A.S.T.M. Rank		Class II — Group 3	Class II — Group 3	Class II — Group 1	Class II — Group 3	Class II — Group 3
% By Wt. Prox Anal	Moisture	2.50	3.00	4.00	2.50	2.50
	VM	37.46	30.26	18.40	29.95	34.50
	FC	55.04	58.74	73.60	60.55	57.00
	Ash	5.00	8.00	4.00	7.00	6.00
	Total	100.00	100.00	100.00	100.00	100.00
% By Wt Ult Anal	Ash	5.00	8.00	4.00	7.00	6.00
	S	1.85	1.07	0.64	0.91	0.91
	H_2	5.18	4.72	4.42	4.80	4.94
	C	77.15	76.72	83.17	79.10	77.87
	Moisture	2.50	3.00	4.00	2.50	2.50
	N_2	1.28	1.51	1.50	1.54	1.30
	O_2	7.04	4.98	2.27	4.15	6.48
	Total	100.00	100.00	100.00	100.00	100.00
Heat Value, Btu/lb		13,900	13,550	14,550	14,000	13,900
(Fuel-Ash), lb/10,000 Btu		0.683	0.679	0.660	0.664	0.676
Theor. Air, lb/10,000 Btu		7.560	7.570	7.610	7.570	7.570
Total H_2O, lb/10,000 Btu		0.353	0.335	0.300	0.326	0.336

TABLE 18-3 Typical as-Fired Analyses of Various Bituminous Coals *(cont.)*

STATE		Washington	West Virginia			
COUNTIES			McDowell, Raleigh, Fayette, Mercer	Preston, Monongalia	Kanawha, Logan, Mingo	Marion, Monongalia, Harrison, Brooke
DISTRICTS OR SEAMS			Pocahontas No. 3, Sewell, Beckley, Fire Creek	Freeport, Kittaning	Cedargrove, Chilton, Coalberg, No. 2 Gas, Winifrede	Pittsburgh
A.S.T.M. Rank		Class II — Group 4	Class II — Group 1	Class II — Group 2	Class II — Group 3	Class II — Group 3
% By Wt Prox Anal	Moisture	5.16	2.50	3.0	2.00	2.50
	VM	35.03	17.11	26.4	34.04	37.56
	FC	48.74	75.39	61.6	57.96	52.94
	Ash	11.07	5.00	9.0	6.00	7.00
	Total	100.00	100.00	100.0	100.00	100.00
% By Wt Ult Anal	Ash	11.07	5.00	9.00	6.00	7.00
	S	0.61	0.74	1.76	0.92	2.26
	H_2	4.89	4.35	4.66	4.97	4.98
	C	69.13	83.53	76.21	78.20	75.11
	Moisture	5.16	2.50	3.00	2.00	2.50
	N_2	1.33	1.74	1.53	1.52	1.45
	O_2	7.81	2.14	3.84	6.39	6.70
	Total	100.00	100.00	100.00	100.00	100.00
Heat Value, Btu/lb		12,300	14,550	13,550	14,050	13,600
(Fuel-Ash), lb/10,000 Btu		0.723	0.653	0.671	0.669	0.684
Theor. Air, lb/10,000 Btu		7.560	7,620	7.580	7.570	7.560
Total H_2O, lb/10,000 Btu		0.400	0.286	0.333	0.333	0.348

TABLE 18-4 As-Fired Analyses of Typical Coals

Type		Anthracite	Bituminous		Subbituminous		Lignite
ASTM RANK		Class I Group 2	Class II Group 1	Class II Group 3	Class III Group 1	Class III Group 2	Class IV Group 1
Origin	State	Pennsylvania	W. Virginia	Pennsylvania	Wyoming	Colorado	N. Dakota
	County	Luzerne	McDowell	Allegheny	Carbon		Mercer
SEAM		Buck Mt	Pocahontas No. 3 Bed	Pittsburgh			
% By Wt Prox Anal	Moisture	11.0	2.0	3.5	12.0	20.8	37.3
	VM	5.5	19.9	35.7	38.4	30.0	27.1
	FC	70.5	70.4	51.8	43.6	43.8	29.7
	Ash	13.0	7.7	9.0	6.0	5.4	5.9
	Total	100.0	100.0	100.0	100.0	100.0	100.0
% By Wt Ult Anal	Ash	13.0	7.7	9.0	6.0	5.4	5.9
	S	0.5	0.7	2.2	0.7	0.6	0.6
	H_2	1.9	4.3	4.8	4.5	3.2	2.7
	C	70.6	80.9	72.8	62.6	57.6	41.1
	Moisture	11.0	2.0	3.5	12.0	20.8	37.3
	N_2	0.8	1.5	1.5	1.2	1.2	0.4
	O_2	2.2	2.9	6.2	13.0	11.2	12.0
	Total	100.0	100.0	100.0	100.0	100.0	100.0
Heat value, Btu/lb		11,230	14,100	13,080	11,030	9,670	6,940
(Fuel-Ash), lb/10,000 Btu		0.773	0.654	0.696	0.852	0.978	1.355
Theor. Air, lb/10,000 Btu		7.76	7.59	7.55	7.48	7.53	7.46
Total H_2O, lb/10,000 Btu		0.250	0.287	0.358	0.473	0.514	0.889

SOURCE: *Useful Tables*, 13th ed., The Babcock & Wilcox Company, New York, 1978.

18-3 HOW TO DETERMINE ULTIMATE ANALYSIS OF COAL FROM PROXIMATE ANALYSIS*

$$H = V\left(\frac{7.35}{V+10} - 0.013\right)$$

(accurate to ± 0.02% for American coals)

$C_{vol} = 0.9\,(V - 10)$ for semi-anthracite
$= 0.9\,(V - 14)$ for bituminous and semi-bituminous
$= 0.9\,(V - 18)$ for lignites
(accurate to ± 2%)

$N = 0.07V$ for anthracite and semi-anthracite
$= 2.10 - 0.012V$ for bituminous and lignite
(accurate to ± 0.5 per cent)

Where V = Volatile matter, % by weight
H = Hydrogen, % by weight
C_{vol} = Volatile Carbon, % by weight
N = Nitrogen in Combustible, % by weight

*Reproduced by permission of The Babcock & Wilcox Company.
NOTE: Reference material used by Babcock & Wilcox was extracted from *Marks' Standard Handbook for Mechanical Engineers,* 7th ed., T. Baumeister, Ed., McGraw-Hill Book Company, New York, 1967.

TABLE 18-5 Fuel Analyses — Ultimate

(Average—Percent by weight—dry basis)

	BAGASSE	WOOD BARK	
		Pine	Hardwood
C	48.1	55.3	49.7
H_2	6.1	5.5	5.4
S	0.1	0.1	0.1
N_2	0.2	0.3	0.2
O_2	43.1	37.3	39.3
Ash	2.4	1.5	5.3
Total	100.0	100.0	100.0
Heat value, Btu/lb			
Dry, MF	8270	9170	8370
Dry, ash-free MAF	8475	9300	8840
Moisture range	45-50%	40-60%	

SOURCE: *Useful Tables*, 13th ed., The Babcock & Wilcox Company, 1978.

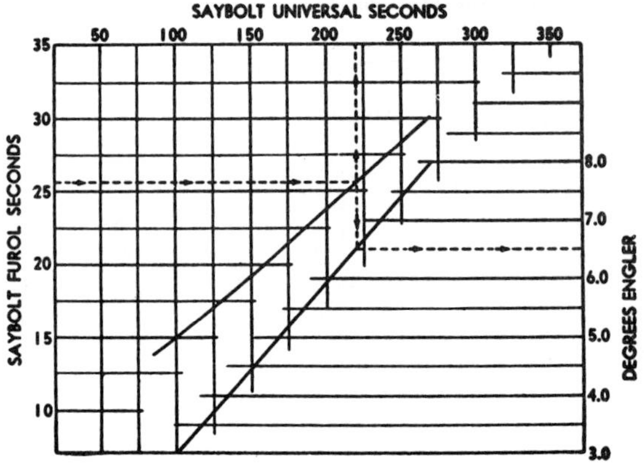

Fig. 18-4 Fuel Oil Viscosity Conversion Chart *(Babcock & Wilcox)*.

18-4 HEAT LOSSES IN STEAM GENERATING UNITS*

(Based on ASME Test Form for Abbreviated Efficiency Test)

Dry refuse per lb of as-fired fuel, lb/lb

$$= \frac{\%\text{ ash in as-fired fuel}}{100 - \%\text{ combustible in refuse sample}}$$

Carbon burned per lb of as-fired fuel, lb/lb

$$= \frac{\%\text{ carbon by weight in fuel sample}}{100} - \left(\frac{\text{dry refuse per lb fuel} \times \text{Btu per lb of refuse}}{14{,}500}\right)$$

Note: If flue dust and ash pit refuse differ materially in combustible content they should be estimated separately.

Dry gas per lb of as-fired fuel burned, lb/lb

$$= \frac{11\,CO_2 + 8\,O_2 + 7\,(N_2 + CO)}{3\,(CO_2 + CO)} \times (\text{lb carbon burned per lb as-fired fuel} + 3/8\text{ S})$$

where: CO_2, O_2 and CO are the percent by volume of carbon dioxide, oxygen and carbon monoxide, respectively in the flue gas; N_2 is the percent by volume of nitrogen, by difference, in the flue gas. S is the pound of sulfur per lb of as-fired fuel from the fuel analysis,or $\frac{\%\text{ sulfur in fuel}}{100}$

1. Heat loss due to dry gas

= lb dry gas per lb as-fired fuel burned $\times$ .24 (tg − ta)
where: .24 = specific heat of gas
tg = temperature of gas leaving unit, F
ta = temperature of air entering unit, F

2. Heat loss due to moisture in fuel

$$= \frac{H_2O}{100} \times (\text{enthalpy of vapor at 1 psia and tg} - \text{enthalpy of liquid at ta})$$

where: H_2O = % moisture in fuel

3. Heat loss due to hydrogen in fuel

$$= \frac{9\,H_2}{100} \times (\text{enthalpy of vapor at 1 psia and tg} - \text{enthalpy of liquid at ta})$$

where: H_2 = % hydrogen in fuel
tg = temperature of gas leaving unit, F
ta = temperature of air entering unit, F

4. Heat loss due to CO in flue gas

$$= \frac{CO}{CO_2 + CO} \times 10{,}160 \times \text{lb carbon burned per lb as-fired fuel}$$

where: CO and CO_2 are per cent by volume of carbon monoxide and carbon dioxide in flue gas
10,160 = Btu generated burning 1 lb of CO to CO_2

5. Heat loss due to unburned combustible

= dry refuse (ash pit + fly-ash) per lb as-fired fuel $\times$ Btu per lb in refuse (weighted average)

Calculations for each of the above five losses will give the Btu per lb for each loss. To determine the per cent loss in efficiency, which is the per cent of heating value of as-fired fuel:

$$\frac{\text{Btu in loss}}{\text{Btu per lb as-fired fuel}} \times 100 = \%\text{ loss}$$

6. Heat loss due to radiation

The per cent loss in efficiency due to radiation may be obtained from the ABMA Standard Radiation Loss Chart on page 67.

7. Unaccounted for losses

These losses include relatively minor losses such as sensible heat in ash or slag, radiation to ash pit, moisture in air, heat pickup in cooling water, etc., generally not measured because the effort is not justifiable. A previously agreed upon amount can be assigned for these losses, if they are not measured.

Unit efficiency as determined by heat loss measurement then becomes the total of the above percentage efficiency losses subtracted from 100%.

*Reproduced by permission of The Babcock & Wilcox Company.

18-5 THERMAL EFFICIENCY OF BOILER*

$$\text{Efficiency} = \frac{W_1(H_1-h_1)+W_2(H_2-h_2)}{C} \times 100\%$$

Where W_1 = Actual initial evaporation in lb per hr.
W_2 = Actual steam reheated (if any) in lb per hr.
C = Gross heating value of fuel burned, Btu per hr.
H_1 = Enthalpy in initial steam in Btu per lb.
H_2 = Enthalpy in reheat steam (if any) in Btu per lb.
h_1 = Enthalpy of feed water in Btu per lb.
h_2 = Enthalpy in steam entering reheater (if any) in Btu per lb.

To calculate the net, or lower heating value from the gross, or higher heating value

$$LHV = HHV - 1040\,\frac{(M + 9H_2)}{100}$$

Where M and H_2 are the percent moisture and hydrogen in the fuel.

NOTE: In the United States the HHV is commonly used in heat balance calculations. In Europe the LHV is most commonly used.

*Reproduced by permission of The Babcock & Wilcox Company.

NOTES

NOTES

Section Nineteen

■

ECONOMICS

TABLE 19-1 Loan Payment Schedule

Years	Monthly Payment Necessary on a $1,000 Loan									
	5%	5½%	6%	6½%	7%	7½%	8%	8½%	9%	9½%
1	$85.61	$85.84	$86.07	$86.30	$86.53	$86.76	$86.99	$87.22	$87.45	$87.68
2	43.87	44.10	44.32	44.55	44.77	45.00	45.22	45.45	45.67	45.90
3	**29.97**	**30.20**	**30.42**	**30.65**	**30.88**	**31.11**	**31.34**	**31.57**	**31.80**	**32.02**
4	23.03	23.26	23.49	23.72	23.95	24.18	24.42	24.65	24.89	25.11
5	18.87	19.10	19.33	19.57	19.80	20.04	20.27	20.50	20.74	20.97
6	**16.11**	**16.34**	**16.57**	**16.81**	**17.05**	**17.29**	**17.52**	**17.76**	**18.00**	**18.23**
7	14.13	14.37	14.61	14.85	15.09	15.34	15.59	15.83	16.08	16.32
8	12.66	12.90	13.14	13.39	13.64	13.89	14.14	14.39	14.63	14.88
9	**11.52**	**11.76**	**12.01**	**12.26**	**12.51**	**12.76**	**13.01**	**13.27**	**13.52**	**13.77**
10	10.61	10.85	11.10	11.36	11.60	11.87	12.13	12.39	12.64	12.90
11	9.86	10.11	10.37	10.62	10.89	11.15	11.41	11.66	11.92	12.18
12	**9.25**	**9.50**	**9.76**	**10.02**	**10.28**	**10.55**	**10.80**	**11.06**	**11.32**	**11.59**

Example: $3100 Loan @ 7% for 10 yrs.

Multiply $\frac{3100}{1000} \times 11.60 = \$35.96$ per month

SOURCE: *Useful Tables,* 13th ed., The Babcock & Wilcox Company, New York, 1978.

TABLE 19-1 Loan Payment Schedule *(cont.)*

	Monthly Payment Necessary on a $1,000 Loan									
Years	5%	5½%	6%	6½%	7%	7½%	8%	8½%	9%	9½%
13	$ 8.73	$ 8.99	$ 9.24	$ 9.51	$ 9.78	$10.05	$10.32	$10.58	$10.85	$11.11
14	8.29	8.55	8.81	9.08	9.35	9.63	9.89	10.16	10.43	10.70
15	**7.91**	**8.17**	**8.44**	**8.71**	**8.99**	**9.27**	**9.56**	**9.84**	**10.10**	**10.38**
16	7.58	7.84	8.11	8.39	8.67	8.96	9.24	9.51	9.78	10.06
17	7.29	7.56	7.83	8.11	8.40	8.69	8.96	9.24	9.51	9.79
18	**7.03**	**7.30**	**7.58**	**7.87**	**8.16**	**8.45**	**8.74**	**9.02**	**9.30**	**9.49**
19	6.80	7.08	7.36	7.65	7.94	8.24	8.50	8.78	9.06	9.35
20	6.60	6.88	7.16	7.46	7.75	8.06	8.34	8.63	8.92	9.22
25	**5.85**	**6.14**	**6.44**	**6.75**	**7.07**	**7.39**	**7.69**	**8.00**	**8.30**	**8.61**
30	5.37	5.68	6.00	6.32	6.65	6.99	7.29	7.60	7.93	8.24
35	5.05	5.37	5.70	6.04	6.39	6.74	7.06	7.40	7.74	8.08
40	**4.82**	**5.16**	**5.50**	**5.85**	**6.21**	**6.58**	**6.90**	**7.25**	**7.60**	**7.96**

TABLE 19-2 Compound Interest Amount of a Given Principal

The amount A at the end of n years of a given principal P placed at compound interest to-day is A = P × x, the interest (at the rate of r percent per annum) is compounded annually, the factor x being taken from the following tables. Values of x.

Years	r = 4	5	6	7	8	9	10	11	12
2	1.082	1.102	1.124	1.145	1.166	1.188	1.210	1.232	1.254
3	1.125	1.158	1.191	1.225	1.260	1.295	1.331	1.368	1.405
4	**1.170**	**1.216**	**1.262**	**1.311**	**1.360**	**1.412**	**1.464**	**1.518**	**1.574**
5	1.217	1.276	1.338	1.403	1.469	1.539	1.611	1.685	1.762
6	1.265	1.340	1.419	1.501	1.587	1.677	1.772	1.870	1.974
7	**1.316**	**1.407**	**1.504**	**1.606**	**1.714**	**1.828**	**1.949**	**2.076**	**2.211**
8	1.369	1.477	1.594	1.718	1.851	1.993	2.144	2.305	2.476
9	1.423	1.551	1.689	1.838	1.999	2.172	2.358	2.558	2.773
10	**1.480**	**1.629**	**1.791**	**1.967**	**2.159**	**2.367**	**2.594**	**2.839**	**3.106**
11	1.539	1.710	1.898	2.105	2.332	2.580	2.853	3.152	3.479
12	1.601	1.796	2.012	2.252	2.518	2.813	3.138	3.498	3.896
13	1.665	1.886	2.133	2.410	2.720	3.066	3.452	3.883	4.363
14	1.732	1.980	2.261	2.579	2.937	3.342	3.797	4.310	4.887
15	**1.801**	**2.079**	**2.397**	**2.759**	**3.172**	**3.642**	**4.177**	**4.785**	**5.474**
16	1.873	2.183	2.540	2.952	3.426	3.970	4.595	5.311	6.130
17	1.948	2.292	2.693	3.159	3.700	4.328	5.054	5.895	6.866
18	**2.026**	**2.407**	**2.854**	**3.380**	**3.996**	**4.717**	**5.560**	**6.543**	**7.690**
19	2.107	2.527	3.026	3.616	4.316	5.142	6.116	7.263	8.613
20	2.191	2.653	3.207	3.870	4.661	5.604	6.727	8.062	9.646
25	**2.666**	**3.386**	**4.292**	**5.427**	**6.848**	**8.623**	**10.834**	**13.585**	**17.000**
30	3.243	4.322	5.743	7.612	10.062	13.267	17.449	22.892	29.960
40	4.801	7.040	10.285	14.974	21.724	31.408	45.258	64.999	93.049

This table is computed from the formula

$$x = [1 + (r/100)]^n$$

SOURCE: *Useful Tables*, 13th ed., The Babcock & Wilcox Company, New York, 1978.

TABLE 19-3 Amount of an Annuity

The amount S accumulated at the end of n years by a given annual payment Y set aside at the end of each year is $S = Y \times v$, where the factor v is to be taken from the following table. (Interest at r percent per annum, compounded annually.) Values of v.

Years	r = 4	5	6	7	8	9	10	11	12
2	2.040	2.050	2.060	2.070	2.080	2.090	2.100	2.110	2.120
3	3.122	3.152	3.184	3.215	3.246	3.278	3.310	3.342	3.374
4	**4.246**	**4.310**	**4.375**	**4.440**	**4.506**	**4.573**	**4.641**	**4.710**	**4.779**
5	5.416	5.526	5.637	5.751	5.867	5.985	6.105	6.228	6.353
6	6.633	6.802	6.975	7.153	7.336	7.523	7.716	7.913	8.115
7	**7.898**	**8.142**	**8.394**	**8.654**	**8.923**	**9.200**	**9.487**	**9.783**	**10.089**
8	9.214	9.549	9.897	10.260	10.637	11.028	11.436	11.859	12.300
9	10.583	11.026	11.491	11.978	12.487	13.021	13.579	14.164	14.776
10	**12.006**	**12.578**	**13.181**	**13.816**	**14.486**	**15.193**	**15.937**	**16.722**	**17.549**
11	13.486	14.206	14.971	15.783	16.645	17.560	18.531	19.561	20.654
12	15.026	15.917	16.870	17.888	18.977	20.140	21.384	22.713	24.133

Formula: $v = [\{1 + (r/100)\}^n - 1] \div (r/100)$
$= (x - 1) \div (r/100)$

SOURCE: *Useful Tables*, 13th ed., The Babcock & Wilcox Company, New York, 1978.

TABLE 19-3 Amount of an Annuity *(cont.)*

The amount S accumulated at the end of n years by a given annual payment Y set aside at the end of each year is $S = Y \times v$, where the factor v is to be taken from the following table. (Interest at r percent per annum, compounded annually.) Values of v.

Years	r = 4	5	6	7	8	9	10	11	12
13	16.627	17.712	18.882	20.140	21.495	22.953	24.522	26.211	28.029
14	18.292	19.598	21.015	22.550	24.215	26.019	27.975	30.095	32.392
15	**20.023**	**21.578**	**23.275**	**25.129**	**27.152**	**29.360**	**31.772**	**34.405**	**37.279**
16	21.824	23.657	25.672	27.887	30.324	33.003	35.949	39.189	42.753
17	23.697	25.840	28.212	30.840	33.750	36.973	40.544	44.500	48.883
18	**25.645**	**28.132**	**30.905**	**33.998**	**37.450**	**41.300**	**45.598**	**50.395**	**55.749**
19	27.671	30.538	33.759	37.378	41.446	46.017	51.158	56.939	63.439
20	29.777	33.065	36.785	40.995	45.761	51.159	57.274	64.202	72.052
25	**41.645**	**47.725**	**54.863**	**63.247**	**73.105**	**84.699**	**98.345**	**114.411**	**133.332**
30	56.083	66.436	79.055	94.458	113.281	136.303	164.489	199.017	241.330
40	95.023	120.794	154.755	199.628	259.050	337.869	442.576	581.810	767.079

Formula: $v = [\{1 + (r/100)\}^n - 1] \div (r/100) = (x - 1) \div (r/100)$

TABLE 19-4 Principal Which Will Amount to a Given Sum

The principal P, which, if placed at compound interest to-day, will amount to a given sum A at the end of n years is P = A × x′, the interest (at the rate of r percent per annum) is compounded annually, the factor x′ being taken from the following table. Values of x′.

Years	r = 4	5	6	7	8	9	10	11	12
2	0.925	0.907	0.890	0.873	0.857	0.842	0.826	0.812	0.797
3	0.889	0.864	0.840	0.816	0.794	0.772	0.751	0.731	0.712
4	**0.855**	**0.823**	**0.792**	**0.763**	**0.735**	**0.708**	**0.683**	**0.659**	**0.636**
5	0.822	0.784	0.747	0.713	0.681	0.650	0.621	0.593	0.567
6	0.790	0.746	0.705	0.666	0.630	0.596	0.564	0.535	0.507
7	**0.760**	**0.711**	**0.665**	**0.623**	**0.583**	**0.547**	**0.513**	**0.482**	**0.452**
8	0.731	0.677	0.627	0.582	0.540	0.502	0.467	0.434	0.404
9	0.703	0.645	0.592	0.544	0.500	0.460	0.424	0.391	0.361
10	**0.676**	**0.614**	**0.558**	**0.508**	**0.463**	**0.422**	**0.386**	**0.352**	**0.322**
11	0.650	0.585	0.527	0.475	0.429	0.388	0.350	0.317	0.287
12	0.625	0.557	0.497	0.444	0.397	0.356	0.319	0.286	0.257

Formula: $x' = [1+(r/100)]^{-n} = 1/x$

SOURCE: *Useful Tables*, 13th ed., The Babcock & Wilcox Company, New York, 1978.

TABLE 19-4 Principal Which Will Amount to a Given Sum *(cont.)*

The principal P, which, if placed at compound interest to-day, will amount to a given sum A at the end of n years is P = A × x′, the interest (at the rate of r percent per annum) is compounded annually, the factor x′ being taken from the following table. Values of x′.

Years	r = 4	5	6	7	8	9	10	11	12
13	0.601	0.530	0.469	0.415	0.368	0.326	0.290	0.258	0.229
14	0.577	0.505	0.442	0.388	0.340	0.299	0.263	0.232	0.205
15	**0.555**	**0.481**	**0.417**	**0.362**	**0.315**	**0.275**	**0.239**	**0.209**	**0.183**
16	0.534	0.458	0.394	0.339	0.292	0.252	0.218	0.188	0.163
17	0.513	0.436	0.371	0.317	0.270	0.231	0.198	0.170	0.146
18	**0.494**	**0.416**	**0.350**	**0.296**	**0.250**	**0.212**	**0.180**	**0.153**	**0.130**
19	0.475	0.396	0.331	0.277	0.232	0.194	0.164	0.138	0.116
20	0.456	0.377	0.312	0.258	0.215	0.178	0.149	0.124	0.104
25	**0.375**	**0.295**	**0.233**	**0.184**	**0.146**	**0.116**	**0.092**	**0.074**	**0.059**
30	0.308	0.231	0.174	0.131	0.099	0.075	0.057	0.044	0.033
40	0.208	0.142	0.097	0.067	0.046	0.032	0.022	0.015	0.011

Formula: $x' = [1+(r/100)]^{-n} = 1/x$

TABLE 19-5 Annuity Which Will Amount to a Given Sum (Sinking Fund)

The annual payment, Y, which, if set aside at the end of each year, will amount with accumulated interest to a given sum S at the end of n years is $Y = S \times v'$, where the factor v' is given below. (Interest at r percent per annum, compounded annually.) Values of v'.

Years	r = 4	5	6	7	8	9	10	11	12
2	0.490	0.488	0.485	0.483	0.481	0.478	0.476	0.474	0.472
3	0.320	0.317	0.314	0.311	0.308	0.305	0.302	0.299	0.296
4	**0.235**	**0.232**	**0.229**	**0.225**	**0.222**	**0.219**	**0.215**	**0.212**	**0.209**
5	0.185	0.181	0.177	0.174	0.170	0.167	0.164	0.161	0.157
6	0.151	0.147	0.143	0.140	0.136	0.133	0.130	0.126	0.123
7	**0.127**	**0.123**	**0.119**	**0.116**	**0.112**	**0.109**	**0.105**	**0.102**	**0.099**
8	0.109	0.105	0.101	0.097	0.094	0.091	0.087	0.084	0.081
9	0.094	0.091	0.087	0.083	0.080	0.077	0.074	0.071	0.068
10	**0.083**	**0.080**	**0.076**	**0.072**	**0.069**	**0.066**	**0.063**	**0.060**	**0.057**
11	0.074	0.070	0.067	0.063	0.060	0.057	0.054	0.051	0.048
12	0.067	0.063	0.059	0.056	0.053	0.050	0.047	0.044	0.041

Formula:
$v' = (r/100) + [\{1 + (r/100)\}^n - 1] = 1/v$

SOURCE: *Useful Tables*, 13th ed., The Babcock & Wilcox Company, New York, 1978.

TABLE 19-5 Annuity Which Will Amount to a Given Sum (Sinking Fund) *(cont.)*

The annual payment, Y, which, if set aside at the end of each year, will amount with accumulated interest to a given sum S at the end of n years is $Y = S \times v'$, where the factor v' is given below. (Interest at r percent per annum, compounded annually.) Values of v'.

Years	r = 4	5	6	7	8	9	10	11	12
13	0.060	0.056	0.053	0.050	0.047	0.044	0.041	0.038	0.036
14	0.055	0.051	0.048	0.044	0.041	0.038	0.036	0.033	0.031
15	**0.050**	**0.046**	**0.043**	**0.040**	**0.037**	**0.034**	**0.031**	**0.029**	**0.027**
16	0.046	0.042	0.039	0.036	0.033	0.030	0.028	0.026	0.023
17	0.042	0.039	0.035	0.032	0.030	0.027	0.025	0.022	0.020
18	**0.039**	**0.036**	**0.032**	**0.029**	**0.027**	**0.024**	**0.022**	**0.020**	**0.018**
19	0.036	0.033	0.030	0.027	0.024	0.022	0.020	0.018	0.016
20	0.034	0.030	0.027	0.024	0.022	0.020	0.017	0.016	0.014
25	**0.024**	**0.021**	**0.018**	**0.016**	**0.014**	**0.012**	**0.010**	**0.009**	**0.008**
30	0.018	0.015	0.013	0.011	0.009	0.007	0.006	0.005	0.004
40	0.011	0.008	0.006	0.005	0.004	0.003	0.002	0.002	0.001

Formula: $v' = (r/100) \div [\{1+(r/100)\}^n - 1] = 1/v$

TABLE 19-6 Present Worth of an Annuity

The Capital C, which, if placed at interest to-day, will provide for a given annual payment Y for a term of n years before it is exhausted is C=Y×w, where the factor w is given below. (Interest at r percent per annum, compounded annually.) Values of w.

Years	r = 4	5	6	7	8	9	10	11	12
2	1.886	1.859	1.833	1.808	1.783	1.759	1.736	1.713	1.690
3	2.775	2.723	2.673	2.624	2.577	2.531	2.487	2.444	2.402
4	**3.630**	**3.546**	**3.465**	**3.387**	**3.312**	**3.240**	**3.170**	**3.102**	**3.037**
5	4.452	4.329	4.212	4.100	3.993	3.890	3.791	3.696	3.605
6	5.242	5.076	4.917	4.766	4.623	4.486	4.355	4.231	4.111
7	**6.002**	**5.786**	**5.582**	**5.389**	**5.206**	**5.033**	**4.868**	**4.712**	**4.564**
8	6.733	6.463	6.210	5.971	5.747	5.535	5.335	5.146	4.968
9	7.435	7.108	6.802	6.515	6.247	5.995	5.759	5.537	5.328
10	**8.111**	**7.722**	**7.360**	**7.024**	**6.710**	**6.418**	**6.145**	**5.889**	**5.650**
11	8.760	8.306	7.887	7.499	7.139	6.805	6.495	6.206	5.938
12	9.385	8.863	8.384	7.943	7.536	7.161	6.814	6.492	6.194

Formula: $w=[1-\{1+(r/100)\}^{-n}]\div[r/100]=v/x$

SOURCE: *Useful Tables*, 13th ed., The Babcock & Wilcox Company, New York, 1978.

TABLE 19-6 Present Worth of an Annuity *(cont.)*

The Capital C, which, if placed at interest to-day, will provide for a given annual payment Y for a term of n years before it is exhausted is $C = Y \times w$, where the factor w is given below. (Interest at r percent per annum, compounded annually.) Values of w.

Years	r = 4	5	6	7	8	9	10	11	12
13	9.986	9.393	8.853	8.358	7.904	7.487	7.103	6.750	6.424
14	10.563	9.899	9.295	8.745	8.244	7.786	7.367	6.982	6.628
15	**11.118**	**10.380**	**9.712**	**9.108**	**8.559**	**8.061**	**7.606**	**7.191**	**6.811**
16	11.652	10.838	10.106	9.447	8.851	8.313	7.824	7.379	6.974
17	12.166	11.274	10.477	9.763	9.122	8.544	8.022	7.549	7.120
18	**12.659**	**11.689**	**10.828**	**10.059**	**9.372**	**8.756**	**8.201**	**7.702**	**7.250**
19	13.134	12.085	11.158	10.336	9.604	8.950	8.365	7.839	7.366
20	13.590	12.462	11.470	10.594	9.818	9.129	8.514	7.963	7.469
25	**15.622**	**14.094**	**12.783**	**11.654**	**10.675**	**9.823**	**9.077**	**8.422**	**7.843**
30	17.292	15.372	13.765	12.409	11.258	10.274	9.427	8.694	8.055
40	19.793	17.159	15.046	13.332	11.925	10.757	9.779	8.951	8.244

Formula: $w = [1 - \{1 + (r/100)\}^{-n}] \div [r/100] = v/x$

TABLE 19-7 Annuity Provided for by a Given Capital

The annual payment Y provided for a term of n years by a given capital C placed at interest to-day is $Y = C \times w'$. (Interest at r percent per annum, compounded annually; the fund supposed to be exhausted at the end of the term.) Values of w'.

Years	r = 4	5	6	7	8	9	10	11	12
2	0.530	0.538	0.545	0.553	0.561	0.568	0.576	0.584	0.592
3	0.360	0.367	0.374	0.381	0.388	0.395	0.402	0.409	0.416
4	**0.275**	**0.282**	**0.289**	**0.295**	**0.302**	**0.309**	**0.315**	**0.322**	**0.329**
5	0.225	0.231	0.237	0.244	0.250	0.257	0.264	0.271	0.277
6	0.191	0.197	0.203	0.210	0.216	0.223	0.230	0.236	0.243
7	**0.167**	**0.173**	**0.179**	**0.186**	**0.192**	**0.199**	**0.205**	**0.212**	**0.219**
8	0.149	0.155	0.161	0.167	0.174	0.181	0.187	0.194	0.201
9	0.134	0.141	0.147	0.153	0.160	0.167	0.174	0.181	0.188
10	**0.123**	**0.130**	**0.136**	**0.142**	**0.149**	**0.156**	**0.163**	**0.170**	**0.177**
11	0.114	0.120	0.127	0.133	0.140	0.147	0.154	0.161	0.168
12	0.107	0.113	0.119	0.126	0.133	0.140	0.147	0.154	0.161

Formula: $w' = [r/100] \div [1 - \{1 + (r/100)\}^{-n}] = 1/w = v' + (r/100)$

SOURCE: *Useful Tables*, 13th ed., The Babcock & Wilcox Company, New York, 1978.

TABLE 19-7 Annuity Provided for by a Given Capital *(cont.)*

The annual payment Y provided for a term of n years by a given capital C placed at interest to-day is $Y = C \times w'$. (Interest at r percent per annum, compounded annually; the fund supposed to be exhausted at the end of the term.) Values of w'.

Years	r = 4	5	6	7	8	9	10	11	12
13	0.100	0.106	0.113	0.120	0.127	0.134	0.141	0.148	0.156
14	0.095	0.101	0.108	0.114	0.121	0.128	0.136	0.143	0.151
15	**0.090**	**0.096**	**0.103**	**0.110**	**0.117**	**0.124**	**0.131**	**0.139**	**0.147**
16	0.086	0.092	0.099	0.106	0.113	0.120	0.128	0.136	0.143
17	0.082	0.089	0.095	0.102	0.110	0.117	0.125	0.132	0.140
18	**0.079**	**0.086**	**0.092**	**0.099**	**0.107**	**0.114**	**0.122**	**0.130**	**0.138**
19	0.076	0.083	0.090	0.097	0.104	0.112	0.120	0.128	0.136
20	0.074	0.080	0.087	0.094	0.102	0.110	0.117	0.126	0.134
25	**0.064**	**0.071**	**0.078**	**0.086**	**0.094**	**0.102**	**0.110**	**0.119**	**0.127**
30	0.058	0.065	0.073	0.081	0.089	0.097	0.106	0.115	0.124
40	0.051	0.058	0.066	0.075	0.084	0.093	0.102	0.112	0.121

Formula: $w' = [r/100] \div [1 - \{1 + (r/100)\}^{-n}] = 1/w = v' + (r/100)$

NOTES

Section Twenty

■

FIRST AID

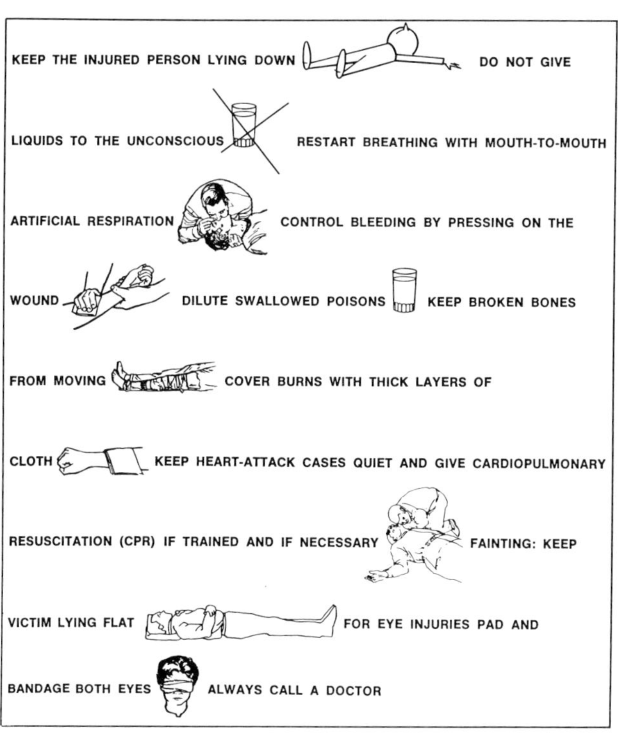

KEEP THE INJURED PERSON LYING DOWN
DO NOT GIVE
LIQUIDS TO THE UNCONSCIOUS
RESTART BREATHING WITH MOUTH-TO-MOUTH
ARTIFICIAL RESPIRATION
CONTROL BLEEDING BY PRESSING ON THE
WOUND
DILUTE SWALLOWED POISONS
KEEP BROKEN BONES
FROM MOVING
COVER BURNS WITH THICK LAYERS OF
CLOTH
KEEP HEART-ATTACK CASES QUIET AND GIVE CARDIOPULMONARY
RESUSCITATION (CPR) IF TRAINED AND IF NECESSARY
FAINTING: KEEP
VICTIM LYING FLAT
FOR EYE INJURIES PAD AND
BANDAGE BOTH EYES
ALWAYS CALL A DOCTOR

Fig. 20-1 General directions for giving first aid *(American Red Cross Poster).*

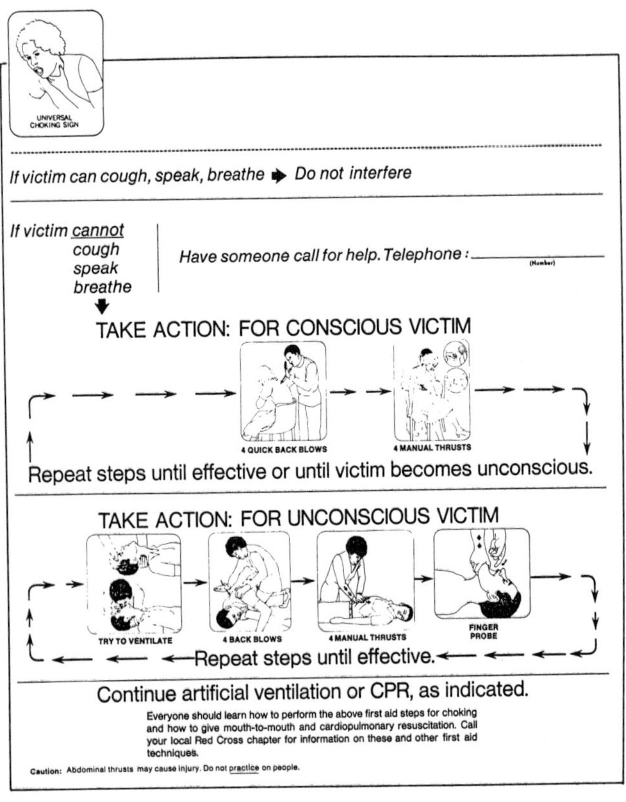

Fig. 20-2 First aid for choking *(American Red Cross Poster)*.

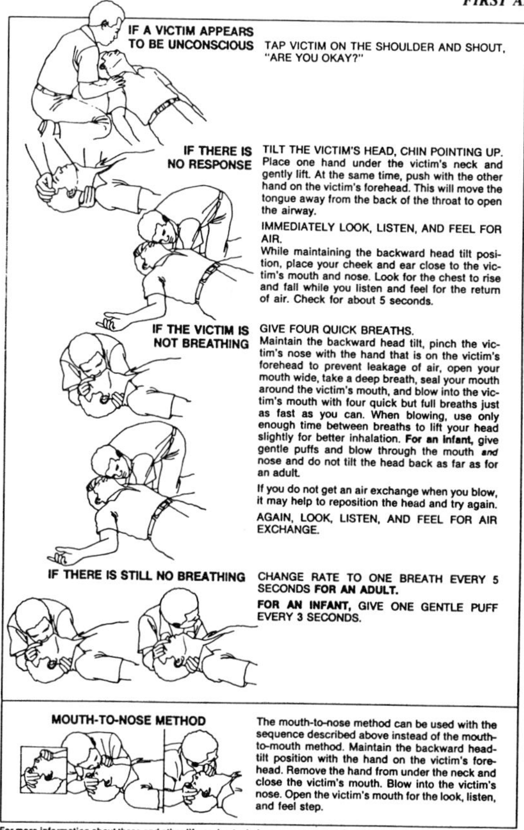

IF A VICTIM APPEARS TO BE UNCONSCIOUS
TAP VICTIM ON THE SHOULDER AND SHOUT, "ARE YOU OKAY?"
IF THERE IS NO RESPONSE
TILT THE VICTIM'S HEAD, CHIN POINTING UP.
Place one hand under the victim's neck and gently lift. At the same time, push with the other hand on the victim's forehead. This will move the tongue away from the back of the throat to open the airway.
IMMEDIATELY LOOK, LISTEN, AND FEEL FOR AIR.
While maintaining the backward head tilt position, place your cheek and ear close to the victim's mouth and nose. Look for the chest to rise and fall while you listen and feel for the return of air. Check for about 5 seconds.
IF THE VICTIM IS NOT BREATHING
GIVE FOUR QUICK BREATHS.
Maintain the backward head tilt, pinch the victim's nose with the hand that is on the victim's forehead to prevent leakage of air, open your mouth wide, take a deep breath, seal your mouth around the victim's mouth, and blow into the victim's mouth with four quick but full breaths just as fast as you can. When blowing, use only enough time between breaths to lift your head slightly for better inhalation. For an infant, give gentle puffs and blow through the mouth and nose and do not tilt the head back as far as for an adult.
If you do not get an air exchange when you blow, it may help to reposition the head and try again.
AGAIN, LOOK, LISTEN, AND FEEL FOR AIR EXCHANGE.
IF THERE IS STILL NO BREATHING
CHANGE RATE TO ONE BREATH EVERY 5 SECONDS FOR AN ADULT.
FOR AN INFANT, GIVE ONE GENTLE PUFF EVERY 3 SECONDS.
MOUTH-TO-NOSE METHOD
The mouth-to-nose method can be used with the sequence described above instead of the mouth-to-mouth method. Maintain the backward head-tilt position with the hand on the victim's forehead. Remove the hand from under the neck and close the victim's mouth. Blow into the victim's nose. Open the victim's mouth for the look, listen, and feel step.
For more information about these and other life-saving techniques, contact your Red Cross chapter for training.

Fig. 20-3 Artificial Respiration *(American Red Cross Poster).*

POISONOUS OR NONPOISONOUS

Poisonous or nonpoisonous, a snakebite should have medical attention. A snakebite victim should be taken to a hospital *as quickly as possible*, even in cases when snakebite is only suspected.

FIRST AID

1. As stated above, *get the victim to a hospital fast.* Meanwhile, take the following general first aid measures:
   - Keep the victim from moving around.
   - Keep the victim as calm as possible, preferably lying down.
   - Immobilize the bitten extremity and keep it at or below heart level.

   If a hospital can be reached within 4 to 5 hours and no symptoms develop, this is all that is necessary.

2. *If mild to moderate symptoms develop, apply a constricting band* from 2 to 4 inches above the bite but NOT around a joint (i.e., elbow, knee, wrist, or ankle) and NOT around the head, neck, or trunk. The band should be from ¾ to 1½ inches wide, NOT thin like a rubber band. The band should be snug, but loose enough to slip one finger underneath. Be alert to swelling; loosen the band if it becomes too tight, but do not remove it. To ensure that blood flow has not been stopped, periodically check the pulse in the extremity beyond the bite.

3. *If severe symptoms develop, incisions and suction should be performed immediately.* Apply a constricting band, if not already done, and make a cut in the skin with a sharp sterilized blade through the fang mark(s). Cuts should be no deeper than just through the skin and should be ½ inch long, extending over the suspected venom deposit point (because a snake strikes downward, the deposit point is usually lower than the fang mark). Cuts should be made along the long axis of the limb. DO NOT make cross-cut incisions; DO NOT make cuts on the head, neck, or trunk. Suction should be applied with a suction cup for 30 minutes. If a suction cup is not available, use the mouth. There is little risk to the rescuer who uses his mouth, but it is recommended that the venom not be swallowed and that the mouth be rinsed.

IF THE HOSPITAL IS NOT CLOSE *(cannot be reached within from 4 to 5 hours)*

1. Continue to try to obtain professional care by transportation of the victim or by communication with a rescue service.
2. *If no symptoms develop*, continue trying to reach the hospital and give the general first aid described above.
3. *If ANY symptoms develop*, apply a constricting band and perform incisions and suction immediately, as described above.

OTHER CONSIDERATIONS

1. *Shock:* Keep the victim lying down and comfortable and maintain body temperature.
2. *Breathing and heartbeat:* If breathing stops, give mouth-to-mouth resuscitation. If breathing stops and there is no pulse, cardiopulmonary resuscitation (CPR) should be performed by those trained to do so.
3. *Identifying the snake:* If the snake can be killed without risk or delay, it should be brought, *with care*, to the hospital for identification.
4. *Cleansing the bitten area:* The bitten area may be washed with soap and water and blotted dry with sterile gauze. Dressings and bandages can be applied, but only for a short period of time.
5. *Cold therapy:* Cold compresses, ice, dry ice, chemical ice packs, spray refrigerants, and other methods of cold therapy are NOT recommended in the first aid treatment of snakebite.
6. *Medicine to relieve pain:* A medicine *not containing aspirin* can be given to the victim for relief of pain. DO NOT give alcohol, sedatives, aspirin, or other medications.
7. *Snakebite kits:* Keep a kit accessible for all outings in snake-infested or primitive areas.

SYMPTOMS

1. *Mild to moderate* symptoms include mild swelling or discoloration and mild to moderate pain at the wound site with tingling sensations, rapid pulse, weakness, dimness of vision, nausea, vomiting, and shortness of breath.
2. *Severe* symptoms include rapid swelling and numbness, followed by severe pain at the wound site. Other effects include pinpoint pupils, twitching, slurred speech, shock, convulsions, paralysis, unconsciousness, and no breathing or pulse.

The information on this poster is based on a report prepared for the American Red Cross by the National Academy of Sciences-National Research Council.

American Red Cross 

Snakebite prevention practices that can eliminate needless illness and worry may be learned in a Red Cross first aid course. Call your chapter to enroll.

Fig. 20-4 First Aid for Snakebite *(The American National Red Cross,* Copyright © 1978; reprinted with permission.)

20-1 FIRST-AID SUGGESTIONS

Wear a Med-A-Lert tag at all times if you have any allergies to a particular medication or have a disease, e.g. diabetes, that requires special attention.

Store the majority of items listed in Table 20-1 in a dust- and water-proof container. Label the container "First-Aid Kit." Place a list of contents on the inside lid of the first-aid kit. Replenish items after each use.

TABLE 20-1 Suggested First-Aid Kit

Quantity	*Item*	*Use*
12	4 x 4 Compresses	Dressings for wounds and burns
12	2 x 2 Compresses	
1 box	Assorted Band-aids	
3	40″ Triangle bandages	Slings; to hold dressings in place
10 yards	1″ Roller gauze or flexible gauze	Slings, open wounds, or dry dressings for burns
10 yards	2″ Roller gauze or flexible gauze	
10 yards	4″ Roller gauze or flexible gauze	
1	2″ Roll elastic bandage	For sprains; to hold dressings in place
5 yards	1″ Adhesive tape	To hold dressings in place
5 yards	2″ Adhesive tape	
2	Small bath towels or large Pampers	Dressings for large wounds or burns
2	Eye pads	For eye injuries
1	Small bar of soap or bottle of liquid soap	Antiseptic for cleansing wounds
1	Single-edge razor blade, individually wrapped	For snake bite
1	Small container with needle, thread, and safety pins	To remove slivers; do emergency repair; fasten dressings
1	Flashlight with extra batteries	
1	Book matches	To sterilize needle (above)
1	Clean sheet, individually wrapped	For dressing large burns; splinting
1	Blanket	For transporting or shock

1	Cup and spoon	For giving fluids or for flushing wounds
1	First-Aid Book	
1	Tweezers	For removing foreign objects, insect stingers
1	Pocket knife	To cut splints
1	Scissors	To cut bandages
1	Small package salt }	For shock, heat exhaustion (to be mixed with water)
1	Small package soda }	
1	8″-long Tourniquet twist stick	As a last resort in bleeding; for snake bite
2	Small plastic bag	Container for ice or snow; for sealing chest wounds
1	1 Bottle antiseptic (Consult physician on selection and follow his or her recommendation.)	For general use

SOURCE: Jackie Pederson, R.N.

NOTES

INDEX

ABOUT THE AUTHOR

Robert O. Parmley is the president of Morgan & Parmley, Ltd., Consulting Engineers. A Registered Professional Engineer in Wisconsin, California, and Canada, he is a member of the National Society of Professional Engineers, Wisconsin Society of Professional Engineers, American Society of Mechanical Engineers, and the American Institute for Design and Drafting. He was the Editor in Chief of the *Standard Handbook of Fastening and Joining* and has published 42 technical articles in such periodicals as *Product Engineering, American Machinist, Machine Design, Assembly Engineering,* and *Plan and Print.*